HANDBOOK OF TELECOMMUNICATIONS ECONOMICS

VOLUME 2

TECHNOLOGY EVOLUTION AND THE INTERNET

Edited by

SUMIT K. MAJUMDAR
University of Texas at Dallas, Richardson, USA

INGO VOGELSANG
Boston University, Boston, USA

and

MARTIN E. CAVE
University of Warwick, Coventry, UK

ELSEVIER

Amsterdam • Boston • Heidelberg • London
New York • Oxford • Paris • San Diego
San Francisco • Singapore • Sydney • Tokyo
North-Holland is an imprint of Elsevier

N·H

ELSEVIER B.V.
Radarweg 29
P.O. Box 211, 1000 AE
Amsterdam, The Netherlands

ELSEVIER Inc.
525 B Street Suite 1900
San Diego, CA 92101-4495
USA

ELSEVIER Ltd
The Boulevard, Langford Lane
Kidlington, Oxford OX5 1GB
UK

ELSEVIER Ltd
84 Theobalds Road
London WC1X 8RR
UK

First edition 2005

Library of Congress Cataloging in Publication Data
A catalog record is available from the Library of Congress.

British Library Cataloguing in Publication Data
A catalogue record is available from the British Library.

ISBN-13: 978-0-444-51423-3
ISBN-10: 0-444-51423-6
ISSN: 1569-4054

The paper used in this publication meets the requirements of ANSI/NISO Z39.48-1992 (Permanence of Paper).

Printed in The Netherlands.

LIST OF CONTRIBUTORS

Martin Cave	Centre for Management Under Regulation, Warwick Business School, University of Warwick, Coventry, UK
Jeffrey Church	Department of Economics, University of Calgary, Calgary, Alberta, Canada
Robert W. Crandall	Economic Studies Program, The Brookings Institution, Washington, DC, USA
Nicholas Economides	Department of Economics, Leonard Stern School of Business, New York University, New York, USA
Gerald R. Faulhaber	Department of Business and Public Policy, The Wharton School, University of Pennsylvania, Philadelphia, Pennsylvania, USA
Neil Gandal	Department of Public Policy, School of Government and Policy, Tel-Aviv University, Tel-Aviv, Israel
Joshua S. Gans	Melbourne Business School, University of Melbourne, Carlton, Victoria, Australia
Damien Geradin	Department of Law, University of Liege and College of Europe, Liege, Belgium
Shane M. Greenstein	Elinor and Wendall Hobbs Professor of Management and Strategy, Management and Strategy Department, Kellogg Graduate School of Management, Northwestern University, Evanston, Illinois, USA
Alok Gupta	Department of Information and Decision Sciences, Carlson School of Management, University of Minnesota, Minneapolis, Minnesota, USA

Dale N. Hatfield	Interdisciplinary Telecommunications Program, University of Colorado at Boulder, Boulder, Colorado, USA
Thomas W. Hazlett	George Mason University, Fairfax, Virginia, USA
Stephen P. King	Department of Economics, The University of Melbourne, Victoria, Australia
Sumit Majumdar	School of Management, University of Texas at Dallas, Richardson, Texas, USA
Bridger M. Mitchell	CRA International, Inc., Palo Alto, California, USA
Milton Mueller	School of Information Studies, Syracuse University, Syracuse, New York, USA
Jeffrey H. Rohlfs	Strategic Policy Research, Inc., Bethesda, Maryland, USA
J . Gregory Sidak	Georgetown University, Law Center, Washington, DC, USA
Pablo T. Spiller	Jeffery A. Jacobs. Distinguished Professor of Business and Technology Chair, Haas School of Business, University of California, Berkeley, California, USA
Padmanabhan Srinagesh	CRA International, Inc., Palo Alto, California, USA
Dale O. Stahl	Malcolm Forsman Centennial Professor. Department of Economics, University of Texas at Austin, Austin, Texas, USA
David N. Townsend	President, David Townsend & Associates, Swampscott, Massachusetts, USA
Ingo Vogelsang	Department of Economics, Boston University, Boston, Massachusetts, USA
Björn Wellenius	Consultant on Telecommunications in Emerging Markets, Potomac, Maryland, USA

Andrew B. Whinston
Hugh Roy Cullen Centennial Chair in Business Administration and Professor of Information Systems, Department of Management Science & Information Systems, McCombs School of Business, University of Texas at Austin, Austin, Texas, USA

Julian Wright
Department of Economics, National University of Singapore, Singapore

CONTENTS OF VOLUME 2 OF THE HANDBOOK

Chapter 15

Institutional Changes in Emerging Markets: Implications for the Telecommunications Sector

PABLO T. SPILLER

Chapter 1

TECHNOLOGY EVOLUTION AND THE INTERNET: INTRODUCTION

SUMIT MAJUMDAR

University of Texas at Dallas

INGO VOGELSANG

Boston University

MARTIN CAVE

University of Warwick

Contents

Handbook of Telecommunications Economics, Volume 2, Edited by S. Majumdar et al.

DOI: 10.1016/S1569-4054(05)02001-4

1. Introduction

The Objective of the Second Volume of the *Handbook of Telecommunications Economics* is to highlight the economic aspects of the evolution of communications technologies beyond the basic fixed-line telephony infrastructure that was covered in Volume 1. In that book, structural, regulatory, and competition policy issues with respect to a well-known technology were covered. The state of technology evolution is still in a flux. Yet, convergence is real. Technological options have increased in a quantum manner, fueled by the creativity of entrepreneurs and policy makers world wide, and it is safe to infer that a process of creative destruction is underway. The current volume, therefore, covers the major technological developments and tracks the changes in these developments, linking them to the ways that both communications can take place and that institutions and policies can evolve.

A 'command and control' institutional structure that has evolved over a century, with the traditional PT&T model, has now to face a battle of ideology with the concept of self-regulation that has driven the growth of the Internet. Yet, there is an extremely vital symbiotic relationship between the two ideologies as the Internet ultimately depends on the last mile connection in the old-established telephone system for it to make its presence felt in the consumer's home.

Following these overarching themes, the initial chapters try to answer the following issues. What are the key communications and network technologies? What are the key economic characteristics of these key technologies? For example, how do bandwagon effects arise in these sectors and how do different communications technologies compete with each other as they are evolving? This part includes the specific evolutionary patterns of some key communications technologies and infrastructures, some of which are broadband applications in the existing backbone, the cable sector, the mobile sector, the satellite sector, and the related behavioral and institutional patterns that are affecting these evolutionary patterns.

After that follows a number of chapters on the Internet, which is not a technology per se but a phenomenon—perhaps akin to a Tsunami. These chapters deal with patterns of diffusion of the Internet, issues of connectivity between Internet service providers, dealt with as the peering phenomenon, issues of quality of service and network pricing in an Internet-driven context, and the issue of Internet domain names.

In the following institutional section, the first two chapters deal with some germane institutional issues in the context of technological convergence—namely mergers between parties, such as AOL-Time Warner, and antitrust issues, such as Microsoft. At the same time, the chapters focuses on regulation vs. competition policies in addressing bottleneck and bandwagon effects and the U.S. and European telecommunications frameworks.

Finally, the closing chapters sum up what is known about the relationship between the evolution of new technologies, particularly communications technologies, and economic performance parameters, such as growth and development, and what policy mechanisms will be required at a country level to enhance the speed of diffusion of new technologies across various country contexts.

2. Evolution of major alternatives to traditional telephone networks

2.1. Emerging network technologies

In the traditional Industrial Organization framework the opening chapter, Chapter 2, "Emerging Network Technologies," by Dale Hatfield, Bridger M. Mitchell, and Padmanabhan Srinagesh characterizes the basic conditions of supply, as it is now evolving, in the telecommunications sector. At the same time, it clearly shows how these basic conditions shape and are themselves shaped by the structure, conduct, and policies in the sector. The authors take the traditional switched telephone network as their starting point to show how emerging technologies will affect message communication.

Following a detailed description of characteristics of traditional voice data and cable television networks the authors discuss the architecture of public switched telephone networks (PSTN), and its limitations in responding to changing demands placed upon it by alternative media and new technologies. In contrast, the architecture and institutional approach of the Internet has led to the emergence of a vision of an evolving network that conveys all applications, whether they be voice, data, images, video, or multimedia. Similar pressures are shown to have resulted on cable TV, commercial wireless, and satellite networks with corresponding responses. Finally, the authors address the effects of positive and negative message externalities that have gained prominence in light of emerging message technologies.

The traditional PSTN is ideally matched to voice traffic. It is based on circuit switching and time division multiplexing to prevent latency of traffic. In contrast, data traffic is based on packet switching and statistical multiplexing, which involves the sharing of transmission capacity. Thus, a data network can deal with bursting data traffic and is error free, but it results in some latency. Because the traditional PSTN cannot handle data and multimedia traffic well, it cannot make use of scale and scope economies from combining these types of traffic. On the traditional PSTN bandwidth is only available in fixed increments. Its dedicated capacity cannot handle bursting traffic. There exist high costs of setting up calls. The access network is largely analog and there is vulnerability to single-point failures. New technologies and more demand for network reliability therefore suggest changes in the PSTN architecture.

New but already established developments in traditional wire line architecture include SONET/SDH, interoffice signaling and intelligent network (IN) features. The signaling network SS7 is packet-switched. Additional service logic and associated databases connected to signaling packet switches form the IN. Thus, the intelligence is contained in the network, while user equipment is dumb. The integrated services digital network (ISDN) access network has not been successful in the U.S. but is now widespread in Europe. The worldwide spread of digital subscriber line (DSL) supports two parallel and quite different networks that have the access lines in common.

In contrast to the PSTN the Internet uses routers with dumb functions. The intelligence is on the edges and in the terminals. In contrast to traditional data networks the Internet is not responsible for correcting errors. Rather, error correction is done by hosts. Voice over Internet Protocol (VoIP) potentially has large capacity requirements. Again, the intelligence is in the terminals. The authors expect that traditional telephone companies will disappear, as the Internet architecture takes over.

The trends in the traditional PSTN are shaped by advances in cable TV and wireless, but these influences are entirely mutual so that new cable TV networks (triple play) look more like the new PSTN and, similarly, cellular networks are evolving into an equivalent structure. The greater similarity of network structures and, therefore, network abilities in performing services along with their parallel existence will increase competitive pressure among the different types of companies, even if competition within the fixed network types is limited by near monopolies in the access networks, while wireless networks support oligopolistic access. The authors predict increased competition at the 'wholesale' level between network components of all networks. In our view, the intensity of such competition increases with the level of player in the network hierarchy. Wholesale competition is strongly affected by unbundling policy. Intermodal competition will be limited by differentiating demand and supply characteristics that can be maintained for some time. Relevant demand characteristics include quality of voice communication, reliability of service, and prompt and fault-free delivery of messages.

The main differentiating factors on the demand side are speed, voice quality, and mobility and on the supply side the relative access costs and usage costs (Vogelsang, 2003). For example, wireless networks have low access and high usage costs, combined with low speed and voice quality, and high mobility. In contrast, fixed networks have high access and low usage costs, which they combine with high speed and voice quality and no mobility. The combination of these factors gives fixed networks a limited niche for high demand customers outside their mobility needs but lets them exit the low demand voice market, while wireless networks will dominate low demand voice markets and compete with fixed networks for non-mobile voice and low-speed data services. Wireless will continue its niche for mobile services.

Positive message externalities, as opposed to pure network externalities, lose importance through reduction in calling prices, and the move to flat rates, and moves toward 'single-party' forms of communication, such as web browsing, file transfers, etc. At the same time, negative message externalities from unwanted calls and messages increase due to low cost of sending multiple messages to many receivers. The authors show in a formal model that, as a result, welfare-optimal calling charges should slightly exceed marginal costs.

2.2. Bandwagon effects

The importance of network effects in telecommunications networks has been appreciated for decades, but recognition of the role such or similar considerations have across the much broader and longer value chain of information and communications industries is more recent. While Jeffrey Rohlfs, in his chapter, focuses on telecommunications, Jeffrey Church and Neil Gandal consider a broader range of illustrations of network effects in their chapter.

Earlier work on bandwagon effects has emphasized herd-like behavior of consumers, but Jeffrey Rohlfs, in Chapter 3, "Bandwagon Effects in Telecommunications," grounds his analysis on special features of network industries. He distinguishes "network externalities"—direct network effects, from 'complementary bandwagon effects,' where consumers benefit from the greater availability of competitively supplied complementary product. He notes that owners of CD players, digital television sets and PCs benefit from the latter, as do users of the Internet—in each case, the greater the spread of 'hardware' ownership or use, the more the 'software' that is available.

Bandwagon effects have implications for the demand curve, creating the possibility that it is upward sloping over a range, expressing the fact that, as more units are sold and bandwagon effects come into play, marginal consumers' willingness to pay increases. Provided a critical mass of consumers is achieved, which depends on the size of the relevant community of interest, sales will grow.

This has a powerful impact on pricing incentives and optimal pricing, and the author exemplifies this by the telephone sector. Whereas call externalities can be internalized (e.g., by alternation of the roles of caller and called party), the same does not apply to subscriptions, and this creates a possible basis for providing subsidies to get people on to the network, thereby benefiting existing subscribers. Devising optimal telephone prices involves the estimation of these externalities and also of the complex demand relationship linking subscriptions and calls. Jeffrey Rohlfs relates these considerations to the historical pattern of telephone prices in the United States, pointing out inefficiency of (but popularity of) low subscription rates and unmeasured local service. He also notes that since the externalities associated with calls to Internet service providers (ISPs), unlike those associated with voice calls, cannot easily be internalized there is a case for providing those at below cost.

The final section of Jeffrey Rohlfs' chapter is devoted to the impact of bandwagon effects on the termination of fixed to mobile calls, an area which, as the chapter "Wireless Communications" by Joshua Gans, Stephen King, and Julian Wright demonstrates, has recently become subject to regulation. Under the calling party pays (CPP) system in Europe (and unlike the receiving party pays (RPP) system in North America), the called party's operator provides exclusive access to that party, for which it can change a monopoly price. If that price is to be regulated at an efficient level, it is appropriate to take account of the network externality associated with subscription. Citing work that he undertook in the U.K., the author shows how this can be done.

2.3. Platform competition in telecommunications

Jeffrey Church and Neil Gandal, in Chapter 4, "Platform Competition in Telecommunications," set out the classification of networks into those which are direct, linking complementary inputs, such as origination and termination assets for a telephone call and those which are indirect, where the effect on consumers is achieved through the supply of complementary products, such as the software available for PCs[1]. These networks then generate externalities, which influence both consumer benefits and the competitive process, leading (possibly) to coordination problems, standards wars, tipping equilibria, and consumer lock-in.

A major strand of the literature, reviewed in this chapter, deals with standards, set either administratively by state organizations or in the marketplaces. Business strategies in standards contests, involving pricing, marketing, and the creation of open standards have been extensively analyzed. As to their efficiency, many results have been found in particular cases—authors noting that 'the tendency in theoretical literature is for the equilibrium to be characterized by *insufficient standardization or too much variety*' (emphasis in original).

A second strand deals with battles for compatibility. A firm owning a platform that is dominant can extend its market power by denying compatibility to equipment produced by rivals—either directly (e.g., refusing to allow a competitor's equipment to be attached to its physical network), or through the exercise of intellectual property rights. Manipulation of interfaces has become a major issue in both communications and computer systems, attracting the attention of regulators and the antitrust authorities—see the Microsoft cases in the U.S. and Europe. Successful use of compatibility weapons can help a firm to extend its dominance from one generation of technology to another. For example, ensuring that a rival's superior software is incompatible with the installed base of hardware assists the maintenance of that hardware dominance into the next generation, by preventing rivals from breaking into either sector.

Much attention has been paid, both theoretically and empirically, to the standard setting process. Cooperative standards come into existence where no firm will earn higher profits by refusing to participate in the common process. Cooperation is thus favored in situations where a common standard will enhance consumer adoption and where no firm has acquired a dominant position in the relevant

[1] Rohlfs' slightly different terminology is described below.

intellectual property. The technologies for wireless local area and wider area networks, such as Wi-Fi and Wi-Max, discussed by the authors in their chapter, provide interesting case studies of cooperative standards setting.

In Europe at least, mobile communications have been subjected to governmental standard setting, with the adoption by the European Union of a harmonized GSM standard for second generation wireless technology. In North America, by contrast, operators were free to choose their own technology. This provides an interesting case study of the costs and benefits of the alternative approaches, although the issue remains unresolved as yet.

Among the examples of platform competition that the authors consider is that between satellite and cable television in the U.S. In this case satellite distribution sought to establish itself in a market dominated by cable, assisted by various legislative and regulatory interventions designed to prevent the new platform from being denied access to programming. The impact on cable TV is analyzed in more detail in Hazlett's chapter in this volume.

2.4. Broadband

Broadband is the area where the three platforms of DSL, based on the existing telecommunications networks, cable, and wireless compete, in a marketplace to which governments attach the greatest importance, many of them seeing broadband diffusion as the key to growth in the knowledge economy.

Robert Crandall, in Chapter 5, "Broadband Communications," identifies these and other more advanced networks, each characterized by different speeds (upstream and downstream) and areas of application, drawing attention to the differences between DSL over copper and the hybrid fiber-coaxial cables on which cable relies.

A review of national experience of broadband penetration in the OECD area has showed, in June 2003, Korea to be an outlier at the top, with the United States and Japan at the head of the second quartile. The presence of a cable network accounts for a high, but normally declining, proportion of subscribers, in most of the countries at the top of the diffusion league, thus demonstrating the advantages which are derived from interplatform competition.

Robert Crandall describes the regulatory framework for the development of broadband in the United States and the European Union[2]. He cites studies by Hausman and others that argue that regulatory requirements on incumbent telephone companies to unbundle network elements are inappropriate because incumbents lack market power. Their effect has been to chill investment by telephone companies, which would otherwise have challenged the cable companies' stronger market position in broadband. At the same time Hazlett, in his chapter in this volume, suggests that cable companies have themselves restricted

[2] See also the chapter by Geradin and Sidak in this volume.

the capacity they devote to broadband, because of the threat of regulation. The Federal Communications Commission (FCC), is however seeking to restrict, if not eliminate, mandatory access by competitors to incumbent's future deployments.

In Europe, some progress has been made in regulating the supply by dominant telecommunications companies of a service known as wholesale broadband access, or 'bitstream,' which comprises of access to carriage on copper loops, digital subscriber line multiplexers (DSLAMs) and backhaul on, for example, an ATM network. The pricing of this service has proved problematic, not least because its costs are uncertain. For this reason, several European regulators have favored an approach known as 'retail minus,' where prices are set, in the style of the efficient component pricing rule (ECPR), on the basis of the access provider's retail price for a broadband service minus its costs of providing those services, which the access sector provides itself. The ERG has mooted in the same document an approach to broadband regulation in which competitors are encouraged progressively to replicate the incumbents assets, starting with backhaul and DSLAMs and possibly culminating in replication of the local loop itself.

The diffusion of broadband is also leading to concerns about the 'digital divide' and to discussion of the extension of universal service obligations to cover broadband. With household take-up rates of the order of 10–25 percent in the more advanced OECD countries, and with access to broadband being increasingly expanded as a result of enabling more and more exchanges to provide asymmetrical digital subscriber line (ADSL), there is little appetite on the part of governments and regulators at this stage to provide subsidized access to all.

The current focus on developing broadband competition is both widespread and understandable. Competitive broadband suppliers based on cable or DSL technologies have the potential, via new voice technologies, to break incumbents' dominance in fixed voice. They might also provide a basis for more competition in the roll-out of fiber to what is sometimes referred to as next generation broadband. As the diffusion of broadband accelerates, the success or failure of attempts to promote competition will become more visible.

Yet, the danger is all too real that the quantity and intensity of regulatory debates in the older economies, in telecommunications terms, such as that of the United States and Europe, will lead them to decline as the newer economies, such as China, India, Japan, and South Korea, bypass the regulatory regimes considered irrelevant and promote broadband diffusion actively so that the largest numbers of persons in the world have eventual Internet connectivity.

2.5. *Cable television*

Thomas Hazlett begins his Chapter, "Cable Television," by noting Negroponte's prediction that while we were born into a world of making voice calls over wires and watching over-the-air television, this would soon be reversed. While the use of cable to deliver television programs has become the dominant mode of transmission in highly cabled countries in North America and parts of Europe, even in the U.S.A. satellite television represents a competing platform, and elsewhere new higher capacity digital terrestrial transmission technologies are being developed.

Nonetheless the market power exercised by U.S. cable companies, and legislative and regulatory attempts to curb it, is a major theme of this chapter. After initially restricting cable growth in favor of traditional broadcasters—as the FCC put it in 1966, restricting the role of cable to complement, rather than substitute for over-the-air and services—from 1976 cable was subject to deregulation. First, the FCC dropped its rules preventing cable systems from carrying any but the closest broadcast TV stations; then the courts struck down regulations on premium movie channels; then, in 1984, Congress passed the Cable Communications Act, which simultaneously outlawed rate-fixing by local authorities and restricted competition in video from telephone companies. This caused prices, quality of service and subscriber numbers to grow.

The rise in prices itself contributed to a re-regulation via the Cable Act of 1992, which led to rate regulation where effective competition was found to be lacking. The author reports a complex series of reactions by cable companies, of which the most prominent were reduced investments, efforts to switch subscribers to less regulated higher tiers of service and overload of regulatory officials by requests for rate increases. Following passage of the 1996 Telecommunication Act, cable rate controls were again abandoned from 1999.

The rapid alteration of policy over rate regulation was accompanied by extensive debate over the degree to which cable operators, particularly multiple system operators (MSOs), were able to exercise market power in other ways, as monopsonists in relation to program suppliers, through vertical integration with the latter and by predation of rival operators or 'overbuilders.' Thomas Hazlett notes the difficulty of establishing whether a large buyer is underpaying for programming in a heterogeneous product market characterized by keen bargaining, and that there appears to be no rule that ownership ties between program makers and cable operators are essential for the success of either.

As noted by Robert Crandall, in his chapter "Broadband Communications," cable operators now have a major alternative revenue stream through the supply of broadband and, in some countries, cable telephony. At the same time DSL and satellite offer increasing competition in the delivery of cable's traditional entertainment services. The widening of the marketplace through convergence will attack market power in previously separated services and should transfer the pressures on cable operators throughout the world.

2.6. Wireless communications

Gans, King, and Wright, in Chapter 7 survey developments in "Wireless Communications." As they note, the growth in mobile telephones used for voice over the last 10 years has been spectacular, reaching by 2004 more than 80 percent inhabitants in many countries in Europe and elsewhere, although lower in North America. But wireless technologies now extend much more widely; 3G, available in an increasing number of countries, provides high bandwidth mobile services. Short range nomadic services, known as Wi-Fi, provide broadband access in tens of thousands of airport lounges, coffee shops and other locations throughout the world; and new fixed broadband technologies provide high speed services over longer ranges, and appear to offer good prospect for providing broadband in less densely populated areas, as well as providing competitive services elsewhere.

By definition, wireless communications services for households and firms depend on spectrum, for which they compete not only with broadcasting services, but also with the whole range of spectrum-using activities, such as closed-group private mobile radio, aeronautical radar and communications, defense, science, such as astronomy, and others.

Balancing these needs is accomplished using one of three methods—the traditional 'command and control' process of allocating bands to particular uses by international agreement and then assigning spectrum to individual firms or organizations by an administrative process; use of market instruments, such as auctions and secondary trading; and creation of unlicensed spectrum to which users satisfying certain conditions or the power of their apparatus can have access.

Governments and regulatory bodies are moving gradually away from the first method in favor of markets and—to some degree—unlicensed spectrum. The spectrum or license auctions described in this chapter are a step in this direction, but by themselves do little to eliminate imbalances in spectrum allocation as the auctioned spectrum is tied to particular uses. To eliminate such imbalances, liberalization of use is required, which itself requires a re-specification of license conditions in the direction of restrictions on the ability of any licensee to impinge upon spectrum utilized by a licensee in an adjoining frequency or geography. Such liberalization has occurred in several countries.

Unlicensed spectrum has come into prominence through its widespread use of Wi-Fi. Because of its short range and low power, users in adjoining sites need not interfere with one another, giving the employed spectrum a zero opportunity cost. In fact, about 9 percent of prime spectrum in the U.K. is set aside for unlicensed uses. A number of new technologies exist, which permit more services to be provided in unlicensed spectrums and several proponents of universal unlicensed spectrum have looked forward to spectrum becoming a plentiful and free resource. Regulators have now to address themselves to the questions of the balance between licensed, and increasingly marketable, spectrum and unlicensed spectrum. It is, however, hard to see how technologies using the latter can compete in, for example, wide-area point to multipoint communications, such as satellite or terrestrial broadcasting, for which an interference-free channel is required to benefit from the low-cost distribution method.

A second key issue discussed by the authors is the scope for competition in voice between fixed and mobile networks. In some low- and middle-income countries without fixed networks, mobile telephony has, temporarily or for longer, filled the gap. The pertinent question for fixed and mobile operators in other countries is whether mobile is increasingly attracting subscriptions and calls from fixed networks. The authors cite chiefly European studies. The U.S. mobile market, where mobile operators provide large, inexpensive call 'buckets,' may be more advanced. Imminent technological change will affect both. 3G mobile networks have the capacity to provide calls on a packet-switched basis, rather than the traditional circuit-switched method. Equally, fixed networks are increasingly moving toward VoIP or Voice over DSL (VoDSL) technologies—the latter using broadband connections.

The rates at which these new technologies are rolled out, themselves governed by competition between fixed and mobile operators and within each group, will determine future trends. Already, mobile revenues have overtaken fixed line revenues in many countries, such as the U.K., even though the volume of fixed voice calls is far greater than that of mobile.

It is sometimes asserted that wireless communications are a regulation-free zone. But, despite the lack of retail price regulation, this is hardly the case, as the authors show. First, regulation through licensing determines the structure of the industry, in a way not encountered in fixed networks. Second, regulators often determine choice of technology, either directly or indirectly by imposing conditions, which only one standard can satisfy. Third, regulations often require number portability, the unlocking, and transferability, of SIM cards and national roaming, with the latter requiring access provision, normally temporary, by a new entrant to an existing network. Finally, regulators increasingly examine and intervene in two wholesale services provided by mobile operators—the termination of calls on the network and the provision of international roaming, which enables a customer of a network in one country to make calls in another country—the subscriber's operator compensating the visited country's operator for the services supplied, and recovering it as retail charges from its own customer.

International wholesale roaming charges in Europe are under scrutiny both by the group of industry-specific regulators known as the European Regulators Group and, in the case of two U.K. operators, by the Competition Directorate of the European Union. Analysis of the market is complicated both by technological changes, which now allow an operator to program its customers' SIM cards so that they will seek a particular network when the customer seeks access in a country, and by the growth of operators active in many countries, or forming cross-country alliances.

Intervention in mobile termination charges is more widespread and has a longer history. In the European Union where the CPP principle, under which the calling pay for termination, all national regulators are investigating the termination, with a requirement to impose remedies where they find dominance—a likely outcome if they declare each operator's mobile termination to be a separate market. As the authors point out, however, if all excess profits on termination are ploughed back into attracting customers buying subscriptions and making outgoing calls, it is the structure of termination charges and outgoing prices rather than the level of profits, which is affected. But this requires very high levels of competition among mobile operators which, given the small number involved, may not exist. Even if it does, it will be inefficient.

Linking Gans, King, and Wright with Thomas Hazlett convinces us that the 'Negroponte switch' whereby the public would consume television programs via land connections and telephone connections via space has probably already taken place.

3. The Internet

Along with mobile telephony, the Internet has emerged as the single most simultaneously creative and destructive influence unleashing the creation of new infrastructures, businesses, business models, and economic concepts in the last decade. In the 10 years that the electronic portals and gateways were made open to the world at large, not only has a wave of Schumpeterian creative destruction changed businesses, but it also has changed the sociology of how communications services are consumed. As the proportion of people using the Internet increases, the proportion of people watching television drop, as documented in a recent study by the Center for the Digital Future. What was once the passive activity of television watching now becomes interactive, conducted via the Internet, and brings about not just economic but social change as well.

In 1994, about two million computers were connected to the Internet, as it was then used mainly by academics, scientists, and corporate researchers. By 2000, about 70 million computers were connected to the Internet. As we write this chapter, the Internet is a ubiquitous presence in the lives of over three-fourths of the population living in the United States. In 2004, over 200 million computers were connected to the Internet. An aspect of writing about the Internet for a handbook that is essentially global in character, and for a phenomenon that permits global connectivity on an unprecedented scale, is that the chapter on the Internet is intrinsically U.S.-centric. The history of how the Internet has evolved has surely affected its geography. Thus, the lessons of the U.S. experience apply generally to other communications infrastructures that are being created around the world.

3.1. The economic geography of the Internet infrastructure

The first among the various chapters dealing with the Internet, Chapter 8 by Shane Greenstein, is titled "The Economic Geography of the Internet Infrastructure." It is a self-explanatory title, and the chapter is an extremely comprehensive review of the origins and diffusion of the Internet. The origin of the Internet is dealt with in a number of other comprehensive histories, with a variety of details about the individuals who were key players also provided, but the diffusion issue has hardly merited attention. The author deals with six broad questions. The first is, why did near geographic ubiquity arise in the United States after commercialization? The second is, why did market forces encourage extensive growth? The third is, has the Internet diffused disproportionately to urban areas? Fourth, is the Internet a substitute or a complement for urban agglomeration and an influence on business location decisions? The fifth question is, which policies mattered and were the effects intended or unintended? Finally, are there lessons for other countries?

In dealing with the first question, given that Internet technologies had been available for around 20 years, by the early 1990s, the role of standardization was critical. By the early 1990s, items of infrastructure, such as hubs and routers, Ethernet cards, and T-1 line were available and hooked together, but the acceptance of the TCP/IP protocol voluntarily by the various players made comprehensive interconnectivity possible. An aspect of the Internet is its overlay structure, with many components being retrofitted to the existing network. The computing, telecommunications, and the network equipment industries were mature and had existed for some time. Since such assets were already available, the standardization of components throughout the system quickly led to the realization of scale economies. The characteristics of the core physical elements of the Internet were malleability, deployment ability, and retrofit ability. These led to high-speed infrastructure creation.

The second issue is the role of market forces and entrepreneurial organizations. The Internet has spawned hundreds of ISPs, at the consumer or the retail end, who quickly provided cheap access to millions of U.S. households. In fact, the growth of such organizations, and the density of competing suppliers, is enormous. The ISP sector has shown the classic patterns inherently noticed in the evolution of any industry. At the initial phase, there are a small number of firms, which establish very strong positions to be followed by, often, thousands of new entrants and then there is a shakeout, as the industry matures, leaving the field to a smaller set of competent players.

The author goes into the demography of the ISPs and the strategy of the ISP firms. Such demography dynamics have important implications for other countries. A key issue is, however, why did the dynamics evolve the way they did? In our assessment, one reason why these firms were able to quickly go into business is because of the pricing model they were able to adopt. Based on the contents of the Computer Inquiry 2 (CI2) report of 1980, Internet services were classified as an enhanced service, avoiding the telephony access charges, even though the authors of CI2 had not at all then foreseen the arrival of the Internet in its present form, and its objective was not to propel the growth of the Internet. Internet Service Providers were, thus, able to provide Internet services to subscribers at the price of a local phone call, which was essentially free since a fixed sum per month permitted customers unlimited local calling. Thus, a pricing model permitting low-cost service supply led to the set up of many ISP businesses that then provided connectivity to the Internet. The grassroots origin of the Internet, as opposed to a top down imposition of the phenomenon, has had major consequences in the diffusion of the Internet within the United States. It has also had major consequences for the growth of Internet-based entrepreneurship, which has been propelled by the architecture of openness that characterizes the Internet phenomenon.

At the grassroots level of the consumer, the role of ISPs is paramount. Equally important, however, is the role of the backbone network, which is the wholesale part of the network as it were, since information originates from a variety of sources and is accessed by the ultimate customer. This information is aggregated at the backbones so that it may be exchanged among the servers and the ISPs. A variety of backbone networks carry traffic, in substantial quantities, and there is interchange of such traffic. The interexchange of traffic between backbone networks leads to the same sorts of interconnection issues that arise in voice networks. These were initially governed by self-regulated cooperative peering arrangements, eliminating the need to monitor traffic.

The sociology of the Internet community created a system that included openness, fairness, and a competitive ISP sector as core values. However, as some backbone networks have gotten larger than others, the question of market power has arisen as bigger networks can discriminate against smaller ones, highlighting the need for possible regulation of the Internet. The author concludes that interconnection issues have not shaped the evolution of the Internet phenomenon, as it has in voice telephony, and even the collapse of WorldCom, one of the world's largest backbones, with over 40 percent market share, did not affect the spatial shape of the sector.

The next major issue has to do with broadband. Internet relies heavily on broadband, and thus this part of the chapter should be read in conjunction with the chapter on broadband by Crandall. From a geographic perspective, the key conclusion that the author reaches is that dial-up is still more widely spread than broadband and broadband is available primarily in dense urban areas. However, as he states, the regulatory environment is in a major flux, and the jury is out on what will happen to universal broadband diffusion, across rural as well as urban areas of the United States.

The comments that we make with respect to policies for broadband diffusion apply to the Internet phenomenon as well, since the Internet is an add-on infrastructure on top of the existing network. The basic telephony infrastructure is still relevant for the Internet. The issue of the last mile still holds, we feel, since there is a very clear symbiotic relationship between the evolution of the Internet, in diffusion as well as quality terms, and the evolution of the fixed-line telephony network as a whole. The Internet does use preexisting infrastructure and is embedded in the communications system of a country as a whole. The embeddedness of the Internet in the national communications infrastructure is a key contingency that policy makers must keep in mind. A holistic policy assessment must link Internet diffusion with the basic telephony regulatory issues.

Because now one can work from practically anywhere, and deliver products or services to any location, the issue of whether the Internet has replaced urban clustering or not becomes highly relevant. It is this area of concern that the author addresses next. In an era of progressive urban congestion, in parts of the world that are salubrious, the location of businesses based on the usage of the Internet infrastructure becomes important. Since the Internet provides a substitute for face-to-face communications, the necessity of locating one's offices at the same place where one delivers products or services no longer holds. Therefore, a new approach to the whole question of spatial considerations in the engendering of economic activity is required. This is an open research area.

In our view, the area of analysis should be further boosted by bringing in the issues of outsourcing and business-to-business (B2B) commerce as the topics of salience in the twenty-first century. Business-to-business commerce growth, responsible as it was for the boom, and subsequent bust, was propelled by the very presence of the Internet, which provided for the possibility of conducting commerce in manners that would have enhanced efficiencies. Outsourcing too, as it is developing, would not have been possible had it not been for the feasibility of information technology enabled services (ITES) to be provided over substantial distance, in many cases these distances being several thousand miles. Both B2B and several variants along the alpha-numeric theme, such as government-to-citizen (G2C) activities that are engendered by the development of e-governance, and outsourcing are two extremely important topics that now become significant and worthy of study given the spatial impact of the Internet.

3.2. Economics of the Internet backbone

The following Chapter 9, by Nicholas Economides "Economics of the Internet Backbone" comprehensively describes the organization of the Internet backbone and reviews issues related to the backbone from a traditional industrial organization perspective. The Internet is akin to a standard telephone system in that there is a local layer of players, the ISPs, and a layer of players similar to the long-distance operators, which are the Internet backbone providers (IBPs). The IBPs operate the high speed hubs permitting data transfer from far-flung servers to the local ISPs who then transmit it to their local customers. It is not surprising that in 1997 the major IBPs were MCI WorldCom, GTE, AT&T, and Sprint. These were the companies then dominating the U.S. long-distance telephony sector as well. In common with the rest of the sector, the fortunes of these firms have reversed in the years since 1997.

In a sense, the economic, strategic, and regulatory issues relating to IBPs and their relationship with the ISPs are similar to those obtaining between a wholesaler and a retailer, or an upstream producer supplying units to a firm that is downstream in the industry value chain. With the Internet, the added complications arise from the fact that a variety of ISPs are connected to each other, so that the network is much richer in terms of connectivity than a standard telephone network. Thus, this complication, in fact, is an advantage since there are many routing choices available for the traffic to pass, and the presence of many parties who provide the data carriage helps bring competitive discipline to pricing. This feature of the Internet, the ability of the ISP to connect with several IBPs, called multihoming, ensures that no ISP is captive to any IBP. Multihoming is also dealt with in some detail in the next chapter, by Alok Gupta, Dale Stahl, and Andrew Whinston.

With respect to the issues of market power that can emerge within the Internet backbone segment, the author asks an important question, and one that has surfaced in various regulatory proceedings given the various telecommunications mergers that have taken place: Can the IBPs exercise market power so that ISPs have to pay more for carriage and eventually pass on these costs to the ultimate consumer? The answer is a resounding 'no' for several reasons. First, there is overcapacity in the sector, similar to the overcapacity in the trunking segment of the long-distance sector, and there is ease of entry and expansion. Following the majors, firms, such as Quest, Level 3, and Williams were also able to create Internet backbones. Second, the Internet operates on public standards and protocols, and these enhance interoperability so that the possible absence of technical connectivity is hardly a moot point. Connectivity is universal. There is also the possible commoditization of the backbone as a whole.

The second major issue that this chapter deals with is whether IBPs can behave strategically by raising prices, as in vertically organized market segments, or degrade quality. Nicholas Economides describes some key features of the Internet as opposing the successful implementation of such strategies. If the price of transit that is the carriage of traffic between an IBP and an ISP for a consideration, increases there simply will be bypass. Other IBPs will be used, or if that is not a feasible strategy, then ISPs will simply provide transit for each other, either commercially for a price or via a peering arrangement whereby no charges are made for inter-ISP traffic transfer, a strategy similar to a knock-for-knock agreement among automobile insurance companies. For similar reasons, an IBP cannot price discriminate either. In a manner similar to the above, service quality degradation is also not a viable strategy for the IBP since the implications are far-reaching and touch every part of the Internet. Capacity is now fungible, and an ISP can easily switch between IBPs who provide the least cost data trunk routing capacity.

The Internet portion of the overall communications infrastructure is extremely competitive, both at the level of the ISPs and the IBPs, in the United States. While there has been concern that IBPs, by virtue of their larger size, and smaller numbers, could exercise market power, this is unlikely. Our view, based on the industrial organization of the Internet, and the descriptions provided in the chapters so far, is that it is a sector that is organized in a horizontal manner, as opposed to the traditional PSTN sector that still displays a vertical structure. This feature of the Internet makes it a sector in which competition is enjoying its full potential, since the opportunities for interconnection available to the various ISPs are considerably greater than those which would be available to local telephone company. The issue that arises is how is the network, which is inherently more complex, managed since so many routing options are available. This is a capacity management issue, which is dealt with in the next chapter by Alok Gupta, Dale Stahl, and Andrew Whinston.

3.3. Pricing traffic on interconnected networks: issues, approaches, and solutions

In Chapter 10 titled "Pricing Traffic on Interconnected Networks: Issues, Approaches, and Solutions," Alok Gupta, Dale Stahl, and Andrew Whinston deal with the question of how to price network resources in the face of congestion. The approaches of the Internet chapters so far have been grounded in economic geography and industrial organization economics. The overarching theme of the chapter is operations management. We have so far discussed the fact that there is substantial capacity available in the Internet backbone. There may be commoditization of the backbone. Nevertheless, traffic growth is equally substantial such that congestion is likely. The chapter highlights that next generation Internet developments, such as Digital Video over IP (DVIP), will create greater traffic bursts that then create congestion problems. Equally, there can be congestion problems in different parts of the network. The growth of ITES and outsourcing will create peak periods for traffic transfer and that too on the particular routes along which the bulk of the outsourcing businesses travel. Since the Internet is going to be the primary network for not just communications needs, but also for conducting business worldwide, the need for pricing schemes is paramount. Management of the Internet capacity via an appropriate pricing scheme, so that resources are optimally allocated, then is an important consideration.

The authors initially describe the optimal network resource allocation methods that are based on advances in the computer science fields. The main congestion problem in computing systems revolves around load balancing, and three main techniques are used. These are: global knowledge, randomization, and feedback. The problem that the authors note with these approaches is that they ignore the economic consequences of load balancing. It is assumed that the value of each specific task is the same. It is, however, eminently feasible that the value of each particular job will vary according to the priorities of the particular user: authors use the case of a teenager downloading large music files versus that of a researcher conducting major computations while connected via the Internet for a remote site. Clearly, the economic and social cost-benefit trade-offs are different.

Given the complexity of the Internet, optimality is difficult to attain; but the reduction of negative externalities, via reduction of congestion, and the attainment of allocative efficiency, so that those users whose valuations are the highest receive the largest allocations of bandwidth, should be goals of a pricing scheme. A dynamic pricing scheme is described by the authors. This is principally a usage-based pricing method wherein two important components of the scheme are an 'opportunity price,' which is charging a user what the other users on the network would have paid for that capacity, and a 'priority price,' which is the price including an additional element paid for faster access or a higher priority. This chapter also demonstrates the superiority of the pricing scheme, relative to other models of Internet usage, based on the results of simulations that they have carried out.

The authors also suggest that in the overall Internet structure, another layer has now emerged. This the overlay network that lies between the ISP and the IBP, and is best understood in the context of transactions that happen in peer-to-peer (P2P) networks. Overlay networks consist of network nodes that perform data aggregation tasks as the various customers, connected to the Internet via their ISPs, conduct their transactions using the P2P framework. The data, for example, could relate to all the users that have contacted a particular file-sharing scheme and who therefore comprise a particular service network. Overlay networks are logical service networks generated on top of the actual physical service network by users' transactions.

The presence of these overlay networks, therefore, changes our assessment of the industrial structure of the Internet. The industrial organization of the Internet has obtained a new component. We now have to account for another link in the chain, as it were, and review the competitiveness issues, if any, for this element of the network. General knowledge would suggest that the existence of overlay networks has given rise to intellectual property rights issues but at this point are unlikely to give rise to market power issues. The authors suggest three, of several possible, applications of overlay networks: these are collaborative work, distributed resource management, and access and task automation using software agents. All of these applications are also clearly ones that are carried out across extreme geographical distances, thanks to the economics of the outsourcing phenomenon. Clearly, dynamic resource allocation in networks of ever increasing complexity, but surprising openness, generates complexities in routing tables' construction as well. Market power considerations are now not so important, but what is important is traffic and operations management. In both cases, appropriate mechanism design remains important.

3.4. Toward an economics of the domain name system

The final chapter on the Internet, Chapter 11 by Milton Mueller is titled "Toward an Economics of the Domain Name System." In a sense, this chapter now takes us down to the grassroots level; starting with the overall landscape of the Internet

(Chapter 8), we have dealt with the forests (Chapter 9) and the trees (Chapter 10). Thus, we have evaluated the Internet phenomenon across the various critical levels of analysis. The chapter by Milton Mueller, that comprehensively deals with an underresearched yet important area, is also historical in analysis and describes the emergence of one among several phenomena, now at best a dozen years old, associated with the Internet. That phenomenon is that of domain names. It has spawned its own market and its own crime—that of 'cybersquatting.' The domain name phenomenon is sufficiently critical that a specialized United Nations agency, the World Intellectual Property Organization (WIPO) is regularly involved in proceedings and is a topic that now engages the minds of the United Nations as a whole.

One of the issues is identity. In particular, unique cyber identity is predicated on the possession of a unique domain name, such as *www.elsevier.com*, and the trade in domain names is a market worth $2.5 billion annually. Because domain names are marketable items, other attributes associated with items that are traded, such as the basic demand and supply conditions, need to be understood. The author remarks, and in our opinion quite validly, that a great deal of policy has been made with respect to domain names without the basics of the domain name as a resource being understood.

This chapter contains, first, a technical description of the domain name system (DNS), which is then followed by sections highlighting the key demand side characteristics and the key supply side characteristics. This characterization is important as it helps define the contours of an industry segment that is actually turning out, as we describe later, to be most critical component of the Internet space in an institutional sense.

On the demand side, the author defines two dimensions associated with the demand for domain names. These are the purely technical demand, for the basic identification routines, and the semantic demand. Semantic demand is the human side of demand, associated with making a domain name appealing or memorable or unique. In a sense, the area of domain name demand analysis merits greater attention using the frameworks that exist in the consumer demand and marketing literatures on product variety. The semantic dimension also has a public policy component to it, since the authentication function associated with the semantic dimension is a signal either of quality or of particular status. Thus, how semantic names are given to individuals, entities, and organizations is not a function of explicit demand itself but also involves a process of implicit certification. In other words, there is a control aspect to the awarding of domain names, and this attribute goes against the *free-spirit* nature that has characterized the Internet phenomenon as a whole.

In a sense, the implicit certification of domain names helps breed trust among the Internet using community, which is now almost the whole world, and the issue of trust in online activities is a very large area of research. The issue of implicit certification, however, does raise another important issue, who chooses the certifiers? This is akin to the question of who regulates the regulators. As we further describe below, it is precisely this issue, over the formation and jurisdiction of the Internet Corporation for Assignment Numbers (ICANN), which has created a substantial turmoil in the Internet community throughout the world.

On the supply side, the author goes into detail about the structure of the domain name supply industry, which has three components: the root servers, the registries, and the registrars. Root servers maintain root zone files, and every top level domain name, such as *www.elsevier.com*, cannot be accessed until it has been verified by the root server and advertised to the public at large via the Web. Having a domain name on the root zone file is the ultimate expression of market entry into the Internet space. The political dimension now becomes important because other than three root servers of which one is in Japan, one in Sweden, and one in the United Kingdom, all of the others are based in the United States, and in the Internet space today capacity is not a constraint but an identity is. Thus, even if there is a great deal of bandwidth and connectivity, and the ability to actually enjoy the benefits of the Internet, there are significant controls over identity. As the author describes it in his chapter, the fundamental problems are now institutional and not technical. It is not a production problem but one of distribution.

The institutional issues that arise also relate to the registries and registrars. This is a vertical market, with registry services coming upstream and registrar services coming downstream. The original domain name registry was Network Solutions, Inc., which was acquired by VeriSign, now the largest registrar. The author highlights the lack of competition in the registry market, and the dominance of the main operator, which was allowed to charge for its services by the National Science Foundation even though at the time there were no alternatives.

This chapter also highlights the relationships between registry and registrar services as a sensitive one for competition policy because now there is substantial growth in the secondary market for names and the role of registrar services become important. If, however, there are limited facilities at the registry level and almost total dominance at the root server level, then there is a bottleneck in the Internet as well, just as there is in the local loop for fixed-line telephony. Only, now the bottleneck is in the market for identity. This chapter discusses a very large number of issues that are worth pursuing, and his chapter ought to open a stream of research and policy debates that will keep the Internet phenomenon in the public eye for the next generation, but from different perspectives relative to that for voice telephony.

4. Institutional considerations

The four chapters in this section concentrate on the benefits and costs of different policies for the telecommunications sector, where the policies include competition policy and regulation, public ownership, privatization, and liberalization.

4.1. Bottlenecks and bandwagons: access policy in the new telecommunications

In his chapter on "Bottlenecks and Bandwagons: Access Policy in the New Telecommunications," Gerald Faulhaber differentiates between two interrelated access concepts, bottleneck access and bandwagon access. These two effects provide for two different potential justifications and policy directions for access policies. The telecommunications sector is an example of an industry with both effects. These concepts are closely related to the differentiation between one-way access, or vertical access and two-way access, also known as horizontal access or interconnection, made in the access-pricing literature. Bandwagons are similar to network effects. In spite of these similarities, the author uses bottlenecks and bandwagons to provide a refreshingly new synthesis of policy insights.

These insights are aptly illustrated by the AOL-Time Warner merger decisions by U.S. Federal Trade Commission (FTC) and the Federal Communications Commission (FCC). This merger addressed both problems with the FTC focusing on the access to cable modems as bottlenecks and the FCC on access to advanced IM services as bandwagons. The author makes clear that bottlenecks and bandwagons present very different types of market failure and require different regulatory and antitrust remedies.

The author's discussion brings out that the difference between a bottleneck or essential facility, and a simple monopoly is by no means trivial. Four properties in combination are conventionally viewed as causing a bottleneck problem. They are: (1) sole ownership of a resource by the bottleneck owner; (2) inability of others to duplicate it; (3) unwillingness of the bottleneck owner to provide the facility on reasonable terms to competitors; and (4) harm from this unwillingness to the competitive process. Property (2) requires natural monopoly or some other barrier to entry (patent, copyright etc.). Property (3) hints at boycott, but would the problem go away if the bottleneck owner charged the monopoly price for use of the bottleneck input? Property (4) hints at a fixed proportions of input that downstream competitors cannot substitute.

Since in the fixed proportions case the monopolist of an input can exercise all market power (both upstream and downstream) in the price of the input, there should be no incentive for boycott. Thus, the bottleneck problem seems to be how to take the market power out of the input monopoly. Hence, the author questions if bottlenecks are really a problem if the input monopoly is earned by innovation or superior efficiency rather than inherited from a monopoly franchise. He further asks if there exists a feasible mechanism to correct the problem, using price and terms of agreement. If not, then competition for the bottleneck may be preferable to bottleneck regulation. This implies the questions, can the essential facility be circumvented by *final consumers* (instead of competitors!), and are new technologies in sight to replace the essential facility? Furthermore, in the absence of regulation, does bottleneck access develop by itself?

In contrast to bottlenecks the monopoly property is not a necessary characteristic of bandwagons but only a possible outcome. A bandwagon example of a sustainable oligopoly is the growth of the Internet backbone through peering. The main property of bandwagons is the mutual interdependence that is absent from bottlenecks. However, the mutual interdependence of bandwagons is commonly not symmetric but rather depends on market shares. As a result, market share conveys to a supplier an independent advantage like demand side economies of scale.

The bandwagon property has been at the core of strategic use of network effects in creating compatibility issues by dominant firms, such as Microsoft. Can therefore a market leader kill competition through her refusal to interconnect? Such effect, known as market tipping, depends on the shape of network effects curve and the market share of the dominant firm. For example, Economides, in his chapter on IBPs, argues that for ISPs tipping may not occur before a market share is close to 100 percent.

The new telecommunications are characterized by a high rate of technical change, a property for which the serial monopoly hypothesis was developed. This hypothesis conjectures that innovators, who monopolize a market, are themselves threatened by the following generation(s) of innovations. Thus, static inefficiency is deemed necessary to generate the next round of innovations. The author argues that, even if one accepts the serial monopoly hypothesis in general, it is difficult to replace a current monopolist if network effects lead to stickiness, say, because of difficult customer switching or barriers to entry. Network effects combined with merger assets would make entry of the next serial monopolist nearly impossible. Also, there is little empirical evidence on the serial monopoly hypothesis in general. To the extent that the serial monopoly hypothesis holds water, the problem of weak intellectual property rights may call for anticompetitive practices as substitutes for such rights. However, the author argues that improved intellectual property rights would be much better than lax antitrust enforcement.

In future, essential facilities in telecommunications networks may become rare because of technical progress and intermodal competition, but bandwagon effects may flourish because of the convergence of the media.

4.2. Antitrust remedies and the institutional design of regulation

The chapter by Damien Geradin and Gregory Sidak, "Antitrust Remedies and Institutional Design of Regulation," contrasts the American and European policy approaches toward the ongoing structural changes in the telecommunications sector. Although the U.S. was for a long time the undisputed leader in the liberalization of the telecommunications sector, it has by now fallen behind other countries in several areas, such as mobile communications. This chapter provides an insightful analysis of the main differences and commonalities in policy that may be responsible for this shift. They observe that the U.S. and Europe differ both in their approach toward the shift from regulation to competition policy tools and in their competition policies. According to the authors, since the break-up of AT&T in 1984, the U.S. has reduced the influence of antitrust policy on the telecommunications sector, while the EU has increased that influence, particularly in its new framework.

While the European approach toward the relationship between competition policy and regulation is based on a systematic design, the U.S. approach is driven by a diversity of interests that work their way through the legislature, regulatory agencies, and courts. It is not that these interest groups are absent in the EU. However, their influence is more contained at the design stage of the general EU policies. As a result, the EU approach, driven by the vision of convergence, has become uniform with respect to the different parts of the telecommunications sector and is geared toward competition on equal terms applied to relevant markets, while the U.S. approach has enormous difficulties reconciling historic differences in the regulation of telephony, cable TV, and wireless services. For example, the different regulatory treatment of broadband access via DSL and cable modem is explained by their original assignment to the telephone and cable sectors.

In spite of these differences, both EU and U.S. policies are shaped by the same vision, which is that competition shall ultimately rule in the entire telecommunications sector. Pockets of natural monopolies and network effects are viewed as ultimately vanishing. In the meantime regulation is there to bring competition to all end-user areas of telecommunications. Thus, regulation has shifted from the protection of end users to the protection of competition, and, as the authors argue, of competitors. This protection can be associated with heavy-handed regulation. Such regulation creates direct administrative and network-related costs. The latter include unbundling and interconnection costs, the establishment of number portability, and the like. In addition, indirect costs arise in the form of distorted or reduced incentives for innovation, investment, and cost reduction. Neither the U.S. nor the EU approach provides for an explicit analysis of such costs. However, considering, for example, the extent of network unbundling required in the U.S. relative to the EU reveals that the implicit judgment of the EU authorities has been that U.S. unbundling has been going too far. As of this writing, though the U.S. seems to be on course in reversing its policy on unbundling, as shown by the FCC's Triennial Order of 2003 and the Revised Unbundling Order of 2005.

Both antitrust remedies that can be *ex ante* and *ex post*, and *ex ante* regulation deal with market failures. Generally, the *ex ante* remedies are more drastic and would therefore have to concern more drastic market failures. The most prominent area of *ex ante* antitrust policies concerns mergers. Mergers have a lot in common with the issues raised by regulation in the presence of competition. In both cases, future market developments have to be evaluated before remedies are applied and in both cases the remedies affect the future competitiveness of the sector. In that sense, both policies are speculative and, according to the authors, need to be evaluated more carefully than *ex post* policies, which correct observed market failures. However, a status quo argument may be applied in favor of *ex ante* policies. Thus, in the case of mergers the remedies, while potentially drastic, are less drastic than divestiture in *ex post* policy would be. Similarly, an interconnection policy imposed on a currently dominant incumbent is less drastic than the same policy on a quickly expanding entrant, who is expected to become dominant in the future. The EU policy takes this difference into account, for example, by excluding emerging markets from regulation.

The EU telecommunications framework is fully in line with the authors' characterization of regulation as *ex ante* and competition policy as *ex post* remedies. In fact, all the regulatory remedies under the framework are explicitly formulated as *ex ante* policies. In contrast the new German telecommunications law (TKG 2004) additionally introduces an intermediate version of regulation, called *ex post* price regulation. Under this approach, dominant firms can set prices on their own but have to announce these to the regulator in advance, who can either simply accept them or postpone their enforcement and initiate a regulatory proceeding. Furthermore, the regulator can collect any abusively excessive amounts collected by the firms in the meantime. This section has been incorporated in the TKG 2004 among others as a way to prevent full-blown regulation of the mobile sector, which so far is not regulated under the market dominance provisions.

In fact, the fear may be justified that, while the EU framework may be able to move end-user markets toward deregulation, the wholesale market regulation may be sticky and may even perpetuate itself for call termination, although intermodal competition between fixed and mobile networks will eventually gain full steam. If one applies *ex ante* price regulation to workably competitive mobile sectors now there would be no reason to abandon it ever. There is no foreseeable increase in replicability that could lead to a gradual softening of this regulation in line with the ladder approach by Cave et al. (2001). Then only two deregulatory options would exist. The first and most straightforward would be to move to a general receiver-pays principle. That would resolve the termination pricing issue and would put the emphasis squarely on the qualitative conditions and location issues of interconnection. The second would be to live with imperfect termination markets.

4.3. Telecommunications and economic development

While a development gap in telecommunications between the high income countries and the developing world is substantial generally and with respect to the digital divide, the chapter on "Telecommunications and Economic Development" by Bjorn Wellenius and David Townsend shows that telecommunications in developing countries can be changing as much as or more than in developed countries. The emergence of China and India as potential world leaders in the telecommunications sector is an example of the latter phenomenon. At the same time, new gaps emerge within the developing world between leaders and laggards.

The chapter demonstrates at many places the strong relationship between institutional change and the evolution of the sector. Countries that have privatized, liberalized, and created stable regulatory institutions have also made greater progress in telephone and Internet penetration and technical developments than the countries continuing government ownership of dominant providers without opening telecommunications markets to competition. Competition, in particular, is a more powerful driver of telecom expansion than privatization alone. Internet services in developing countries expand best if left largely unregulated and with open access to the public networks. Protecting established network providers often conflicts with the introduction of more modern and cheaper services. This holds particularly for VoIP that is forbidden in a substantial number of countries, in order to protect dominant network providers from an erosion of long-distance and international telephone call charges. However, VoIP will eventually emerge into a formidable competitor for established networks, whether forbidden or not.

The authors point out the consequences of improved telecommunications sectors for the economies as a whole, as improved telecommunications sectors enable developing countries to export services thereby replacing white-collar workers in high income countries. This suggests that countries that have not yet reformed their telecommunications sectors should see their opportunities from joining the reform bandwagon. At the same time, as the next chapter by Spiller demonstrates, the reform ability itself depends on a country's institutional endowment so that reform in countries with inadequate institutional endowments may not lead to the desired outcomes. It would therefore be important to learn about reforms that either failed in the sense that they had to be abandoned before completion or in the sense that, although completed, they did not yield the desired outcomes.

Universal service policies in developing countries usually mean something quite different from such policies in high income countries. In high income countries, universal service concentrates on the access of individuals or household to basic or advanced services. In contrast, in developing countries universal service policies largely concern the connection of villages or remote areas to modern communications, including Internet access. Both have in common an urban-to-rural cross-subsidization pattern along with a similar political-economy drive.

4.4. Institutional changes in emerging markets: implications for the telecommunications sector

Pablo Spiller's chapter on "Institutional Changes in Emerging Markets: Implications for the Telecommunications Sector" builds on and extends a framework he has developed earlier. This chapter starts from the observation that telecommunications utilities are characterized by large and sunk investments, by the presence of economies of scale and scope and by products that are mass consumed. The combination of these features has traditionally led to state ownership and/or government regulation of the sector and has made pricing in the sector a politically sensitive issue. The author particularly emphasizes the short-term political profitability of expropriating the sunk costs in favor of low prices for the masses, requiring safeguards against government opportunism in order to achieve investment in this sector. State ownership is then interpreted as the result of the inability of the polity to commit not to expropriate private investments in the sector. Viewed this way, continued state ownership would not necessarily reduce investment if the alternative is a private sector suffering from massive government interference.

Central to the author's analysis is the institutional endowment of a country that determines what governance and incentive structures for the telecommunications sector are feasible and optimal for a country. The institutional endowment includes different political, legal, and economic institutions and different populations, standards of living, and geography. In order to prevent regulatory opportunism, the endowment needs to restrict a regulator's discretion. At the same time, the telecommunications sector, in particular, requires flexibility of regulations to adapt to a quickly changing environment. Overcoming this tension and achieving the right balance between restrictions and flexibility is the difficult task of sector reform. The author uses the U.S. and U.K. as two different, largely successful models of institutional endowments, where the U.S. takes a decentralized and the U.K. a centralized approach.

The main feature of the U.S. regulation is that telecommunications regulators have discretion to reform within statutory and due process limits, but they are monitored first by a court and then by legislators who step in with new legislation, when regulators and courts, in interpreting the current law, significantly deviate from what the legislative body wants. Through the Telecommunications Act of 1996 this only occurred very late in the U.S. telecommunications reform process, after competition in long-distance services and telecommunications equipment and the divestiture of AT&T took place without legislative interference. In contrast, in the U.K. the telecommunications reform process started with legislation, the subsequent judicial reviews have been much less pronounced, government rather than regulators took the principal regulatory decisions and licenses were a major instrument of limiting regulatory discretion.

The author takes these observations as the starting point for developing theoretical ideas about regulatory governance. He first concentrates on federal systems and brings out that the conditions for a strong tradition of judicial review of regulatory decisions are that the division of powers is real and that the judiciary cannot be manipulated by political parties. This chapter further hypothesizes that independent agencies will be favored in countries where the legislative and executive branches are not necessarily dominated by the same party. Under those circumstances, as the legislators cannot hope to implement the regulatory laws themselves they have to rely on courts to interpret the laws so that regulatory agencies cannot deviate too much from legislative intent.

Interestingly, in Germany, a country with no tradition of regulatory discretion, fairly independent telecommunications regulation was introduced in 1998, along with the liberalization of the telephone sector. This led to an unprecedented explosion of administrative lawsuits challenging regulatory decisions. In Germany, the judicial control was part of the existing regulatory endowment, but its major applications to the telecommunications sector had to wait until a new regulatory institution was introduced that involved discretionary decision making.

It is expected that the new telecommunications framework of the European Union, discussed extensively in Chapter 13 by Geradin and Sidak, will increase the discretionary powers of the national regulatory authorities vis-à-vis their own governments and legislatures. This again can be interpreted under the author's framework from the separation of two main institutional decision makers, in this case the European Commission on the one hand and the individual country legislators and their governments on the other. It will be interesting to see how this plays out. For unified government systems spiller particularly considers the properties of contract based regulation, showing how this can lead to commitment, provided there is an independent judiciary to uphold contracts. Under such a system reform can be achieved through contract renegotiations.

While Chapter 14 on "Telecommunications and Economic Development" ended on the note that it is the institutional choice that allows technical progress, this note could mark the beginning and is at the center of the current chapter. In fact, the statement could be reversed. Technical progress and market growth substantially reduce the governmental and regulatory expropriation problems, while at the same time reducing governmental and regulatory abilities to finance and manage the sector. In light of this, technical progress, market developments and the institutions that govern the telecommunications sector mutually influence each other.

5. Conclusions

The chapters describe the continuing evolution of a sector from both technological and institutional perspectives. Central to the evolution of the sector is the Internet, which, while available in component form much earlier, has emerged as a systemic phenomenon in the last decade. With its coming, a number of allied phenomena have emerged. The first is that old countries and economies, in communications sector terms, such as the United States and Europe, are still struggling with old questions, such as interconnection and the existence of market power. On the other hand, new countries and economies, particularly those in Asia, have leapfrogged ahead both technologically and in terms of future economic progress through a process of bypass of the regulatory regimes that were seen to be not conducive to the spread of technology.

Next, the emergence of new technologies and technological systems, often like the Internet via a self-organizing process, has created new types of networks, such as overlay networks, and new types of bottlenecks. The necessity for domain names has created a new industry segment that is not at all well understood, and yet this segment can be critical in the years to come. While the availability of technological options means that capacity will no longer be a constraint, identity can now be a constraining feature and the provision of identity a source of market power with its attendant conflicts.

Understanding the industrial organization of several new sectors, the advances in institutional and mechanism design that must, of necessity, accompany the formation and analysis of new sectors, learning about new business models, their applicability, newly evolving operational concerns, and the role that new geographic locations play, all are a set of agenda items that will occupy the field for years to come. In this spirit, we present the chapters in this volume of the Handbook as continuations of and precursors to an interesting set of debates that will remain in progress for a considerable period of time.

References

Cave, M., S. Majumdar, H. Rood and T. Valletti, 2001, The Relationship between Access Pricing Regulation and Infrastructure Competition, Report to OPTA and DG Telecommunications and Post, March.

Vogelsang, I., 2003, The German Telecommunications Reform: Where did it come from, where is it, and where is it going? Perspektiven der Wirtschaftspolitik, 4(3), 313–340, September.

Chapter 2

EMERGING NETWORK TECHNOLOGIES

DALE N. HATFIELD

Interdisciplinary Telecommunications Program, University of Colorado at Boulder, Boulder, USA

BRIDGER M. MITCHELL

CRA International, Palo Alto, USA

PADMANABHAN SRINAGESH

CRA International, Palo Alto, USA

Contents

Handbook of Telecommunications Economics, Volume 2, Edited by S. Majumdar et al.

DOI: 10.1016/S1569-4054(05)02002-6

1. Introduction

The traditional public switched telephone network described in Volume 1 was optimized to handle ordinary, one-to-one voice conversations in a technically efficient manner. Its architecture reflected the requirements for human voice communications—the amount of transmission capacity needed for an individual call, the frequency and duration of those calls, and similar factors. The technologies used in the network have changed dramatically over time: switching, for example, has evolved from manual switchboards, to electro-mechanical switches, to computerized switches. Still, the basic architecture has remained remarkably stable. More recently, however, that architecture has been under intense pressure to evolve due to fundamental changes in the marketplace—increasing competition on the supply side and rapidly changing technology and new services and applications on the demand side. This chapter describes how the traditional public switched telephone network is changing and suggests several ways that emerging technologies are likely to affect message communication.

Our discussion proceeds in several sections. Section 2 contains basic background material for the other sections. We describe the characteristics of three major types of signals and include a brief, high-level discussion of how traditional voice, data, and video (cable television) networks evolved in response to those characteristics.

Section 3 reviews the architecture of the traditional public switched telephone network (PSTN). We provide a more detailed discussion of how changes in demand and the need to remain competitive with other alternatives are impacting the PSTN and describe the substantial disadvantages or limitations of its architecture in responding to those changing demands.

The heart of the chapter is Section 4, which explains in detail how the fundamental architecture of the traditional PSTN is changing to overcome these technological limitations and respond effectively to new demands and competitive conditions. We describe the Internet Protocol (IP) and architecture of the Internet and propose an ultimate vision of the evolved network in which all applications—voice, data, image, video, and multimedia—are conveyed to an all-digital, packet-switched, broadband, and low-latency network.

Although the focus of this chapter is on the traditional wireline telephone network, other major telecommunications platforms are under similar pressures. How these other networks evolve has competitive implications for the traditional telephone network. Section 5 briefly describes how cable television, commercial wireless networks, and satellite networks are responding to pressures similar to those on the traditional wireline telephone network.

In the final section, we examine some economic issues and consider the relevance of both positive and negative message externalities in light of emerging messaging technologies.

2. Voice, data, and entertainment video signals

2.1. Signal characteristics

The signals of voice telephony, computer-to-computer data, and entertainment video have very different characteristics. The networks that were designed to meet these different requirements have fundamentally different architectures. We briefly examine these different characteristics, which are summarized in Table 1.

Basic signal type. There are two fundamental types of signals—analog and digital. As humans, we communicate in an analog world and voice telephony signals and entertainment video signals are basically analog signals. In contrast, digital computers and data communications signals are basically digital[1].

Bandwidth of a signal is the measure of information content per unit of time[2]. Unfiltered, the human voice has a bandwidth of several thousand cycles per second (hertz). Long experience backed by scientific measurements, however, has established that the bandwidth of a voice signal of a only few thousand hertz is adequate to preserve fidelity for ordinary conversations[3]. The bandwidth required for data communications is highly variable, ranging from the few bits-per-second (bps) necessary to carry a 'mouse click' from a computer client to a server, to hundreds of thousands or even millions of bps needed to download a video clip from the server or to allow two mainframe computers to exchange large files quickly.

Holding time refers to the duration of a call (in telephony) or of a session (in data communications). Telephone calls have holding times of several minutes while holding times for data-communication sessions vary widely—from a few seconds to hours. Holding times for entertainment video are not clearly defined, but single channels—absent 'channel surfing'—are typically viewed for about ten minutes at a time.

[1] There are also two types of networks—analog and digital. To send digital signals over an analog network, a device called a *modem* (an acronym for the term modulator-demodulator) converts digital signals to analog signals for conveyance over the network; a second modem reverses the conversion at the other end. To send analog signals over a digital network, a *codec* (an acronym for the term coder-decoder) converts analog signals to digital signals and then back again at the other end.

[2] The concept of bandwidth is used in two, interrelated, ways: (1) the range of signal frequencies that a communications circuit or channel will pass; and (2) the range of signal frequencies that are occupied or utilized by a particular type of signal, say an ordinary analog voice or entertainment video signal. The amount of information that a circuit or channel can carry per unit of time depends upon the bandwidth available. Although bandwidth is an analog concept, it has been adopted for use in the digital world where bandwidth is measured in bps.

[3] How much bandwidth to provision for a particular service is a fundamental tradeoff in telecommunications network design. If the bandwidth of the information source is greater than the bandwidth of the circuit or channel, then some information (e.g., quality) will be lost in transmission. However, while providing greater bandwidth improves the quality ('fidelity') of the transmission, it usually comes at an added cost.

Table 1
Characteristics of different types of signals

	Voice telephony	Data (computer)	Entertainment video
Basic signal type	Analog	Digital	Analog
Bandwidth	Few thousand hertz	Highly variable	Few megahertz
Holding times	Several minutes	Varies widely	Tens of minutes
Activity	'Continuous'(see Note 4)	Bursty	Continuous
Delay tolerance	Very little	Some (few seconds)	Good (one-way)
Error tolerance	Good	Variable	Fair
Symmetry	Symmetrical	Often asymmetrical	Asymmetrical (one-way)

Activity refers to actual end user consumption of the available bandwidth during the call or session. Ordinary voice conversations are essentially continuous—one party to the conversation or the other is always talking and consuming the available bandwidth on the connection[4]. Entertainment video signals also have continuous activity during the session. Data communications signals, in contrast, are often quite bursty—once a session is established, the activity is not continuous: a person checking e-mail may pause for long periods to read individual messages before generating a response, or an ATM machine with a dedicated connection to a distant computer may only be used sporadically[5].

Delay or *latency* refers to the time taken for information (e.g., an analog voice syllable or a packet of data) to travel from one point in the network to another. Contributors to latency include the irreducible propagation delay due to the speed of light, various processing delays such as the time it takes to convert an analog signal to a digital signal and compress it for transmission over a digital network, and congestion delays produced by queuing at routers in a packet-switched network. Two-way voice conversations are particularly intolerant of delay and one-way delays of more than about 150 milliseconds or round-trip delays of more than about 300 milliseconds significantly reduce the perceived quality of the voice signal[6]. Bandwidth and latency are two fundamental measures of network performance.

Many data communications applications are more tolerant of delay. For example, a delay of 1 second before a web page starts to download may be perfectly

[4] In an ordinary telephone call about 20 percent of the time represents normal gaps and pauses in the conversation. Sophisticated statistical multiplexing systems take advantage of these gaps to pack additional conversations into each direction of transmission.

[5] Various compression techniques can be used to reduce the bandwidth required for digital signals. Some compression techniques reduce the transmission (bit) rate when little change is occurring in the original (i.e., uncompressed) signal and increase it when large changes are occurring. This can produce 'bursty' signals as well.

[6] Variations in delay ('jitter') can also be troublesome in certain applications.

acceptable. Likewise, a delay of a few seconds may be tolerated when making a debit or credit card transaction at an ATM machine or retail store. However, long delays can be problematical in certain types of data communications applications. For example, some data communications protocols send a block of information and then wait for the distant computer to acknowledge error-free reception before sending the next block. If the roundtrip delays are large relative to the time it takes to send the block, then the transmitting node will spend much of its time waiting for the acknowledgement to be returned. Long delays may also be unacceptable in certain highly interactive games, such as those played over the Internet. In contrast, for traditional entertainment video, because of the one-way nature of delivery reasonable delays are not a problem—it doesn't make a practical difference if a video frame from a hockey game arrives at a television receiver a second or more after it is created at an arena.

Error tolerance—the frequency with which an incorrect bit of data is received reflects the susceptibility of the different types of signals to digital bit error rates. The corresponding analog term would be susceptibility to noise or distortion. Because of the redundancy in human voice communications, users are able to understand a voice message even in the presence of some noise and distortion or, equivalently, in the presence of errors in the digital signal conveying the voice communications. Data communications applications, on the other hand, are often completely intolerant of errors. For example, a single bit in error in a large computer program being downloaded can render the program inoperative. Similarly, errors in an e-commerce message transferring millions of dollars electronically could be disastrous to the institutions involved. Entertainment video signals are somewhat more tolerant of noise or bit errors. While 'snow' or other defects in the received signals are annoying, they are not as devastating as uncorrected errors in a computer communications applications. However, as television picture quality has improved (e.g., with the movement toward high definition television), transmission imperfections become more noticeable and, hence, less tolerable.

Connectivity refers to the linkages between the source of information and its destination. In traditional voice telephony, the connection is typically one-to-one—one user communicating with the other user. Likewise, in some traditional data communications networks, one computer or other digital device communicates with one other computer or digital device at a time. In contrast, in the delivery of entertainment video one source device simultaneously communicates with multiple receiving devices. Over-the-air telecast or cable delivery of a popular sporting event is an example of such a connection.

Symmetry refers to the evenness—or lack of it—in the information flows between the two end points of a connection. Traditional two-way voice communications are symmetrical and require the same bandwidth in each direction. Traditional data communications, on the other hand, are often asymmetrical. For example, a small home office user may communicate a relatively small amount of information upstream toward the Internet (e.g., 'mouse clicks') and receive vast amount of information downstream. At the other extreme, a large computer server 'farm' may receive downstream from the Internet only a few 'mouse clicks' and yet deliver vast amount of information upstream to the Internet. Since traditional entertainment video delivery is a one-way service, it is totally asymmetric.

2.2. Network architectures

A network is a set of interconnected nodes for transmission of voice, data, or entertainment video signals. Because of the differences in these types of signals, the networks designed to carry those signals differed in their fundamental architectures. Each network was optimized to handle a particular type of traffic and captured economies of specialization.

2.2.1. Traditional telephony

The traditional voice telephone network used a star topology that combined circuit switching with transmission that initially used frequency division multiplexing and later, with the conversion to digital transmission, time division multiplexing. The advantages of a star topology, circuit switching, and time division multiplexing for handling ordinary voice traffic are apparent when the characteristics of the signals are taken into account. Consider a small community with 100 households, each with a telephone and desiring to be able to communicate with the other household. One configuration would be to permanently wire each telephone to every other telephone in a *full mesh* topology. The number of wire pairs necessary to construct a full mesh network is given by:

$$M = \frac{N \times (N - 1)}{2}$$

where M is the number of transmission paths required and N is the number of nodes (in this case, the number of telephones) to be interconnected. The number of paths required goes up exponentially with the number of telephones; to interconnect just 100 telephones would require nearly 5000 transmission paths. Clearly, a full mesh topology would be very costly and very quickly become impractical. Moreover, because of the statistical nature of telephone calling, most of the links between individual telephones would be lightly, or perhaps never, used if the two households were not acquainted.

The local telephone network therefore evolved using a star topology in which each home was connected to a central switchboard or central office, thereby reducing the number of transmission paths to N. When two people at different locations wished to communicate with each other, the calling party requested that the central office connect the pair of wires serving him to the pair of wires serving the called party. The star topology and circuit switching dramatically reduce the number of transmission paths required and increase the utilization of those paths—but at the added cost of a switch. This trade-off between switching and transmission costs is one of the most fundamentals in telecommunications.

Local telephone networks could achieve a similar economy by using a star topology to connect central offices in large communities to each other so that a person served by one central office could complete a call to a person served by a distant central office. To reduce costs, a second level of switching was added to the local network. In this arrangement, each local switch or 'end office' is connected to the tandem switch rather than directly to each of the other end offices in the community. Reflecting the same switching-versus-transmission trade-off, this topology reduces the number of transmission paths or trunks required between end offices and increases their average utilization[7]. Thus, the traditional U.S. local telephone network represents a two-level hierarchy: customers are connected in a star topology to end offices and the end offices are connected in a star topology to one or more tandem offices[8].

One additional topic—multiplexing technology—will complete our discussion of the architecture of the traditional telephone network. One way of meeting the requirements for trunks between central offices would be to install individual wire pairs to create the transmission path; that is, there would be one voice conversation per pair of wires. As noted above, a voice conversation requires less than 4 kHz of bandwidth for reasonable transmission quality. Over short distances, however, an ordinary pair of wires has a significantly greater bandwidth and microwave radio links and fiber optic transmission links have bandwidths that are thousands or, in the case of fiber, even millions of times greater than the bandwidth required for a single voice call.

Multiplexing combines multiple signals onto a common transmission path and thereby reduces the number of individual transmission facilities needed to carry a given number of calls. There are two fundamental types of multiplexing for signals—frequency division and time division. In frequency division multiplexing, each conversation gets a fraction of the available bandwidth all the time, whereas in time division multiplexing each conversation gets all the available bandwidth but only for a fraction of the time. Traditionally, analog networks have employed frequency

[7] In actual practice, there may be sufficient traffic between two particular end offices to justify direct ('high-usage') trunking between those two offices. If all of the high-usage trunks are busy, an individual call can still be routed via the tandem office. This is referred to as alternative routing.

[8] Instead of sizing interoffice trunks to handle the maximum number of simultaneous calls that would occur, *traffic engineering* takes advantage of the statistical nature of telephone calling. Only enough trunks are installed to reduce the probability of a call between the two end offices being blocked to a small amount, say 1 percent. This illustrates a fundamental tradeoff between cost (in terms of added trunks) and performance (in terms of the probability of encountering a 'trunks busy' indication). The same cost-performance principle applies when considering not just direct trunking between end offices but the more complex local tandem network as well.

division multiplexing while digital networks have used time division multiplexing, although this is simply a matter of convenience, not of technical necessity[9].

One of the earliest forms of digital multiplexing allows the transmission of 24 two-way voice conversations over two wire pairs—one for each direction of transmission. This allowed a telephone network, for example, that needed 24 two-way voice trunks between two central offices to reduce the number of pairs of wires required from 48 to 4[10]. This illustrates another fundamental trade-off in telecommunications engineering—the multiplexing tradeoff. Multiplexing reduces the cost of installing additional physical transmission facilities—whether wire, fiber or radio—but requires additional expenditure for electronics[11].

It is no accident that the architecture of the traditional public switched telephone network is ideally matched with the characteristics of voice communications traffic that are summarized in Table 1. For example, it supplies just the right amount of bandwidth (4 kHz in analog terms, or 64 Kbps in digital terms) for a voice call. Moreover, because it utilizes circuit switching and time division multiplexing, once the call is established there is essentially no latency or delay between the originating and terminating ends. Traffic engineering and multiplexing allow efficient utilization of circuits based on the statistical characteristics of the traffic and efficient use of transmission facilities when voice traffic is being handled. However, as we will discuss in some detail at the beginning of Section 3, this traditional architecture has significant limitations when presented with the need to serve other emerging needs.

2.2.2. *Data networks*

The characteristics of data communications traffic, summarized in Table 1, led to a quite different network architecture for traditional data communications networks, one based on packet switching and statistical multiplexing in addition to circuit switching and time division multiplexing. Data networks incurred some degree of latency or delay in order to achieve fewer errors and greater efficiency in the use of individual transmission links. In packet switching, the unit of

[9] However, modern optical networks employ wave-division multiplexing, a form of frequency division multiplexing, to transmit digital signals.

[10] In the interoffice portion of the network, technical considerations normally require four-wire rather than two-wire transmission using one pair of wires in each direction of transmission. In the example, without multiplexing, 48 pairs of wires would be required to handle 24 two-way voice conversations. The T1digital multiplexing that is widely used in the U.S. operates at a transmission rate of 1.544 Mbps on each pair of wires in each direction and provides sufficient capacity to transmit 24 voice conversations, each digitized (using a codec as described in Note 1) at a rate of 64 Kbps. In Europe, the corresponding E1 digital multiplexing scheme operates at a transmission rate of 2.048 Mbps, providing for up to 30 voice conversations.

[11] There is also a tradeoff between bandwidth and transmission quality (Note 3). For example, music transmitted over an ordinary 4 kHz analog voice channel or a digitized (and uncompressed) 64 Kbps voice channel sounds poor because very low frequencies and very high frequencies are lost. Adding bandwidth to improve the quality comes at an added cost, especially in long-haul transmission.

information (in the form of binary digits or bits) to be transmitted is divided up into packets—groups or 'chunks' of bits—and the address of the receiving computer (in binary digits or bits) is included in the header of each packet. The packet is then sent over the network in store-and-forward fashion where each node (packet switch) reads the address and forwards the packet to another node closer to the recipient. If, instead of a packet-switched network, dial-up connections or dedicated connections between the terminal and receiving computer were used, the links would often be idle between the bursts of data communications. In packet switching, the links between packet switches carry traffic from multiple users so when one user is not transmitting packets others can utilize the available capacity. This sharing of transmission capacity is referred to as *statistical multiplexing*. Occasionally, however, the number of packets being sent will exceed the capacity (bandwidth) of a link or overwhelm the ability of the packet switches to process the packets, leading to queues and congestion delays.

In the very early days of data communications, the data terminals had almost no processing capability and the analog circuits used for data transmission (with relatively rudimentary modems) produced relatively high errors. Given the importance of error-free performance in data applications, the responsibility for error detection and correction was assigned to the network where some of the processing power in the packet switches performs the functions. Thus, a packet switch would store a copy of each packet that it sent to another node until it received an acknowledgement that the forwarded packet had been received error-free. When receiving a packet, the switch would validate the packet, and if an error was detected, it would request the sending switch to resend it. In an era of high-error rates, without such link-by-link checks for errors a significant portion of the network capacity would have been used simply to forward erroneous packets.

The extra processing at the packet switches to control errors added to the latency in a traditional data communications network. However, since modest delays—up to as much as several seconds in some important applications—are acceptable in data applications, this seeming disadvantage was more than compensated by nearly error-free performance and the efficiency gains realized from the statistical multiplexing of bursty data traffic. In short, traditional data communications networks were optimized to reflect the special characteristics of data communications traffic and the state of the network technology at the time. However, these delays meant that a network that used traditional packet switching and statistical multiplexing was unsuitable for the handling voice communications[12].

[12] Specialized forms of statistical multiplexing were developed for carrying voice conversations, especially on costly international circuits. These systems used digital techniques in specialized terminals to rapidly switch to another active conversation during the normal pauses of a conversation (e.g., when one party to a conversation is listening rather than talking). One prominent example of such a system was known as time assignment speech interpolation (TASI), developed in the late 1950s.

2.2.3. *Entertainment video*

The traditional networks used to deliver entertainment video signals to residential consumers operated on a broadcast (one-to-many) basis reflecting the characteristics summarized in Table 1[13].

In a traditional cable television system, the television signals to be delivered are received from 'off-the-air' broadcasts at the cable television 'headend' or are delivered to that point by terrestrial microwave or geostationary satellite systems. The collection of signals is then inserted into the coaxial cable, using frequency division multiplexing to separate the individual channels. The coaxial cable lines spread out from the headend in a 'tree and branch' topology. In this architecture, all the video signals are available everywhere in the network and all the channels are available to all of the customers. In contrast to the telephone network, there is no dedicated (unshared) facility that runs from the headend to each customer. Instead, when a customer is added to the cable network a small amount of the signal on the cable is extracted from the cable and fed to the customer's cable-ready television set or set-top box. In this case, the switching of signals occurs at the edge of the network, in the set-top box, and not in the network itself.

If most customers want to view the same set of video signals, this *bus topology* is extremely efficient. In a star topology used in the traditional telephone access network, in contrast, hundreds or even thousands of copies of the same program would be delivered over individual lines to residences in a neighborhood when a single shared line (the coaxial cable) would suffice. As cable networks have expanded their offerings to include individualized services—data communications (Internet access), telephony, and the more individualized video programming (e.g., video on demand)—the architecture of cable television networks have evolved further into a more star-like topology.

3. Traditional circuit-switched wireline architecture: limitations in the face of new demands

We have described the basic architecture of the traditional voice telephone network, observing that the traditional voice network's use of a star topology with circuit switching and time division multiplexing was ideally matched with the characteristics of voice traffic. It provides just the right amount of dedicated bandwidth for the duration of a call and, once the call is established, there is a minimal amount of end-to-end latency.

[13] Our focus here is on traditional cable television systems, which use coaxial cable as the medium of transmission, rather than over-the-air broadcast systems. The bandwidth of coaxial cable is significantly greater than ordinary twisted-pair copper cable used in traditional local telephone networks and is able to simultaneously transport scores of analog television channels.

However, this architecture is under increasing pressure to evolve. Changes in the marketplace include: (1) new demands and, in particular, customers' needs for the more efficient handling of data, image, and video traffic—multimedia signals—as well as voice signals; (2) the desire of the incumbent local exchange carriers to realize economies of scale and scope associated with handling a wider range of traffic and to respond to cable television competitors whose upgraded networks are capable of efficiently handling multimedia traffic; (3) rapid changes in underlying potentially enabling technologies (e.g., the increasing capacity and falling cost of computer processing and fiber optic transmission systems) that facilitate the evolution away from the traditional architecture; and (4) increased concerns about network reliability as firms, government agencies, and other institutions intensify their reliance on electronic communications.

The limitations or disadvantages of the traditional circuit switched wireline architecture include the following:

First, bandwidth is normally available in only fixed increments. The traditional telephone network is designed to deliver just the amount of bandwidth needed for a normal voice conversation—a 4 kHz analog channel or a 64 Kbps digital channel. There is no way of supplying varying amounts of bandwidth ('bandwidth on demand') that changes smoothly in response to the rapidly varying demands associated with data communications or compressed video[14]. These constraints are inherent in circuit switching and time division multiplexing—technologies that provide a fixed amount of bandwidth that can only be varied in pre-set multiples on a longer-term basis.

Second, dedicated capacity cannot efficiently handle bursty traffic. Once a call is established, the bandwidth is reserved exclusively for that conversation or session and cannot be shared with other users in moments when the channel is idle.

Third, the traditional voice network incurs high overhead due to the time and processing needed to set up and take down the call. Before the actual conversation can take place, information must be exchanged between the end user and the network and between network switches to establish the link-by-link connections[15]. This includes, for example, assigning a particular time slot in a time division multiplexer for the duration of the call. The overhead contributed by analog dual tone, multifrequency (DTMF or touchtone) dialing on ordinary subscriber lines is particularly onerous, even if the required tones are computer/modem generated. The overhead requirements make the traditional circuit switched telephone

[14] With packet switching and statistical multiplexing, the bandwidth consumed can be varied by simply transmitting more or fewer packets in a given amount of time. This provides short term 'bandwidth on demand.' Above a certain maximum rate of course, excessive delays (latency) and/or dropped or lost packets will result.

[15] *Signaling* is the exchange of information between the subscriber and the circuit switched network or between switching machines in the network necessary to establish the end-to-end connection.

network inefficient for handling brief (short holding time) messages. For example, circuit switching is poorly suited for downloading a web page that contains information from multiple servers at different sites.

Fourth, the traditional voice network also suffers from limitations on end-user signaling, such as the ability of end users to exchange signaling messages between themselves before the actual conversation path (connection) between the two is established. On the traditional analog loop, subscriber signaling is done 'in-band' using the same audible frequency range as the voice conversation. This seriously limits signaling that is possible during the conversation itself (i.e., receiving the calling number of an incoming call during a conversation or instructing the network to deal with the call by, e.g., routing it to a voice mail system). Another limitation is that the traditional telephone instrument is capable of generating only a very limited number of signaling messages. An ordinary telephone generates only 10 digits and 2 special characters (* and #)[16]. Finally, as pointed out above, with the current telephone network architecture, it is not possible for two customers to exchange signaling messages in advance of setting up the conversation path[17]. When telephone instruments were electro-mechanical devices with limited functionality this was not a serious limitation because, for example, an unattended instrument would not be able to respond to an incoming signaling message in any event. With today's more advanced devices (e.g., powerful personal computers), end-to-end signaling could enable the development of more sophisticated services. We give an example below.

Fifth, even though the interoffice portion of the network is nearly all digital (i.e., voice signals are carried in digital rather than analog format), much of the access network, especially in residential areas, still employs analog transmission[18]. Relying on analog technologies can degrade performance and necessitates the use of modems to transmit digital data signals.

Sixth, the access network that is used to serve many small businesses and nearly all residences uses twisted-pair copper wire as the transmission medium. The bandwidth of a twisted-pair copper wire depends on distance but cannot exceed several hundred kilohertz with typical wire (local loop) lengths. This ultimately

[16] One additional piece of signaling information is available on a traditional analog loop. When the subscriber goes 'off-hook' by lifting the handset, a switch in the telephone set closes. The resulting current flow to the central office—or its absence—is used as a simple signal (e.g., to indicate to the central office switching equipment that the subscriber desires to place a call, or is already engaged in a call). During a conversation switch, hook 'flashes,' which briefly interrupt the current, can be used for signaling, but this is crude at best.

[17] After the call is set up, with a DTMF telephone it is possible to send signaling messages and, indeed, this is the capability subscribers use daily to access their voice mail systems or check their bank balances using automated response systems.

[18] With a traditional analog loop, the voice signal is normally converted from the analog format to the digital format at a port on the switch in the serving central office; the loop itself operates in the analog mode but the balance of the network is typically digital.

limits the maximum rate at which digital signals can be transmitted over an ordinary loop[19].

Seventh, as the traditional telephone network developed, it evolved from analog, frequency division multiplexing to digital, time division multiplexing. In North America time division multiplexing was organized in a hierarchical fashion with 24 64-Kbps (DS-0) voice channels multiplexed into a 1.544-Mbps bit stream (DS-1), and then with 28 1.544-Mbps DS-1 channels multiplexed into a 44.736-Mbps (DS-3) bit stream[20]. In order to maintain accurate timing across the network, these multiplexed systems operated asynchronously. As a result, adding, dropping, or cross-connecting individual DS-1 streams out of a higher rate circuit is a relatively cumbersome and costly multistep process that constrained the architecture of the traditional public switched telephone network and tended to exacerbate certain reliability issues.

Eighth, the traditional network is especially vulnerable to single-point failures. As described earlier, it is largely a two-level star topology in which customers are connected to local central offices (end offices) and the end offices, in turn, are then connected to the local tandem office[21]. This means that if the multipair cable connecting a set of residential customers or a large business customer to the end office is cut, service is lost. Likewise, if the cable is cut between the end office and the local tandem office, customers served by that end office would be cut off from the balance of the network. Moreover, this vulnerability to single-point failures was exacerbated by the utilization of asynchronous transmission which had the effect of concentrating the multiplexing/demultiplexing function at a few locations (e.g., at an existing tandem office).

The traditional telephone network was optimized to handle ordinary voice traffic and it did so in a generally efficient and satisfactory manner. Most of the limitations we have described become of concern only when the network is asked to provide services it was not originally designed for.

[19] Filtering associated with the analog-to-digital conversion process limits dial up modem speeds to approximately 50 Kbps in order to conserve the use of network resources to just that is necessary to convey voice calls with acceptable sound quality. As described later, the additional bandwidth on the individual wire pairs—beyond that needed for voice communications—can be used for other purposes (e.g., the provision of Internet access).

[20] The DS-1 bit stream at 1.544 Mbps utilizes the T1 transmission technology (Note 10). Note that the bit rates at DS-1 and higher are greater than the sum of the tributary rates because of additional bits are needed for overhead functions. In Europe and elsewhere the digital multiplexing hierarchy is different (Note 10).

[21] In practice, high-usage trunks may also be employed between certain central offices in a metropolitan area (Note 7). Thus, the local transport portion of the network is sometimes referred to as a partial (rather than full) mesh topology.

4. Evolution of the traditional wireline architecture

The architecture of the traditional wireline telephone network is evolving to overcome certain technological disadvantages and limitations associated with its roots in supplying ordinary telephony services. In this section we take up, in turn, developments in the technology of interoffice transport, interoffice signaling and the Intelligent Network, and the end-user access network. We then examine the architecture of the Internet and consider the technological trend toward a future all-digital, packet-switched broadband network.

4.1. Interoffice transport facilities

Interoffice transport facilities are the transmission links that connect end offices to each other and to their associated tandem offices. Because of the technical and economic advantages of digital transmission and its synergy with digital switching, the transformation from analog to digital technology in interoffice transport has long been underway and has largely been accomplished in most countries[22]. A second widespread change in the interoffice transport facilities is the replacement of twisted-pair copper wire or microwave radio by fiber optic cables. The latter have tremendous capacity and the practical effect is that, except for certain remote areas or under-developed countries, lack of bandwidth is generally not a technical problem except in the access portion of the local network.

The change to digital technology and fiber optic cables has been accompanied by the shift from asynchronous to synchronous transmission technology. The specific development is known as the synchronous optical network (SONET) in North America and the synchronous digital hierarchy (SDH) in Europe and elsewhere. The synchronous transmission standards define how the billions of bits flowing down a fiber optic cable are organized, the physical interfaces to the cables (thus promoting interoperability among equipment provided by different vendors), and certain overhead functions that facilitate rapid restoration of service when failures occur.

A major advantage of the SONET/SDH time division multiplexing technology is that it allows single-stage multiplexing and demultiplexing. This, in turn, allows more flexible network topologies, including *ring* topologies. In the latter arrangement, the end offices/tandem offices and other points of traffic concentration are connected to a fiber optic cable that is configured in a ring. In normal operation, the traffic flows around the ring in one direction and traffic is added or dropped by add-drop multiplexers (ADMs) at each location according to its origin and destination.

[22] The conversion from analog to digital technology is occurring in all portions of the telecommunications and related industries. For an in-depth discussion of the advantages of digital transmission, storage and processing of information, see, for example, Bellamy (2000).

The advantage of the ring topology is that, in the event of a cable cut or other failure, the traffic that would normally flow across the failure point can be rerouted in the other direction around the ring and still reach a node beyond the break. The redundancy inherent in the ring, and the associated network management capabilities of the SONET/SDH technology, virtually eliminates the single-point-of-failure problem associated with the traditional star architecture used in the interoffice portion of the local telephone network[23].

Used together, digital transmission, fiber optic cable, SONET/SDH technology, and more robust topologies overcome most, if not all, of the technological issues associated with the interoffice portion of the traditional local wireline telephone network. These interoffice technologies are capable of handling not only traditional circuit-switched voice traffic, but also of providing a robust platform for the handling of packet-switched data traffic as well.

4.2. Interoffice signaling and the intelligent network

As noted earlier, the traditional public switched telephone network is connection-oriented. Before end user information can be exchanged between the calling and called parties, signaling sets up a dedicated path or connection between the two end points, a link-by-link connection that must be maintained for the duration of the call. During the conversation (exchange of end-user information), it is as if a pair of wires were dedicated to connecting the two end points (e.g., telephones). To set up and take down a circuit-switched call, signaling information must be exchanged between the end users and the network (subscriber loop signaling) and also between the switches in the network (interoffice signaling). The exchange of the signaling information is logically separate from the exchange of end-user information that takes place after the connection is established.

There are two major categories of signaling—in-band and out-of-band. Signaling information can be carried in-band as audible tones over the same path that will eventually carry the actual conversation. For example, on an ordinary analog telephone line, the called number is signaled by a sequence of audible DTMF tones[24]. With out-of-band signaling, the signaling information is separated and moved outside the audible bandwidth and carried either on the same circuit or on a separate network. The out-of-band signaling information associated with a group of interoffice trunks between the two switches can be aggregated and carried on a common transmission path between the switches, a technology called common channel interoffice signaling (CCIS).

[23] The ring topology can also include end-user customer locations where traffic volumes and reliability concerns justify it.

[24] At one time, similar multifrequency (MF) tones were used for signaling between switching machines in the interoffice network.

Establishing a connection between two switches requires an exchange of messages which are inherently digital (e.g., the telephone numbers of the calling and called parties). Such bursty digital messages are typical of data communications more generally. Thus, over time, a specialized, packet-switched technology was developed to carry signaling messages between circuit switches in the traditional voice telephone network. The modern version of this technology is known as Signaling System No. 7 (SS7). The SS7 signaling technology provides faster call setup time, more efficient trunk utilization, and, with the addition of computer processing power as explained below, support for a plethora of advanced, network based services. The packet switches, which serve a collection of switching machines or service switching points (SSPs), are called signaling transfer points (STPs).

Today, the traditional public switched telephone network consists of two logically distinct 'sub-networks'—the conversation sub-network and the signaling sub-network. In the interoffice portion of a modern telephone network, the actual conversations are carried on a circuit switched sub-network using time-division multiplexing, while the signaling information is carried on a specialized packet switched sub-network using statistical multiplexing. Both sub-networks typically share the same underlying interoffice transmission facilities that use digital, time-division multiplexed, SONET/SDH-based technology described above[25].

At a conceptual level, a circuit switch consists of two basic parts—the switching matrix and the service logic. The switching matrix or 'fabric' makes the connection between an incoming port and an outgoing port on the switch. These ports may be connected to subscriber lines or to interoffice trunks. The service logic or 'intelligence' decides, based, for example, on the digits dialed by the customer and information stored in a routing table, which incoming line or trunk to connect to which outgoing line or trunk. In early implementation if the call is destined for another switch, the service logic in the originating switch communicates with the service logic in the terminating switch to establish the necessary link between the two switches (e.g., to assign a particular trunk or time slot for the duration of the call). In the early days of telephony, the services rendered by a switch were relatively simple and the necessary service logic resided in the mind of a telephone operator who manually made the connections between lines and trunks in response to spoken information from the subscriber or another operator. Operators and manual switchboards were eventually replaced by automated switches that used electro-mechanical devices for both the switching fabric and the service logic, and later electro-mechanical switches were replaced by computer-based switches in which the service logic resided locally in software rather than in hardware.

[25] Packets are typically carried on an underlying time-divison multiplexing (TDM) network by disassembling each packet into sequential groups of bits and transporting the groups of bits in the time slots associated with a particular circuit. This process is then reversed at the receiving end.

These computer-based switches, called stored program control switches, could provide a much richer set of services compared to their electro-mechanical predecessors. For example, the more powerful logic allowed a subscriber to instruct the switch to automatically redirect calls from one line or telephone number to another (call forwarding) or, if the called line was busy, to forward the call to a voice messaging machine (voice mail). In the era of manual and then electro-mechanical switches, and even in the early days of stored program control switches, the service logic was tightly tied to the switching matrix or fabric of individual switches. This meant that in order to create a new service, the software (and any associated database) had to be modified in each individual switch. The development of the sophisticated and specialized packet-switched SS7 signaling network, however, allowed most of the service logic to be moved out of the switch and be accessed remotely. This separation of the service logic and the switching matrix is referred to as the intelligent network concept.

In the intelligent network, a computer containing service logic (called a service control point) and an associated database is connected to the signaling network's packet switch (STP) that serves a number of circuit switches (SSPs). When a subscriber connected to one of the local circuit switches triggers a particular service, the call is briefly suspended and the local switch sends a signaling message to the service control point via the SS7 network asking, in effect, how the call should be handled. The service control point uses its service logic to decide and returns the answer via the SS7 network.

Perhaps the best example of an Intelligent Network-based service is toll-free '800-number' or 'free-phone' service. When a subscriber dials an 800-number, the particular dialed number acts as a trigger, causing the local circuit switch, in essence, to suspend the call briefly and forward the called 800-number and the subscriber's calling number to the service control point over the SS7 network. The service control point uses the 800-number to interrogate a database that associates the 800-number with an ordinary (routable) telephone number. The 800-number is translated into an ordinary telephone number and returned over the SS7 network to the originating switch. The originating switch then handles the call in the normal way[26]. A simple extension of this basic 800 service would be a feature that would allow the subscriber to the 800-number service (the called party) to specify different routings depending on the time-of-day or the geographic location of the calling party.

Using combinations of called number, calling number, time of day, information requested from and entered by the calling party (e.g., 'Please enter your account number'), and information stored in the network (e.g., location information on

[26] Although the call is routed as an ordinary call once the number translation has occurred, the billing is handled differently to reflect that the called party (the subscriber to the 800-number service) pays for the call.

their stores supplied by 800-number subscriber), extremely flexible services can be created[27]. Indeed the intelligent network-type architecture provides a platform for a host of services including personal number calling, local number portability, virtual private networks, wireless network roaming, and prepaid calling cards. Because the service logic and associated databases can be centrally located, changes to that service logic and those databases only have to be made at one location rather than at each local switch—a significant advantage with some services.

The intelligent network architecture allows the service logic (the intelligence) to be located at various places in the network. For example, consider an abbreviated dialing service that enables a consumer to dial a short sequence of digits, say *13, to indicate that he wants to call the thirteenth person on a personal list of frequently called people. The service logic and associated memory (database) could be located in the handset itself, if it had the requisite intelligence. That is, when the customer dialed *13, the handset would intercept the digits, translate the sequence into a regular number and transmit it to the local central office in the normal way. A second alternative would be to attach the service logic and associated database to the local central office. When the customer dialed *13, the equipment in the central office would recognize that a special service was being requested by this customer and do the necessary translation based on service logic contained in the stored program control switch and information previously supplied by the calling subscriber and stored locally. The third alternative would be similar, except that the service logic and associated database would be located remotely from the local switch and accessed via the packet based signaling SS7 network using the intelligent network-type architecture. Note that in the second alternative, it could be the local telephone carrier offering the service and in the third alternative it could be a long distance carrier[28].

The optimal location for the intelligence or service logic associated with a given service depends on a host of factors such as the cost of computer processing and electronic storage, the cost and latency of moving signaling messages from one point to another, operational costs associated with updating the logic and data in numerous locations versus a handful of locations, customer preferences for

[27] Within the constraints on end-to-end signaling (Note 17), it is also possible to supply similar services using logic located at a subscriber's location. For example, once the call is set up, a customer can enter a particular digit to have his or her call routed to a particular department within a company. As discussed later, with more 'intelligent' end user devices, it is even possible for the customer to speak the number and have it recognized by the equipment at the terminating end of the call. Here, though, the focus is on services that are offered using logic and associated databases located within the provider's network—not on the customer's premises.

[28] Perhaps a more familiar example is voice mail, where the service can be provided in an end user device (i.e., in an answering machine on the customer's premises) or in the public switched telephone network itself.

storing certain information at locations they control versus on computers controlled by others, and the impact of failures.

The location and control of the intelligence or service logic has enormous competitive consequences. In the traditional public switched telephone network, the intelligence—the service logic—resides within the network itself and is under the control of the operator. The network assumes that end user equipment has limited functionality ('dumb terminals') and the public switched telephone network operator controls access to the packet switched interoffice signaling (SS7) network. As we will discuss later, in the much different architecture of the Internet the network itself is 'dumb' and the intelligence (or service logic) resides in devices (hosts, e.g., personal computers and servers) that are located at the edge of the network, outside the control of the network provider. In a sense, the Internet is like the telephone network 'turned inside out.' Moreover, in the Internet, both the signaling information and the end users' information content are carried over a common, general purpose, packet switched network.

4.3. The access network

Our discussion to this point has concentrated primarily on the steps that the operators of the traditional public switched telephone network have taken to reduce or eliminate the limitations associated with the interoffice network. While large customers in most developed countries typically have access over high capacity, digital transmission facilities, residential users are often served over traditional analog loops that employ in-band (DTMF) signaling with its associated limitations. This is true even though the local central office switch itself is usually digital. That is, the voice (or voice-band data) signal is carried in the analog format over the local loop and the necessary analog-to-digital and digital-to-analog conversions occur at the central office.

Beginning in the 1960s, however, the telephone industry developed and later began to deploy a new digital access technology—the integrated services digital network (ISDN). It takes advantage of the fact that ordinary twisted-pair copper wire has capacity (bandwidth) that exceeds that necessary to transmit a single voice conversation. Two versions of ISDN were developed—basic rate interface (BRI) and a higher capacity primary rate interface (PRI). With basic rate ISDN, the net transmission rate is 144 Kbps in each direction and it uses TDM to provide two voice (or data) channels of 64 Kbps each plus a separate (out-of-band) signaling channel at 16 Kbps. With primary rate ISDN, the net transmission rate is approximately 1.5 million bits per second in each direction and is channelized to provide 23 voice (or data) channels at 64 Kbps each along with a separate 64-Kbps signaling channel[29]. Because of the higher data rate associated with primary rate

[29] Basic rate ISDN is a worldwide standard. The Primary rate ISDN described in the text is the North American standard. In Europe and other parts of the world, PRI provides 30 voice (or data) channels at 64 Kbps each, plus a signaling channel at 64 Kbps.

ISDN, digital repeaters must be employed at regular intervals to regenerate the digital signal. The network operator tightly controls access to the signaling channel, and the end user generally cannot use the signaling channel as a general-purpose packet network to, for example, send his own message over that channel to another end user.

While ISDN overcomes many of the limitations associated with the traditional access network, providing faster and more flexible out-of-band signaling and more robust digital transmission, it has not been particularly successful in the U.S. marketplace[30]. Reasons advanced for this lack of success include not having uniform standards early in the rollout process, the high cost of the digital customer premises equipment compared to mass produced analog equipment, the high cost of terminal adapters necessary to accommodate legacy analog equipment during the transition to ISDN, and substantial improvements in analog modem technology that reduced the advantages of the per-channel ISDN transmission rates of 64 Kbps in data communications applications. More fundamentally, with ISDN, user content is still circuit switched and it is voice-oriented in terms of its channelization. Hence, for increasingly important data communications applications, ISDN does not provide the advantages of statistical multiplexing, always-on connectivity, and 'bandwidth-on-demand' that can be realized with packet switching technology.

More recently, the telephone industry has deployed a new access technology—digital subscriber line (DSL). Like ISDN, DSL is basically a multiplexing technology that exploits the fact that, over limited distances, an ordinary twisted-pair copper wire has a bandwidth that significantly exceeds that occupied by a single voice conversation[31]. One popular version of the technology, asymmetrical digital subscriber line (ADSL), allows the simultaneous transmission of analog voice and digital data signals over a single local loop. The high speed, two-way, digital signals do not interfere with a voice conversation being carried on the same local loop because analog voice and digital data signals occupy different frequency bands. With DSL technology, frequency division multiplexing is used to transmit relatively high rate, two-way digital signals in the otherwise unutilized bandwidth that lies above the frequencies that convey the analog voice conversation. The combined transmission rate of the two-way digital signals is limited by the maximum bandwidth of the twisted-pair copper wire which, in turn, depends on its length, the distance between the customer location and the serving central office and other factors. This version of DSL is asymmetric because more digital

[30] In the U.S., ISDN has had some commercial success in commercial rather than residential applications including access to call centers; in Europe, it has achieved considerable success even in the residential space.

[31] It is the filtering associated with the analog-to-digital conversion process at the central office, or as described later, in a remote multiplexing terminal closer to the customer, that limits the information carrying capacity of the twisted-pair copper cable in the loop plant.

capacity is supplied in the downstream direction (toward the customer from the central office) rather than in the upstream direction[32].

When DSL technology is deployed in the access portion of the network, a filter at the customer's location splits the analog voice conversation from the two-way data signals and routes the voice signals to ordinary analog customer premises equipment. The digital signals in turn are sent to a modem-like device that connects to the customer's personal computer or other equipment. In the central office, at the other end of the DSL loop the analog voice signals are split off and connected to the regular circuit switch and the data signals are split off, aggregated in a device called a DSL Access Multiplexer, and connected to the data network (i.e., the Internet). Note that, in the local loop, the capacity of the high data rate (broadband) digital signals can, in turn, be further divided using either time division multiplexing, statistical multiplexing, or both[33]. In addition to the economic advantages of allowing the provision of broadband services on existing loops, this access network technology has the advantage that, after arriving at the central office, ordinary voice telephone traffic is processed in the traditional way using circuit switching and time division multiplexing while the data traffic can be handled using packet switching and statistical multiplexing.

Digital subscriber line technology effectively supports two parallel networks—one optimized for voice and other optimized for data—over a common 'last mile' path to the home or small business. Hence, unlike ISDN, the parallel data network has the advantages of statistical multiplexing and always-on connectivity, and of providing bandwidth on demand.

The traditional twisted-pair copper cables used in the access network can be supplanted by electronic technology and ultimately fiber optic cables. A single copper feeder cable running from a central office to a particular neighborhood may contain several hundred individual wire pairs. By using digital TDM the number of wire pairs required to serve the neighborhood can be reduced, an increasingly attractive alternative with the falling cost (and size) of the needed electronic equipment. These digital loop carrier (DLC) systems then use ordinary analog lines to complete the connection from the multiplexing equipment (called remote terminals) in the neighborhood to the individual customer locations. Thus, the analog-to-digital conversions take place at the remote terminal rather than in the central office. The multiplexed connection between the remote terminal and the central office requires fewer wire pairs (hence the term 'pair-gain system').

[32] There are also symmetrical versions of DSL that do not include simultaneous carriage of an analog signal, including high bit-rate digital subscriber line (HDSL).

[33] For example, two home computer users can simultaneously access the Internet using simple local area networking techniques in the residence and statistical multiplexing based on packet switching and the IP over the broadband connection. As we discuss in more detail later, this broadband connection can also be used to provide Voice-over-the-Internet Protocol (VoIP) services at satisfactory quality if sufficient bandwidth is available and the latency and/or dropped packets due to congestion are properly controlled.

The copper cables in the feeder portion of the access network can be replaced by fiber optic cables running from the central office to the remote terminals, offering both the potential economic advantage associated with the using fiber-optic technology and reducing the length of the ordinary copper wires serving the customer. This, in turn, can alleviate the normal distance limitations associated with the use of DSL technology and/or increase the digital transmission rates of existing DSL customers.

However, as we now discuss, the ultimate vision is to have both voice and data (including image and entertainment video signals) carried on a common, broadband, digital connection to the home or small business.

4.4. Architecture of the Internet

In contrast to the circuit-switched, connection-oriented, TDM technologies of the traditional telephone network, the Internet is based on packet switching and statistical multiplexing technologies using protocols referred to as the IP suite. Computers connected to the Internet are known as *hosts* and each host has a unique IP address. In the Internet, addressed packets of information (IP packets) are sent through the network from one router (packet switch) to the next router in store-and-forward fashion. The router performs a 'dumb' function—it examines the address information and very rapidly forwards the packet to another router topologically 'closer' to the intended recipient. Unlike earlier packet switched data networks, Internet routers are not responsible for correcting errors in the information contained in the packets, for lost packets, or for assuring that the packets are delivered in the right sequence. Instead, the hosts (e.g., personal computer 'clients' and 'servers') at the edge of the network take on this responsibility.

In concept, the Internet is comprised of a series of logical layers. At the IP *packet layer*, the Internet, unlike the traditional telephone network, is connectionless. This means a packet of information can be sent from one end user or third-party service provider to another without first establishing a connection through a call setup procedure. Several sessions or 'calls' can exist between the end points, but these are established in the *transport layer* between blocks of complementary software that resides in the host computers that are controlled by the end users or third-party service providers and are individually identified by a port number.

The creation of a session or connection does not have to involve the network itself; the Internet simply forwards IP packets based on the address information in the packets. What is established, then, is not a 'real' connection that consists of identifiable, physical capacity dedicated to the session for its duration. Rather, the connection is a relationship between transport layer software residing in hosts that allows packets of information to be moved or transported from one port to another. In the case of the most common transport layer protocol, the Transmission Control Protocol (TCP), the software detects packets with errors and requests repeats, delivers packets to the application in the order that they were sent, and eliminates duplicate packets[34].

At the third, *application layer* of the Internet, other software can, in turn, use these virtual connections and other services at the transport layer to provide a great variety of user-oriented applications and services—for example, to exchange a collection of e-mail messages between an e-mail client and an e-mail server or to stream music from a digital audio music server to a music client in an end-user device. Given sufficient bandwidth and processing power, a single host or client can have multiple concurrent sessions, enabling the end user to download e-mail and listen to streaming music simultaneously over a single physical connection.

The openness of the Internet is summarized by the *end-to-end principle*—the ability to send an IP packet containing signaling information or data, image, video, and even voice media content from one host to another host, each located somewhere at the edge of the network, without interference by the underlying service provider. The end-to-end principle, combined with the creation of independent protocol layers, represents the critical architectural features incorporated in the Internet. Because the protocol layers are independent, an application is not tightly coupled to the characteristics of the underlying network and the application software resides in hosts (e.g., in personal computers and servers) that are at the edge of the network outside the control of the underlying service provider. In essence, the Internet architecture provides a powerful general-purpose packet-switched network—a platform that is capable of handling both signaling information and media content on an end-to-end basis. In contrast, the traditional public switched network combines a specialized, tightly controlled packet-switched sub-network having relatively limited bandwidth for signaling with a relatively inflexible circuit switched sub-network for handling media content[35]. These fundamental architectural differences between the traditional public switched telephone network and Internet have far-reaching implications in terms of both the ability to support multimedia applications and in the control over the development of such applications.

Several examples will illustrate. The ability to send sophisticated signaling messages between end user devices before a connection is established allows the creation of a number of advanced services. For instance, one end user can send another end user a signaling message that says 'I would like to talk to you' and the other end user's device can respond automatically with a signaling message that says 'I am unavailable at the moment; would you like to leave a message on my voice mail system'? The calling party can then make the decision whether or not to place the call to an answering machine or voice mailbox. (In the traditional public

[34] Because IP and TCP are often used together, the Internet suite of protocols is often referred to as TCP/IP.

[35] As explained before, traditional circuit-switched networks cannot smoothly supply varying or 'elastic' amounts of bandwidth—bandwidth on demand—to end users.

switched telephone network, there is no way of exchanging information between end user devices before the session or call is established; a caller does not know whether she will reach an answering machine before the call itself is established[36].) The called end user's Internet device can also be programmed to return a new address (thus providing call forwarding) or to provide an e-mail address and an invitation to the calling party to send a text message instead. Again, we emphasize that such services can be developed in software in end user (or third party) devices without the knowledge or cooperation of the underlying Internet service provider.

As a powerful general-purpose packet switched network or platform, the Internet (as well as business intranets based on the IP suite) is capable of handling any combination of sound, data, image or video traffic. To continue the above example, end users can negotiate, say, the audio quality of the connection (e.g., by using more or less compression or bandwidth) or the combination of media to be used (e.g., voice and video or voice and still images) before a conversation is established. Furthermore, the quality or combination of media and number of connections can be renegotiated during the conversation or session. For the most part, these features are impossible to duplicate on a traditional circuit switched network using time division multiplexing and legacy forms of signaling.

From an economic perspective, the general-purpose network also allows the capture of economies of scale and scope associated with carrying all types of traffic—and combinations of traffic—on a common platform. Thus, it not only provides a powerful platform for the creation of interactive services that employ animation, streaming music and video, and other special effects, but it also provides economic efficiencies by, for example, carrying voice and data on common rather than separate networks.

Finally, as touched upon earlier, in the Internet architecture, the network itself is dumb while the intelligence or service logic and associated databases reside in hosts at the edge of the network. Because the Internet lowers the cost and increases the flexibility associated with accessing intelligence and because the end-to-end principle ensures that end user devices will be able to access that intelligence wherever it resides, the Internet fundamentally changes the trade-offs associated with both where that intelligence is located and who has control over it. With the ability to create applications residing in end-user and third-party devices at the edge of the network, the Internet—in contrast to the closed, proprietary data networks that preceded it—facilitates the development of revolutionary new services such as the Worldwide Web, Instant Messaging, and VoIP. It empowered developers ranging

[36] End user devices connected to the PSTN (e.g., answering machines) traditionally did not have the intelligence to handle such signaling messages. However, as voice telephony switches to packet-switched networks, telephones and computers are becoming more alike. This means that the devices have the necessary intelligence as well as greatly expanded capabilities for entering information (e.g., keyboards and tablets), processing information (e.g., running games), and displaying information. This trend is perhaps most clearly seen in wireless communications. It stands in stark contrast to the relatively simple telephone sets of the past.

from teenagers working in basements to computer scientists working in the best-equipped laboratories. The intelligence and associated databases can be concentrated in only one or a handful of servers, or distributed in a decentralized fashion in millions of end user devices, or any combination of the two extremes. For example, the software associated with interactive games can be located in a handful of third-party provided servers accessed by less sophisticated end user devices, or, alternatively, in a personal computer or specialized game console that uses the Internet mainly as a medium of communications among players.

The implications of the Internet for telephony are similarly revolutionary. The intelligence or service logic and associated databases necessary to provide voice telephone service over the Internet (VoIP) can be concentrated in a few servers, or it can be distributed in a highly decentralized fashion on end user devices. The Internet provides the basic connectivity among calling and called parties—using the dumb network functionality—while all the intelligence necessary to provide services associated with traditional telephone service (e.g., call forwarding, abbreviated dialing, calling card, and 800-number services) resides in end user or third party provided devices (or some combination of the two) at the edge of the network[37]. A customer can access the intelligence, and hence the services, wherever a connection to the Internet is available with sufficient bandwidth.

Providing telephone service over the Internet has many implications. Service creation is no longer limited to just the underlying carrier. And the variety of services that are now regarded as constituting voice telephony can be separated from the network that provides the underlying basic connectivity. Also, the individual services (beyond basic connectivity) that comprise voice telephony service can further be unbundled so that a customer can choose among providers of different elements or can, with the more distributed arrangements, in effect self-provide. Finally, in the extreme, if essentially all of the intelligence or service logic comes to reside in end user devices, telephone companies as we know them today would disappear[38]. Stated another way, voice telephony could become simply another software application on end user devices[39].

[37] Of course, for many years, telephony services provided on the traditional PSTN will necessarily have to coexist with telephony services providing via VoIP. Gateways between the two types of networks will be needed to translate PSTN signaling protocols (e.g., SS7) and circuit switched/time division multiplexed voice on one side into Internet signaling (e.g., SIP) and packet-switched/statistically multiplexed voice with different coding on the other side (and vice versa).

[38] Certain functions, such as the administration of numbering resources, will continue to require centralized management and it is likely that there will be some other services that are more efficient to provide on a centralized basis. The extreme scenario is useful as a way of understanding the implications of current trends.

[39] Indeed, some suppliers of VoIP are arguing that they are not 'service providers' but merely suppliers of equipment or software that allow end users to make the functional equivalent of telephone calls. Suppliers of equipment or software are not usually subjected to telecommunications regulation.

In the past, the principle challenge to providing telephony over packet-switched networks has been to provide adequate quality for services that are sensitive to latency or variations in delay (jitter). Indeed, such delays were accepted in order to obtain the other advantages of packet switching, such as bandwidth on demand and statistical multiplexing, in data communications applications that were not particularly sensitive to small delays. Delays in two-way voice telephone service and certain highly interactive data services are principally produced by a lack of adequate bandwidth and/or processing delays in the packet switches themselves[40]. Because of the normal statistical variations in the amount of traffic offered to the network, the delay or latency varies over time.

Conceptually, the latency for delay-sensitive applications can be reduced in three different ways. First is the 'brute force' method of adding greater bandwidth and faster switches to the network to reduce delay even at peak traffic times. The falling costs of bandwidth (at least in the interoffice portion of the network) and processing power make this a viable solution. Second, priority can be given to packets associated with delay-sensitive applications such as two-way voice at the expense of occasionally delaying packets associated with less time-critical applications such web browsing[41]. The third method involves reserving sufficient capacity along the route to ensure that latency encountered by packets associated with a particular session or call does not exceed some maximum amount[42]. Because network capacity must be reserved in advance, this method requires a call setup phase and the service supplied to delay-sensitive traffic becomes connection oriented. There are, of course, tradeoffs in making a connectionless packet-switched network look more like a connection-oriented network with capacity reservations. The tradeoffs include both the time it takes to reserve the capacity (the call setup time in circuit switched network terms) and the loss of the statistical multiplexing advantage when reserved capacity goes unused[43]. All three methods of reducing latency are being adopted in varying degrees in both the public Internet and in the managed networks that utilize the IP suite.

[40] Compare a toll road, where traffic can be backed up because there are not enough lanes to carry the number of vehicles (analogous to bandwidth) and/or the toll-takers do not collect the charges fast enough (analogous to processing delays in the packet switches).

[41] In contrast to circuit switching, where the necessary capacity is reserved in advance during the call setup phase of the connection, this method does not completely eliminate the possibility of excessive latency for delay sensitive applications. (Excessive delay—in the form of no dial-tone—can arise in a circuit-switched network if resources are not available.) However, with adequate bandwidth and extremely fast packet switches, giving priority to certain packets may be adequate to reduce the latency to acceptable levels.

[42] An approach based on a combination of technical solutions including multiprotocol label switching (MPLS) is described in Xiao, Telkamp, Fineberg, Chen and Ni (2002).

[43] With very short call setup times, the capacity need not be reserved on a continuous basis for both directions of the conversation as in conventional circuit switching. That is, the capacity can be reserved only when bursts of speech occur.

As described earlier, in the traditional telephone network, the principal constraint on bandwidth has been in the access network. This limitation is being relieved by the deployment of DSL technology and by installing fiber optic feeder cable from central offices toward end users. In addition, as will be described in more detail in a following section, cable television systems are installing two-way digital coaxial cable that provides a second source of high-bandwidth access. Combined with the techniques for reducing latency just described, it now appears feasible to handle all types of traffic—voice, data, image, video and combinations thereof, on a common network—or more properly—a network of networks.

4.5. The network of the future

Based on the descriptions contained in the prior sections of this chapter, the trend is towards an architecture where all applications—audio/voice, data, image, and multimedia—are conveyed to an all-digital, packet-switched, broadband, low-latency network of networks that uses common, open standards and protocols (i.e., the IP suite). Using the modularity and layering associated with modern communications protocols, this network, in turn, will increasingly rely on fiber optic transmission facilities for the needed bandwidth and on wireless technologies to extend the network to allow users to communicate anyplace, anytime, in any mode or combination of modes. In this architecture, traditional voice conversations will become just another digitized, compressed, and packetized bit stream along with data, image and video traffic and with the intelligence or service logic defining the equivalent of traditional voice services being distributed throughout the network and not necessarily controlled by traditional telephone carriers.

5. Evolution of cable, wireless, and satellite networks

The dominant trend in the traditional public switched telephone network is towards an architecture wherein all applications—audio/voice, data, still image, video, and multimedia—are conveyed to an all digital, packet-switched, broadband, low latency, network of networks that uses common, open standards and protocols (i.e., the IP suite). This trend also affects the evolution of cable television, wireless, and satellite networks.

5.1. Cable television

The 'tree and branch' architecture of cable television networks is evolving rapidly toward a star-like configuration, one better suited for delivering more individualized voice, data, and entertainment video signals. The newer hybrid fiber-coax (HFC) architecture is a star composed of separate runs of fiber optic cable from the cable headend to individual nodes serving clusters of homes in relatively small geographic areas. At each node optical signals are converted into electrical signals, which are then distributed to individual homes in a bus topology using conventional coaxial cable and two-way amplifiers. In this arrangement, the entire bandwidth of the coaxial cable is shared among only a few hundred homes. Its capacity is divided into non-interfering frequency bands or channels that simultaneously deliver traditional analog entertainment video programming, digital television signals (including more individualized 'pay-per-view' programs), and broadband, two-way data communications (high data rate access to the Internet using cable modems). The broadband data capability, using the IP suite, is capable of handling voice, data, image, and some video traffic on an integrated basis. In time, it is expected that most of the remaining analog signals on the cable (entertainment video programming) will be digitized, compressed and packetized and carried in the same integrated manner.

Viewed from afar, the traditional cable television and traditional public switched telephone networks look increasingly similar. Both employ fiber optic cable in a star topology to reach nodes that serve clusters of residences (or connect fiber directly to high volume and high value customers such as large businesses) and from there rely, respectively, on their traditional coax or twisted-pair copper cable facilities to connect to individual customers. Both platforms provide IP-based, packet switched, high-data rate (broadband) and low latency access services while maintaining support for certain legacy applications. Cable television networks use cable modems and the HFC technology; telephone networks use DSL technology over shorter runs of twisted-pair copper cable. And both platforms can eventually migrate the legacy services to the more advanced packet-switched, broadband technology and, eventually, to fiber-optic cable that connects all the way to the home.

5.2. Wireless

Interestingly, wireless networks—both those serving fixed subscriber units and mobile subscriber units—are also evolving towards a similar architecture. The trend in wireless access systems is to deploy nodes (base stations) throughout a large geographic area and to connect them back to a central location (a mobile switching center in the case of cellular mobile radio systems) using fiber-optic or other broadband (microwave) facilities in a star topology[44]. The balance of the distance between the node or base station and the fixed or mobile subscriber is spanned using wireless (radio) technology. Through the use of relatively low power transmitters and antennas close to the ground at the nodes, the geographic coverage and interference range associated with each base station is limited. This,

[44] In actual practice, for the reasons stated earlier, a ring network may be used for the backhaul from the base station to the central location, rather than a star. However, this does not change the more generic notion of having high capacity facilities serving local nodes that, in turn, connect to coaxial cable, twisted-pair cable, or radio to create the balance of the connection.

in turn, allows the same scarce radio spectrum to be reused many times in a given geographic area with concomitant improves in spectrum efficiency.

Just as the cable television network is evolving with separate cable runs to each cluster of end users, this reuse of the radio spectrum has the advantage of allowing more capacity to be delivered to each subscriber. The actual amount of bandwidth depends on the amount of spectrum that is assigned to a particular provider. This can range from an amount sufficient to support only narrowband applications up an amount sufficient to support true broadband applications.

Beyond the changes in topology, the evolution of mobile telephone (cellular) services has followed a path similar to those outlined above for traditional cable and telephone networks. First-generation (1G) cellular systems employed analog, narrowband, and circuit switched technology that was optimized for voice. The second-generation (2G), digital systems were still narrowband and employed circuit switching; although optimized for voice, they were somewhat more 'data friendly[45].' The third generation (3G) cellular systems, designed for data services, are digital, packet-switched, and offer variable bandwidth[46].

In summary, all three networks—telephone, cable, and cellular—are evolving to a similar structure with fiber-optic cable (or another high capacity facility) providing transport to a node near the subscriber and with coaxial cable, twisted-pair copper cable, and radio, respectively, going the remaining distance and supporting IP-based, packet switched, high-data rate (broadband) and low latency access services.

5.3. Satellite

Communications satellites, which have played a key role in the telecommunications industry but have not been discussed earlier, are also evolving in similar ways. The commercially important satellites are those in geostationary orbits above the equator[47]. A basic satellite network consists of the satellite itself plus associated earth stations that communicate with each other using the satellite as a relay point.

Terrestrial microwave radio signals can only travel a few tens of miles because the very high frequency signals are blocked by the curvature of the earth; longer distances require microwave 'repeaters' placed at regular intervals to relay the

[45] Second generation cellular systems are also being upgraded to so-called 2.5G technology, utilizing packet switching techniques albeit with bandwidths less than that associated with 3G systems.

[46] While the focus here is on traditional cellular systems and their evolution, the connection from the node (base station) to the subscriber can also be made using other technologies in both licensed and unlicensed (license-free) bands. 'Wi-Fi' systems, which use unlicensed spectrum, are being rapidly rolled out in many countries as a way of providing high speed Internet access in homes, businesses, and public spaces.

[47] Low Earth Orbiting Satellites (LEOS)—non-geostationary orbit satellites—have achieved only limited success in commercial applications and will not be discussed here.

signal from one terrestrial antenna tower to the next. In their most basic form, microwave repeaters consist of: (1) a receiver (and associated antenna) which collects incoming signals in one frequency range coming from the previous microwave tower; (2) a device which processes and then translates the signals to a different frequency range; and (3) a transmitter and associated antenna which amplifies the translated signal and aims it toward the next tower in the chain.

A satellite system dramatically reduces the distance limitation associated with terrestrial microwave by putting the repeater (transponder) far out in space so that it is visible over a large fraction of the earth's surface. A geostationary satellite is about 35,000 kilometers (22,000 miles) above the earth.

The earth stations associated with early generations of geostationary satellite systems were very large, costly installations that used analog techniques and frequency division multiplexing to provide interoffice transport of large volumes of telephone calls or to extend network television signals between major centers. Higher power satellites, additional spectrum, and improvements in earth station designs have allowed for smaller, more easily installed earth stations. These advances initially allowed entertainment television signals to be distributed efficiently to hundreds of individual cable television head-ends scattered across a wide area and, eventually, to directly broadcast signals to individual receive-only earth stations ('dishes') owned by consumers.

This satellite-based system architecture matches closely the characteristics of entertainment video summarized in Table 1. A single earth station (uplink) can use the geostationary satellite to simultaneously deliver, on a one-way basis, scores of television signals to millions of individual households. Using the satellite receiver (dish) the consumer can choose from scores of different channels. These direct broadcast satellite systems use digital transmission and advanced compression technology. They are ideal for the point-to-multipoint situation in which many customers want simultaneous access to the same set of video signals.

As with traditional cable television network architecture, satellite networks are constrained when it comes to two-way data communications (Internet access), voice telephony, and the delivery of more individualized video programming (video on demand).

The irreducible round-trip delay (latency) for a radio signal to reach the satellite and return to the earth—about one-quarter of a second—is not a defect when delivering entertainment video programs because it doesn't make a practical difference if a video frame from a hockey game arrives at a television receiver a second or more after it is created at an arena. However, the same latency makes satellite systems less desirable for voice (except in more isolated or difficult to serve areas where no terrestrial facilities are available) and largely unsuitable for certain two-way data applications (e.g., highly interactive games).

A satellite system with very wide geographic coverage is also less suited to deliver more individualized signals. For example, there are hundreds of terrestrial broadcast television channels in the U.S. To provide program offerings comparable to a cable television system, all these local channels would need to be carried on the satellite. But doing so would exceed the capacity of the satellite system and waste radio spectrum resource by spreading signals over large geographic areas where there may be few viewers interested in receiving most of the local stations. This constraint can be partially alleviated by using satellite antennas that transmit a narrow spot beam, allowing different sets of signals or channels to be delivered to various regions using the same radio spectrum. This technology is similar to the reuse of spectrum accomplished in cellular mobile radio systems and in cable systems that use the HFC architecture. However, spot beams are less satisfactory for two-way services. For example, in order for two consumers served by different spot beams to communicate, two satellite relays or hops are required (exacerbating the latency problem) or switching within the satellite is required (with associated increases in the cost and complexity of the satellite).

These technical challenges have led satellite service providers and telephone companies to combine capabilities and create packages of services to compete with the bundled offerings of cable television companies. Whether the satellite operators can develop a standalone, integrated package that supports all applications—audio/voice, data, still image, video, and multimedia on a common platform remains to be seen. Nevertheless, satellites' unique ability to serve isolated and other hard to serve areas would seem to guarantee them an important segment of the market even if certain compromises must be made.

6. Economic issues in the telecommunications sector raised by converging technologies

We now turn to an assessment of several of the larger economic issues that emerge from the major trends in telecommunication network technology we have identified. In broad terms, the convergence of previously quite distinct telecommunications networks for voice telephony, cable television, and data services toward a common technological platform—one with broadband digital, packet-switching, low latency, and open standards—will greatly increase the competitive pressures throughout the telecommunications sector. At the same time, technical characteristics will enable each modal type of network to maintain differentiating characteristics that convey advantages in serving particular demands. And a third set of issues will also gain increased importance—those associated with message communication, the proliferation of unwanted messages, and the preservation of identity and integrity in anonymous digital networks.

6.1. *Increased possibilities of competition*

6.1.1. *End-to-end competition*

The fundamental trend in each of the major telecommunications networks—voice telephony, mobile telephony, cable television, and data communications—toward a common technological platform greatly increases the possibilities that formerly disparate networks will be in direct competition to offer consumers many of the same services. This evolution is clearly apparent in mobile, broadband, and messaging networks.

Mobile telephone networks have matured rapidly, to the point that mobile service substitutes for primary fixed telephone service in meeting the demands of a growing number of subscribers. As mobile operators fill in gaps in coverage and as technological improvements in speech encoding and antennas continue, mobile operators will compete more effectively for an increasing share of the fixed voice telephone market.

The market for broadband data access to the Internet, once supplied entirely by dedicated digital T1 telephone lines, is now vigorously contested by telephone operators with DSL and by cable television operators with digital cable modem service. In addition, fixed wireless access (Wi-Fi) may complement or compete with both.

Messaging, when conveyed in digital form, is a service that is readily transported over any sort of network. Wireline telephone, mobile telephone, and data networks can all strive to supply voice/fax/e-mail messages over an integrated service.

6.1.2. *Network components competition*

Technological convergence similarly increases the substitutability of individual components of the networks and thus heightens competition at the 'wholesale' level of the telecommunications sector. In a world with a mixture of analog and digital, circuit- and packet-switching, and both narrowband and broadband network technologies, any interchange of components usually required additional interface technology—analog-to-digital modems, protocol conversion, and the like.

At the ends of the network, end-user access can be provided by DSL, coaxial cable, or fixed wireless loops. These access facilities can interconnect with a fixed telephony network or a data communications network, which may be provided by unaffiliated firms. Moving in from the ends of the network, switching need not be provided by the incumbent local wireline telephone network operator, but can be supplied by a long-distance, cable television, or competitive local exchange carrier.

Perhaps the most far-reaching substitution of network components may occur with the maturation of VoIP, when voice telephony is packetized and transported over data networks. As voice traffic migrates to data networks, routers and ATM switches will substitute for traditional voice circuit switches and personal computers and corporate LAN infrastructures (with associated servers) may substitute for traditional telephones and PBXs.

The opportunities for components competition are significantly affected by regulatory policy. Requirements that local wireline telephone networks unbundle their network elements and provide them to competitors at prices based on forward-looking economic costs have further enabled the entry of competing local service suppliers who rely on the incumbent network for key components. However, other (cable, mobile wireless, and satellite) networks have not been subjected to similar unbundling rules.

6.1.3. Imperfect intermodal substitutes

The convergence of basic technologies is creating very substantial opportunities for the major types of telecommunications networks to compete in supplying end-user services. Nevertheless, important technical differences between the various modes of wireline, wireless, cable television, and data communications networks will likely continue to differentiate telecommunications services in some applications and for particular consumers.

First, services supplied to end-users by different modes vary in characteristics that are highly valued by some consumers and for some applications. The quality of voice and video communication, the year-round reliability of service, and the assurance of prompt delivery of messages, will continue to differentiate networks. Geographic coverage also distinguishes network providers—wireline telephony is effectively ubiquitous, but the mobility provided by payphones is inferior to that afforded by wireless providers. And communication capacity varies substantially, with cable and high-speed fixed wireless links providing high bandwidth that is unavailable, or far more costly, over mobile and twisted pair telephone links.

Second, technologically integrated networks allow some operators to ensure quality of service and to introduce innovations that competitors, using interconnected, disparate equipment and technology, have difficulty supplying. For example, unbundled elements of an incumbent's local telephone network can enable entrants to provide competing local services. Nevertheless, dependent on a rival organization and numerous technical interfaces, the entrants may be unable to provision service, repair faults, and create new service offerings that are fully comparable in the eyes of consumers. Similarly, a video service provider who maintains end-to-end control of its traffic is able to avoid quality degradation that can occur if packets are routed over the public Internet.

Third, network operators can use technology strategically to differentiate their service offerings from competitors. Proprietary protocols encourage customers with common interests to purchase from the same supplier and enable an operator to capture some of the gains from network effects. Open protocols and interfacing standards, on the other hand, weaken this grip and may foster faster growth through easier entry of new suppliers and the shift of intelligence to devices at the edges of networks.

6.1.4. Message communication issues

The development of multiple messaging technologies increasingly provides an individual consumer with a plethora of communications channels. Messages can be delivered over different devices, including the telephone, fax, and computer. The devices and accounts have a variety of addresses—fixed line voice and fax telephone numbers at more than one location, mobile phones, and multiple e-mail accounts.

The multiplication of messaging channels creates a need for tools to manage and unify messaging services. Early technologies of this sort include a single personal telephone number (700 service), 'intelligent' message attendants that can route calls to the currently-active device and multimedia agents that can receive and retrieve voice and text messages from dissimilar devices. As these technologies mature, it is likely that unified messaging service will emerge to integrate functionalities most demanded by consumers.

6.1.4.1. Trends in message communication. Telecommunications messages, like other messages, are consumed by two (or more) persons. The message—a letter, telephone call, e-mail message, or fax—is a local public good that directly affects the utility of both parties to the communication. Whether it is a synchronous, interactive telephone call or a message delivered to a voice-mail, fax, or e-mail account, a message is jointly consumed. Thus, message communication involves externalities that can broadly be classified in two categories. The first is network externalities, or network effects, in which the value of subscribing to a network depends on the ability to exchange messages with others, and hence on the number and identities of other subscribers to the network. Network effects in telecommunications have been considered in Volume 1 by Liebowitz and Margolis. The second category of externalities is message (or call) externalities, in which the value of a message (or telephone call) depends, not solely on the value to the person initiating the message, but also on the utility it provides to the recipient.

Although most message communications have been assumed to be of positive value to both parties, one consequence of recent trends in telecommunications technology has been explosive growth in the volume of negative 'message externalities.'

A second thrust of technology trends is the growth in 'single-party' forms of communications. Web browsing, file transfers, streaming audio and video, and other uses of the Internet provide utility to the initiating user. While other parties are certainly involved in supplying the content of such communications, the utility obtained by those other parties resembles the consumption of other goods supplied in economic markets—the suppliers of those services realize value through direct or indirect sale of the consumed services, often promoted by advertisements and other forms of marketing. Consequently, message externalities are generally of less significance for this category of uses.

6.1.4.2. Message externalities. A typical telephone call begins with one person (the calling party) placing a telephone call to another person (the called party). The call is established when the calling party picks up the phone—in most wireline networks, this is the moment at which usage-based charges start to accumulate. Subscribers who also have caller ID or a privacy manager service can identify the calling party and then decide whether to answer the call or not. Thus, the call is established jointly by decisions made by both the calling and the called parties, reflecting the expectations of both that the call will be valued.

Once the call is established, the call proceeds as long as both parties to the call wish to continue the communication, and the call concludes when one party, or both, decides to hang up. The call will be terminated when, for either person, the additional benefit of continuing the conversation is less than the total cost of the call to that party—the price of the call to that party plus the opportunity cost of the party's time. The duration of a call (and hence the calling volume) is determined by the party that disconnects, often in agreement with the other party. Calling volume is not unilaterally determined by the caller. Rather, the monthly calling volume between any two subscribers is determined by the demand of the party with the smaller demand at the given prices, irrespective of who initiates individual calls[48]. Additionally, both the caller and the called party consume the same volume of calling.

Both parties to a call obtain benefits (generally positive, but possibly negative for the called party) from the call. With few exceptions, the literature assumes, often implicitly, that the call externality is positive, even though most demand analysis has also assumed that consumption decisions can be satisfactorily modeled by representing the volume of calls as a function of the price to the call initiator alone[49]. This assumption might initially appear to be reasonable since calling parties will not make calls that have negative value to them, and called parties are likely to hang up on unwanted calls. However, the popularity of

[48] For these reasons it is misleading to regard the calling party as the 'cost causer' of all calls or as the decision maker with regard to call volume.

[49] For example, analyses of mobile call termination rates for the U.K. regulator (Oftel) have alternatively assumed (a) that both parties to a call obtain positive benefits, and (b) that the volume of fixed-to-mobile calls depends only on the price of fixed-to-mobile calls, while the volume of outbound mobile calls depends only on the per-call charge paid by mobile subscribers. Compare Rohlfs (2002a) with Armstrong (2002, pp. 2–3).

'do-not-call' lists and other efforts undertaken to avoid unwanted calls suggest that there is a substantial disutility generated by such calls.

Negative Call Externalities. That some calls create negative utility to those that receive them has been recognized for some time. Rohlfs, for example, noted that 'the value of the [telecommunications] service to others would probably be lessened if a large number of life insurance salespersons subscribed to the service to solicit other subscribers[50].' The implication is quite clear: receiving unwanted calls at dinnertime from importuning salesmen has disutility for many consumers. More than 50 million U.S. subscribers made their preferences clear by registering their home telephone numbers on a national do-not-call list established by the Federal Trade Commission and the Federal Communications Commission to block unwanted telemarketing calls. Similar arrangements were instituted earlier in Australia[51] and the U.K.[52]. In Europe, the EC requires that all member states implement similar mechanisms to stop unwanted calls.

Laws prohibiting some types of calls do not fully protect subscribers from receiving any unwanted calls. Consumers continue to demand a variety of complementary services, including unlisted numbers, answering machines, and even devices specifically designed to block calls from telemarketers[53]. While the high penetration rates of telephone service in advanced countries attest to the value subscribers place on connectivity, the high demand for, and expenditure on, a range of legal and technical tools to block unwanted calls suggests that most subscribers demand restricted—not unrestricted—connectivity. The implications for network architecture are a physical network that reaches all locations, combined with a virtual network that mirrors the restrictive connectivity demanded by subscribers.

Technology trends, particularly advances in computer telephony integration (CTI), undoubtedly account for much of the increased volume of calls undesired by one party. Computerized predictive dialers, cross-referenced databases of telephone numbers and subscriber characteristics, and call centers[54] have increased the productivity of telemarketing employees as viewed by marketing firms, if not by call recipients. Nevertheless, telephone marketing requires some labor input as well as payments for telephone network usage and therefore has a minimum expected cost per completed message, limiting the volume of undesired calls[55].

[50] Rohlfs (1974).

[51] http://www.aca.gov.au/consumer/fsheets/consumer/fsc16.pdf.

[52] See http://www.tpsonline.org.uk/tpsr/html/default.asp.

[53] Telezapper (http://www.telezapper.com) and PrivateTime PT1000 wireless phones (see http://www.3g.co.uk/PR/June2002/3595.htm) are two such devices.

[54] See Yarberry (2002).

[55] With the increasing migration of call centers to low labor cost countries such as India, the economic costs of these calls has been falling.

Internet e-mail, in contrast, has much lower incremental costs per completed message and multicasting technology has made marketing via unsolicited e-mail (spam) an increasingly costly annoyance to consumers and businesses.

Unwanted messages therefore impose a cost not only on the calling party, but also on the party called. But the corresponding incremental private cost of sending unwanted e-mail has become vanishingly small. Computer programs generate the mailing lists used by spammers, and computers transmit the e-mail messages with little human intervention. Since most spammers have access to flat-rate Internet accounts, the additional transport costs to the sender of sending spam are zero, and the additional cost of adding another e-mail address to a list is also negligible. Consequently, unlike telemarketing, the supply of spam is not effectively constrained by private marginal costs.

Until recently, regulations on traditional telemarketing calls did not apply to spam e-mail. In these conditions, the volume of spam (more generally, unwanted communications including spam, pop-up advertisements, viruses, and other intrusive communications) is likely to rise significantly, spurring the demand for services that filter out these communications. This has created a demand for new technologies (sometimes called middleware) at the application layer. In 2003, the U.S. passed limited legislation[56] requiring that unsolicited commercial e-mail be labeled as such and must include opt-out instructions, and prohibiting deceptive subject lines and false headers. The FTC has been authorized to establish a 'do not mail' registry. Eight bills addressing spam (one to wireless phones) are pending[57].

Positive call externalities. Despite the exploding volume of unwanted telephone and e-mail messages, the great majority of (at least telephone) message communication provides positive benefits to both parties. Yet it is not apparent that the volume of messaging arising from conventional price structures is efficient. For most messages only one party to a call faces a positive price (although both incur the opportunity costs of time engaged in conversation). Some analysts of traditional telephone services have suggested that when both parties are members of a community of interest and benefit from communicating with each other, it should be relatively easy for the parties to internalize the call externality, for example, by agreeing to alternate calls or for a parent to reimburse costs to a child[58]. When this is the case the effective prices paid after internalization are relevant for determining demand.

When informal arrangements are impractical, technological mechanisms provide some alternatives. For business subscribers seeking to attract calls from customers, toll-free 800-number services had grown to account for more than 40% of the traffic on AT&T's long-distance network by 1992[59]. The total number

[56] CAN-SPAM Act (S 877).
[57] http://www.spamlaws.com/federal/summ108.html#s877.
[58] See Rohlfs (2002a), p. 2.
[59] http://www.800voicemailstore.com/toll-free-history.htm.

of toll-free numbers assigned increased from about 4 million in 1993 to about 23 million in 2003[60]. Still, the available internalization mechanisms are limited by transactions costs—personal 800 numbers, collect calls, and charge-backs based on detailed call accounting record are all more costly than a directly dialed call. Moreover, 900 numbers, which allow one party to a call to be reimbursed for the opportunity cost of its time in addition the cost of the call, are quite expensive and have been abused, raising the cost of use to both parties.

For e-mail messages, file exchanges and streaming media, where computers mediate both sides of a transaction, it may be easier to support price structures that enable call internalization mechanisms. Technology for direct debits from one party to another (PayPal) can be more easily used to internalize the call externality, allowing individuals to reallocate the cost of the communications (including the opportunity cost of time and the costs of any goods traded as a result of the communications) in real time, on a transaction-by-transaction basis. These opportunities for electronic commerce will require the development of concomitant technologies (authentication, privacy, security, legal protections for fraud, etc.).

6.1.4.3. Formal results. Both prices and technology can help to filter out unwanted messages while ensuring that the efficient volume of desirable messages can be made. In Appendix A, we present a formal analysis of a communications network in which some messages create positive value for both parties while others create negative values for one party to the message. We show, under fairly general conditions that optimal prices will generally exceed marginal costs and that the optimal mark-up over marginal cost will be smaller when technologies for blocking unwanted messages are more widely adopted. Finally, we draw out the implications of successful filtering strategies for the deployment of new technologies, and hence convergence.

The key insight from the model is to categorize pairs of subscribers into two classes. The first consists of pairs of subscribers that together obtain a positive value from their communications when the usage charges equal marginal usage cost. The second class consists of all other subscriber pairs. Pairs of subscribers in the first category have the incentive (and several available techniques) to internalize the call externality and achieve the optimal calling volume. For this category of subscribers, provided that internalization is possible, any usage charges (levied on one or both parties) that exactly recovers the usage cost will be socially optimal. For other subscriber pairs, a usage volume of zero will be socially optimal. In general, this will require a usage charge on the calling party that exceeds marginal cost.

When a single usage charge is applied to the calls of both categories of subscribers, there is a trade-off between setting the price equal to marginal cost (to facilitate optimal calling volume among members of the first category) and

[60] FCC, Trends in Telephone Service, 2003. Table 18.3.

setting prices above marginal cost (to reduce unwanted calls). In general, the trade-off will be resolved by setting usage charges above the marginal costs of usage, but applying a smaller mark-up when the volume (or nuisance value) of unwanted calls is smaller.

6.2. Growth and technology

The explosive growth of data communications is one of the primary drivers of convergence. With rapid growth, operators are more easily able to transition from embedded products and a legacy architecture serving specific applications to the state-of-the-art products and converged architectures. A common approach adopted by carriers is to 'cap and grow'—to limit expenditure on embedded products, start installing new products, and replace embedded units with the new products over a chosen transition period. Equipment vendors adopt a complementary strategy of reducing support (e.g., stocking spare parts), and eventually discontinuing all support, for products that are no longer demanded by major carriers. If new regulations or spam-fighting technologies effectively reduced the volume of unwanted e-mail that is transmitted, the growth in data traffic would be reduced and the process of convergence may be considerably slowed. According to one report, the volume of spam doubled over a 6-month period, and 40% of all e-mail traffic at that time was spam[61]. The removal of nuisance traffic (including other annoyances such as pop-up banners) could have a substantial effect on the need for carriers to expand their networks, and legacy architectures might enjoy a longer life than would otherwise be the case.

7. Public policy puzzles

The converged networks of the future will, we postulate, be digital, packet switched, statistically multiplexed, low latency, and broadband. A 'dumb' network focused on transporting signals reliably and efficiently will form the core, with the intelligence (service creating capabilities) residing mostly in client and server hosts at the edge of the network. Control of services will increasingly pass from the traditional service provider (who heretofore owned the links and nodes comprising the physical network and largely defined and created services) to a highly diverse group of service and equipment suppliers. These networks will be, and in many cases already are, capable of providing many combinations of voice, data, still image, and video services—so-called 'rich media.'

This evolution will pose fundamental and far-reaching challenges to public policy. Pressures on traditional regulatory classifications and jurisdictions are already being felt in the evolving organization of public regulation, in the pursuit

[61] CNSNews.com (2003).

of traditional social goals (such as universal service) in the telecommunications sector, and in competition and innovation. The challenge to policymakers and regulators is to reorganize their institutions to reflect the process of convergence that is now well underway. We conclude this chapter with a number of puzzling questions thrown up by this challenge.

7.1. Regulatory organization

Telecommunications law, public policy and regulatory bodies have traditionally been organized along two major dimensions: technology and geography. These distinctions are rapidly losing meaning and creating inconsistent and unintended outcomes.

Over-the-air broadcasting, cable television, wireline telephony, and wireless telephony have previously been distinguished by key differences in technology and have been regulated by different legislation and operational departments and bureaus. As telecommunications technologies converge, these traditionally non-competing segments provide alternative means of delivering services that were earlier separated into distinct technological 'stove-pipes.'

Federated oversight of economic and social regulation has also been divided between predominantly local and more global authorities. In the U.S., jurisdiction is shared between 50 state commissions and a national regulator (FCC). In Europe, community-wide policy is set by a single commission, with implementation effected at the nation state level by national regulators.

With technological convergence, these industry/technological/organization distinctions will become increasingly ineffective. Today's telephone service can be provided over the 'dumb' network using intelligence, service logic, and associated databases that are distributed around the world. When this occurs the traditional telephone carrier disappears—customers (or companies integrating such services) can assemble their own telephone services (e.g., gateway services, directory services, call forwarding services or whatever) on a mix and match basis.

Distance as an organizing distinction also vanishes. A 'telephone' with its IP address can be anywhere in the world, connected to the Internet; the service provider has no knowledge of the instrument's location, as IP addresses do not have geographic significance. In such an environment pricing based on distance, and distinctions between local, state toll, and domestic and international long distance traffic lose all meaning. For example, one VoIP provider offers its subscribers a 'virtual' telephone number associated with the city of their choice. Calls originated to the virtual number from wireline telephones located in that city are treated as local calls, and billed as such by the local telephone carrier serving the calling party, even though the called party might be located in a different country.

In the U.S., for example, what is the future role of state regulators in such an environment when the 'service' may be provided via a server in Stockholm and it is impossible to know where a call originates and terminates? In the European Union, what is the role of the individual member states in this distributed environment?

7.2. Social goals

Funding social objectives from sectoral revenues becomes problematic when there is no telephone service provider in the traditional sense. When VoIP software applications are widely distributed over hosts at constantly changing locations around the world, the concept of a 'provider' is tenuous and likely beyond the legal jurisdiction of a regulatory agency. On the other hand, the underlying physical assets (such as cables, switches, and spectrum) that undergird the service applications are geographically fixed and easy to identify. Taxation to fund social goals could then shift to the assets of these providers. Is this a desirable public policy?

Who would be eligible to receive subsidies from universal service funds collected on this basis—only providers of physical infrastructure who contribute to the fund, or also service providers? If the latter, for what collection of services should they be paid? Similarly, if particular social goals are to be pursued—such as providing access to persons with disabilities—on whom does the regulator impose those obligations? For example, today, ordinary physical telephones are supposed to be accessible—will a softphone displayed on a computer monitor also have to be accessible? If so, who will be required to provide this capability:—the software vendor? the computer vendor?

How will a VoIP provider provide enhanced emergency (E911) services when there is no fixed physical connection associated with the subscriber? Since IP addresses carries no location information, how will emergency technicians be dispatched in response to an emergency call from a caller who cannot speak? Should the partial solutions that are currently available be evaluated by any regulatory bodies, and will regulations be required? Should VoIP providers be required to offer at least the degree of security that mobile providers are now required to provide? How will these regulations affect intermodal competition?

Convergence poses new tensions between private information and public security. How can lawful wiretapping for law enforcement and national security purposes be supported in a distributed environment? On whom does the obligation fall?

7.3. Competition and innovation

Technological convergence of formerly dissimilar networks aggravates the familiar but difficult issues of inter-carrier compensation. So long as traffic could be identified by location, compensation for transport and termination could discriminate according to state, national, and international boundaries in the interest of serving social goals other than economic efficiency. But with increasing volumes of telephone traffic carried via VoIP, the ability of national regulators to maintain high settlement rates for international calls is rapidly being eroded.

The providers of 'last mile' broadband facilities—local telephone carriers and cable systems—control the access to end-users. These vertically integrated providers of both basic transport and service applications may be able to discriminate against services provided over the Internet, when those services (such as streaming video) compete (in the case of cable) with their core service applications. What if the telephone service offering of the cable company has access to the quality-of-service functions associated with the cable modem protocol, but other VoIP providers do not? Should such discrimination be of concern to regulators? Or will the consumers' choice of alternative physical networks in a converged environment provide sufficient competitive discipline?

Vertically integrated incumbents may take other steps to discourage intelligence from migrating to the edge of the network. A local telephone carrier can, for example, deliver IP traffic to the end user, but terminate the local loop with a box on the side of the house that replaces the analog network interface device (NID). The subscriber still has a simple analog telephone set with limited functionality—DTMF signaling, etc., while the intelligence in the box converts the signaling to IP and digitizes, compresses, and packetizes the analog signal. In this way, the network operator thwarts the 'end-to-end' transparency of the Internet and controls the service definition/creation. Similarly, subscribers to mobile (cellular) services may not be able to get their own IP address and access to the Internet in such a way that they—or third parties—can develop their own services or applications.

Would such discrimination harm consumers? Or are these forms of competition desirable if they promote innovation and if consumers have a choice of multiple physical networks that provide somewhat different, but competing bundles of multimedia services?

Convergence may heighten the tension between open standards and proprietary services. The Internet's explosive growth and rapid innovation is frequently attributed (in part) to the open IP standard. However, most of the computers that are attached to the Internet, whose evolution is necessary for convergence, are based on proprietary standards, such as the Windows operating system and the applications that are increasingly bundled with it. How should the balance between open standards and proprietary systems be struck to best benefit consumers?

More generally, while there is widespread agreement that intelligence is migrating to the edge of the network, where is the edge anyway? Does the customer control the edge device and the intelligence residing in it, or does the carrier?

Might this device be controlled by computer companies who seek to extend their business into multimedia communications? Should consumers have the option of self-provision of telephone service? And, who will decide?

Appendix A

A1. Personalized prices

Preferences for each pair of subscribers (i, j) are described by inverse demand functions $V_{ij}(Q)$ and $V_{ji}(Q)$, where Q is the number of minutes per month of ij communications, regardless of which subscriber initiates any specific call. This representation of preferences incorporates the assumption that utility is derived from the content of the communications and not by the identity of the subscriber who initiates the call. $V_{ij}(Q)$ defines the marginal value placed by subscriber i on Q minutes of calls with subscriber j during a period (e.g., a month) and captures the value to subscriber i of calls with j, net of i's opportunity cost of time. Consequently, $V_{ij}(Q)$ can be positive or negative. We assume that $dV_{ij}(Q)/dQ < 0$.

The corresponding ordinary demand function of subscriber i for communications with j is denoted $Q_{ij}(p)$, the volume of calling that i will demand if he has to pay a price p for *all* of the calls between i and j[62]. If j pays for some of those calls, i may well demand a greater calling volume. In general, $Q_{ij}(p)$ may not be equal to $Q_{ji}(p)$. $Q_{ij}(p)$ and $Q_{ji}(p)$ denote the desired demands of each subscriber in the pair if each paid for all of the calls, while Q represents the realized volume of calling, which is always the same for both parties. In some cases considered below, these three quantity variables take on identical values, and in other cases they do not.

Consider, first, a subscriber pair for whom $V_{ij}(0) > m$, $V_{ji}(0) > m$, and thus $V_{ij}(0) + V_{ji}(0) > m$, where m is the marginal cost of a call. For this case, economic efficiency requires a positive call volume between these two subscribers. If the pair does not coordinate its calls, additional assumptions would be required to establish the pair's calling patterns. Subscriber i would demand no more than $Q_{ij}(0)$ (if j paid for all calls) and no less than $Q_{ij}(p)$ (if i paid for all calls), and subscriber j would demand no more than $Q_{ji}(0)$ and no less than $Q_{ji}(p)$. The actual calling volume will be the smaller of the demands of i and j. Without additional assumptions, there is little more to be said about the pair's calling patterns.

[62] Traditionally, for direct-dial calls, p is the price paid by the caller. The called party pays nothing. In the current formulation, we assume that the price for all ij calls is paid by i, regardless of which party originated individual calls.

Now, assume that pairs of subscribers are related in some way (e.g., through family ties, friendship or a business relationship) and that each such subscriber in a pair can adjust his calling behavior to internalize the call externality. Such arrangements encompass agreements by parents to pay for their children's telephone bills, including providing children with mobile phones; implicit agreements regarding the relative frequency with which each subscriber originates calls, and the use of 800 and 900 numbers as well as collect calls. In the first case, privately constructed prices (with the children paying a zero price) replace market prices that did not meet the needs of the subscriber pair. In the remaining cases, subscribers may select 800 and 900 number services to obtain market prices that are closer to the optimal prices described below, facilitating the choice of optimal calling volumes. Rohlfs[63] and Brown and Sibley[64] conclude that the message externality can readily be internalized by the callers agreeing to share the cost of calling.

When internalization is feasible, both subscribers in any subscriber pair will demand the same volume of calls (i.e., $Q_{ij} = Q_{ji} - Q$). If this condition did not hold, the utility of the pair could be increased by an alternative arrangement in which the subscriber demanding the greater volume of calling offered an incremental subsidy to the other subscriber. When the caller pays a price p, the optimal volume of calls consumed by a subscriber pair ij is the solution to:

$$\max_{Q_{ij}} \left\{ \int_0^{Q_{ij}} V_{ij}(Q)\mathrm{d}Q + \int_0^{Q_{ij}} V_{ji}(Q)\mathrm{d}Q - pQ_{ij} \right\}$$

The first-order condition is:

$$V_{ij}(Q_{ij}^*) + V_{ji}(Q_{ij}^*) - p = 0.$$

As with other shared or public goods, the optimal Q_{ij}^* is the quantity at which the sum of both parties' marginal values is equal to the market price p, which is the total marginal cost to both parties. Using the implicit function theorem:

$$\frac{\partial Q_{ij}^*}{\partial p} = \frac{1}{V_{ij}' + V_{ji}'}.$$

Incentive-compatible or 'personalized' prices are required to ensure that both individuals will demand the same, optimal quantity:

$$p_{ij} = V_{ij}(Q_{ij}^*);\ p_{ji} = V_{ji}(Q_{ij}^*);\ p_{ij} + p_{ji} = p.$$

63 Rohlfs (2002b), Annex A, p. 2.
64 Brown and Sibley (1986), p. 197.

At these personalized prices, $Q_{ij}(p_{ij}) = Q_{ji}(p_{ji}) = Q^*_{ij}$. The prices are incentive compatible since each subscriber's demand, given the internalized price, is equal to the realized volume of calling. It can be easily shown that if all subscriber pairs satisfied the assumptions made above, marginal cost pricing (i.e., $p = m$) would be efficient.

At the efficient volume of calling the personalized price for one member of the pair can be negative. In practice, such negative prices are quite common. Businesses that bill customers directly for time spent on the telephone (e.g., lawyers speaking with clients or technical support staff assisting customers) are essentially paying a negative price for time spent on the telephone. Also, 900 numbers, where the telephone company performs the billing and collection for calls accepted by a subscriber, are another example of a market price that internalizes the message externality by offering negative prices to one of the subscribers in the pair.

For subscriber pairs for whom $V_{ij}(0) + V_{ji}(0) < m$, the analysis is significantly different. For these subscriber pairs, the optimal number of calls is 0. If each subscriber in each such pair was aware of the preferences of the other, or if internalization mechanisms were in place, optimally no calls would occur at the marginal-cost price. But with imperfect information, unwanted (nuisance) calls can occur. For example, if i's expected value of calling j, $V_{ij}(0) > p$, and if subscriber i does not know whether j will value a call or not, i might decide to place a call to j. This kind of unwanted call can occur, for example when a telemarketer places a call to a subscriber, not knowing whether that subscriber will value the call and purchase the marketed product (internalizing the positive message externality in the manner described above) or will hang up in disgust. No doubt a large proportion of marketing calls (and similar e-mail messages) are unwanted in this sense. As long as laws cannot successfully block all unwanted calls[65], prices can be useful in reducing the number of such calls that get made. The market price charged to the caller can be used to offset inefficiencies that result from the inability of the parties to arrive at the efficient calling arrangement.

Consider subscriber pairs for which $V_{ij}(0) + V_{ji}(0) < m$, $V_{ij}(0) > 0$, and $V_{ji}(0) < 0$. We assume for simplicity that subscriber i attempts to call subscriber j, who hangs up. The volume of calling is assumed to be $KQ_{ij}(p)$, where $K < 1$ is taken to be an exogenous parameter in this model. That is, at price p, telemarketer i would like to spend $Q_{ij}(p)$ minutes with potential customer j, but j hangs up as soon as he

[65] In the U.S., charitable and political organizations are exempt from requirements placed on telemarketers even though many subscribers view calls from such organizations as unwanted. In addition, the transactions costs of getting on a 'do-not-call' list, and the costs of reporting telemarketers who violate the law can be significant. Since a fundamental purpose of many unwanted calls is to identify individuals with specific characteristics, it may be impossible to devise perfect blocking schemes without an overly intrusive discovery process.

realizes that he is receiving an unwanted call. The duration of the average call is assumed to be only K times the duration desired by the telemarketer.

An optimal price for these subscriber pairs is any price greater than $V_{ij}(0)$ (and hence m), since such a price will result in a demand of zero for calls of this type. While marginal cost pricing was shown to be optimal when callers are related and all calls satisfied the condition $V_{ij}(0) + V_{ji}(0) > m$, it is not necesarily optimal when this condition is not met. Having considered the two cases in isolation, we turn to a consideration of optimal pricing when only *some* subscriber pairs are related satisfy the condition $V_{ij}(0) + V_{ji}(0) > m$.

A2. Optimal uniform pricing of outgoing calls

In this section, we characterize the (uniform) optimal price per outgoing call. That is, we focus on the traditional pricing arrangement in which the caller is charged for the call, while the called party is not charged. Subscriber pairs are assigned to one of two groups: group A, for which $V_{ij}(0) + V_{ji}(0) > m$; and group B for which $V_{ij}(0) > m$ but $V_{ij}(0) + V_{ji}(0) < m$. We can neglect other subscriber pairs because they will not make calls to each other at prices at, or above, marginal cost.

Summing over all pairs in group A and in group B, total welfare is:

$$W = \sum_{A} \int_{0}^{Q^*_{ij}(p)} [V_{ij} + V_{ji}]\mathrm{d}Q + \sum_{B} \int_{0}^{KQ_{ij}(p)} [V_{ji} + V_{ji}]\mathrm{d}Q - m\left[\sum_{A} Q^*_{ij} + K\sum_{B} Q_{ij}\right]$$

The first-order condition is:

$$\frac{\partial W}{\partial p} = \sum_{A}[V_{ij} + V_{ji}]\frac{\partial Q^*_{ij}}{\partial p} + K\sum_{B}[V_{ij} + V_{ji}]\frac{\partial Q_{ij}}{\partial p} - m\left[\sum_{A}\frac{\partial Q^*_{ij}}{\partial p} + K\frac{\partial Q_{ij}}{\partial p}\right]$$

Since pairs in group A can internalize the call externality, for group A, these pairs W make Q^*_{ij} cells that $V_{ij} + V_{ji} = p$, and

$$\sum_{A}(V_{ij} + V_{ji})\frac{\partial Q^*_{ij}}{\partial p} = p\sum_{A}\frac{\partial Q^*_{ij}}{\partial p}$$

so that we obtain:

$$p - m = \frac{-K\sum_{\mathrm{B}}[V_{ij} + V_{ji} - m]\partial Q_{ij}/\partial p}{\sum_{\mathrm{A}} \partial Q^*_{ij}/\partial p}.$$

The right-hand side of the equation is positive since $V_{ij} + V_{ji} - m$ is assumed to be negative for each subscriber pair in group B. Therefore, the solution of the first-order condition requires $p > m$.

There is considerable evidence that many customers prefer restricted connectivity to related subscribers over broad connectivity to all subscribers, including those from whom they are likely to receive unwanted calls. If subscribers are able to internalize call externalities with other callers within their desired calling community, yet unable to internalize call externalities with subscribers outside that community, optimal usage prices for originating calls should exceed marginal cost. By the same token, as more effective internalization mechanisms (such as do-not-call lists and anti-spam technology) are developed, the need for prices to exceed marginal cost diminishes. As is clear from the equation for optimal prices, as $K \to 0$, $p \to m$.

References

Armstrong, M., 2002, Call termination on mobile networks, Nuffield College, 11 April.
Bellamy, J. C., 2000, Digital Telephony, Third Edition, New York: John Wiley & Sons.
Brown, S. J. and D. S. Sibley, 1986, Public utility pricing, Cambridge, UK: Cambridge University Press.
CNSNews.com, 2003, Spam epidemic provokes lawmakers, Regulators, April 30.
Rohlfs, J. H., 1974, A theory of interdependent demand for telecommunications service, Bell Journal of Economics and Management Science, 5, 16–37.
Rohlfs, J. H., 2002a, Annex A: Network externalities and their internalization with respect to the UK Mobile Market Network, 19 April.
Rohlfs, J. H., 2002b, A model of prices and costs of mobile network operators, Bethesda, MD: Strategic Policy Research, May 22.
Xiao, X.-P., T. Telkamp, V. Fineberg, C. Chen and L. M. Ni, 2002, A practical approach for providing QoS in the Internet backbone, IEEE Communications Magazine, 40, 56–62.
Yarberry, W. A., 2002,Computer Telephony Integration, Boca Raton: Auerbach Publications.

Further reading

Arnbak, J., 2000, Regulation for next generation technologies and markets, Telecommunications Policy, 24, 477–487.
Elixmann, D., U. Schimmel and A. Metzler, 2003, Next generation networks' and challenges for future competition and regulatory policy, WIK Discussion Paper No. 248, November.
Elixmann, D. and M. Scanlan, 2002, The economics of IP networks—Market, technical and public policy issues relating to Internet traffic exchange, Report, Eu Commission (DG Info Soc), Brussels.

Engebretson, J., 2002, Next generation Internet, America's Network, February 2.

Engebretson, J., 2001, Softswitch Ecpo & Conference—The Holy Grail of Class 5 Replacement, America's Network, December 7.

Giordano, S., M. Potts, M. Smirnov and G. Ventre, 2003, Advances in QoS, IEEE Communications Magazine, 41, 28–69.

Marcus, J. S., 1999, Designing wide area networks and internetworks—A practical guide, Reading (Mass): Addison-Wesley.

Sinnreich, H. and Johnston, A. B., 2001, Internet communications using SIP—Delivering VoIP and multimedia services with Session Initiation Protocol, Networking Council Series, New York: Wiley.

Smith, C. and Collins, D., 2002,3G wireless networks, New York: McGraw-Hill Telecom Professional.

Uebele, R. and Verhoeyen, M., 2001, Strategy for migrating Voice Networks to the Next Generation Architecture, Alcatel Telecommunications Review, 2nd Quarter, 85–90.

Wydrowski, B. and M. Zukerman, 2002, QoS in best-effort networks, IEEE Communications Magazine, 40, 44–49.

Chapter 3

BANDWAGON EFFECTS IN TELECOMMUNICATIONS

JEFFREY H. ROHLFS*

Strategic Policy Research

Contents

* The author thanks MIT Press for permission to use portions of the author's book, *Bandwagon Effects in High-Technology Industries* (MIT Press, 2001). The author thanks Fred Bellemore, John Haring, Geoffrey Myers, Peggy Rettle, and Chip Shooshan for helpful comments. The author additionally thanks Fred Bellemore, Peggy Rettle, and Jordan Schneider for research contributions to this chapter. The author is, of course, solely responsible for any remaining errors.

Handbook of Telecommunications Economics, Volume 2, Edited by S. Majumdar et al.

DOI: 10.1016/S1569-4054(05)02003-8

1. Introduction[1]

Bandwagon effects are defined as increases in a consumer's utility as others consume the same product or service that she does. Bandwagon effects are pervasive in the high-technology sector of the economy and in telecommunications, in particular.

Bandwagon markets have dynamics that differ from those of conventional products and services. They are quite difficult to get started, because bandwagon benefits are initially limited to those generated by a small user set. Many would-be bandwagons never get underway[2]. Once enough consumers have gotten on a bandwagon, however, the bandwagon benefits may be huge. Highly successful bandwagon products and services include the telephone, facsimile machines, compact disc players, videocassette recorders, personal computers, television, and the Internet[3].

Bandwagon effects continue to operate even after the start-up problem has been solved and the market has reached maturity. In particular, the social value of a bandwagon product or service includes bandwagon benefits, as well as what the marginal consumer is willing to pay. Consequently, at cost-based prices demand may equilibrate at a level below the optimum. For that reason, pricing bandwagon products and services below marginal cost in order to stimulate demand may be economically efficient.

This issue has special relevance for public policy where prices of bandwagon services are regulated, in particular, telephone services and mobile termination charges. In such markets, bandwagon effects must be considered, together with marginal costs and demand elasticities, in order to determine the economically efficient price structure.

Several chapters in this handbook discuss the start-up problem with respect to bandwagon effects in particular network industries[4]. In this chapter, we focus on economically efficient pricing in mature bandwagon markets. Historically,

[1] Parts of the first three sections of this chapter are taken from Rohlfs (2001).

[2] The Picturephone comes to mind. Digital audiotapes and digital compact cassettes are other examples. The color television bandwagon took over a decade to get underway in the U.S. More recently, digital HDTV is another (would-be) bandwagon that is struggling to get underway.

[3] See Rohlfs (2001) for discussion of bandwagon effects relating to all these products and services.

[4] Some other studies of start-up problems in bandwagon markets are as follows: Farrell and Shapiro (1993) examine the dynamics of bandwagons using a series of models to study adoption time and bandwagon competition, growth and abandonment. Hong and Konrad (1998) offer an explanation of bandwagon effects in voting behavior by hypothesizing uncertainty aversion, and demonstrate the effect polling and bandwagons have on outcomes. Abrahamson and Rosenkopf (1997) investigate the function that social network structures have in bandwagon adoption and posit that slight idiosyncratic network characteristics of a network can result in large bandwagon effects. Strategic Policy Research (2002) examined the start-up problem for broadband distribution (e.g., DSL and cable modem) in the U.K.

the most important application of such pricing has been to promote universal telephone service. That issue was discussed in Volume I of this handbook[5]. In this chapter, we consider historical pricing of telephone services and the following two recent issues involving pricing of mature bandwagon services:

- Local-usage charges for calls to Internet service providers; and
- Regulated prices for fixed-to-mobile calls, in countries where the calling-party pays for such calls.

These topics are addressed after discussion of the relevant theory of bandwagon demand and economically efficient pricing.

2. Theory of bandwagon demand

The importance of bandwagon effects has increased dramatically in recent years as a result of a number of high-technology innovations, including those discussed in this Handbook. Not surprisingly, therefore, most of the economic work on bandwagon effects has been done relatively recently.

Nevertheless, bandwagon effects were first analyzed half a century ago—by Harvey Leibenstein in 1950. In Leibenstein's analysis, bandwagon effects are purely psychological—a quirk in the utility function. That is, a consumer may feel better doing the same as others do. Leibenstein compared bandwagon effects to 'snob' effects, whereby a consumer feels better doing as other do not.

In contrast, bandwagon effects in high-technology industries often have a basis that goes beyond what is in the consumer's head. In particular,

- Consumers of a product or service that uses a telecommunications network may benefit from being able to communicate with more persons as the set of users expands. These effects are known as 'network externalities.'
- Consumers of a product (e.g., hardware) may benefit from greater availability of competitively supplied complementary products (e.g., software) as the set of users of the product expands. We denote these effects as 'complementary bandwagon effects'[6].

Examples of services that are subject to network externalities are the telephone, fax, and the Internet. Users of these services benefit from being able to communicate with more persons as the user set expands. For these, consumers benefit from increases in output. These benefits can be characterized as external demand-side economies of scale. They are external because they relate to the size of the entire user set—not to the size of a single user organization.

Although network externalities and internal supply-side scale economies both benefit consumers as output expands, the two effects are easily distinguished. The

[5] For example, see Riordan (2002) and Armstrong (2002). Downes and Greenstein (1998) consider Internet-related services in relation to this issue.

consumers' benefits from network externalities derive from observable effects on the demand side of the market; e.g., the sending of more faxes by each user. The consumers' benefits from internal supply-side scale economies derive from observable effects on the supply side of the market (e.g., cost savings through automation and centralization). Bandwagon theory, as described in this chapter, is useful for modeling network externalities. Internal supply-side scale economies can, however, be modeled without recourse to bandwagon theory.

Complementary bandwagon effects involve complex interactions between supply and demand. The nature of those effects depends on the extent of vertical integration within the supplying industry.

One possibility is that a fully integrated supplier provides the hardware and all the complementary software, which the user organization needs and does not write itself. This industry structure is common for special-purpose equipment (e.g., some telecommunications switches).

Under this industry structure, complementary bandwagon effects are simply internal supply-side scale economies. As output expands, the supplier can spread the cost of writing software over more units of production. Consequently, writing more and better software is likely to be profitable. Users then benefit from greater availability of software. Since users benefit (indirectly) from the expansion of the user set, one can view this phenomenon as a bandwagon effect. Actually, it is no different from other kinds of internal supply-side scale economies that benefit users as output expands. Consequently, where supply is fully integrated, one does not need bandwagon theory to model complementary bandwagon effects (notwithstanding their name).

Bandwagon theory is, however, useful for modeling these effects where supply of hardware and software is not integrated. In that case, the bandwagon effects are *external* economies of scale. The economic consequences of such economies differ considerably from those of *internal* supply-side scale economies, which apply to output of a single firm.

[6] These benefits are greatly increased if the products of all suppliers are compatible and therefore 'interlinked' (using the term from Rohlfs (2001)). There is an extensive literature on the issues of compatibility and adoption. The March 1992 issue of the *Journal of Industrial Economics* (Vol. 40, No. 1) presents a symposium on compatibility (all of the 1992 cites listed below). Matutes and Regibeau (1992) show model compatibility and bundling where a user combines components to form an agreeable product. Farrell and Saloner (1985) demonstrate that whether or not excess inertia traps an industry in an obsolete technology depends on the existence of complete information. Katz and Shapiro (1992), in contrast to excess inertia, present conditions where a new technology exhibits insufficient friction, stranding users of existing technology. Arthur (1989) examines the competition between technologies for adopters, where increasing returns magnify random events and lead to 'lock-ins' and multiple equilibria. Farrell and Saloner (1992) find that converters can actually make outcomes even more inefficient than the market one because they decrease the chance of standardization occurring. Other relevant work includes Church and Gandal (1992) and Economides and Salop (1992).

Examples of complementary bandwagon effects where supply is *not* fully integrated are the following:

- Owners of CD players and VCRs benefit as other consumers buy products using the same technical standard, because a wider range of programming will become available;
- Owners of PCs benefit as others buy compatible systems, because a wider range of applications programs will become available;
- Owners of digital television sets benefit as others buy such sets because a wider range of digital programming will become available; and
- Internet users benefit as a wider range of information becomes available online.

The properties of bandwagon demand are largely the same for all three types of bandwagon effects discussed above: Leibenstein's concept, network externalities, and complementary bandwagon effects. The same demand model can be used for all three. One simply posits that the utility that each individual derives from a service depends (for whatever reason) on others' consumption of that service. In particular, we assume in the pure bandwagon model that as the set of users expands, the service becomes more valuable to some persons and does not become less valuable to anyone.

Telemarketers have come to represent an important exception to the pure bandwagon model[7]. Since modern technology has lowered telecommunications costs, telemarketers can often operate profitably, even if they make only one sale in 100 attempts. They can and do disregard the bother and inconvenience that they cause the other 99 persons, since they are not required to pay that cost. The other 99 persons may well feel that they would be better off if telemarketers unsubscribed to telephone service.

Much the same phenomenon occurs with respect to other bandwagon services. Junk faxes appear on the office fax machine every morning. E-mail provides the means for low-cost, pervasive proliferation of junk mail, which is usually bothersome and sometimes offensive. E-mail has also recently provided the means for spreading computer viruses[8].

For these reasons, real-world services and products generally do not conform precisely to the pure bandwagon assumptions, described above. Nevertheless, the pure bandwagon model is often a reasonable approximation. It provides useful insights with regard to services and products, which embody substantial (although not pure) bandwagon effects.

In our formal bandwagon demand model, we assume that each consumer makes a simple yes–no decision about whether to subscribe to a particular service that embodies bandwagon effects. This assumption is not really restrictive, because the

[7] For discussion of this issue, see Noam (1995).
[8] See Krebs (2003) for further discussion of this problem.

same individual could make yes–no decisions about whether to purchase successive units of the service.

2.1. Equilibrium user sets

For some services, a reasonable bandwagon model could define each user's demand as a function of the number of other users. In general, however, demand in a bandwagon model also depends on who those users are; in particular, how strong is the community of interest between existing subscribers and the marginal subscriber.

Because communities of interest are often important, the basic theoretical concept in the bandwagon model is not the number of users, but the user *set* (i.e., those consumers who have for some reason—perhaps animal spirits—chosen to consume the bandwagon service). A user set is in equilibrium if and only if:

- No consumer who has chosen to consume the service would be better off not consuming it—given the other users of the service; and
- No consumer who has chosen *not* to consume the service would be better off consuming it—given the users of the service[9].

2.2. Multiple equlibria

A key result of bandwagon demand theory is that there are generally multiple equilbria, which may differ substantially. If perfect cooperation were possible among all members of society, the best (maximum) equilibrium user set could always be attained. In general, if all members of that user set could agree to consume the bandwagon service, they would thereby all benefit.

In reality, of course, perfect cooperation is not possible; nor is anything remotely resembling perfect cooperation. Indeed, one can reasonably assume that cooperation is limited to the following:

- An organization may be able to make decisions, taking into account bandwagon benefits within the organization; and
- Small groups (in most cases, probably 2 persons or 2 organizations) can jointly make decisions to consume a bandwagon service[10].

Suppose that the initial user set consists of all individual entities and small groups (mainly pairs) of entities that can justify purchasing the service, even if others don't purchase it. This set may possibly be the null set. For example, the

[9] This analytical approach follows Rohlfs (1974).

[10] In technical economic terms, bandwagon effects are external technological economies in consumption. The two types of cooperation described above are ways to internalize the externality. Larger groups are unable to internalize the externality, because (in Coase's terminology) the transactions costs are too large.

bandwagon service may have little utility within an organization, and the bandwagon benefits for small user sets do not justify the (high) price. In that case, the initial user set is an equilibrium user set, and there is no tendency for demand to grow.

For many services, however, the initial user set is not the null set. In that case, it is probably not an equilibrium user set and demand will expand until an equilibrium user set is reached. We refer to the resulting equilibrium as the 'demand-based' equilibrium user set. This equilibrium derives solely from demand adjustments—assuming that suppliers simply offer the service at some price[11].

The demand-based equilibrium user set may, unfortunately, be far from optimal. There may exist a much larger equilibrium user set wherein all consumers are better off. The original subscribers would all be better off with a larger equilibrium user set because of bandwagon benefits. Also, if new subscribers make voluntary decisions to subscribe, they are presumably better off as well. The ideal in this regard is the 'maximum equilibrium user set'[12].

To the extent that the maximum equilibrium user set is larger than the demand-based equilibrium user set, ordinary demand adjustments do not provide a path to the optimum. This situation is the start-up problem for bandwagon products and services.

2.3. *Demand as a function of price*

All our previous discussion of equilibrium user sets is based on the assumption that the bandwagon service is offered at some fixed price. Let us now consider how demand varies as price varies.

In bandwagon models, there is a general tendency for demand to be negatively related to price. In particular, starting from any given equilibrium user set, equilibrium demand declines as price rises; it rises as price declines. Nevertheless, the level of demand depends on the starting point as well as price. A larger equilibrium user set may quite possibly be reached with a higher price and a larger initial user set than with a lower price and a smaller initial user set.

Even though it is generally unrealistic to assume that demand depends only on the number of subscribers, we now analyze precisely that model. Albeit unrealistically simple, the model does provide useful insights about pricing of bandwagon products.

Suppose that the maximum price, which an individual would be willing to pay for the service (her 'reservation price'), depends only on the number of subscribers.

[11] For further discussion of the equilibrium adjustment process, see the Mathematical Appendix of Rohlfs (2001).

[12] The Mathematical Appendix of Rohlfs (2001) describes the properties of the maximum equilibrium user set. It also provides a mathematical comparison of the various user sets we have been discussing.

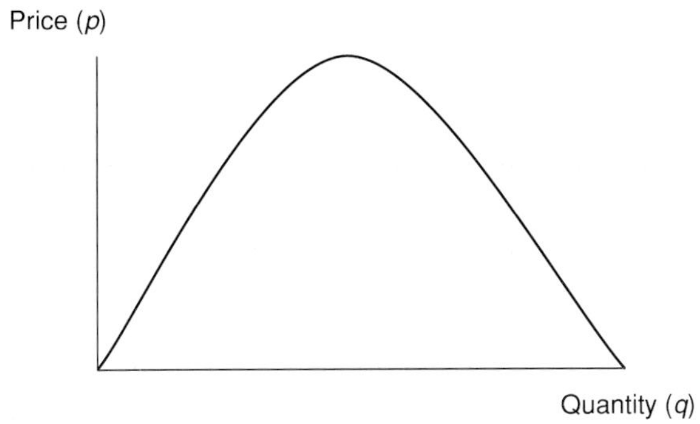

Fig. 1. Equilibrium Demand Versus Price: Simplest Shape.

The simplest shape for the inverse demand function (price as a function of quantity) is depicted in Figure 1[13]. The quantity variable in the figure (q) indicates the number of subscribers who have joined the user set. We assume that the first subscribers to join are those who value the service the most, *for any given q*. Then additional subscribers who value the service progressively less (for any given q) join. The variable for price (p) indicates the price at which the user set with q subscribers is an equilibrium user set. For any given q, p in Figure 1 (as in any inverse demand function) is the reservation price of the marginal subscriber.

Because additional subscribers value the service progressively less, ordinary (nonbandwagon) demand curves are necessarily downward sloping. The reservation price of the marginal subscriber declines as q increases.

For bandwagon services, however, there is an additional consideration. As more subscribers join, bandwagon benefits increase the value of the service to each subscriber. For services in the start-up phase, this bandwagon effect usually dominates the normal tendency toward downward-sloping demand. That is, the reservation price of the marginal subscriber *increases* as additional subscribers join. Thus, we have the anomalous result that the demand curve in the left part of Figure 1 is upward sloping!

The downward sloping part of Figure 1 has properties similar to those of ordinary (nonbandwagon) demand curves. For any given price, demand tends to gravitate toward the demand curve, which indicates the equilibrium at that price.

The upward-sloping part of Figure 1, however, consists of unstable equilibria. Demand has no tendency to gravitate toward that part of the demand curve. On the contrary, if demand is to the left of the upward-sloping curve, it tends to

[13] See the Mathematical Appendix of Rohlfs (2001) for a mathematical description of demand in Figure 1.

contract toward zero. If demand gets past the upward sloping part of the curve, it tends to expand until it reaches the downward-sloping part[14]. Thus, for any price, the upward-sloping part of the demand curve can be considered the 'critical mass' at that price. In this example, the service cannot be successful unless critical mass is somehow achieved. If critical mass is achieved, the service can take off, and a positive level of demand can be sustained.

Once critical mass is achieved, demand becomes subject to positive feedback. The feedback starts when enough users join the bandwagon to exceed critical mass. The initial user set then provides sufficient bandwagon benefits to induce additional users to join. When the additional users join, more bandwagon benefits are generated, inducing still more users to join. In the model illustrated in Figure 1, this positive feedback continues until the maximum equilibrium user set is achieved.

Figure 1 is the simplest possible bandwagon demand function. The economics become more complicated when we take into account communities of interest among individuals. In that case, a tight community of interest might have its own critical mass. Its achieving critical mass may or may not suffice to allow other groups that have some community of interest with the original group, to achieve critical mass.

In the simplest model (Figure 1), the service is either a market success or failure, depending on whether critical mass is achieved. In more complex models, there are gradations of market success, depending on the number and sizes of community-of-interest groups that achieve critical mass[15].

2.4. *Metcalfe's law*

A principle often ascribed to network services is Metcalfe's Law; *viz.*, the value of a network goes up as the square of the number of users. The law is named for Bob Metcalfe, the inventor of Ethernet and the founder of 3Com[16].

The assumption underlying Metcalfe's Law is that the value that any user A derives from a communications link with any other user B is the same for all users A and B. For a user set of n users, each user enjoys the value of n links[17]. The sum of the values derived by all n users is therefore the value of n^2 links.

[14] Further discussion of the stability properties of the demand curve is given in the Mathematical Appendix of Rohlfs (2001).

[15] For further discussion of more complex bandwagon demand curves, see the Mathematical Appendix of Rohlfs (2001).

[16] Shapiro and Varian (1999) quote Metcalfe as saying that the law was ascribed to him by George Gilder, but he is willing to take credit for it.

[17] One needs to include the link between the user and herself in order for the law to work out exactly. In reality, the self-link may have little or no value, but the treatment of the self-link makes relatively little difference for large n.

In reality, the values that pairs of users derive from the links connecting them are likely to differ considerably. Some pairs of users will derive much more value from links than do other pairs.

In general, one would expect that the first users would be those who, together with the persons with whom they communicate, derive the most value from links. Then, users for whom this value is progressively less would follow. Under these circumstances, the value of a network increases *much less than proportionately* to the square of the number of users. Indeed, the total value of a bandwagon service may go up less than proportionately to the number of users—let alone the square of the number of users. Such an outcome could occur, for example, if a service in equilibrium has a low price, but some inframarginal users value it very highly. Because the price is low, we know that the value of the service to marginal users must also be low. That value, in turn, may be low because marginal users do not or would not use the service very much. Under these circumstances, it is likely that existing users would derive only small bandwagon benefits as marginal subscribers join. The sum of these two low values (benefits to marginal subscribers and bandwagon benefits to existing subscribers) may be far less than the average value of the service to inframarginal users (for communicating with one another). This point is developed more rigorously and in more detail in the Mathematical Appendix of Rohlfs (2001) and in Swann (2002). A large user set would be even less valuable if it attracted negative-value users, such as telemarketers.

The fundamental insight underlying Metcalfe's Law—that bandwagon effects raise the value of a service to each user—is certainly correct—indeed, the main point of this chapter. Nevertheless, the precise statement of the law is unlikely to hold, even approximately, in practice.

An important possible exception is internal networks (e.g., a local area network, LAN) within an organization. It is, for example, quite plausible that a LAN is four times as valuable for an organization that has twice as many connected computers. This is, perhaps, the application that Metcalfe had in mind.

Internal networks are bandwagon products in a sense. Nevertheless, the managers of the organization can internalize the network externality by simply taking bandwagon effects into account in making their decision whether to set up the network. The theory developed in this book applies to external bandwagon effects that cannot be internalized so easily. It is neither necessary nor appropriate to use that theory to analyze decisions involving (purely) internal networks.

3. Pricing of mature bandwagon services

Pricing initially below cost may be a necessary component of the start-up strategy for a new bandwagon product or service. Additionally, even in a mature bandwagon market, below-cost pricing may be economically efficient. In particular, there may be individuals such that:

- The benefit *to the individual* of consuming the bandwagon service does not justify incurring the marginal cost, but
- The benefit *to society*, including bandwagon benefits that would be enjoyed by other users, does justify incurring the marginal cost.

Under these circumstances, economic efficiency would be improved if the individual consumed the service; but she will choose not to do so.

In formal economic terms, consumption of a bandwagon service involves external technological economies. The benefits are external (to the subscription decision), because an individual will presumably base her subscription decision on the expected benefits to herself. She cannot be expected to decide on the basis of total social benefits—including benefits to existing subscribers. For this reason, the number of individuals who choose to subscribe may be fewer than optimal. Corrective pricing can, if properly implemented, compensate for this market failure and improve economic efficiency and aggregate economic welfare. In this context, seminal work by Artle and Averous (1973), Squire (1973), and Littlechild (1975) analyzes the nature of, and necessary adjustments owing to, consumption externalities.

Externalities also exist with respect to calling. In particular, both the caller and the called party often benefit from a call,[18] but only the caller pays. It is possible that callers may decline to make certain calls where the benefits to the caller do not justify the cost but the total benefits (including the external benefits to the called party) do, indeed, justify the cost.

3.1. Internalization of externalities

Market participants can often internalize externalities themselves (especially calling externalities), even if corrective pricing is not implemented[19]. Examples of how market participants in network industries can themselves internalize externalities are as follows:

1. Persons may contribute toward the cost of telephone service (both subscription and calls) for others with whom they have a substantial community of interest. For example, parents may pay for telephone service for their children who no longer live in the parents' home, or sons and daughters may pay for mobile service for their aging parents.
2. Two persons may each agree to call the other half the time;
3. Persons or businesses may accept collect calls under certain circumstances;

[18] An exception that has become increasingly important in recent years is telemarketing calls. Such calls presumably benefit the telemarketer but are often a nuisance (a source of disutility) to the called party.

[19] Liebowitz and Margolis (1994) emphasize the effectiveness of the market in internalizing network externalities.

4. A person may be subject to social pressure to subscribe so that others can call her;
5. Persons or businesses may obtain toll-free numbers;
6. A business may bill its customers explicitly for calls made on their behalf; and/or
7. Businesses presumably price their services so as to recover their total telecommunications costs (subscription and usage), as well as other costs, from their customers. Consequently, in aggregate, the beneficiaries bear the costs; so they are not external.

Where externalities are internalized in such fashion, corrective pricing is unnecessary. Indeed, it is counterproductive, because the same externality thereby gets internalized twice. As a result, economic efficiency is likely to be diminished.

In general, it is reasonable to assume that users can internalize a large portion of calling externalities, even if corrective pricing is not implemented. Calling inherently involves only two persons, and those two persons have many ways to internalize the externality, as described above. Nevertheless, uninternalized calling externalities may still be significant—especially with regard to calls from persons outside the called party's usual community of interest.

The externality with respect to subscription is often not easy to internalize. If the external benefits to a single existing subscriber are sufficiently large, he may contribute to the cost of the marginal subscriber's service, as in (1), above. Often, however, the benefits to any single subscriber are not very large, but the benefits to *all* existing subscribers are quite large. In such cases, there may be no practical way for existing users to induce nonsubscribers to join. In Coase's terminology, the transactions costs are too high[20].

3.2. *Budget constraint*

Prices in some bandwagon markets (e.g., telephone service) are subject to regulation. In the remainder of this section, we examine the principles of efficient pricing, as they apply to such markets.

As a matter of practicality (and often also as a matter of law), regulated prices must afford the regulated firm the opportunity to recover its costs, including the cost of capital. Otherwise, the regulated firm would have neither the ability nor the incentive to attract capital. The unavailability of capital would generally lead to a suboptimal outcome in the long run, since new investment is typically needed to accommodate growth and to modernize the firm's physical plant.

For this reason, the problem of efficient regulatory pricing is usually posed subject to the budget constraint that total revenues cover total costs. The solution to the problem, so posed, was worked out by in Ramsey (1927). In the absence of

[20] See Coase (1960).

cross elasticities of demand, the Ramsey's rule is that percentage mark-ups over marginal costs are inversely proportional to the (direct) demand elasticities. That is:

$$\frac{Xp_i - mc_i}{\eta_i} = \frac{p_j - mc_j}{\eta_j} \quad \text{for all } i,j \tag{1}$$

$$\text{and} \quad \sum_i p_i q_i = c \tag{2}$$

where:

p_i = price of service i;
mc_i = marginal cost of service i;
η_i = the direct elasticity of demand for service i with respect to its price;
q_i = quantity of service i; and
c = total cost.

This rule can be adapted to apply to the case where cross elasticities of demand are nonzero. The rule, given in Baumol and Bradford (1970) with credit to Boiteux (1951) is as follows:

$$\frac{p_i - mc_i}{mr_i - mc_i} = \frac{p_j - mc_j}{mr_j - mc_j} \quad \text{for all } i,j. \tag{3}$$

Equation (3) is then used, in conjunction with Eq. (2), to determine economically efficient prices.

Equation (3) can be interpreted in the usual way for constrained optimization. The numerator is the marginal contribution to the objective function, which is the sum of consumer and producer surplus. That objective function reflects the difference between aggregate willingness to pay and costs. The denominator of Eq. (3) is the marginal contribution to profits (i.e., the marginal contribution to meeting the budget constraint). The ratio of the numerator to the denominator must be the same for all services.

Rohlfs (1979) showed that the usual Ramsey's rule can be used in this case, if 'superelasticities' are used instead of direct elasticities. That is, the following is used in conjunction with Eq. (2):

$$\frac{p_i - mc_i}{\xi_i} = \frac{p_j - mc_j}{\xi_j} \quad \text{for all } i,j \tag{4}$$

where:

ξ_i = the superelasticity of service i.

Superelasticities are, in turn, defined as follows:

$$\xi_i = \frac{1}{\sum_j r_{ji}\phi_{ji}} \text{ for all } i \tag{5}$$

where:

r_{ji} = ratio of revenues derived from the *j*th service to those derived from the *i*th service; and
ϕ_{ji} = flexibility of the *j*th price with respect to *i*th quantity.

'Flexibility' is the multivariate reciprocal of elasticity; i.e.,

$$\phi_{ji} = \frac{q_i}{p_j} \frac{\partial p_j(q_1, \ldots, q_n)}{\partial q_i} \tag{6}$$

ξ_i is the inverse of a weighted sum of the ϕ_{ji}. If there is only a single market, the superelasticity is the same as the direct elasticity. If there are multiple markets, ξ_i takes account of cross elastic effects as well as direct effects.

3.3. Externalities

A useful concept for optimal pricing with network externalities is the 'externality factor,' developed in Rohlfs (1979). The externality factor is the ratio of total social value (including external benefits) to the private value (excluding external benefits). The externality factor can be an explicit parameter in determining the optimal degree of corrective pricing in the presence of externalities.

Absent cross elasticities, the equation for economically efficient pricing with externalities is as follows:

$$\frac{e_i p_i - mc_i}{mr_i - mc_i} = \frac{e_j p_j - mc_j}{mr_j - mc_j} \quad \text{for all } i, j, \tag{7}$$

where e_i = the externality factor for service *i*.

Equation (7) is used in conjunction with Eq. (2). Equation (7) can be interpreted in the same way as Eq. (3). The denominators of the two equations are the same. The numerator of Eq. (7) is the marginal contribution to the objective function—aggregate willingness to pay less aggregate costs—taking externalities into account.

3.4. Externalities and cross elasticities

The optimization problem becomes considerably more complex where there are externalities *and* cross elasticities of demand. This case is especially important because externalities themselves generate cross elastic effects. In particular, as the price of subscription falls and the user set expands, existing users call the new subscribers. These calls contribute to the cross elasticity of usage with respect to the price of subscription. They also reflect externalities to the extent that the benefits to the callers are not fully internalized by the marginal subscribers.

3.4.1. *Gross and net externality factors*

We have defined the externality factor as the ratio of social value to private value. The social value includes the cross-elastic bandwagon benefits of usage. Indeed, these benefits may well account for most of the externalities.

Nevertheless, there may be additional bandwagon benefits not related to usage. In particular, the option to call a particular subscriber may have considerable value, even if that option is rarely exercised[21].

Under these circumstances, it is useful to distinguish the 'gross externality factor' (as defined above, e_i for service i) from the 'net externality factor' (n_i for service i). The net externality factor is the gross externality factor less the external benefits associated with usage. The value of n_i exceeds unity to the extent that there are external benefits (e.g., option value) that are not associated with usage.

Consumer surplus, measured as the area under demand curves, includes the surplus associated with external bandwagon benefits of usage, because those benefits engender willingness to pay for usage. The area under demand curves does not, however, include the bandwagon benefits reflected in the net externality factor, because those benefits are external to market demand functions. They reflect utility that inframarginal subscribers derive without any incremental consumption.

The surplus integral that includes all bandwagon benefits is as follows:

$$s = \int \sum_i (n_i p_i - mc_i) dq_i, \tag{8}$$

where s is evaluated along some path from the origin to the final market equilibrium.

Note that the bandwagon benefits associated with usage would be double counted if e_i were substituted for n_i in Eq. (8).

[21] One can argue that in theory even the option value is related to usage. That is, there may be few calls, but they have extremely high value (e.g., in emergencies). From an empirical perspective, however, it is impractical to measure the extremely high value of such calls. Consequently, postulating an option value may be the best way to model them in practice.

3.4.2. *Path dependence of consumer surplus*

Externalities affect inframarginal consumers, as well as marginal consumers. In particular, bandwagon effects increase the willingness to pay of inframarginal consumers as consumption by others increases. The optimization conditions therefore inherently depend on the *global* properties of the market–demand functions. They do not depend solely on the local properties, which (in turn) depend only on marginal consumers.

Samuelson (1965) observed that the marginal conditions for Pareto optimality (absent externalities) could be derived without reference to global consumer surplus. He argued that use of the marginal conditions is preferable to using global consumer surplus, which may be path dependent and hence ill defined[22].

In the present case, however, we have no alternative but to deal with the global properties of market demand functions. Since we utilize a variant of consumer plus producer surplus as our objective function, we must deal with the problem of possible path dependence of the surplus line integral s in Eq. (8).

In modeling demand in the absence of externalities in a sector that is a small part of the total economy, it is generally reasonable to assume that the matrix of partial derivatives (not elasticities!) of demands with respect to prices is symmetrical. That is, if:

$$q_i = q_i\ (p_1, \ldots, p_n) \qquad i = 1, \ldots, n \tag{9}$$

then

$$\frac{\partial q_i}{\partial p_j} = \frac{\partial q_j}{\partial p_i} \quad \text{for all } i, j \tag{10}$$

The inverse demand functions corresponding to Eq. (9) are:

$$p_i = p_i\ (q_1, \ldots, q_n) \quad \text{for } i = 1, \ldots, n, \tag{11}$$

where p_i is the willingness of the marginal consumer to pay for service i.

If the matrix of partial derivatives of demand is symmetrical, so is the matrix of partial derivatives of inverse demand.

That is,

$$\frac{\partial p_i}{\partial q_j} = \frac{\partial p_j}{\partial q_i} \quad \text{for all } i, j. \tag{12}$$

[22] See the section on "Why Consumer's Surplus is Superfluous" in *Foundations of Economic Analysis* (Atheneum, 1965, originally published 1947). See also Willig (1976).

The matrices of partial derivatives of demand and inverse demand are both symmetrical for a single consumer if Slutsky income effects are zero—generally a reasonable assumption when one is analyzing a small part of the total economy[23]. In that case, one can also assume, as a reasonable approximation, that the matrices of partial derivatives of aggregate demand (and aggregate inverse demand) are symmetrical. This assumption ensures that the consumer surplus line-integral is path-independent and, hence, well defined.

The 'buy-through' model of demand for a telecommunications service leads to symmetrical cross partials of demand in the absence of externalities. In that model, the willingness to pay for subscription is the surplus that the consumer would derive from all types of usage if she subscribed. A key property of the buy-through model is that demand for subscription is equally affected by:

- A change in the monthly subscription price; or
- A change in usage prices that when multiplied by the usage of a marginal subscriber yields the same monthly amount as the change in the subscription price.

Either of these changes has the same effect on the surplus that a marginal subscriber would derive from subscription and usage, considered together.

Even where the matrix of cross partials of demand is symmetrical in the absence of externalities (as in the buy-through model), externalities cause that matrix to be asymmetrical. This effect can be clearly seen in the buy-through model:

- In that model without externalities, the partial derivative of demand for a particular type of usage with respect to the subscription price reflects (only) the usage of the marginal subscriber.
- Externalities increase the absolute value of that cross partial, because existing subscribers will call marginal subscribers after the latter join the network.
- The absolute value of the partial derivatives of subscription demand with respect to usage charges will increase equally only if the externality is fully internalized (i.e., only if the marginal subscriber, in making her subscription decision, takes full account of the surplus of those who call her).

Given that full internalization of the externality is unlikely—especially with respect to subscription decisions—the matrix of partial derivatives of demand will be asymmetrical.

3.4.3. Externalities with respect to subscription

In this subsection, we consider economically efficient pricing for the (most important) case of externalities with respect to subscription. We initially assume no externalities with respect to usage. We also assume the buy-through model,

[23] Slutsky income effects relate to the effect of price changes on the consumer's real income. For price changes in a small sector of the total economy, the effect that price changes in that sector have on real income is disproportionately small.

conditional on any given number of subscribers. Given these assumptions, we can prove the following theorem:

Theorem: If consumer surplus is integrated first over subscriptions (holding all usage at zero) and then over the various types of usage in arbitrary order, the resulting value reflects aggregate willingness to pay.

Proof: The first step of the proof is to demonstrate that willingness to pay for usage is properly evaluated, contingent on the final number of subscriptions. This point follows from the economic logic. At the final equilibrium, each consumer's willingness to pay for any quantity of usage depends on the final number of subscribers—with all of whom she can communicate.

Secondly, we must demonstrate that one can integrate over the various types of usage in arbitrary order. This point follows directly from our symmetry assumption. For any given number of subscribers—in particular the final number—the surplus integral is path independent and on any path reflects aggregate willingness to pay.

Thirdly, we must demonstrate the appropriateness of integrating over subscription, conditional on all quantities of usage being zero. This point follows from the buy-through model. We separately integrate over all types of usage—conditional on the final number of subscribers. Thus, we have separately taken account of all the consumer surplus associated with all usage. That surplus should not be counted again. The subscriber may also reap benefits that are not related to usage. That surplus is reflected in the net externality factor, and also should not be counted again. It follows that subscription must be integrated, assuming zero quantities, in order to avoid double counting.

End of Proof

3.4.4. *Usage externalities*

The previous subsection considered the case of externalities with respect to subscription but not with respect to usage. The results of that subsection do, however, hold even if the net externality factors of usage differ from unity—so long as the externalities do not generate cross elasticities.

3.4.5. *Solution*

The solution for economically efficient prices with externalities and cross elasticities is as follows:

$$\frac{\frac{\partial s}{\partial q_i}}{mr_i - mc_i} = \frac{\frac{\partial s}{q_j}}{mr_j - mc_j} \quad \text{for all } i,j, \tag{13}$$

where s is integrated along the path described above.

3.5. Nonuniform pricing

The preceding discussion all assumed a single (uniform) price for each product or service. In reality, firms with significant market power typically have the ability to price-discriminate (i.e., charge different prices to different consumers for the same product or service). Price discrimination, where feasible, greatly changes the optimization problem[24].

The extreme case is the perfectly discriminating monopolist. Such a monopolist can achieve the Pareto optimal outcome in which all consumers consume the optimal amounts, but the monopolist (if unregulated) extracts all the surplus.

Of course, perfect discrimination is not feasible in practice. A common form of price discrimination that is feasible is to offer consumers a choice of two-part tariffs. In mobile telecommunications, for example, subscribers can choose from a variety of pricing plans with different amounts of usage included in the monthly charge.

Under certain conditions, Eqs. (13) and (2) can be applied to find optimal nonuniform prices with externalities, cross elasticities and a budget constraint. We assume a given set of pricing plans, with given quantities of usage in the various plans[25]. The first step is to formulate demand equations in which subscription is unbundled from the usage packages (which then do not include subscription). At the end, the optimal price of subscription is added to the optimal price of each unbundled usage package (without subscription) to get the optimal prices of the bundled packages. As before, we must assume that the matrix of partial derivatives of inverse demands for (unbundled) usage is symmetrical—conditional on the number of subscribers.

The ability to use nonuniform pricing always relaxes the budget constraint, compared to the alternative of economically efficient pricing with uniform pricing. If the firm has sufficient ability to price discriminate, it can set marginal prices at economically efficient levels (assuming no budget constraint). Whatever costs are not recovered from marginal prices would then be recovered from inframarginal consumers and not affect consumption.

If the firm cannot price discriminate sufficiently to achieve this result, marginal prices will differ from the economically efficient levels. Nevertheless, a higher level of surplus will be attainable, compared to uniform pricing.

[24] See Brown and Sibley (1986) for a full discussion of optimal nonuniform pricing.

[25] Brown and Sibley (1986) address the more general problem in which the entire price schedule is optimized.

4. Historical pricing of telephone services

4.1. Promotion of universal service

Historical pricing of telephone service was similar in some respects in many countries. Long-distance and (especially) international calls were priced far above marginal costs. At the same time, the monthly charge for access to the network was often below marginal cost.

Such pricing was intended to promote 'universal service'[26]. In the context of economic analysis, the benefits of universal service are the bandwagon benefits of increased telephone penetration.

In reality, the policy, as implemented, was almost surely massively inefficient—notwithstanding bandwagon effects. Rohlfs (1979) showed that because demand for long-distance and international services is much more elastic than demand for local services,[27] the economically efficient price structure is as follows:

- Long-distance and international services should be priced slightly above their respective marginal costs; and
- Local services should then be priced so that total revenues equal total costs.

To be sure, the marginal costs of local service and network externalities affect the level of economically efficient prices. These effects are, however, largely balanced by the need to meet the budget constraint. For example, suppose that the combination of low marginal costs for local services and network externalities would justify a low price, without a budget constraint. In that case, marginal-cost pricing would probably lead to a large deficit, and large mark-ups would be required to meet the budget constraint. Because demand for local services is so inelastic, it follows from the results of the previous section that those services would bear the largest percentage mark-ups by far. The end result is not much different from the case in which the marginal costs of local service are much higher and there is no network externality.

A related consideration is that lowering access prices yields little benefit in terms of promoting universal service—again because the demand for access to the network is so inelastic. Greater economic value can be created, with the same effect on the budget constraint, by moving prices of services with more elastic demand closer to their costs[28]. For this reason also, the network externality cannot justify the historical structure of costs.

[26] See Mueller (1997) for an irreverent history of the universal-service policies in the U.S.

[27] A result that Taylor (2002) confirms still obtains.

[28] Wenders (1987) estimates the potential gains from a move to fully efficient telecommunications pricing to be tens of billions of dollars annually. At the same time, because the relevant markets are so large, very small taxes on service could, in principle, generate substantial amounts of revenue to finance subsidies to the poor and maintain universal service. See Haring (2002).

This historical price structure has come under great pressure in recent years as long-distance and international markets have been opened to competition. With competition, long-distance and international prices that are far above marginal costs are not sustainable[29]. In the long run, those services cannot yield the contributions necessary to keep the price of access to the network below the economically efficient level.

Although the necessity to rebalance telecommunications rates may seem obvious to economists, the transition has been politically difficult. Rebalancing probably favors most consumers, but some (in particular, those who make few long-distance or international calls) are made worse off. Those consumers believe they have a right to the benefits that they enjoy under the historical rate structure, which has been in effect for decades. They exert strong pressure against any change in that rate structure. Regulators have been reluctant to resist that pressure. The result has been progress that, while constructive, has been exceedingly slow. In the U.S., almost 20 years after the AT&T divestiture, telecommunications prices are still not close to being fully rebalanced[30].

4.2. *Local usage charges*

In one respect, the historical telephone rate structure evolved quite differently in the U.S. than in Europe and many other countries. In the U.S., residents generally pay no local-usage charges whatever. Instead, they pay a flat monthly fee that covers the costs of all local calls. In many parts of the country, businesses also do not pay for local usage[31].

In contrast, charges for local usage have been the norm in Europe and many other countries. Such charges were historically assessed on the basis of pulses but are now often assessed on a per-minute basis.

The absence of local usage charges in the U.S. certainly did not result from lack of effort by the Bell System to impose them. On the contrary, the Bell companies launched a long and aggressive campaign during the 1960s and 1970s to impose local-usage charges. State regulators were not, however, persuaded that such charges were in the public interest—apart from some optional measured service plans.

During the 1960s and 1970s, one could make a colorable argument that because of measurement costs, local-usage charges were not efficient[32]. Since then,

[29] See Monson and Rohlfs (1993) for discussion of this point as it applies to local exchange carriers.

[30] The imbalance is evidenced by the disparity between local interconnection prices and interexchange access charges.

[31] Local exchange tariffs of major local telecommunications holding companies in the U.S. are posted on those companies' Websites. Further, the FCC regularly issues surveys of local telephone rates in 95 urban areas of the U.S. Those studies can be found on the Industry Analysis Division's Federal-State Link Web page.

[32] Mitchell's (1978) Optimal Pricing paper made an important contribution in this area.

however, measurement costs have declined dramatically. Today, the efficient telecommunications rate structure would almost surely embody significant local-usage charges.

For economic efficiency, the price of local usage should cover its marginal cost, since usage externalities are probably largely internalized. Large mark-ups on local-usage charges may also obviate large mark-ups for access to the network and thereby promote universal service. In addition, if large mark-ups for access to the network are politically impossible, local-usage charges ameliorate the inefficient overpricing of long-distance and international services to meet the budget constraint.

5. Local-usage charges for calls to Internet service providers

For two principal reasons, the Internet has greatly changed the economics of local-usage charges: (1) the demand for minutes of Internet calls is much more sensitive to price than is demand for minutes of voice calls; and (2) calls to the Internet involve complementary bandwagon effects.

5.1. Sensitivity of demand to price

One would expect demand for calls to ISPs to be much more sensitive to price than demand for voice calls. Internet users are often on line dozens of hours per month. At current levels of usage charges, the bill at the end of the month can be quite large. Furthermore, most Internet users can probably cut back substantially on Internet calls with much less inconvenience than cutting back substantially on voice calls. The U.K. regulator Office of Telecommunications (OFTEL, now Ofcom, Office of Communications) took this price sensitivity into account in mandating that a flat-rate Internet access call origination product be offered[33].

This price sensitivity was confirmed by the econometric study Haring et al. (2001). Haring et al. found that flat rate pricing increases number of dial-up Internet subscribers by 30 percent and average usage per subscriber by 30 percent. Thus, the effect on total dial-up Internet usage is about 60 percent.

In contrast, the elasticity of demand for local voice usage has usually been measured as approximately –0.1[34]. Since the constant elasticity model is untenable when applied to price reductions to zero, let us assume that for price reductions, demand is linear and that the elasticity is –0.1 at current prices. The stimulation in voice usage from decreasing price to zero would then be only 10 percent—one-sixth the estimated stimulation for calls to ISPs.

[33] See OFTEL (2001).

[34] Taylor (1994).

Table 1

Fixed Termination Charges

Country	Mobile-to-fixed termination rate (US$/minute)	Fixed-to-fixed: peak (Euro/minute)
Austria	0.017	
Denmark	0.008	0.009
Finland	0.0157	0.0143
Germany	0.008	0.0107
Italy	0.009	0.0114
Korea	0.009	
Norway	0.004–0.0097	
Sweden	0.0079–0.0119	0.009
Spain	0.009	0.0116

Note: ITU (2000b).

The high sensitivity of Internet usage to price is also confirmed by the 2.4-fold increase in AOL's average usage per customer when it eliminated usage charges. This increase marked a 137.5 percent increase in usage per customer. By comparison, since the change to a flat rate, usage has increased between 17 and 21 percent per year[35].

Since demand for minutes of calls to ISPs is quite sensitive to price, the efficient percentage mark-up over marginal cost is small—even in the absence of externalities.

5.2. *Costs of fixed termination*

This subsection and the next develop estimates of the costs of local usage, including calls to ISPs. Given those cost estimates, we then examine the actual percentage mark-ups embodied in the prices of those calls in various countries.

Table 1 shows charges for fixed termination in OECD countries for which the charges are supposed to be cost oriented. Cost-oriented interconnection charges include all directly attributable costs plus a contribution to common and overhead costs. They are not, however, supposed to include any subsidy to other services.

Fixed termination charges generally apply where the interconnector delivers the call to a tandem office. The fixed operator then completes the call to the called party. Fixed termination includes a single occurrence of tandem switching, a single interoffice transmission link, and a single occurrence of local switching[36].

[35] See Haring et al. (2001) at 8. See also OECD (2001).

[36] The single occurrence of local switching may include switching by a remote switch, as well as the host switch. The costs of the joining link to the interconnector and the loop to the subscriber are non-traffic sensitive costs, properly recovered through fixed monthly charges.

5.3. Costs of local usage

The average local call, including a call to an ISP, may cost more or less than fixed termination. It is unlikely to cost twice as much. In particular, the most costly type of local call is routed through a tandem switch. Such calls involve twice as many local switching occurrences and interoffice transmission links as does fixed termination—two of each instead of one of each. But they involve the same number of tandem switching occurrences as does fixed termination; namely, one. Thus, even these most costly of local calls probably cost less than twice the cost of fixed termination.

Other types of local calls cost considerably less. In particular:

- A significant percentage of local calls originate and terminate at the same local switch. They involve a single local switching occurrence, the same as fixed termination. But unlike fixed termination, they involve no interoffice transport and no tandem switching. The cost of this type of local call is much less than the cost of fixed termination, let alone twice the cost.
- Some local calls are transported directly to the terminating local office, without use of a tandem switch[37]. Those calls involve twice as many local switching occurrences as does fixed termination—two instead of one. But they involve only a single interoffice transmission link—the same as fixed termination. And they involve no tandem switching, while fixed termination does.

The median fixed termination charge in Table 1 is about US$ 0.009 per minute. It follows from the above that the cost of local usage is probably considerably less than US$ 0.018 per minute.

5.4. Bandwagon effects

We observed earlier that network externalities associated with usage are probably largely internalized, because they inherently involve only two persons. Calls to ISPs, however, involve a different kind of externality; namely, complementary bandwagon effects. In particular, as Internet usage increases, so does the incentive to put information on line. The increased incentives apply to both commercial suppliers and to many who put information on line for personal reasons.

The magnitude of this externality is difficult to assess precisely. Nevertheless, we do observe the following:

- Internet usage has been much lower in Europe than in the U.S. (Table 2 below).
- The number of Internet hosts is much higher in the U.S. than in Europe (Table 3 below).

[37] Direct routing of calls to the terminating local office is very common in the U.S., but less common in some other countries.

Table 2

Internet Subscribers

Country	Internet subscribers per 100 inhabitants
United States	18.2
EU	9.9

Note: Data as of January 2000. See OECD, Telecommunications database, June 2001.

Table 3

Internet Hosts

Country	Number of Internet hosts per 100 inhabitants	
	1997	2001
EU	12.25	40.79
United States	56.51	218.75

Note: OECD (2002), OECD calculations based on Netsizer (www.netsizer.com).

It is certainly plausible—indeed, likely—that a significant part (though certainly not all) of the difference in growth of hosts is related to the lower level of usage in Europe. Although the Internet is an international resource, many Internet hosts seek a national audience. On average, they are likely to attract a larger audience if Internet usage is higher. Thus, the incentives to establish the host are greater.

A further point is that online information performs a valuable economic function. The economy functions better with a freer flow of information. To the extent that high charges for local usage diminish the amount of online information, they can significantly retard economic growth.

For these reasons, one can make a sound case for having usage charges significantly less than marginal cost for calls to ISPs.

5.5. Comparison with actual usage charges for calls to ISPs

Table 4 shows actual usage charges for calls to ISPs. In some countries the usage charge is zero, which may well be economically efficient. Also, some countries, notably the U.K., have established regimes with no usage charges for calls to ISPs[38]. Still other countries have usage charges significantly less than US$ 0.01 per minute.

[38] See OECD (2000).

Table 4

Local Usage Charges

Country	Local usage charge (US $/minute)
Australia	0.00175
Austria	0.03625
Belgium	0.04943
Canada	0.00000
Czech Republic	0.05833
Denmark	0.02999
Finland	0.00861
France	0.00000
Hungary	0.06733
Iceland	0.01685
Ireland	0.01848
Italy	0.01157
Japan	0.01088
Korea	0.01407
Luxembourg	0.00000
Mexico	0.00000
The Netherlands	0.02640
New Zealand	0.00000
Poland	0.04903
Portugal	0.02660
Spain	0.02643
Sweden	0.01953
Switzerland	0.03098
Turkey	0.00664
United Kingdom	0.03089
United States	0.00194
OECD Average	0.02003

Note: See OECD (2000).

Nevertheless, in many countries, local usage charges for call to ISPs are substantially above US$ 0.02 per minute and are therefore far above costs. Indeed, the OECD average in Table 4 slightly exceeds US$ 0.02 per minute. Such pricing is economically inefficient and retards economic growth.

6. Charges for fixed-to-mobile calls

In mobile as well as fixed telephony, the pricing structure that evolved in the U.S. differs considerably from that in Europe and in many other countries. The U.S. opted for a structure under which mobile subscribers pay airtime for both outgoing and incoming calls. Europe and much of the rest of the world (with the notable exception of Canada) adopted a structure under which calling party pays ('CPP') for fixed-to-mobile ('FTM') calls[39].

The economic effects of these different regimes depend on the nature of the network externalities associated with FTM calling. The externality is usually positive; that is, both the caller and the called party benefit from the call[40]. The relative benefits, however, differ considerably from call to call. Furthermore, there are calls—for example, from telemarketers—which have no value, perhaps even negative value, to the called party.

We observed in Section 3 that usage externalities are largely internalized, especially within the subscriber's usual community of interest. Usage externalities are unlikely, however, to be fully internalized with respect to persons outside the subscriber's usual community of interest. They are decidedly not internalized with respect to unwanted telemarketing calls. Because externalities are not fully internalized, the two charging regimes for FTM calling have led to very different market outcomes.

Regardless of the regime, the caller is the one who decides whether a potential call gets made. Where the externalities are uninternalized, that decision depends on whether the benefits exceed the costs for the caller. The costs include the usage charges under CPP, but not if the receiving party pays ('RPP') airtime for incoming calls.

At the other end, the called party has historically had to decide whether to answer a call without knowing whom it was from. Under RPP, answering the call means incurring airtime charges. A common strategem for subscribers under these circumstances has been to limit the distribution of their mobile telephone numbers.

Obviously, that strategem has reduced the potential benefits of mobile telecommunications. Mobile telephony has served for outgoing calls and for incoming calls from the subscriber's usual community of interest. Potential calls from outside that community of interest could not, however, be made. Potential callers were thereby made worse off, as well as possibly the subscriber, herself.

Apart from this consideration, CPP works especially well with prepaid service, for which the subscriber usually pays no monthly charges whatever. For the one-time cost of purchasing a mobile telephone, a prepaid subscriber can be continuously in touch with whoever wants to call her. Unless she originates calls, she pays no usage charges. The prepaid subscriber can thereby enjoy the full (external) benefits from receiving calls, while not having to pay for unwanted calls or those that have minimal value to her.

For this reason, prepaid mobile subscription can be regarded as an important part of a modern solution to promoting universal service. It allows an individual to be connected to the network for very low cost, apart from outgoing-usage charges.

[39] Hausman (2002) offers a synopsis of U.S. and European cellular telephone pricing at 579–582.

[40] DeGraba (2003) discusses the implications of this point for interconnection pricing.

Table 5
Mobile Subscribers in Countries with CPP and RPP

Countries w/CPP	Mobile subscribers per 100 inhabitants (2001)
France	60.53
Germany	68.23
Italy	83.94
Japan	58.78
United Kingdom	77.04
Countries w/RPP	
Canada	36.19
United States	45.08

Note: See ITU (2004).

Prepaid service can also be, and is, offered in conjunction with RPP. That option has, however, proved to be much less popular than in conjunction with CPP. The reason is that the subscriber cannot keep her costs down by simply making few calls. She must also pay for the calls that she receives.

The advantages of CPP, as described above, have contributed to higher mobile penetration in countries that have CPP. In particular, penetration is substantially higher in all five of the G-7 countries that have CPP than in the U.S. and Canada, which have RPP (See Table 5 above).

Notwithstanding this history, the disadvantages of subscribers' paying for airtime have declined significantly, as a result of technological progress. In particular:

- Caller ID allows the mobile subscriber to see who is calling before answering the call; and
- Mobile-usage charges have declined dramatically, especially in the increasingly popular bulk-calling packages. Consequently, mobile subscribers no longer need be so concerned about incurring usage charges on unwanted calls and those that have little value to the subscriber.

6.1. Charges under calling-party-pays

Most countries, especially industrialized countries, now have multiple suppliers of mobile telephony[41]. The retail market for mobile telephone services in these countries is often quite competitive.

[41] See ITU (2000c).

In contrast, the market for call termination on a specific mobile network (from another network, either mobile or fixed) is more complex. In order to complete a call to a particular mobile telephone, the person that originates the call must connect to the *exclusive* mobile operator that provides the service to the called party.

Competition among mobile operators could conceivably exert downward pressure on the call termination price. Nonetheless, under CPP, once a person decides to subscribe to a certain mobile network, the selected network has a certain degree of market power *vis-à-vis* those who want to communicate with her. The party that originates the call often has no satisfactory alternative to paying the price that the operator of the receiving party charges. This market power has become known as the 'terminating access monopoly'[42].

Given this imperfection of the market, the market outcome depends critically on the regulatory policy. There are three general possibilities: (1) the fixed operator is not subject to any restrictions with regard to negotiations with mobile network operators ('MNOs'); (2) MNOs can establish whatever termination charges they choose on their networks, and the fixed operator is forced to pay them (and presumably pass them on to its fixed subscribers); and (3) the call termination charge is subject to regulation.

Case 1. The fixed operator is unrestricted.

If there is only one fixed operator and multiple mobile operators, the fixed operator will have greater leverage in the negotiations and will probably be able to negotiate a mobile termination charge that does not fully cover the costs of mobile termination. The leverage is all the greater if the fixed operator is affiliated with an MNO. In that case, the operator gains a competitive advantage in the mobile market if the other MNOs do not acquiesce to the terms that it imposes and no interconnection agreement is reached. The fixed operator may additionally have the ability and incentive to subsidize its mobile service (with anticompetitive consequences) in order to stimulate lucrative fixed origination.

Apart from these anticompetitive possibilities, mobile termination charges that are below cost do not necessarily reduce the economic profits of MNOs in the long run. If the market to attract mobile subscribers is effectively competitive, MNOs will earn approximately zero economic profits in any event.

Below-cost charges do, however, reduce economic efficiency. They encourage calls whose value does not justify their costs. In addition, they force MNOs to recover the deficit from their subscribers in order to cover total costs. Such charges repress demand for mobile subscription and are likely to lead to fewer mobile subscribers than would be optimal.

Case 2. MNOs can set whatever termination charges they choose and the fixed operator is forced to pay them.

[42] See Laffont and Tirole (1994).

Research studies done recently by OFTEL and the U.K. Competition Commission ('CC') found clear signs of competition among mobile operators in the U.K. to attract subscribers, measured by the increase in the coverage and the capacity (and, therefore, in the quality) and the existence of many new price packages. Nevertheless, both government organizations determined that competition in mobile termination was not effective and ordered reductions in the termination charges[43].

In addition, data collected from countries where MNOs are allowed to charge whatever price they choose for termination of FTM calls indicate that competition in the retail market does not necessarily translate into cost-oriented charges for the *call termination*. The CC determined that it is not probable that the competitive pressure directed to attract users has much effect on termination charges. The data analyzed by the CC indicated that the majority of mobile users assigned low priority to incoming call charges.

Excessive termination charges in the mobile network have been widely noted. For example, the International Telecommunications Union ('ITU') observed that the interconnection charges among cellular networks are substantially lower than the interconnection charges between fixed and mobile networks in most countries[44]. One can reasonably presume that interconnection charges between MNOs are not less than marginal cost—unless the bargaining leverage of the two MNOs is quite unbalanced. The MNO with the higher fraction of incoming traffic would be unlikely to agree to such a charge. It follows that the much higher mobile termination charges for FTM traffic are far above marginal cost. Moreover, concerns about excessive charges have led to corrective regulatory measures. In July 1998, the European Commission carried out a series of comprehensive investigations regarding mobile termination charges for FTM calls in 14 member countries.

Case 3. Termination charges for the mobile network are subject to regulatory restrictions.

Given that alternatives (1) and (2) both have serious disabilities; regulation of mobile-termination charges appears to be the best alternative. Such regulation should ideally mandate prices that maximize economic efficiency.

6.2. Economically efficient mobile termination charges

The first possibility to consider is setting mobile termination charges equal to marginal cost. Such pricing would be economically efficient if no budget constraint had to be met and there were no externalities.

[43] More information about these decisions is available on the Website of the Competition Commission (www.competition-commission.org.uk).

[44] ITU (2000a).

In reality, two types of externalities must be considered; namely, the externality with respect to FTM usage and the externality with respect to mobile subscription.

As previously discussed, the usage externality is probably largely internalized. Nevertheless, part of the externality may be uninternalized, especially with respect to persons outside the subscriber's usually community of interest. Usage externalities can be modeled by setting n_i at the appropriate level (greater than unity) for FTM usage in Eq. (13). This externality lowers the economically efficient price.

The externality with respect to mobile subscription is probably larger and more important than the usage externality. To take account of this externality, one must carefully consider how the FTM price is likely to affect mobile subscription.

The cross-elasticity of mobile subscription with respect to the FTM price is probably small in absolute value. As previously discussed, studies have indicated that mobile subscribers pay little heed to that price in their subscription decision. It is precisely because this cross-elasticity is small in absolute value that regulation is justified.

The FTM price can, however, affect mobile subscription indirectly by affecting the prices that MNOs charge their subscribers. In particular, an increase in the mobile termination charges makes subscribers more lucrative for MNOs because the subscribers receive calls that are more lucrative. Consequently, one would expect MNOs to respond by competing more aggressively for subscribers; in particular, by offering them more attractive pricing plans.

As previously discussed, one can reasonably assume that competition for mobile subscribers is effective in most countries. Consequently, one would expect them not to earn sizable monopoly rents.

Under these circumstances, economically efficient pricing can be modeled subject to the budget constraint that total revenues of MNOs cover total cost. In this optimization, the regulator sets the price of mobile termination at the optimal level. MNOs set the prices to mobile subscribers. Competition ensures that the budget constraint is satisfied.

This process would lead to economically efficient prices if only a single rate element were charged to mobile subscribers. In reality, however, mobile pricing plans are highly complex and involve many rate elements. Under these circumstances, MNOs can be expected to use incremental revenues from mobile termination charges to pursue their own agendas (subject to the competitively imposed budget constraint). They are not likely to choose the prices that maximize economic efficiency.

Equation (13) can be applied to this problem to get an estimate of the economically efficient level of mobile termination charges. In reality, however, a somewhat lower level of economic efficiency is likely to be achieved, as MNOs pursue their own agendas.

One can reasonably conclude that the solution to Eq. (13) overestimates the optimal mobile termination charge. That level is based on balancing the loss of surplus that derives from fixed subscribers' paying more than marginal cost against the efficiency gains that the MNOs are expected to provide through price reductions. If the MNOs provide less efficiency gain because they pursue their own agendas, the trade-off changes in favor of a lower mobile termination charge.

6.3. Setting mobile termination charges in practice

The OFTEL was recently involved in setting mobile termination charges with a goal for economic efficiency, taking into account network externalities. Using Eq. (13), OFTEL estimated optimal prices[45]. However, OFTEL's analysis recognized that MNOs may pursue their own agendas.

A thorny and contentious issue in this proceeding was the estimation of cross elasticities of demand. The OFTEL believed that the available econometric estimates of those elasticities, supplied by the MNOs, were inaccurate and unreliable. For that reason, in the optimization [Eq. (13)], it used estimates of cross elasticities that were based largely on *a priori* reasoning.

Rohlfs (1979) reasoned that if all communication links were equally valuable to the two parties and no external benefits were internalized, the (gross) externality factor would be two. What is relevant for efficient pricing are the relative values of communications links to the marginal subscriber and to existing subscribers with whom she communicates. If they are worth more to the subscriber, then the externality factor before internalization would be less than two. If they are worth more to those that call her, the externality factor before internalization would exceed two.

It is difficult to say whether on average a mobile communications link benefits a marginal subscriber more or less than existing subscribers, including both mobile and fixed subscribers. One can, however, observe that calls to mobile subscribers typically significantly benefit both parties to the communication. The mutual benefit is evidenced by the costs that both parties voluntarily incur to enable the communication. The calling party pays for the call. The mobile subscriber undergoes the inconvenience of keeping the mobile telephone battery charged and otherwise making herself available to receive calls. Consequently, the benefits to marginal and existing subscribers are not likely to be too unequal.

At the same time, as discussed in Section 3, the opportunities to internalize the externality are considerable, especially within the marginal subscriber's usual community of interest. Internalization probably does not suffice to drive the externality factor down to unity, because of transactions costs. It does seem likely, however, that internalization drives the externality factor below two—especially since it may not have been much above two before internalization (if, indeed, it was above two).

The OFTEL assumed that the (gross) externality factor is in the range of 1.3 to 1.7. It then reasoned that the nonusage externalities (option values) were a

[45] That formula was developed by the author in connection with consulting for OFTEL on this issue.

small part of the total externality. In particular, OFTEL assumed that the net externality factor is in the range of 1.05 to 1.1. The remaining externalities are then related to usage.

A further step in OFTEL's analysis was to reason that MNOs have the ability and incentive to internalize some of the externalities through their nonuniform pricing structures. As discussed in Section 3, the optimal use of nonuniform pricing always relaxes the budget constraint. In this particular case, MNOs can take into account the external benefits that marginal subscribers bestow on the MNO's existing subscribers. They can additionally reason that even if their price cuts are met by other MNOs, all will benefit from the network externalities generated by new subscribers. The use of nonuniform pricing by MNOs reduces the efficient mark-up (over marginal cost) of FTM usage charges.

The OFTEL then estimated cross elasticities based on the above considerations. It applied Eq. (13) and also considered a model in which MNOs pursue their own agendas. The analysis supported a moderate mark-up for mobile termination above average incremental cost.

6.4. *A possible dual regime*

We noted earlier that CPP has had historically significant advantages over RPP, but those advantages are declining. Still, however, CPP retains the advantage of allowing prepaid subscribers to obtain access to the network at very low cost.

A disadvantage of CPP is that regulation of mobile-termination charges is necessary for efficient pricing. The charges are unlikely to cover costs unless the fixed operator is constrained. The rates are likely to be excessive if the fixed operator is constrained but MNOs are allowed to charge whatever they want.

A dual regime combining both approaches is also possible (though the author knows of no country in which it has been adopted). Under the dual regime, there would be two classes of mobile telephone numbers: one for CPP and one for RPP.

Such a regime has the advantage of internalizing both directions of usage externalities that the users cannot internalize themselves. That is, callers who are unwilling to pay usage charges could call a RPP number (if it has been disclosed to them); mobile subscribers who are reluctant to pay airtime on incoming calls could publicize only their CPP number. If network externalities are more effectively internalized in these ways, mobile subscription and usage would likely increase and all telephone subscribers (mobile and fixed) would then enjoy greater bandwagon benefits.

The dual regime also gives MNOs more flexibility. In particular, if they are dissatisfied with the regulated mobile termination charges, they could offer especially attractive terms to subscribers who have only a RPP number. In that way, if the regulatory regime is not working well, a larger and larger fraction of the mobile market could migrate to the fully unregulated RPP regime.

7. Conclusions

Bandwagon effects are pervasive in high-technology industries, and in telecommunications in particular. They are of two types:

1. Network externalities, whereby existing subscribers benefit from being able to communicate with a larger user set; and
2. Complementary bandwagon effects, whereby purchasers of the base product (e.g., hardware) benefit from the greater availability of competitively supplied complementary products (e.g., software) as the user set expands.

A key property of bandwagon markets is the start-up problem, which arises because bandwagon benefits are initially limited to those generated by a small user set. Several chapters in this *Handbook* discuss the start-up problem in specific contexts.

In this chapter, we considered economically efficient pricing in bandwagon markets. We derived the general conditions for optimality and discussed three applications:

1. Historical pricing of telephone service to promote universal service;
2. Local usage charges on calls to Internet service providers; and
3. Pricing of mobile termination for fixed-to-mobile calls.

In all three cases, externalities affect economically efficient prices. Nevertheless, other considerations, especially relative elasticities of demand, are at least equally important in determining optimal mark-ups above marginal costs.

References

Abrahamson, E. and L. Rosenkopf, 1997, Social network effects on the extent of innovation diffusion: A computer simulation, Organization Science, 8(3), 289–309.

Armstrong, M., 2002, Theory of access pricing and interconnection, in M. E. Cave, S. K. Majumdar and I. Vogelsang, eds., volume 1, Chap. 8, 295–384.

Arthur, W. B., 1989, Competing technologies, increasing returns, and lock-in by historical events, The Economic Journal, 99, 116–131.

Artle, R. and C. Averous, 1973, The telephone system as a public good: Static and dynamic aspects, Bell Journal of Economics and Management Science, 4(1), 89–100.

Baumol, W. J. and D. F. Bradford, 1970, Optimal departures from marginal cost pricing, American Economic Review, 60(3), 265–283.

Boiteux, M., 1951, Le 'Revenue Distribuable' et les pertes economiques, Econometrica, 19(2), 112–133.

Brown, D. and J. Sibley, 1986, The Theory of Public Utility Pricing, Cambridge, MA: Cambridge University Press.

Church, J. and N. Gandal, 1992, Network effects, software provision, and standardization, Journal of Industrial Economics, 40(1), 85–103.

Coase, R., 1960, The problem of social cost, Journal of Law and Economics III, 1–44.

DeGraba, P., 2003, Efficient intercarrier compensation for competing networks when customers share the value of a call, Journal of Economics and Management Strategy, 12(2), 207–230.
Downes, T. A. and S. M. Greenstein, 1998, Do commercial ISPs provide universal access? in S. Gillet and I. Vogelsang, eds., Proceedings: Telecommunications Policy Research Conference.
Economides, N. and S. C. Salop, 1992, Competition and integration among complements, and network market structure, Journal of Industrial Economics, 40(1), 105–123.
Farrell, J. and G. Saloner, 1985, Standardization, compatibility, and innovation, Rand Journal of Economics, 16(1), 70–83.
Farrell, J. and C. Saloner, 1992, Converters, compatibility, and the control of interfaces, Journal of Industrial Economics, 40(1), 9–35.
Farrell, J. and C. Shapiro, 1993, The dynamics of bandwagons, in J. W. Friedman, ed., Problems of Coordination in Economic Activity, Kluwer Press, 149–184.
Haring, J., 2002, Telecommunications, in The Concise Encyclopedia of Economics at http://www.econlib.org/library/Enc/Telecommunications.html.
Haring, J., J. H. Rohlfs and A. Briceño, 2001, The Effect of Pricing Structure on Residential Internet Demand: Mimeo Strategic Policy Research.
Hausman, J. A., 2002, Mobile Telephone, in M. E. Cave, S. K. Majumdar and I. Vogelsang, eds., Volume 1, Chap. 12, 563–604.
Hong, C. S. and K. Konrad, 1998, Bandwagon effects and two-party majority voting, Journal of Risk and Uncertainty, 16(2), 165–172.
International Telecommunications Union ('ITU'), 2000a, Workshop on connection between fixed and mobile phones Document WFMI/04, 29, September 14.
ITU, 2000b, Regulatory Survey 2000; Mobile Interconnection Section, Fixed-Mobile Interconnection Workshop: Briefing Paper, September 14.
ITU, 2000c, World Telecommunication Development Report.
International Telecommunications Union, 2004, Mobile Cellular Subscribers per 100 inhabitants, downloaded March 31, 2004 from www.itu.int/ITU-D/ict/statistics/at glance/cellular01.pdf.
Katz, M. L. and C. Shapiro, 1992, Product introduction with network externalities, Journal of Industrial Economics, 40(1), 55–83.
Krebs, B., 2003, A Short History of Computer Viruses and Attacks, www.washingtonpost.com/ac2/wp-dyn/A50636-2002 June 26, February 14.
Laffont, J. and J. Tirole, 1994, Access pricing and competition, European Economic Review, 38, 1672–1710.
Liebowitz, S. J. and S. E. Margolis, 1994, Network externality: An uncommon tragedy, Journal of Economic Perspectives, 8(2), 133–150.
Littlechild, S. C., 1975, Two-part tariffs and consumption externalities, Bell Journal of Economics, 6(2), 661–670.
Matutes, C. and P. Regibeau, 1992, Compatibility and bundling of complementary goods in a Duopoly, Journal of Industrial Economics, 40(1), 37–54.
Mitchell, B. M., 1978, Optimal pricing of local telephone service, American Economic Review, 68(4), 517–537.
Monson, C. S. and J. H. Rohlfs, 1993, The $20 Billion Impact of Local Competition in Telecommunications, Strategic Policy Research.
Mueller, M. L., Jr., 1997, Universal Service: Competition, Interconnection, and Monopoly in the Making of the American Telephone System, Washington, DC: MIT Press and The AEI Press.
Noam, E. M., 1995, Privacy in Telecommunications, Part III, New Telecommunications Quarterly, 3(4), 52–69.
OECD, 2000, Working Party on Telecommunications and Information Service Policies, Table A.1 of Local Access Pricing and E-Commerce.
OECD, 2001, Understanding the Digital Divide.
OECD, 2002, Information Technology Outlook 2002.

Office of Telecommunications, 2001, Determination relating to a dispute between British Telecommunications and Worldcom concerning the provision of a Flat Rate Internet Access Call Origination product ('FRIACO').

Ramsey, F. P., 1927, A contribution to the theory of taxation, Economic Journal, 37, 47–61.

Riordan, M. H., 2002, Universal residential telephone service, in M. E. Cave, S. K. Majumdar and I. Vogelsang, eds., Handbook of Telecommunications Economics, Volume 1, Chap. 10, 424–473.

Rohlfs, J. H., 1974, A theory of interdependent demand for a communications service, Bell Journal of Economics and Management Science, .

Rohlfs, J. H., 1979, Economically Efficient Bell System Pricing, 1978, Bell Labs Economic Discussion Paper #138, presented at the 7th Annual Telecommunications Policy Research Conference, Pennsylvania, Skytop.

Rohlfs, J. H., 2001, Bandwagon Effects in High-Technology Industries, MIT Press.

Samuelson, P. A., 1965, Foundations of Economic Analysis, Atheneum, New York.

Shapiro, C. and H. R. Varian, 1999, Information Rules, A Strategic Guide to Network Economy, Harvard Business School Press.

Squire, L., 1973, Some aspects of optimal pricing for telecommunications, Bell Journal of Economics and Management Science, 4(2), 515–525.

Strategic Policy, Research, 2002, Propelling the Broadband Bandwagon, prepared for the United Kingdom Office of Telecommunications and the Office of the e-Envoy, released September 4, 2002.

Swann, G. M. P., 2002, The functional form of market effects, Information Economics and Policy, 14(3), 417–429.

Taylor, L. D., 1994, Telecommunications Demand in Theory and Practice, Boston: Kluwer Academic Publishers.

Taylor, L. D., 2002, Customer Demand Analysis, in M. E. Cave, S. K. Majumdar and I. Vogelsang, eds., Volume 1, Chap. 4, 97–142.

Wenders, J., 1987, The Economics of Telecommunications: Theory and Policy, Ballinger.

Willig, R. D., 1976, Consumer's Surplus without Apology, American Economic Review, 66(4), 589–597.

Chapter 4

PLATFORM COMPETITION IN TELECOMMUNICATIONS

JEFFREY CHURCH

University of Calgary, Calgary, Canada

NEIL GANDAL

Tel Aviv University, Tel Aviv, Israel

Contents

Handbook of Telecommunications Economics, Volume 2, Edited by S. Majumdar et al.

DOI: 10.1016/S1569-4054(05)02004-X

1. Introduction

In this chapter, we consider the economics of platform competition in telecommunications. Platform competition occurs when different, sometimes incompatible, technologies compete to provide telecommunications services to end-users. Battles between competing technologies have been an important feature of telecommunications in the last twenty or so years. Examples of platform competition in telecommunications include wireless vs. wireline networks; competing wireless options, such as satellite vs. cellular; and, within cellular, different digital standards. Other examples include competing incompatible instant messaging offered by AOL, MSN, and Yahoo; and direct broadcast satellite vs. cable networks—to say nothing of the possibility of video delivered over the local phone network. A more broadly defined view of telecommunications widens motivating examples to include battles in consumer electronics between audio and video formats, as well as operating systems for personal computers.

These platform wars have become more significant, and more prominent, given the pace, nature, and magnitude of technological change in telecommunications, and information technologies, more generally. The ongoing convergence between telecommunications, entertainment, and computing is driven by fundamental changes and advances in the ability to manipulate information. All of these industries involve the manipulation, transmission, and/or reproduction of information, which in the broadest sense is anything that can be represented as (reduced to) a stream of binary code. The value of that manipulation depends often on how the information is gathered and distributed. The efficient gathering and distribution of information—perhaps after processing—typically occurs on a network. Competition between platforms in telecommunications, or technologies in information intensive industries typically involves more than just competition between differentiated products—in the interesting cases, and the concern of this chapter, it involves competition between technologies that are not only differentiated, but also are competing networks.

In this chapter, we view platform competition through the lense of network economics to develop an understanding of the determinants of its outcome and adoption patterns by consumers. In Section 2, we define network industries and distinguish between types of networks. We introduce the concept of network effects and discuss how the presence of network effects affects demand for a platform. In this section, we also discuss settings in which network effects give rise to network externalities. In Section 3, we distinguish between standards wars, battles over compatibility, and cooperative standard setting leading to battles on the network—rather than between networks. Section 4 considers firm strategies in standards wars and raises the possibility that competition between incompatible technologies may not result in an efficient standard. Section 5 discusses battles for compatibility between dominant networks and competitors, while Section 6 discusses the economics of cooperative standard setting. Section 7 discusses the role of regulation in insuring compatibility (interconnection) and setting standards. Section 8 contains a number of case studies drawn from telecommunications that illustrate the principles of network competition discussed in the preceding sections.

2. Network industries

The defining feature of network industries is that products consumed are systems of components: the ultimate 'good' demanded is comprised of a group of complementary products that provide value when they are consumed together. It is most often the case that the components in and of themselves have very little, if any, value. In order to be of value they must typically be consumed as part of a system. For the complements to work together as a system requires standards to insure compatibility. In this context a 'standard' refers to the set of technical specifications that enable compatibility between products.

It is common to distinguish between two different types of networks: direct networks and virtual or indirect networks. In the first, the system consists of similar products linked together in a network. In the second, the system consists of a unit of hardware and a unit of software[1].

2.1. Direct networks

Direct networks require horizontal compatibility, which is typically achieved through some sort of common standard across the products chosen by consumers who have joined the network. The archetype is a communication network, where compatibility allows adopters[2] to communicate with each other. An adopter's link to the network has no value except to facilitate the transmission of information to, and from, other adopters. The value of a link to an adopter depends on the systems that can be created by combining their link with links of other adopters.

The classic example of a communication network is a telephone exchange. Consider a very simple example where n subscribers (adopters) are connected to a switch. The creation of a phone call from subscriber i to subscriber j involves combining two complements: subscriber *i's* link to the switch and subscriber *j's* link to the switch. Notice that in this example, it is connection with the switch that insures compatibility with all other connected links. Two other examples of direct networks—where consumption benefits arise because of horizontal compatibility among adopters—are the communications standards built into facsimile machines and the Internet. Direct networks need not consist of, or depend on, cables in the ground, they may consist of individuals who adopt a similar word-processing

[1] Modeling issues are briefly addressed in the appendix.

[2] We sometimes referred to consumers who join the network as subscribers or adopters.

program and derive benefits from being able to swap files with others (the horizontal compatibility here is the ability of the software to recognize and read files produced by others) or the network created by speakers of a common language (the horizontal compatibility here is the ability of both parties to comprehend and speak a common language)[3].

2.2. *Virtual (indirect) networks*

When the network is indirect, systems consist of one unit of hardware and one unit of software. The two components interact, or are combined, to provide consumption benefits: a unit of hardware or a variety of software typically has no, or relatively very little, stand-alone value. In the case of indirect networks what is important is vertical compatibility—compatibility between hardware and software. In the interesting cases, the unit of hardware is compatible with many different varieties of software.

Examples abound in consumer electronics: televisions and programming, compact disc players and compact discs, video game systems and video games, FM radios and FM radio stations, video-cassette recorders and prerecorded programming, digital music formats and digital music players, satellite radio and satellite radio channels. This 'hardware–software' paradigm is not restricted to consumer electronics and the hardware need not literally be hardware. Other examples are operating systems (hardware) and application programs (software); credit cards (hardware) and the stores that accept them (software); natural gas powered vehicles (hardware) and natural gas filling stations (software); browsers (hardware) and Websites (software); yellow pages (hardware) and yellow page listings (software); ATM cards (hardware) and ATM teller machines (software). In the last example, for instance, the value of an ATM card depends on the number of ATM machines at which it can be used. These examples suggest the identification of the software good as the component for which there are many possible varieties and the hardware good as the component for which there is a unit demand for a single variety. In the multiplicity of systems that can be created the hardware good does not change and it can be interpreted as the component, which allows access to the software varieties.

2.3. *Network effects*

Both types of networks are often characterized by a network effect. A network effect exists if the value of joining a network by buying compatible products is increasing in the number of other adopters who (ultimately) join the network by purchasing compatible products. With positive network effects, 'bigger is better.' In both cases the underlying source of the network benefit is the same: the larger

[3] See Church and King (1993) and Gandal (2002a).

the number of adopters the greater the possible number of systems (combinations of complementary components) an adopter can create.

Recall that a direct network consists of complements linked together to form a network. In this case, horizontal compatibility allows for interconnection of the product purchased by a consumer with that of others. In a direct network the number of systems that can be created by a subscriber equals the number of other adopters: in a telephone exchange with n consumers, subscriber i can create $n - 1$ systems. The greater the total number of subscribers, the greater the number of systems.

When the network effect is indirect, consumption benefits do not depend directly on the size of the network (the total number of consumers who purchase compatible products) per se. Rather, individuals care about the decisions of others because of the effect that has on the incentive for the provision of complementary products. Users of Macintosh computers are better off the greater the number of consumers who purchase Macs because the larger the number of Mac users, the greater the demand for compatible software, which, if matched by an appropriate supply response—entry by software firms—will lead to lower prices and a greater variety of software, which makes all Mac users better off.

As in the direct network effects case, when there are indirect network effects consumers benefit from the adoption by others of compatible hardware because it allows them to consume a wider variety of systems. In this case consumption benefits flow from creating systems consisting of one unit of hardware and one unit of software, where the unit of hardware is compatible with many different varieties of software. If consumers value variety, then they will demand multiple systems, each involving one unit of the hardware good and a different variety of software[4]. The advantage of more adopters of hardware to an existing subscriber arises if an increase in hardware adoption induces the production of more software varieties since existing adopters will then benefit from being able to create more 'two-component' systems.

2.4. From network effects to network externalities

Both direct and indirect network effects can give rise to network externalities. An adoption externality arises when the network benefits of existing adopters increases with the increase in size of the network (i.e., with the addition of another adopter). In both cases, the source of benefit to inframarginal adopters from adoption at the margin is the creation of new systems by existing subscribers. The marginal adopter does not internalize the marginal external benefit when making their adoption decision, leading to underadoption.

[4] It is a rare individual indeed who listens to only one compact disc on the stereo, uses only one application program on the PC, or plays only one video game on the video game console.

In the case of a direct network, when an additional individual joins a network of n individuals, in addition to the n potential types of systems that are open to the new individual, the link of the new subscriber creates new systems for the n inframarginal, or existing, adopters. The addition of a new individual to an n individual network creates n new systems—combinations of complements that can be connected by existing subscribers to create a new good. It is this creation of new systems for existing subscribers/adopters that is the benefit to existing subscribers of network expansion.

As in the direct network effects case, when there are indirect network effects, consumers benefit from the adoption by others of compatible hardware because it allows them to consume a wider variety of systems. Inframarginal adopters of hardware benefit when there is an increase in hardware adoption if it induces an increase in the production of more software varieties, providing them with the option of creating more two-component systems.

The issue is that marginal consumers do not account for the effect that extending the hardware network will have on the variety of software, and thus the benefit inframarginal consumers receive from being able to consume additional software varieties. It is the number of *different* software varieties that is important, not the quantity (or price) of a particular software variety. That is, adoption externalities in the setting with indirect network effects are the result of *variety effects,* not price effects[5]. The manner in which inframarginal consumers benefit from indirect effects is identical to the manner in which they benefit when there are direct effects—the ability to create new systems of complementary products. Network externalities that arise in settings with indirect network effects have the same microfoundations as network externalities that arise in settings with direct network effects.

2.5. *Implications for consumer demand*

For products characterized by network effects the decision by consumers regarding which network to join—often referred to as the technological adoption decision of a consumer—will depend not only on relative product characteristics and prices, but also the expected size of the network. Moreover, the current size of the network, or its installed base, will often be used as a signal or indication to consumers of its future size. In the case of a direct network, the size of the installed

[5] See Church, Gandal, and Krause (2003). They show that the requirements for indirect network effects to give rise to an adoption externality are threefold: (i) increasing returns to scale in the production of software, (ii) free-entry into software, and (iii) consumer preferences for software variety. Under these circumstances, the marginal adopter does not take into account the benefits that accrue to inframarginal adopters from the response of the software industry to an increase in hardware sales. When there are increasing returns to scale and free-entry into the production of differentiated software, the key response to an increase in hardware sales is an increase in software variety, which benefits inframarginal consumers. These three conditions are both necessary and sufficient.

base is usually measured by the number of adopters. In the case of indirect networks the size of the network can also be measured by the number of adopters of compatible hardware, but in many instances the relevant installed base is often the number of complementary software varieties available.

The role of the size of the existing installed base in determining the size of the network in the future arises because positive network effects give rise to positive feedback effects. Consumers will appreciate that a larger installed base today will make the network more attractive to adopters in the future, while the expectation that the network will be attractive to consumers in the future insures strong adoption in the present. These positive feedback effects create a strong tendency for 'the strong to get stronger' in a virtuous cycle—the greater the installed base, the greater network benefits, the more attractive the network to adopters, the greater adoption, the greater the installed base, etc. When there are competing networks and one network experiences positive feedback effects it is often the case that its competitors experience negative feedback effects: 'the weak get weaker' in a vicious cycle. The smaller the installed base, the smaller the network benefits and the less attractive the network, the greater the incentive to abandon the network, the smaller the installed base, etc.

Unlike economies of scale there is no reason for diseconomies from network effects. They may become small, but they will not be negative. Network effects are similar to demand side scale economies, but not identical since expectations matter. Consumption benefits are increasing not only in the number of consumers who adopt at time t—as with demand side scale economies—but also with the number of consumers who join the same network in the future.

2.6. *Expectations and competition between networks*

As we noted in the previous section, the value of joining a network when there are positive network effects depends on the ultimate size of the installed base of the network. This means that the expectations of consumers today, regarding the growth of a network, will be an important determinant of their adoption decision. And since a larger installed base today will contribute to growth in the future, the current size of the installed base will inform those expectations. The central role of expectations and their dependency on the current installed base has a number of important implications for competition in network markets[6]. These are:

[6] The seminal contributions on competition in settings with network effects are a series of papers by Farrell and Saloner (1985, 1986a,b) and Katz and Shapiro (1985, 1986). Katz and Shapiro (1994) and Besen and Farrell (1994) are excellent surveys on the economics of network industries. In what follows we draw liberally on all of these papers, following Church and Ware (1998). David and Greenstein (1990) provide a comprehensive survey of earlier work, while Farrell and Klemperer (2005) provide a detailed survey of more recent work. Gilbert (1992), Gandal (1995), Matutes and Regibeau (1996), Gandal (2002b), and Stango (2004) provide selective reviews of the literature.

(i) coordination problems, (ii) tipping/standardization, (iii) multiple equilibria, and (iv) lock-in[7].

2.6.1. Coordination problems

Typically a consumer must invest in a connection (direct network) or hardware (indirect case) to join a network. If this investment is sunk, the potential for coordination problems arise. Consumers make these investments with the expectation that the network will grow and a certain level of network benefits will be realized. If the expected growth in the network is not realized, perhaps because it is abandoned by future generations of consumers, then its early adopters will be stranded on an 'orphan' technology. In such a case the expected benefits associated with the sunk investment and membership on the network are not realized.

Uncertainty over being stranded makes consumers reluctant to join new networks whose installed base is small. The possibility of being stranded arises because of a coordination problem: consumers would be willing to join a new network if they knew that others were also willing to join, but because no one is presently on the network they do not believe that others are willing to join, and hence they, and others, do not. A similar coordination problem can affect the adoption prospects of an indirect network. Consumers would be willing to buy hardware if sufficient software variety was available or expected to be available. Software suppliers would find it profitable to support a hardware technology if it is adopted, but are reluctant to do so until consumers demonstrate that a market exists by adopting the hardware. The coordination difficulties between consumers and suppliers of software—or more generally complementary products—are, for obvious reasons, known as the 'chicken and egg' problem. The coordination problem in indirect networks is further complicated because hardware might not be introduced without the expectation of support from suppliers of complementary software. Firms will therefore have an incentive to try and minimize the risk to consumers of being stranded.

2.6.2. Standardization

Products characterized by network effects are highly susceptible to 'tipping' or standardization. When a network market tips, consumers adopt or join only one network, they do not support multiple competing networks/technologies. Competition between incompatible networks can easily result in the 'winner taking all' where one technology becomes a de facto standard because all consumers adopt it. This occurs when network effects are particularly strong. When network effects are strong, if one system or technology can establish an initial edge in the size of its installed base, this provides an effective (and correct) signal to all consumers—present and future—that all consumers will also adopt this system. Small initial

[7] Our discussion here follows Church and Ware (1998).

leads in market penetration can result in the creation of a very large sustained advantage and exclusive adoption. Firms whose technology is incompatible with their rivals in markets with strong network effects have an incentive to compete aggressively when their technology is introduced in order to establish an installed base advantage. Tipping and de facto standardization is more likely the stronger network effects are relative to the extent of consumer heterogeneity. If consumer preferences are relatively heterogeneous, then product differentiation considerations can trump network effects and multiple incompatible differentiated networks can coexist.

For instance, if consumers observe that video rental outlets typically stock predominantly video cassettes in the VHS format then they will tend to expect continued supply of VHS tapes in the future, providing them and others with incentives to purchase VHS compatible video cassette recorders. Consumers' expectations become self-fulfilling as video rental outlets—and eventually film studios–respond by reducing their library of Beta tapes and specialize in VHS. The standardization on the VHS format in North America is an example where product differentiation between the two standards was fairly minimal, but network effects were strong. This experience can be contrasted with the personal computer market where network effects have contributed to the dominance of the PC standard (Intel and Windows), but strong product differentiation has enabled Apple to carve out a niche in publishing and graphic design.

2.6.3. *Multiple equilibria*

Multiple equilibria are possible, since the equilibrium outcome will depend on the expectations of consumers. For example, suppose there are two new competing technologies and that network effects are relatively strong. In these circumstances, depending on the expectations of consumers, the equilibrium outcome might be standardization on one of the new technologies or both technologies failing. The latter might be the case if consumers cannot coordinate on a standard and, due to concerns over stranding, play it safe by not adopting either technology. Firms will have an incentive to influence and coordinate the expectations of consumers.

2.6.4. *Lock-in*

Consumers can become locked in to a network because of switching costs. Two types of switching costs 'lock-in' consumers to a network. When adoption entails a sunk cost and networks are incompatible, consumers that switch to another network will need to make an investment in the connection or hardware of the new network. Second, consumers might forgo network benefits if the installed base of the new network is not as large. This disadvantage might only be temporary, but it might be permanent if the installed base advantage of the incumbent network is expected to persist in the future.

Lock-in means that consumers find it costly to switch to a competing network *ex post* and consequently, it makes them subject to opportunism by a network provider. The opportunism can take a number of forms including raising prices or lowering quality. In particular, promises by a network provider to expand its network by charging low prices or providing lots of inexpensive software in the future are not necessarily credible.

3. Battles for standards, compatibility and adoption

Following Besen and Farrell (1994), it is useful to distinguish between four different situations. These are:

- a standards war between two or more incompatible standards (Section 4);
- battles over compatibility (Section 5);
- standard setting by voluntary agreement (Section 6); and
- mandated standards by the state (Section 7).

In a standards war, two or more incompatible systems compete against each other. Examples include: VHS vs. Beta; Visa vs. American Express; Linux vs. Windows; and X-Box vs. PLAYSTATION. Standards wars typically arise between sponsors of closed systems. Sponsors have proprietary rights in their technology, often intellectual property rights, that prevents or limits competing firms from producing compatib products. The lack of competition on the network defines a closed system. Sponsors of a closed system wage a standards war in the hope of becoming a de facto standard (i.e., a monopolist)[8]. The focus of competition will be on trying to create an installed base advantage and creating consumer expectations that its technology will win the standards war.

Battles over compatibility arise typically when the competition between incompatible standards has been resolved. The creation of a de facto standard means that competition between technologies/networks (internetwork competition) is not possible, and instead the focus shifts to competition on the same network (intranetwork competition). However, in order for there to be competition on the network, products of competing suppliers have to be compatible. The sponsor of the dominant network will have an incentive to limit and disrupt the ability of rival firms to produce compatible products, preferring to protect its profits and monopoly by maintaining incompatibility.

Firms that have developed, or are in the process of developing, incompatible technologies can forestall a standards war by agreeing to a common standard. Under a common standard all firms agree to produce compatible products, replacing internetwork competition for competition on a single network. Standard setting by agreement arises when firms forecast that competition on a single

[8] As we discuss below a system sponsor might open up its system to competition in order to mitigate lock-in and convince consumers that it will win the standards war. In this case it shares the monopoly with its intra-network competitors if its technology wins the standards war.

standard is likely to be more profitable than the expected profits from a standards war. In particular, firms will find it more profitable to agree to a common standard when consumers' expectations are fragmented or uncoordinated. In these circumstances a diversity of options with no clear winner may make consumers reluctant to adopt any of the competing technologies, and as a result the next generation of technology fails. Competing firms that have developed next generation networks can avoid this fragmentation by cooperatively agreeing to a standard. This can be done formally through national or industry standard setting bodies, or less formally when firms simply agree to a common standard. Common standards are implemented by firms agreeing on the technical specifications for interfaces to insure compatibility and making the technology embodied in the standard accessible to all.

A particular concern of standard setting bodies is to make sure that all of the technology embodied in a standard is licensed. Failure to do so can create a situation, where a firm with an unlicensed technology embedded in the standard, can end up in control of the standard with considerable market power if the standard is adopted widely. For instance, the code of practice covering intellectual property rights of the International Telecommunications Union-Telecommunications Standardization Sector (ITU-T) specifies that patent holders not willing to waive their patent rights or to negotiate licenses with reasonable terms on a nondiscriminatory basis will not have their technology incorporated in a standard[9].

4. Standards wars

Firms in a standards war engage in a number of strategies aimed at credibly convincing consumers that their technology will become the de facto standard, or at the very least, have a larger installed base. Typically the strategies followed by firms do this by either (i) directly affecting the expectations of consumers, or (ii) by exploiting the link between expectations and the size of the current installed base by making investments in the size of their installed base. The extent to which firms are willing to make investments to enhance the size of their network depends on their ability to capture the benefits from doing so, which in turn depends on being able to restrict access to their network by competitors.

4.1. Strategies in standards wars

Here we discuss a number of strategies that firms can, and have used, in order to influence the expectations of consumers and/or create a larger installed base[10].

[9] The ITU-T Patent Policy can be found online at "http://www.itu.int/ITU-T/dbase/patent/patent-policy.html."

[10] This section follows Church and Ware (1998).

4.1.1. Penetration pricing

Firms that adopt penetration pricing set an intertemporal pattern of pricing that promotes adoption of a product early in its life cycle in order to build up its installed base[11]. Firms following a penetration pricing strategy strategically lower their price, perhaps below marginal cost, in order to convince consumers to adopt their technology and build their installed base. The investment, through low prices, is recouped in the future when the firm's technology becomes a de facto standard or has a sufficient installed base that it provides consumers with considerable network benefits. These network benefits give it room to raise its prices to future adopters: its installed base gives it a competitive advantage over rival platforms, perhaps even deterring entry. The use of penetration pricing is a way for a firm to (partially) internalize the externality and transfer (through lower prices) some of the benefit to consumers today: subsidies to encourage adoption today are financed through higher prices in the future.

4.1.2. Advertising and marketing

Promotional efforts will be aimed at providing information to shape the expectations of consumers regarding the relative size of the installed bases of competing networks. Credible information that a network is ahead of its rivals—both in terms of its installed base or its current rate of adoption—can be particularly effective in changing or reinforcing the expectations of consumers.

4.1.3. Insurance

There are a variety of mechanisms through which firms can reduce the risk that consumers will be stranded. These included sophisticated pricing contracts where the ultimate price paid depends on the size of the network[12], or the firm retains ownership of the connection (direct) or hardware good (indirect) and recovers its cost through service fees or short term leases[13]. By reducing the risk of lock-in, insurance strategies make it easy for consumers to leave a network, thereby encouraging them to join.

4.1.4. Second sourcing and open standards

Second sourcing occurs when firms license their products to other suppliers to create competition[14]. Like insurance schemes it is a means for firms to reduce the risk of lock-in, but rather than provide early adopters with protection from being stranded on a small network, second sourcing is a means to create an installed base and protect them from opportunism in the future if the technology is successfully established. When second sourcing creates competition, it signals to

[11] See Farrell and Saloner (1986b) and Katz and Shapiro (1986).
[12] See Dybvig and Spatt (1983) and Thum (1994).
[13] Katz and Shapiro (1994, p. 103).
[14] See Farrell and Gallini (1988) and Shepard (1987).

consumers that prices will be low now and in the future. This assures consumers today that the network will continue to grow.

A very aggressive form of second sourcing is the creation of an open standard. An open standard exists when its sponsor does not enforce its intellectual property rights. Under an open standard the technical specifications for compatibility are freely available to any firm for incorporation into their products. As a means to create competition among suppliers an open standard can be used when networks are either direct or indirect. When networks are direct, it creates competition among suppliers of substitute products that are horizontally compatible. In the case of indirect networks, the open standard can encompass the hardware good, or more typically, it is used to promote third-party supply of complementary software. In both cases, the strategy can be particularly effective since it creates competition on a standard today, and if credible, it is a means to commit to lower prices, product differentiation, innovation on the standard, and variety (in the indirect network case) in the future, mitigating concerns over lock-in for early adopters and contributing to growth in the size of the network.

4.1.5. Signaling

A firm can also create credible incentives to continue to compete for adoptions in the future, thereby ensuring the growth of its network, and signal these incentives to adopters today by making valuable assets hostages[15]. It can do this by creating and maintaining a reputation for not stranding consumers by eliminating support and sales for a technology. Developing such a reputation can be very valuable if a firm produces multiple products or introduces new generations of technology. Alternatively, it may invest in large sunk expenditures whose recovery depends on growth in the size of the network. By doing so it sends the signal to adopters that it expects the network to grow, otherwise it would not have made the investments.

4.1.6. Product preannouncements

A firm preannounces its product when it informs consumers about the future availability of its products[16]. If the announcement is credible—consumers actually believe that the firm will introduce its product as announced—the effect can be to induce consumers not to join a competing network, but to wait for the firm's product. They are likely to wait if the preannounced product is compatible with their existing network and/or its quality is greater than that of competing networks presently available. A preannouncement that induces consumers to wait limits the growth of the installed base of competing networks.

[15] Katz and Shapiro (1994, p. 104).

[16] See Farrell and Saloner (1986b) and Dranove and Gandal (2003).

4.1.7. Investments in complementary software

In indirect networks, often the relevant installed base is complementary software, and hardware firms can make strategic investments to increase the supply of complementary software[17]. Hardware firms can do this by vertically integrating and supplying software and/or by subsidizing third-party suppliers, either by underwriting some of their costs or instituting support programs that lower the costs of third-party developers. In hardware/software industries what typically influences adoption decisions by consumers is the *relative* number of software titles available, both today and in the future. Hence hardware firms can increase the relative size of their installed base by not only increasing software available for their system, but by also reducing the variety of software available for competing systems. This typically involves contractual restrictions on third-party software developers' freedom to provide software for other systems or foreclosure. Foreclosure involves acquisition of third-party software and elimination of the supply of software compatible with rival systems.

4.2. Standard wars and efficiency

Perhaps the central focus in the literature on standards wars has been on determining whether market processes, such as a standards war, can be relied upon to govern standard selection. There are two issues: (i) is standardization efficient, and if so (ii) is the correct standard chosen. Whether standardization is efficient depends on a trade-off between product diversity and network effects. Standardization maximizes the benefit from network effects but typically results in a reduction in variety for consumers. For standardization to be socially optimal requires network benefits to become more important as consumer preferences become more diverse. The tendency in the theoretical literature is for the equilibrium to be characterized by *insufficient standardization or too much variety*[18].

For instance in Church and Gandal (1992a) the bias against standardization arises from the incentive of software firms to reduce competition. Software firms can support one of two competing differentiated hardware systems. The decision of which system to support depends on expected profits and there are two effects associated with joining a network. An increase in software support increases demand for that hardware (the demand effect), which leads to an increase in software profits as the size of the network expands. On the other hand, an increase in the number of software varieties increases competition on that network, which

[17] See Church and Gandal (1992b, 1996, 2000) for discussions of strategic investment in software by hardware firms, and Whinston (1990), more generally, for a discussion of when tying software to hardware can lead to profitable monopolization of hardware and software.

[18] For the physical networks case, see Farrell and Saloner (1986a). For the virtual network case, see Chou and Shy (1990) and Church and Gandal (1992a). All of these papers show that market forces often result in suboptimal standardization. Markovich (2001) examines the trade-off between standardization and variety in a dynamic setting using numerical methods. Unlike the other papers in the literature, she finds that there can be excessive standardization in equilibrium.

ceteris paribus, decreases profits (the competitive effect). Church and Gandal show that the competitive effect dominates the demand effect and the equilibrium is characterized by excess variety—both networks are supported by software firms—when consumers' valuation of software is relatively large (implying significant network effects) compared to the extent of hardware differentiation.

However, as Katz and Shapiro (1994) observe, an additional advantage of having multiple networks not considered formally in the theoretical literature is that multiple competing networks have an option value. Preserving multiple networks and selecting a standard after technical and demand uncertainty is resolved makes it more likely that the optimal standard will be implemented. Too early a choice may preclude a subsequent change to a superior standard[19]: total surplus would rise if there was a switch to an alternative.

Excess inertia occurs when a technically superior new standard is not able to replace an existing standard even if total surplus would ultimately be greater with a change in standards. Excess inertia might arise because of the advantage network effects provide for a prevailing platform with an installed base. Rather than evaluate their choices on the basis of price and quality, consumers will consider these and network benefits. An installed base advantage might be difficult for a new technology to overcome, deterring its entry and adoption, despite its technical superiority. Because of lock-in to the existing standard the transition to a socially superior technology is not made. Farrell and Saloner (1985) highlight the importance of coordination problems among consumers. Despite the technical superiority of a new technology, consumers will be reluctant to bear the costs and risk of adopting a new technology if they don't think that others will also adopt[20]. The difficulties associated with coordination are amplified when competing standards are introduced simultaneously[21].

[19] See Katz and Shapiro (1994, p. 106).

[20] Choi (1994) shows that uncertainty over the quality of technologies in the future has a similar effect. Early adopters have insufficient incentives to wait for the uncertainty to be resolved because they do not internalize a forward externality. A forward externality arises because when consumers today make their adoption decision before the quality of the standard is revealed rather than waiting, they deny consumers in the future the possibility of coordinating on a better standard. This forward externality is stronger than the backward externality—that adopters tomorrow do not consider the cost of stranding early adopters—leading to the result that there is a tendency for early adopters to move too soon, locking-in an inefficient standard.

[21] Rysman (2003) considers explicitly a dynamic model with indirect network effects—where consumers can wait before adopting—the possibility that competing standards will result in a delay in adoption. His results are similar to Church and Gandal (1992a): the equilibrium depends on the relative strength of the network and competitive effect. When the network effect is strong, the market tips and there is standardization earlier. When the competitive effect is relatively strong, however, the software firms support both of the competing hardware standards, and consumers respond by delaying their adoption decision until one of the standards reaches a critical mass of software support, and then the market standardizes. The adoption delay equilibrium is not efficient. Postrel (1990) attributes, in part, the failure of quadraphonic sound in the 1970s to competing standards.

On the other hand, if the expectations and preferences of consumers are less 'fragmented,' it is more likely the case that a new technology is adopted too easily[22]. Insufficient friction arises when a change to a new technology is socially inefficient. This can be the case since new generations of consumers have socially excessive incentives to switch to a new superior technology: they do not take into account that they strand previous generations on the old standard when they switch, thereby limiting the growth of the network benefits of consumers on the old technology[23].

The work of Katz and Shapiro (1986, 1992) suggests that strategic behavior by sponsored new technologies limits the ability of existing standards and contributes to *insufficient friction*. In particular, the theme of Katz and Shapiro's work is that sponsors of new superior technologies can limit the ability of existing standards to grow their network through penetration pricing. The ability of a firm to engage in penetration pricing depends on the extent to which they can finance below cost pricing today through higher prices in the future. Higher prices in the future are possible if the firm is able to develop an installed base advantage. However, the extent to which it can raise prices from an installed base advantage depends on the quality and price of competing alternatives in the future. As a result the technology expected to be superior tomorrow—either higher quality or lower cost—will have an advantage today, since by providing a more attractive option to consumers in the future *without* an installed base advantage, it is in a better position to limit the ability of the current technology sponsor to finance penetration pricing to attract consumers in the present than the current technology sponsor can limit it. The advantage that a sponsored technology has from penetration pricing gives it a large advantage over socially preferred, but nonproprietary existing standards,

[22] Fragmented expectations can arise in the analysis of Farrell and Saloner (1985) because consumers' have incomplete information; they don't know the preferences of other adopters. Expectations are much less likely to be fragmented when consumers know the preferences of others and are able to coordinate on a Pareto Optimal alternative. A Pareto Optimal alternative is more likely to exist when consumers' preferences are relatively homogenous.

[23] See Katz and Shapiro (1986, 1992, 1994). Choi and Thum (1998) obtain the opposite result (excess inertia not insufficient friction) in a model very similar to Katz and Shapiro (1986) except that first generation consumers can wait and the new technology is not available for first-generation consumers. The latter assumption negates the possibility of penetration pricing and insufficient friction. Choi and Thum (1998) find that consumers today who have the option of waiting for a superior technology tomorrow, that is not available today, tend to adopt the prevailing technology too often and too soon, rather than waiting for the arrival of the superior technology. This is true when the technologies are supplied competitively. Sponsorship of the new technology exacerbates the tendency to excess inertia, since early consumers know that if they wait they will face monopoly pricing from the sponsor of the new technology. The negative externality is that early generations of consumers by moving too early force later generations to either adopt an inferior technology to get network benefits, or forgo network benefits for the superior technology.

since without a sponsor the incentives to engage in penetration pricing are limited. The firm that lowers the price today to build up the installed base may not be able to benefit from the installed base in the future if it faces competition from other suppliers on the same standard. The differential ability of standard sponsors (existing versus new) to utilize penetration pricing contributes to 'insufficient friction' or 'excess momentum' as one incompatible technology replaces another[24].

Excess momentum is less likely with indirect networks, especially if the installed base of complementary software is controlled by the sponsor of the existing hardware standard. In the case of indirect network effects it is easier for an incumbent to strategically manipulate the installed base than in the case of direct network effects. In the latter, it often takes time for consumers to arrive in the market and the installed base can only grow as consumers adopt through time. In the case of the former, hardware firms can invest in software/complementary products. Indeed, Church and Gandal (1996) show that existing hardware sponsors have an incentive to strategically overinvest in software in order to deter entry of a competing standard, resulting in a bias in favor of the incumbent's standard. This bias results in either insufficient standardization (too much variety) or standardization on the wrong technology (the incumbent's). The result is an example of a raising rivals' cost strategy. It is profitable for an incumbent to strategically invest in software varieties—given that it will be a monopolist—but the same level of investment in an installed base is not profitable for an entrant given that it will have to share the market with the incumbent.

5. Battles for compatibility

Firms with an ownership interest in a platform that is dominant or has, through universal adoption, become a de facto standard have an incentive to restrict compatibility to preserve market power and profits. Because it is a de facto standard, the installed base of such a technology is often a sufficient barrier to entry to exclude entrants whose products are incompatible. Nonsponsors can only participate in the market if they gain access to the network (i.e., design compatible products). The issue of compatibility is not exogenous: either by design or exercise of property rights, incumbent firms may be able to block, or reduce, compatibility.

It is unlikely that the sponsor(s) of a network with a large installed base will grant compatibility. Doing so enhances intranetwork competition—and in the absence of strong network competition—provides very little benefit to the system sponsor. Compatibility eliminates the installed base advantage of the

[24] See Katz and Shapiro (1986, 1992).

incumbent, reducing its market power and profits. Dominant firms can attempt to frustrate and disrupt compatibility by denying compatibility and by making frequent and unannounced changes in product standards to introduce incompatibility.

5.1. Denying compatibility

Depending on the circumstances, sponsors of standards can deny compatibility with competitors by: (i) exercising their physical property rights to deny interconnection and therefore insure that competitors' products/services are not on their network and (ii) asserting their intellectual property rights and refusing to license the standard (interface technology/knowledge required for compatibility) or, in the case of indirect networks, the installed base of software.

It is useful to distinguish between cases when changes in product standards make competitors incompatible and when they make complementary products supplied by third-parties incompatible. The economics of the former are relatively straightforward: creating incompatibility enhances or preserves the firm's market power by excluding competitors. The analysis of the latter is more difficult since in general a greater variety of complementary software (typically from independent software sources) increases the value of the standard and, *ceteris paribus*, the higher the price the monopolist supplier of the hardware can charge[25].

However, two circumstances can be identified where the supplier of the standard has an incentive to restrict compatibility of complements. These are: (i) closing up an open standard and (ii) intergenerational leverage.

5.2. Restricting compatibility of complementary products

5.2.1. Closing an open system

In these cases the dominant firm or system sponsor begins with an 'open' system, which allows second sourcing or third party provision of complementary products. In some cases, second sourcing is in fact actively encouraged as the system sponsor recognizes that the credible commitment to low prices and a wide variety of complementary products it creates provides a competitive advantage in the market for the system or hardware good. Once, however, the standard is established or sales of the hardware good are of lesser importance—perhaps because the other standard has lost the standards war—the incentives of the system sponsor change. In the case in which its network becomes a standard it has an

[25] See Whinston (1990).

incentive to close up its system and engage in second degree price discrimination[26]. When it has lost the standards war—so that hardware sales are negligible—but there is still demand for complementary products from its (stranded) installed base, monopolization of complementary products will be profitable. Competitors can be excluded and complementary goods markets monopolized in two ways.

First, it can render third party complementary products incompatible, unnecessary, or inferior by manipulating interfaces. Second, a sponsor of a standard can exclude suppliers of complementary products by tying supply of its proprietary standard to supply of its complementary goods. This can be done either by: (i) contractual terms where the tying arrangement is explicit, (ii) de facto bundling where the proprietary standard is not available as a separate product, of which a special case is, (iii) a technological tie (the products are not physically available separately).

Whether it engages in changing its interface standards or tying the effect is to close up its system and monopolize the supply of complementary products. Not only can this result in a substantial lessening of competition in the market for complementary products, but it clearly reduces the incentive for innovation by suppliers of complementary products.

5.2.2. *Intergenerational leverage*

This occurs when a dominant firm acquires control over the supply of its installed based of complementary products in order to deny access to competing network technologies. Control of the supply of complementary products—or being able to effectively insure that they are incompatible with other hardware technologies—provides the sponsor of the current standard an avenue to forestall entry of a better hardware technology. Through its control of the installed base of software the sponsor of the existing standard may be able to manage the transition to the next generation by insuring compatibility only between its software and its hardware technology[27]. By enforcing incompatibility between its installed base

[26] Monopolization of complementary products through tying allows the system sponsor to price discriminate based on the intensity of use for complementary products. Sales of the complementary products indicate the intensity of use and if intensity of use reflects benefits, sales of complementary products can be used to price discriminate (i.e., extract more surplus from those who realize substantial benefit). This is done by raising the price of the complementary product above marginal cost (i.e., competitive levels). In order to raise the price of complementary goods, the system sponsor must exclude alternative suppliers of complementary products by tying. See Tirole (1988) or Church and Ware (2000) for additional details. Saloner (1990) and Greenstein (1990) discuss the dilemma that the incentive to engage in second-degree price discrimination presents for system sponsors. Second-degree price discrimination and the supply of a wide variety of components depend on standardized interfaces. Raising the price of components above cost and standardized interfaces provide the incentive and opportunity for rival firms to enter and compete in the complementary product markets.

[27] The implicit assumption is that the new technology does not find it profitable to create its own installed base of software. See Church and Gandal (1996) and our earlier discussion in Section 4.0 for why this may be the case.

of software and higher quality hardware introduced by rivals that reflects recent technical advances, the current monopolist may be able to monopolize the next generation of hardware. The effect is not only to maintain its monopoly, but to also reduce the incentives for innovation by rivals for its tying product—its monopoly hardware[28].

6. Cooperative standard setting

Cooperative voluntary standard setting occurs when suppliers agree to compete 'in the market' rather than 'for the market' by making their products compatible. Compatibility is achieved by agreeing to a standard. In doing so, firms suppress competition between networks in favor of competition on a network. The focus of competition is not on building an installed base advantage, but instead price—and depending on the extent of the standard—product features. The more specific and detailed the standard, the less room there is for firms to engage in product differentiation and the more important price competition. Moreover, it is more likely the case that firms will follow a variety of product line strategies, with some that might have offered complete systems in a standards war, instead focusing on a subset of (compatible) components when there is a common standard[29].

Cooperative standard setting requires agreement among potential suppliers. It cannot be imposed unilaterally. The incentive for competing sponsors of incompatible standards to develop and introduce a common standard depends on their expectations of the likely alternatives and their profits. The alternatives are: (i) a standards war that results in de facto standardization; (ii) multiple competing differentiated networks (if network effects are not strong enough to trump the advantages of product differentiation); and (iii) fragmentation of adopter's expectations with the result that none of the competing technologies is viable.

If the likely outcome is understood by suppliers to be fragmentation of consumers' expectations when there are competing standards, then agreeing to reach a common standard will not be difficult. However, agreeing to set and adopt a common standard is not the same thing as reaching agreement on the details of the standard. Disagreement over the details of the standard can arise for a number of reasons, including: (i) preference differences among consumers; (ii) preference differences among sponsors stemming from proprietary rights, first-mover or other experience advantages; (iii) uncertainty and consequently disagreement,

[28] For related formal analyses, see Choi and Stefanadis (2001) and Carlton and Waldman (2002). See also Rubinfeld (1998).
[29] This point is made by Shapiro and Varian (1999, p. 233).

over future developments in the industry; and (iv) asymmetries of information and strategic bargaining[30].

It is often the case that the determination of a voluntary standard has attributes similar to that of the 'battle of sexes game.' Each firm would like to have standardization on its technology, but prefers standardization on the technology of a rival to none at all. Farrell and Saloner (1988) examine the incentives for firms to achieve coordination through voluntary standard setting (standardization committees) where the structure of firm payoffs are similar to that of the battle of the sexes and firms have the option of forgoing negotiations and starting a (possibly unsuccessful) standards war[31].

A firm will be reluctant to agree to a common standard if its expected profits from a standards war exceed its expected profits from agreeing to, and competing on, a common standard[32]. This will more likely be the case when: (i) failure to standardize does not have a significant adverse effect on consumer adoption because of uncertainty and concerns over stranding; (ii) the firm is well positioned to win a battle of standards, perhaps because of its ability to easily implement the strategies discussed in Section 4.0 to increase its installed base and favorably influence consumers expectations; (iii) the firm has sufficient ability to 'harvest' the monopoly rents associated with its technology becoming the standard (that is, it has sufficiently strong intellectual property rights and/or is able to follow the strategies discussed in Section 5.0 to restrict competition on its network); (iv) it has a sufficient competitive advantage *vis-à-vis* competitors that in winning the standards war, its profit dissipation is restricted; and (v) competition on a standard will be particularly dissipative with respect to profits, perhaps because the products of firms will be relatively undifferentiated due to standardization.

In many industries, including telecommunications, there are established institutions, which provide forums for the discussion and determination of industry standards[33]. For example, the ITU-T has 13 study groups, which make

[30] See Besen and Farrell (1994, pp. 124–126) for a discussion of the commitments and concessions often made by firms when negotiating a standard.

[31] They use a simple model in which two firms prefer their own incompatible standard to that of a rival, but also prefer standardization to incompatibility. Belleflamme (2002) is an extension of Farrell and Saloner (1988), where players have to choose between an existing standard and a standard known to be superior in the future. Belleflamme compares two standard setting processes: unilateral adoption to create a bandwagon and negotiation. Unilateral adoption leads to excessive and early adoption of the existing standard, negotiation to excessive and late adoption of the evolving standard. However, if firms can choose the standard setting process, the outcome almost always maximizes the sum of payoffs (i.e., is efficient).

[32] de Palma and Leruth (1996) present a formal model in which firms noncooperatively determine whether to compete with incompatible networks, or instead both agree to a common standard and produce compatible products. Compatibility requires the cooperation of both firms and it will not be forthcoming if one of them has a high enough prior that they will dominate in a standards war.

[33] See Shapiro and Varian (1999), Grindley (1995), Katz and Shapiro (1999), and Farrell and Saloner (1992b) for more extensive discussion and references on formal standard setting.

recommendations on standards in all fields of telecommunications, ranging from the assignment of telephone numbers to protocols for data transmission. As of 2004, the ITU-T had more than 2700 recommendations in force. The standards set by the ITU-T are called recommendations because the ITU-T cannot force compliance. The value of interconnection between telecommunications networks, however, is a powerful incentive for firms to comply with its recommendations.

There are two attributes of formal standard setting bodies that are a source of both their strength and weakness. Formal standard setting bodies are typically open to all participants and work on consensus. The power of consensus is that it insures that standards are open and that any proprietary technology incorporated into a standard is available to all on 'fair, reasonable, and nondiscriminatory' terms. Both their open nature and the inclusive process by which official standards are set often provide them with considerable credibility *vis-à-vis* consumers, encouraging adoption.

On the other hand, because of both of these 'open' attributes the process of negotiating open standards can be very slow, ineffectual, and inflexible. The process is likely to be slow because any interested participants are welcome to participate and they will often have divergent interests. Firms will have an incentive to participate because the outcome of negotiations on standards will often be an important determinant of their profitability. Their interests will diverge because the competitive position of each firm *ex post* will depend on the details of the standard adopted.

Moreover, unless the decision of the standard setting body is adopted by (most) suppliers, the standard set will be irrelevant. An important determinant of the effectiveness of the standard is whether it is incentive compatible: Will firms party to the standard honor their commitment to compatibility? Modifications to a standard may not be timely, or even possible, due to high transaction costs associated with reaching consensus. As a result official standards are more likely to become irrelevant when market circumstances are subject to frequent change.

7. Mandated standards

Another alternative to market mediated standards is the setting of standards by regulators. The presence of an official standard setting body with the power to impose standards, such as a regulator, distinguishes voluntary standard setting from mandated standards[34].

The factors underlying voluntary standard setting by firms discussed in the previous section are industry and firm profitability. However, the importance of

[34] One of the few papers that explicitly models the behavior of a regulator to impose a mandatory standard is Cabral and Kretschmer (2004). They examine how uncertainty over the preferences of consumers between competing standards affects when and which standard is adopted.

adoption of a common standard for consumers in some industries may result in de jure standard setting by a regulator or other apparatus of the state. An important public policy issue arises when firms are unable, or unwilling, to reach an agreement on a common standard. Intervention in standard setting by the state is typically motivated by the presumption that a standards war in such circumstances is inefficient. The existence of a standards battle and the absence of a voluntary agreement on a common standard suggests that at least one sponsor of an incompatible technology believes that it is well positioned to win and profit from a standards war. Besen and Saloner (1989) suggest that these circumstances are when the private value of winning a standards war is high and network effects are significant[35]. If network effects are large then the impact of standardization on demand—both adoption and market size/growth—will likely be significant. And it is precisely in these circumstances when the costs of a standards war may make intervention preferable. The key benefit of mandated standardization is the creation of a credible standard, which encourages adoption, avoiding the costs associated with a market process.

7.1. Advantages of mandated standards

There are a number of costs associated with a standards war. These costs may mean that it is preferable to instead have a mandated common standard[36].

7.1.1. Failure to standardize

Perhaps the most important cost that might arise from a standards war is that in the standards war a de facto standard is not established. As a consequence network benefits are not maximized. If the failure to standardize fragments the expectations of consumers sufficiently, then none of the competing technologies might be adopted. A mandated common standard would maximize network benefits.

7.1.2. Stranding

If there is a winner in the standards war, then some consumers and some producers of complementary products will be orphaned. Those consumers and producers who made sunk expenditures on losing standards will not realize the anticipated benefits from those expenditures. A common standard adopted early enough would have forestalled these type of mistakes by consumers and producers of complementary products.

[35] See also Grindley (1995), Chapter 3.

[36] For more detailed discussion, see Rohlfs (2003), Shapiro and Varian (1999), and, especially, Grindley (1995). Our presentation in this and the next section follows Grindley.

7.1.3. Duplication of development and promotion expenditures

Competing standards also entail duplication of development and promotion programs. To the extent that these expenditures are sunk, the expenditures of losers in the standards war will be economically wasteful. Moreover, given the nature of these expenditures and the winner takes all nature of the competition, it is difficult for firms to reduce the extent and risk of these expenditures through a staged introduction of their products.

7.1.4. Inefficient standardization

As discussed in Section 4.0, the outcome of a standards war may not be efficient. The result could be standardization on the wrong technology or excessive standardization. In the later case welfare would be higher if a variety of standards were available. In the short-run this advantage arises because of the benefits of product differentiation: the greater the selection of products the less the costs from a mismatch between the preferences of consumers and the characteristics of products available. In the long-run the advantage of competing products extends to include competition in innovation.

7.1.5. Market power

The winner of a standards war may be able to exercise considerable market power without concern regarding entry because of the barrier to entry created by its installed base. Moreover, its market power may be particularly enduring if it is able to frustrate attempts by entrants to produce compatible products, or it is able to use its present installed base to extend its monopoly to include the next generation of technology[37].

7.2. Advantages of market standards

The costs associated with mandated standards arise because of the nature of the process. The process is likely to be inclusive and open, suggesting as with voluntary standards, that standard setting will be methodical, slow, and costly. Moreover, the process is made more difficult by the fact that the standard setting body will likely be: (i) remote from the market; (ii) much less well informed than market participants; and (iii) have its own policy agenda. Its remoteness means that it is likely to have a focus on technical criteria rather than market acceptance. Because of asymmetries of information, firms typically know more about costs, quality, and potential technological progress than regulators providing firms with an opportunity to strategically influence regulators and making it difficult for regulators to set efficient standards. If the standard setting body takes into account other policy objectives besides choosing an efficient standard, the result could be a standard that is irrelevant, or one which does not have credibility with adopters.

[37] See the discussion in Section 5.0 supra.

A danger with mandatory standard setting is that these considerations will result in setting a standard too early (i.e., setting a standard without relevant information) that would be gained by waiting. Finally, as with any administrative process, rent-seeking behavior with its attendant inefficiencies will be induced by the prospect of mandated standards.

There are also some benefits of using markets to determine standards, which are lost when a standard is mandated. It is important to remember that strategic behavior designed to enhance a firm's installed base is only contemplated because the winning firm has property rights in its network and thus finds it profitable to attempt to internalize the network externality. If the network was nonproprietary—as in the case of a common mandated standard—then firms will have significantly less incentive to make investments in increasing the size of its installed base. As a result, the coordination problems inherent in network industries may mean that no technology is successfully adopted or that the network size is substandard and consumers do not benefit as much as they could from network effects.

In addition, using the market to determine standards has other advantages relative to a mandated standard. These include: (i) the resolution of the standards war and determination of the standard is often quick and the standard definitive; (ii) quality and costs of competing standards is done post development and any trade-off is evaluated in the market; and (iii) depending on the outcome of the standards war, there is not necessarily the same loss in variety.

Hence, despite the costs associated with market determined standards, competition in the market is probably preferable to mandated standards in most cases–with two exceptions. The first is, when network effects are significant and the presence of multiple competing standards suggest the strong possibility that fragmented expectations on the part of consumers will prevent adoption of a new, superior technology. The second is more obvious and of considerable relevance to telecommunications as we discuss in the next section.

7.3. Mandated standards in telecommunications networks

New networks and platforms are unlikely to compete successfully against the local networks of the incumbent local telephone company without regulatory intervention regarding interconnection. The reason is the installed based advantage of the incumbent carrier. In the absence of interconnection, it would be very difficult for a competing telecommunications provider to compete: call completion would only be possible if both parties subscribed to the same network. Consumers would be unlikely, in these circumstances, to unsubscribe from the incumbent network and subscribe to the network of either a new wired-line network or a new platform such as wireless.

To counter this, virtually every regulator which has introduced competition has mandated and regulated the terms of interconnection between the existing local telecommunications network and the networks/platforms of new entrants. Regulators do so in order to eliminate the installed base advantage of the incumbents. Without mandatory interconnection—a form of standardization—competition in telecommunications would be a nonstarter, whether it was from competing telecommunications networks or other platforms/technologies (e.g., wireless, cable, or voice over internet protocol (VoIP))[38].

8. Case studies

This section provides a number of case studies drawn from telecommunications and information services that illustrate the principles of network competition discussed in the previous sections.

8.1. Competition in the mobile cellular industry

In most settings in which network effects are present, compatibility across platforms or its absence is a key determinant of the success or failure of a particular technology. In the case of wireless telecommunications, however, interconnection and the availability of the relevant infrastructure can be a substitute for compatibility. An individual subscribing to any one of the wireless technologies (analog, AMPS, CDMA, GSM, TDMA, and iDEN) in the U.S. can make calls to and receive calls from someone else subscribing to any one of the other standards (or to and from the wired-line network) as long as there is: (i) interconnection between networks; and (ii) the relevant infrastructure is in place. In the U.S. (and several other developed countries), interconnection has been achieved by standard interconnection protocols.

[38] See Noam (2002) for a discussion of the nature, history, rationale, and importance of interconnection policies in creating and sustaining competition in telecommunications. The issue of mandatory interconnection has similarities to the use of converters and adaptors to create compatibility between incompatible networks. For an examination of the effects of converters, see Farrell and Saloner (1992a) or Choi (1996). Farrell and Saloner show that converters can result in less compatibility and excessive variety of networks since the private costs of ignoring network benefits are mitigated by the converter *ex post*. Choi (1996) shows that the existence of converters can harm the adoption prospects of a new technology competing against an existing network with an installed base. The presence of converters reduces the risk and costs of being stranded, thereby, encouraging consumers to adopt the prevailing standard and not the new technology. Choi also shows that this effect is welfare improving since it reduces the extent of insufficient friction. These results mirror Choi (1994), who finds that *ex post* standardization policy, standardization policy after first generation consumers have adopted a technology, encourages them to adopt an inferior technology knowing that they will be protected from being stranded if it is efficient to standardize after they have adopted. The policy is *ex ante* inefficient if the harm from eliminating stranding in the future is less than the adverse effects of encouraging early adoption.

Hence, at first glance, competition within the wireless industry resembles competition within the market, rather than competition for the market, that is, competition between compatible technologies. Nevertheless, interconnection does not completely eliminate the importance of network effects in mobile telecommunications competition. This is because networks typically charge different prices for on-net and off-net calls[39]. (An on-net call is a call that originates and terminates on the same network, while an off-net call originates on one network and terminates on another network.) Typically on-net prices are lower than off-net prices. This creates what is sometimes referred to as tariff-mediated network effects[40].

Lower on-net prices and the induced tariff-mediated network effects mean that standardization may occur on one platform, despite interconnection. Hence, competition in mobile networks embodies some aspects of competition between incompatible networks. A key difference between competition in 'interconnected' telecommunications networks and other networks is that in a 'variety' equilibrium, interconnection insures that all consumers can indeed call each other.

The mobile telecommunications industry also provides an opportunity to examine the benefits of standard setting vs. market competition[41]. Since 1994, Europe and North America have taken divergent approaches in the market for wireless for voice and data services. The European Community mandated a harmonized standard, GSM, in the second generation (2G) bands. In contrast, the North American approach has been to allow the market to decide (i.e., operators have been free to choose) among four digital wireless standards: CDMA/IS-95, GSM, TDMA, and iDEN[42]. In the case of 3G, a standards war between Wideband CDMA (WCDMA) and CDMA2000 is in a nascent stage.

An interesting question is whether mandated standards have led to faster adoption of mobile technology. Several recent papers empirically examine whether, other things being equal, early penetration rates for mobile networks were lower (or higher) for countries with multiple incompatible digital standards[43].

[39] This is only relevant in calling party pays (CPP) systems, which exist in most European countries. In the U.S., this issue does not arise.

[40] See Laffont, Rey, and Tirole (1998a,b).

[41] Funk and Methe (2001) provide an excellent overview of standards development in wireless and the role mandated standardization in a region/country played in creating an installed base, which transformed the regional/national standard into a global standard.

[42] Other developed countries have enabled market competition as well. See Gandal, Salant, and Waverman (2003) for further discussion.

[43] Gruber and Verboven (2001) and Koski and Kretschmer (2002) estimate logistic diffusion models for mobile telecommunications. These papers find that early diffusion of second-generation mobile telephony was faster in Europe where a single standard (GSM) has been in use, than in other countries (like the U.S.), where multiple standards coexist. According to Cabral and Kretschmer (2004), current diffusion levels are quite similar between the U.S. and Europe. There is clearly a need for additional empirical work on this issue.

8.2. Instant messaging[44]

Before the internet, there were online computer services. Online computer services like America Online (AOL) and Compuserve were accessed by subscribers through dial-up modems. Once connected, subscribers 'surfed' proprietary content provided by their online service and had access to e-mail accounts. AOL introduced instant messaging (IM) for its customers in the late 1980s. Instant messaging allowed its subscribers to engage in nearly real-time text-based dialogue. Its popularity soared in 1996 when AOL introduced its 'buddy list.' The buddy list feature enabled subscribers to determine if other subscribers were online, thereby allowing them to determine who was available to exchange messages. In 1997, AOL introduced AIM, which allowed free Web-based access for non-AOL subscribers to its instant messaging network.

In 1999, a number of firms, including Microsoft and Yahoo! introduced competing IM services, but by this time the installed base of IM subscribers was estimated at 30 million. Without interoperability between the instant messaging networks, the prospects for competing services were not very good. However, their attempts to design their services to interoperate with IM and AIM were blocked by AOL. The issue of AOL's refusal to allow competitors access to its installed base raised concerns for the two relevant regulatory agencies, the Federal Trade Commission (FTC) and the Federal Communications Commission (FCC) when AOL and Time Warner agreed to merge in January 2000 in a deal that was the largest merger ever at the time.

Although AOL offered basic (text-based) instant messaging service before the proposed merger, there were emerging instant messaging services, such as voice over IP, the exchange of pictures, and streaming video, which required broadband capabilities. AOL gained significant broadband capabilities with its acquisition of Time-Warner. These advanced messaging services would use the same directory as text-based instant messaging, and hence the network effects associated with IM and AIM would also be available to advanced instant messaging offered only by AOL. In essence, advanced instant services offered by AOL would start with a significant installed base, providing it perhaps with an insurmountable advantage over rivals. To mitigate this possibility, the FCC imposed as a condition for approval of the merger that AOL must offer interoperability with other providers of advanced instant messenger services before it is allowed to offer advanced instant messaging services itself.

While this decision came out of a merger case, the decision to require interoperability has antitrust implications for other settings with network effects. Should Internet backbone providers for example, be required to interconnect with other backbone providers? There, clearly, are strong network effects in this case as

[44] This case is based on Faulhaber (2002).

well. Currently there is no such policy and interconnection relies on private agreements[45].

8.3. *The 56K modem standards war*[46]

Network effects arise in modem markets because compatible modems are required to transfer data between the sending and receiving parties. Hence, there are direct network effects, similar to those inherent in e-mail or telephone networks.

In September 1996, US Robotics (3COM) submitted a proposed 56K standard to the ITU, the X2[47]. In November 1996, Lucent and Rockwell agreed to make their chipsets interoperable, using a different standard called K56flex. The K56flex and X2 standards were, however, incompatible. If a consumer used one standard while her Internet service provider (ISP) used a different standard, the data transmission speed was not 56K, but rather that of the previous technology, 33K.

Both products came to the market in early 1997: demand for higher speed modems was thought to be significant because of the rise of the World Wide Web and the need to download and display graphic intensive files. The standards war featured extensive efforts by both sides to manipulate the expectations of adopters, with exaggerated claims of dominance made by both sides. However, rather than tip the market, the consensus is that it instead engendered confusion among consumers and ISPs, delaying adoption and retarding sales. Augereau, Greenstein, and Rysman (2004, p. 13) estimate that by October, just over 50% of ISPs had adopted 56K modems, but waiting for the standards war to work itself out, none of the seven largest ISPs nor most consumers upgraded.

Hence, the industry appealed to the ITU to set a standard. In April 1997, the ITU set up a committee to determine a 56K standard. In February 1998, the V.90 standard was approved by the ITU, a standard based on both the X2 and K56flex technologies, but incompatible with both. The V.90 standard was established quite quickly. The Chairman of the relevant ITU Study Group noted, "This is the shortest period of time ever taken for an ITU-T modem Recommendation to

[45] Cremer, Rey, and Tirole (2000) examine a dominant Internet backbone provider's incentives to 'degrade' the quality of its connection with rival backbone providers. The main concern of the FCC, the Antitrust Division of the Department of Justice in the United States, and Directorate General IV, the antitrust enforcement agency of the European Commission, in the merger between WorldCom and MCI was that it would create just such a dominant Internet backbone provider. The merged firm's market share would be in excess of 50%. By degrading the quality of its interconnection with smaller backbone providers, refusing to interconnect, or charging a price for interconnection, it was alleged that MCI/WorldCom could isolate its installed base from smaller rivals, creating differential network effects, which would disadvantage them and increase its market power and profits. The merger was allowed to proceed subject to MCI divesting its entire Internet business to Cable & Wireless for $1.75 billion. See Kolasky (1999, pp. 602–604) for details.

[46] This section draws heavily from Augereau, Greenstein, and Rysman (2004). See also Shapiro and Varian (1999, pp. 267–270).

[47] 56K means that the maximum speed of the modem was 56,000 bits per second.

achieve 'determination' approval status, and demonstrates a commitment by the ITU-T to respond quickly to urgent market needs[48]." If a standard had not been agreed upon quickly, it's quite possible that both the competing 56K standards would have failed.

Interestingly, Augereau, Greenstein, and Rysman attribute the failure of the market to tip, despite large network effects, to the strategic behavior of the ISPs. As suggested by the models of Church and Gandal (1992a) and Rysman (2003), small competing ISPs in local markets, which upgraded introduced product differentiation and reduced competition by adopting the standard not adopted by their rivals.

8.4. Satellite vs. cable television (CATV)

Vertical integration between program producers/packagers (i.e., cable networks), and multiple system cable operators was quite common in the U.S. cable television industry[49]. As a consequence, satellite services and others seeking to offer video-to-the-home services in competition with incumbent CATV operators often experienced difficulty in acquiring programming.

An example is the consent decree entered into the early 1990s by the U.S. Department of Justice in the Primestar case. The Department of Justice alleged that the terms establishing Primestar (in 1990) were designed to restrain competition in the market for multichannel subscription television by restricting the access to programming controlled by the cable companies to high-powered direct broadcast satellite (DBS) operators. Primestar was a joint venture among subsidiaries of seven of the major cable television companies and a subsidiary of General Electric. Those seven cable companies had ownership interests or controlling interests in most of the major cable channels. The General Electric subsidiary operated the only available medium-power direct broadcast satellite. The small dish size and low installation costs of high-powered DBS made it a viable alternative to cable in urban areas.

The consent decree effectively prohibited the seven cable companies from acquiring exclusive distribution rights to the major cable channels or preventing the cable channels from supplying competing distributors[50]. Concerns over the potential for cable system incumbents to deter competition from alternative distributors underlies the 1992 *Cable Act* in the U.S.[51] This *Act* and subsequent Federal Communication Commission rules are intended to prevent programming

[48] See "Agreement reached on 56K Modem standard," available at http://www.itu.int/newsarchive/press_releases/1998/04.html, accessed May 2, 2004.

[49] For detailed discussion and econometric analysis, see Waterman and Weiss (1996).

[50] United States versus Primestar Partners, L.P. et al., Proposed Final Judgement and Consent Decree 58 Federal Register 60672 (November 17, 1993).

[51] 47 U.S.C. Section 548.

suppliers from favoring affiliated cable systems over competing distributors in the supply of programming[52].

Today, direct broadcast satellite (DBS) and cable are the two main platforms in multichannel programming. Cable's market share in 2001 was approximately 75%, while satellite services had a 19% market share[53,54]. Given that cable services held a virtual monopoly before 1992, the 1992 bill seems to have achieved its purpose.

Until recently, satellite provision was perhaps somewhat less desirable because it typically did not offer local channels. However, in 1999, the *Satellite Home Viewer Improvement Act* (SHVIA) was enacted in the U.S. Under the relevant FCC rules, satellite carriers can carry local TV broadcast channels, though access by a satellite carrier to a local TV channel is subject to consent by the channel. If a satellite carrier has chosen to carry one local TV broadcast channel, it must carry all that ask[55].

8.5. DVD vs. DIVX standards war[56]

In April 1997, a consortium of hardware makers and motion picture studios introduced DVD as a replacement for videotapes. The DVD forum wanted to avoid the VHS-Betamax 'format war' in the videocassette market. Hence the DVD consortium decided that DVD would be an 'open format.' Hence, all DVD machines would play all DVD discs.

Circuit City, a major electronics retailer in the U.S., introduced a competing format called Digital Video Express, or DIVX in September 1997. In addition to DVD features, it was possible to purchase DIVX discs for a short time period. This is similar to renting movies. In the end DIVX failed. The DVD vs. DIVX standards war highlights an important consideration about the choice between compatibility and incompatibility.

DIVX was 'one-way' compatible with DVD in the sense that DIVX players could play DVD discs, but DVD players could not play DIVX discs. The idea was similar to second sourcing and the goal was to convince potential adopters that

[52] For a nice summary of the provisions in the Act and the FCC's rules, see Federal Communications Commission, Annual Assessment of the Status of Competition in the Market for the Delivery of Video Programming Tenth Annual Report 5 January 2004 at pp. 91–92.

[53] This corresponds to17 million households in the U.S. receiving satellite multi-channel services. Of these, 16 million of these receive DBS, the remaining million the C-band. There are two major satellite companies in the U.S., Direct TV and EchoStar, both offering DBS services.

[54] SBCA online http://sbca.com/government/competition.htm "Status of Competition in the Multi-channel Video Marketplace."

[55] See FCC NEWS Press Release, September 5, 2001 "FCC AFFIRMS RULES FOR SATELLITE CARRIAGE OF LOCAL TV STATIONS" and Federal Communications Commission, "FACT SHEET Satellite Home Viewer Improvement Act of 1999" December 2000, online at http://www.fcc.gov/mb/shva/shviafac.html.

[56] This discussion draws liberally from Dranove and Gandal (2003) and (2004).

there would be sufficient software available for the DIVX format. Despite the 'one way' compatibility DIVX failed.

Circuit City's choice of one-way compatibility insured potential adopters that purchasers of DIVX machines would not be orphaned. Our earlier discussion suggests that this is a sensible strategy. But there is a difference between one-way compatibility in an 'indirect network' and full interoperability in a 'direct' network. In the case of one-way compatibility in a 'hardware/software' system, software vendors may choose to release their software in the form that is compatible with the incumbent technology since it reaches BOTH audiences. This will mean that very little software will be written specifically for the entrant's technology. In such a case, few consumers will buy the entrant's product, unless the entrant's technology is clearly superior. DIVX failed in part because there was little software written exclusively for its technology.

Appendix: Modeling Issues[57]

In settings with direct network effects, authors typically employ a utility function of the form:

$$U_{ij} = a_i + N_j^b, \ 0 < b \leq 1. \tag{1}$$

U_{ij} is the utility to consumer i from network j. This utility depends on a standalone benefit (a_i), which can differ among consumers (and can be equal to zero). The second term represents the network benefit (or network effect), where N_j is the *expected size* of the network and 'b' represents the strength of the network effect. The restriction $0 < b \leq 1$ means that the marginal benefit of an additional user on the network is positive, but decreasing or constant in the size of the network. Although the framework is quite simple, N_j (the expected size of the network) is endogenous. This makes it difficult to analytically solve all but the simplest models.

In settings with virtual (or indirect) network effects, the typical utility function is of the form:

$$U_{ij} = c_i + M_j^d, \ 0 < d \leq 1. \tag{2}$$

Here, the utility to consumer i depends on the standalone benefit (c_i) and the number of compatible software varieties available for hardware j (denoted M_j). Again the standalone benefit can be zero. In this case, utility does not depend directly on the number of consumers who join the network. The number

[57] This section draws heavily from Gandal (2002b).

of compatible software varieties, however, does depend on and is increasing in the number of consumers who adopt hardware technology j. In other words, $M_j = f(N_j)$, $M_j'(N_j) > 0$, so the reduced form (or equilibrium) utility from (2) does increase in the number of consumers that join the network. The modeling complexity is even greater in settings with virtual network effects because there is an extra set of agents (software firms, in addition to hardware firms and consumers). Additionally, both the number of software varieties and the number of consumers on each network are potentially endogenous.

There are two basic approaches to handling expectations[58]. In the fulfilled expectations approach, consumers' expectations are correct. Although this is probably the most satisfactory approach, it leads to models that are quite difficult to solve analytically[59]. An alternative approach is to assume that consumers have myopic expectations, that is, consumer utility is based only on the network size at the time of purchase. This assumption makes it easier to analytically solve the model and hence allows the models to be more sophisticated. The tradeoff is that myopic expectations are less satisfactory from a modeling standpoint. Since these two assumptions have quite different implications, it makes it difficult to compare results across settings, unless the results are robust to both of these 'extreme' cases.

Timing issues are important as well. This is especially true in the case in which there are indirect network effects. In such cases, there is interdependence between the hardware adoption decisions of consumers and the supply decision of software manufacturers. Do consumers purchase hardware before software firms choose the hardware technology for which to write software, or do software firms first choose which technology to supply software for? This is the chicken and egg problem. The theoretical literature typically assumes either that consumers first purchase software, or that software firms first choose their preferred network[60].

References

Augereau, A., S. Greenstein and M. Rysman, 2004, Coordination vs. differentiation in a standards war: 56k modem, NBER Working Paper 10334.

Belleflamme, P., 2002, Coordination on formal vs. de facto standards: A dynamic approach, European Journal of Political Economy, 18, 153–176.

Besen, S. and J. Farrell, 1994, Choosing how to compete: Strategies and tactics in standardization, Journal of Economic Perspectives, 8, 117–131.

[58] This discussion is based on Matutes and Regibeau (1996).

[59] Of course, due to improvements in computing, models can be solved numerically relatively quickly.

[60] In reality, the process probably involves some 'give and take' (i.e., some software firms choose to make their software available) for a particular technology, then some consumers make purchases, etc. Gandal, Kende, and Rob (2000) develop a theoretical model and use it to estimate the feedback from hardware to software and vice versa in the CD industry.

Besen, S. and G. Saloner, 1989, The economics of telecommunications standards, in R. Crandall and K. Flamm, eds., Changing the rules: Technological change, international competition, and regulation in communications, Washington, DC: The Brookings Institution, 177–220.

Carbal, L. and T. Kretschmer, 2004, Standards battles and public policy, Mimeo: New York University.

Carlton, D. and M. Waldman, 2002, The strategic use of tying to preserve and create market power in evolving industries, RAND Journal of Economics, 33, 194–220.

Choi, J. P., 1994, Irreversible choice of uncertain technologies with network externalities, RAND Journal of Economics, 25, 382–401.

Choi, J. P., 1996, Do converters facilitate the transition to a new incompatible technology, International Journal of Industrial Organization, 12, 825–835.

Choi, J. P. and C. Stefanadis, 2001, Tying, investment, and the dynamic leverage theory, RAND Journal of Economics, 32, 52–71.

Choi, J. P. and M. Thum, 1998, Market structure and timing of technology adoption with network externalities, European Economic Review, 42, 225–244.

Chou, C. and O. Shy, 1990, Network effects without network externalities, International Journal of Industrial Organization, 8, 259–270.

Church, J. and N. Gandal, 1992a, Network effects, software provision and standardization, Journal of Industrial Economics, 40, 85–104.

Church, J. and N. Gandal, 1992b, Integration, complementary products and variety, Journal of Economics and Management Strategy, 1, 651–675.

Church, J. and N. Gandal, 1996, Strategic entry deterrence: Complementary products as installed base, European Journal of Political Economy, 12, 331–354.

Church, J. and N. Gandal, 2000, Systems competition, vertical merger, and foreclosure, Journal of Economics and Management Strategy, 9, 25–51.

Church, J., N. Gandal and D. Krause, 2003, Indirect network effects and adoption externalities, CEPR Working Paper 3738.

Church, J. and I. King, 1993, Bilingualism and network externalities, Canadian Journal of Economics, 26, 337–345.

Church, J. and R. Ware, 1998, Network industries, intellectual property rights, and competition policy, in N. Gallini and R. Anderson, eds., Competition Policy, Intellectual Property Rights and International Economic Integration, Calgary: University of Calgary Press, 227–285.

Church, J. and R. Ware, 2000, Industrial Organization: A Strategic Approach, San Francisco: McGraw-Hill.

Cremer, J., P. Rey and J. Tirole, 2000, Connectivity in the commercial internet, Journal of Industrial Economics, 48, 433–472.

David, P. and S. Greenstein, 1990, The economics of compatibility standards: An introduction to recent research, Economics of Innovation and New Technologies, 1, 3–41.

de Palma, A. and L. Leruth, 1996, Variable willingness to pay for network externalities with strategic standardization, European Journal of Political Economy, 12, 235–252.

Dranove, D. and N. Gandal, 2003, The DVD vs. DIVX standard war: Empirical evidence of network effects and preannouncement effects, Journal of Economics and Management Strategy, 12, 363–386.

Dranove, D. and N. Gandal, 2004, Surviving a standards war: Lessons learned from the life and death of DIVX, in K. Tomak, ed., Advance in the Economics of Information Systems, Hershey, PA: Idea Group, forthcoming.

Dybvig, P. and C. Spatt, 1983, Adoption externalities as public goods, Journal of Public Economics, 20, 231–247.

Farrell, J. and N. Gallini, 1988, Second-sourcing as a commitment: Monopoly incentives to attract competition, Quarterly Journal of Economics, 103, 673–694.

Farrell, J. and P. Klemperer, 2005, Coordination and lock-in: Competition with switching costs and network effects, in M. Armstrong and R. Porter, eds., Handbook of Industrial Organization, Volume 3, Amsterdam: North-Holland, forthcoming.

Farrell, J. and G. Saloner, 1985, Standardization, compatibility, and innovation, RAND Journal of Economics, 16, 70–83.

Farrell, J. and G. Saloner, 1986a, Standardization and variety, Economic Letters, 20, 71–74.

Farrell, J. and G. Saloner, 1986b, Installed base and compatibility: Innovation, product preannouncements, and predation, American Economic Review, 76, 940–955.

Farrell, J. and G. Saloner, 1988, Coordination through committees and markets, RAND Journal of Economics, 19, 235–252.

Farrell, J. and G. Saloner, 1992a, Converters, compatibility, and control of interfaces, Journal of Industrial Economics, 40, 9–35.

Farrell, J. and C. Shapiro, 1992b, Standard setting in high definition television, Brookings Papers on Economic Activity: Microeconomics, 1, 1–93.

Faulhaber, G., 2002, Network effects and merger analysis: Instant messaging and the AOL-Time Warner case, Telecommunications Policy, 26, 311–333.

Funk, J. and D. Methe, 2001, Market- and committee-based mechanisms in the creation and diffusion of global industry standards: The case of mobile communication, Research Policy, 30, 589–610.

Gandal, N., 1995, A selective survey of the literature on indirect network externalities, Research in Law and Economics, 17, 23–31.

Gandal, N., 2002a, The effect of native language on internet usage, CEPR Working Paper 3633.

Gandal, N., 2002b, Compatibility, standardization, and network effects: Some policy implications, Oxford Review of Economic Policy, 18, 80–91.

Gandal, N., M. Kende and R. Rob, 2000, The dynamics of technological adoption in hardware/software systems: The case of compact disc players, RAND Journal of Economics, 31, 43–61.

Gandal, N., D. Salant and L. Waverman, 2003, Standards in wireless telephone networks, Telecommunications Policy, 27, 325–332.

Gilbert, R., 1992, Symposium on compatibility: Incentives and market structure, Journal of Industrial Economics, 40, 1–8.

Greenstein, S., 1990, Creating economic advantage by setting compatibility standards: Can 'physical tie-ins' extend monopoly power? Economics of Innovation and New Technology, 1, 63–83.

Grindley, P., 1995, Standards, strategy, and policy, Oxford: Oxford University Press.

Gruber, H. and F. Verboven, 2001, The evolution of markets under entry and standards regulation—The case of global mobile telecommunications, International Journal of Industrial Organization, 19, 1189–1212.

Katz, M. and C. Shapiro, 1985, Network externalities, competition, and compatibility, American Economic Review, 75, 424–440.

Katz, M. and C. Shapiro, 1986, Technology adoption in the presence of network externalities, Journal of Political Economy, 94, 822–841.

Katz, M. and C. Shapiro, 1992, Product introduction with network externalities, Journal of Industrial Economics, 40, 55–84.

Katz, M. and C. Shapiro, 1994, System competition and network effects, Journal of Economic Perspectives, 8, 93–115.

Katz, M. and C. Shapiro, 1999, Antitrust in software markets, in J. Eisenach and T. Lenard, eds., Competition, Innovation and the Microsoft Monopoly: Antitrust in the Digital Marketplace, Boston: Kluwer Academic Publishers, 29–82.

Kolasky, W., 1999, Network effects: A contrarian view, George Mason Law Review, 7, 577–616.

Koski, H. and T. Kretschmer, 2002, Entry, standards and competition: Firm strategies and the diffusion of mobile telephony, ETLA Discussion Paper 827.

Laffont, J. J., J. Tirole and P. Rey, 1998a, Network competition I: Overview and nondiscriminatory pricing, RAND Journal of Economics, 29, 1–37.

Laffont, J. J., J. Tirole and P. Rey, 1998b, Network competition II: Discriminatory pricing, RAND Journal of Economics, 29, 38–56.

Markovich, S., 2001, Snowball: The evolution of dynamic markets with network externalities, Mimeo.

Matutes, C. and P. Regibeau, 1996, A selective review of the economics of standardization: Entry deterrence, technological progress, and international competition, European Journal of Political Economy, 12, 183–206.

Noam, E., 2002, Interconnection practices, in M. Cave, S. Majumdar and I. Vogelsang, eds., Handbook of Telecommunications Economics: Structure Regulation and Competition, Amsterdam: Elsevier, 387–421.

Postrel, S., 1990, Competing networks and proprietary standards: The case of quadraphonic sound, Journal of Industrial Economics, 39, 169–185.

Rohlfs, J., 2003, Bandwagon Effects in High Technology Industries, Cambridge, MA: MIT Press.

Rubinfeld, D., 1998, Antitrust enforcement in dynamic network industries, Antitrust Bulletin, XLIII, 859–882.

Rysman, M., 2003, Adoption Delay in a Standards War, Boston: University Mimeo.

Saloner, G., 1990, Economic issues in computer interface standardization, Economics of Innovation and New Technologies, 1, 135–156.

Shapiro, C. and H. Varian, 1999, Information rules, Cambridge, MA: Harvard Business School Press.

Shepard, A., 1987, Licensing to enhance demand for new technologies, RAND Journal of Economics, 18, 360–368.

Stango, V., 2004, The economics of standards wars, Review of Network Economics, 3, 1–19.

Thum, M., 1994, Network externalities, technological progress, and the competition of market contracts, International Journal of Industrial Organization, 12, 269–289.

Tirole, J., 1988, The Theory of Industrial Organization, Cambridge, MA: MIT Press.

Waterman, D. and A. Weiss, 1996, The effects of vertical integration between cable television systems and pay cable networks, Journal of Econometrics, 72, 357–395.

Whinston, M., 1990, Tying, foreclosure, and exclusion, American Economic Review, 80, 837–859.

Chapter 5

BROADBAND COMMUNICATIONS

ROBERT W. CRANDALL

The Brookings Institution

Contents

Handbook of Telecommunications Economics, Volume 2, Edited by S. Majumdar et al.

DOI: 10.1016/S1569-4054(05)02005-1

Communication technologies that provide high-speed, always-on connections to the Internet for large numbers of residential and small-business subscribers are commonly referred to as 'broadband' technologies. *High-speed* is an imprecise term—it simply means much faster than dial-up connections. A dial-up connection, using a 56-kilobit per second (Kbps) modem, typically transfers data at about 40 Kbps. The U.S. Federal Communications Commission (2000) defines high speed as a connection that provides at least 200 Kbps in one direction. Much higher speeds are generally available in modern digital subscriber line (DSL) or cable modem services[1]. *Always-on* refers to an Internet connection that is immediately available to the user and does not require that he (or she) log on to the service for each use. With an always-on or always-available connection, the delay from the time that a user goes to the computer and clicks on a web page URL to the time when the request for information is delivered to the remote server is measured in milliseconds.

In this chapter, the literature on the development of mass-market broadband services for residential and small-business subscribers is reviewed. There is no discussion of traditional high-speed services for medium and large businesses, such as DS-1 or DS-3 services, or various high-speed packet-switched services, such as frame relay. Nor is there any discussion of 'special access' or 'leased lines' that are used by large businesses or long distance carriers to originate or terminate large numbers of voice/data calls is included.

1. The technology

It is often asserted that broadband technologies have been available for as much as two decades, but that regulated telecommunications carriers were slow to deploy them. However, there was little demand for broadband connections until a large number of subscribers had access to the Internet, and the cost of broadband services would have been prohibitive 20 years ago. As recently as 1994, less than one-fourth of U.S. households had a computer and far fewer had Internet access. In this environment there was not likely to be much demand for broadband connections at any price.

1.1. Digital subscriber line

Telephone networks are typically constructed with a mix of fiber optics and copper wires. The copper wires typically extend from the central office to the subscriber's premises, but in areas of low-population density or new development,

[1] The FCC retains its current definition for advanced telecommunications capability as infrastructure capable of delivering a speed of 200 Kbps in each direction. In Japan and Korea, broadband services are often delivered at speeds of 8–20 Mbps.

fiber optics may radiate out from the central office to remote terminals from which copper wire is extended to the subscriber. Digital subscriber line modems are used to transmit data over these copper wires at far higher rates than is possible over voice-grade connections because they can take advantage of capacity in the copper wire that is not used for voice communications. There are a number of different DSL technologies available that may be employed by telephone companies under various conditions. These technologies, shown in Table 1, offer a variety of speeds for the downstream and upstream paths.

The higher-speed services shown in Table 1 often require considerable installation expense at the customer's premises because 'splitters' must be installed to divide the lower-frequency voice signal from the higher-frequency data signal or dedicated lines must be installed for the high-speed data connection. Even those not requiring such installations may require substantial network upgrades to reach large numbers of subscribers.

Most DSL services are asymmetric digital subscriber line (ADSL) services, offering different transmission capacities for downstream and upstream communications. Higher speeds are offered for downloading information or even audio or video programming than for communicating upstream. DSL is designed to share the copper loop used for ordinary telephone calls. By splitting the signals on the loop at the central office or remote terminal and routing the lower-frequency portion, generally from 0 to 4 kHz, over the traditional voice network and the higher frequencies to a separate modem or digital subscriber line access multiplexer (DSLAM), a telephone company can deliver a high-speed access service for Internet connections over the same line as it delivers voice services[2].

As Table 1 shows, DSL service is limited by the quality and length of the copper lines that extend to the subscriber's premises. Digital subscriber line connections can be restricted or even made unworkable if the copper loop is impaired by various devices that are used to enhance the quality of voice communications, such as loading coils. Moreover, the ability of copper telephone loops to carry high-speed data signals declines with distance.

The various DSL services shown in Table 1 are designed for different uses, offering a variety of range and capacity alternatives as well as technical standards. Thus, the telephone industry lacks the simplicity of a single standard enjoyed by the cable industry. Typically, ADSL permits downstream speeds of from 500 kbps to 8 Mbps and upstream speeds of up to 640 Kbps. The communications speed actually achieved depends on the quality and length of the copper line connecting the subscriber. Asymmetric digital subscriber line modems adjust the operating rate to deliver the highest possible speed over a specific loop, but they do not operate reliably over copper loops longer than about three miles (5000 m).

[2] For a more detailed exposition, see Jackson (2002).

Table 1
The varieties of digital subscriber line (DSL) service

DSL type	Description	Data rate	Distance limit	Application
IDSL	ISDN digital subscriber line	128 Kbps	18,000 ft on 24 gauge wire	Similar to the ISDN BRI service but data only (no voice on the same line)
CDSL	Consumer DSL from Rockwell	1 Mbps downstream; less upstream	18,000 ft on 24 gauge wire	Home and small-business service not requiring a splitter; similar to DSL Line
G. Lite or DSL Lite	'Splitterless' DSL without the 'truck roll'	From 1.544 Mbps to 6 Mpbs, depending on the subscribed service	18,000 ft on 24 gauge wire	The standard ADSL; sacrifices speed for not having to install a splitter at the user's home or business
HDSL	High bit-rate digital subscriber line	1.544 Mbps duplex on two twisted-pair lines; 2.048 Mbps duplex on three twisted-pair lines	12,000 ft on 24 gauge wire	T1/E1 service between server and phone company or within a company; WAN, LAN, server access
SDSL	Single-line DSL	1.544 Mbps to 2.048 Mbps downstream and upstream on a single duplex line	12,000 ft on 24 gauge wire	Same as for HDSL but requiring only one line of twisted-pair
ADSL	Asymmetric digital subscriber line	1.544–6.1 Mbps downstream; 16 to 640 Kbps upstream	From 1.544 Mbps at 18,000 ft to 8.448 Mbps at 9000 ft	Used for Internet and Web access, motion video, video on demand, remote LAN access
RADSL	Rate-adaptive DSL from Westell	640 Kbps to 2.2 Mbps downstream; 272 Kbps to 1.088 Mbps upstream	Not provided	Similar to ADSL
VDSL	Very high digital subscriber line	12.9–52.8 Mbps downstream;1.5 to 2.3 Mbps upstream	12.96 Mbps at 4500 ft to 51.84 Mbps at 1000 ft	ATM networks; fiber to the neighborhood

Source: http://www.everythingdsl.com/types/index.shtml.

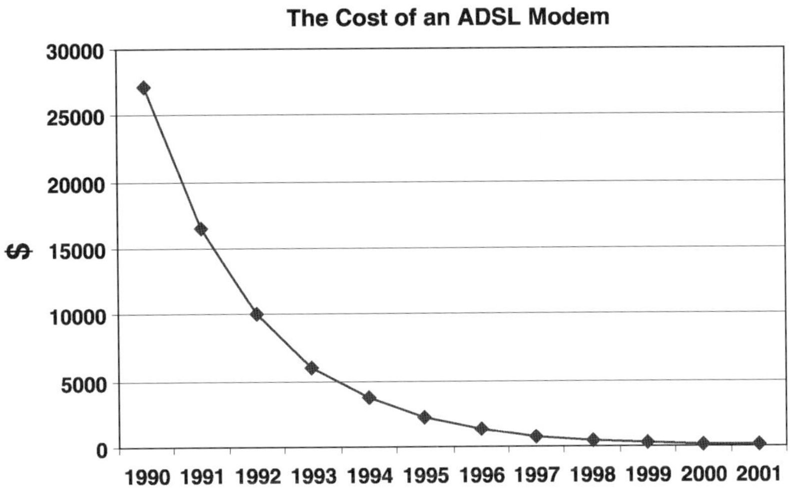

Fig. 1. The cost of an ADSL modem. Source: Based on Jackson (2002).

As the Internet has grown, the cost of providing ADSL service has declined, in large part because of the declining cost of electronics. The cost of ADSL modems has fallen substantially as Figure 1 shows. In 1992, the cost of a modem was approximately $10,000; 9 years later it was in the range of $100. This cost trend helps to explain why telephone companies were initially slow to deploy DSL. Faulhaber (2002a), for example, suggests that the incumbent U.S. telephone companies should have begun to deploy DSL before 1998, but the high cost of modems may have made such deployment uneconomic. Similar conclusions obviously apply for cable modems and for cable television company deployment of broadband.

1.2. Cable modems

Broadband services can also be offered over traditional cable television networks that employ a hybrid fiber-coaxial cable (HFC) technology. In the 1990s, a cable industry research consortium, CableLabs, developed a standard for data communication over cable systems, Data Over Cable System Interface Specification (DOCSIS)[3]. DOCSIS has had a major impact on the cable modem market, allowing multiple manufacturers to compete in the supply of compatible equipment and reducing the risk to consumers of being saddled with orphaned equipment. More than 50 manufacturers now supply DOCSIS cable modems—including firms, such as Motorola, Cisco, Toshiba, and 3Com[4].

[3] The DOCSIS project was renamed as the Cable Labs(r) Certified(tm) Cable Modem project.

[4] See Jackson (2002) for details.

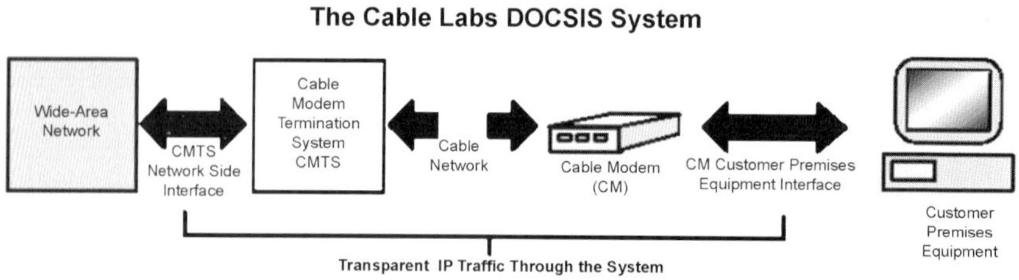

Fig. 2. The Cable Labs DOCSIS System. Source: Cable Labs SP-RF1-106-010829 as reproduced by Jackson (2002).

Figure 2 provides a schematic diagram of a DOCSIS system. Each subscriber has a cable modem at its premises, and the cable company installs a cable modem termination system (CMTS) located at its head-end. The cable modems and the CMTS use the preexisting cable network much as dial-up modems use the telephone network.

Cable modems can receive data at speeds as high as 40 Mbps and transmit at speeds up to 10 Mbps. The next generation of cable modems may be able to transmit upstream at 30 million bits per second[5].

To offer high-speed access, a cable television system must dedicate some of the system's capacity, typically the equivalent of a single TV channel, to the cable modem service. One TV channel provides about 40 Mbps of downstream capacity that is shared by subscribers. This sharing of capacity creates a fundamental problem for cable systems. The downstream data from the head-end to the subscriber is transmitted over the cable through its tree and branch architecture, much as a television signal is distributed over the system. DOCSIS includes an encryption element, the modern version of a secret codes, to prevent anyone other than the subscriber from reading the data in these broadcast packets. In a large cable system with 100,000 subscribers, a downstream capacity of 40 Mbps would provide an average capacity of just 400 bits per second per subscriber if all subscribers were connected. If as many as 5 percent were connected, the capacity would still only be 8 Kbps.

The solution to this network sharing problem is to provide different data streams to a number of remote neighborhood nodes. For example, if a cable system serves an average of 1000 homes per fiber node and no more than 5 percent of homes are using the downstream capacity at one time, the average downstream capacity would be 800 Kbps, an acceptable speed for web-browsing service for most subscribers. The data capacity is not rigidly divided among the subscribers, but the capacity resides in a pool and is allocated dynamically as users need to communicate. Notice however, if many consumers choose to use the Web in a fashion that requires continuous transmission—say, downloading a host of music

[5] This description is drawn from Jackson (2002).

files—the shared capacity will quickly be exhausted. Improving system performance or speed will then require the allocation of additional capacity to the cable modem service.

1.3. Fiber to the home

The highest speed broadband connections are available through fiber to the home (FTTH) technology, which is now being installed in Japan and in the United States. Fiber to the home technology utilizes fiber optics to distribute signals from the telephone central office to the final subscriber and back again. The capacity of a single fiber is substantial; therefore, several FTTH architectures split the fiber so that 16–64 households share the capacity of a single fiber. If the splitting is accomplished at a 'passive' network node that cannot vary the capacity allocation across subscribers, the network is described as a passive optical network (PON). If the node at which the splitting occurs has electronics that permit the allocation to vary with network usage, the network is an active star network. Each subscriber to these networks would be able to receive service at speeds of 10–100 Mbps or even higher if demand warrants.

In Japan, NTT is offering a FTTH service to businesses and residences in dense areas at a monthly price of between US$ 30–70. By August 2002, NTT had enrolled nearly 100,000 subscribers in this service, which offers speeds of up to 100 Mbps, but most subscribers were businesses. By the end of 2004, FTTH had 2.4 million subscribers in Japan[6] and about 700,000 subscribers in the United States[7].

1.4. Wireless access

A third form of broadband access is provided by various wireless technologies. The earliest of these technologies was fixed wireless, utilizing a central wireless transmission facility that radiates signals to dispersed customers with antennas. These services have been used to provide services to medium-sized and even large businesses, but with limited economic success. Several U.S. companies acquired spectrum in the MMDS and LMDS bands in order to offer a mass-market broadband service. Neither of these services has been widely subscribed, nor is it clear that they are competitive with wire-based systems, particularly if they are required to pay auction-based spectrum license fees. An recent analysis by Wanichkorn and Sirbu (2002) suggests that they are only competitive with cable modems and DSL at very low-population densities.

[6] CIAJ, "The challenges for Japan's telecommunications industry," July 2005, available at www.ciaj.or.jp/e/japanmarket/.

[7] U.S. Federal Communications Commission (2005).

Satellite systems are the most widely used of the high-speed wireless options today. Two U.S. firms, DirecWay and StarBand, provide two-way satellite-based Internet access. These systems provide service at data rates of about 400–500 kilobits per second. Thus far, satellite firms appear to be targeting customers in rural areas who will not be offered DSL or cable modem service. Because geostationary satellites have large footprints, they are not well suited for delivering broadband services to large numbers of customers. The next generation of satellite systems will use spot beams to overcome this deficiency and therefore, may be able to offer higher data rates to millions of customers due the greater system capacity afforded by spot beams.

The newest wireless technology that is gaining attention is the 'Wi-Fi' technology that uses unlicensed spectrum, the 2.4 GHz spectrum in the U.S., to reach wireless local area networks (LANs). Using an IEEE standard called 802.11[8], users can access the Internet at thousands of low-power wireless network locations throughout the country, including university campuses, airports, coffee shops, hotels, and even their own homes. This technology supports data rates at up to 54 megabits per second, but it is shared access and no one user would typically be able to achieve this rate[9]. This technology is spreading rapidly, primarily to allow people to gain access from numerous locations through their laptop or notebook computers or through other wireless appliances.

While Wi-Fi solves the problem of distributing Internet services the last few yards to the subscriber, it is not a technology that can substitute for DSL or cable modem service unless high-speed connections are widely available from the wireless LAN to the Internet backbone. Users typically only need to sign up with a Wi-Fi service provider and deploy a Wi-Fi PC card on their laptop to surf the Web or access e-mail, but they must be in relatively close proximity to a 'hot spot' to access the Wi-Fi network. Lehr and McKnight (2003) suggest that the current cellular service providers could integrate Wi-Fi into their networks, either as a substitute for 3G or as a complement to it. This would allow a virtually ubiquitous Wi-Fi service in countries with highly-developed cellular infrastructure.

Broadband service is now also available on commercial wireless ('cellular') networks at speeds of as much as 500 Kbps over '3G' networks. In the United States, Verizon Wireless has begun to offer a broadband access service using an Evolution-Data Optimized (EV-DO) 3G wide-area network based on the CDMA technology. DoCoMo, Japan's leading cellular company, introduced 3G services capable of delivering 384 Kbps in 2001, but it found limited consumer demand for such services until early 2003. By the end of 2003, it had enrolled approximately 5 percent of its 40 million customers in its new high-speed service.

[8] 802.11 comes in 4 flavors-in chronological order they are: 802.11, 802.11b, 802.11a, and 802.11g.

[9] Even if they could, relatively few coffee shops install a 54 Mbps or 11 Mbps upstream connections to the Internet.

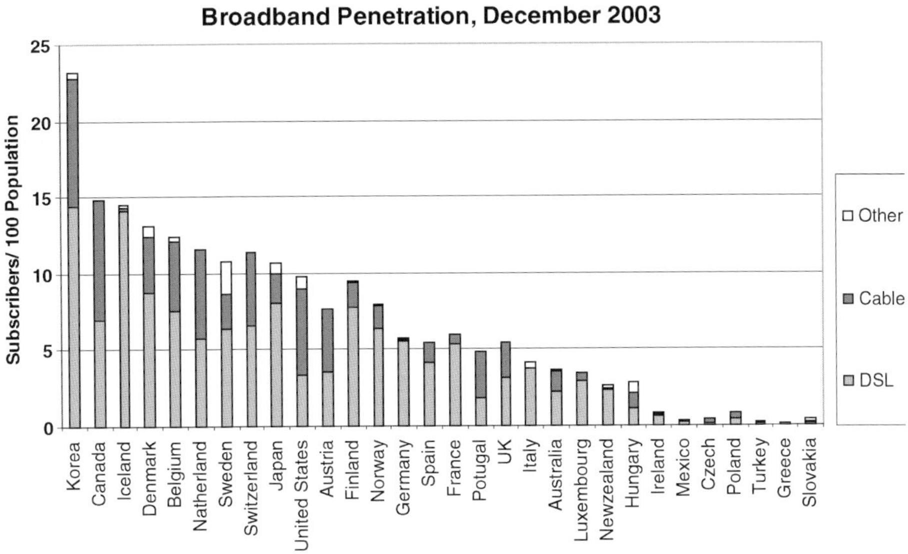

Fig. 3. Broadband penetration, December 2003. Source: OECD (2004).

Vodafone, the world's largest mobile wireless operator, began to launch its 3G services in Europe in early 2004.

2. Broadband diffusion

In its first few years, broadband grew rapidly in some countries, but not in others. Before examining the possible reasons for these differences, it is useful to provide data on the extent of broadband penetration in the developed world.

2.1. Cross country comparisons

Figure 3 shows the data compiled by the Organization for Economic Cooperation and Development as of December 2003. Figure 4 shows that there is a weak relationship between income per capita and broadband penetration, with the obvious outliers being Korea, which deserves special attention, and Luxembourg. North America, Scandinavia, and the Benelux countries tend to have much higher broadband penetration than other OECD countries. The United Kingdom, Australia, and New Zealand lag badly despite their early liberalization of telecommunications and their obvious initial advantage in using the Internet because of their use of the English language.

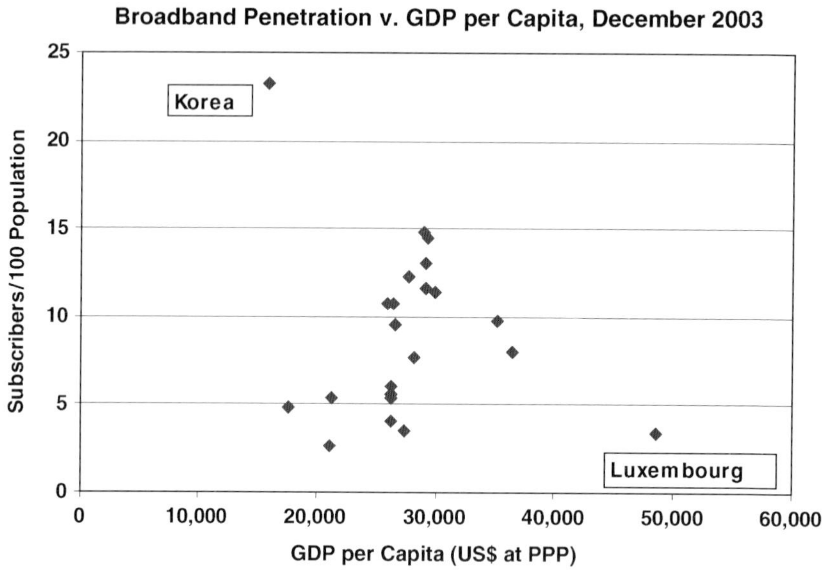

Fig. 4. Broadband penetration vs. GDP per capita.

It is also useful to array these countries' broadband penetration in terms of technology. Korea has the most aggressive policy of extending fiber optics to new residential buildings, particularly the large apartment complexes that have been built in recent years. About two-thirds of Korean subscribers use DSL and one-third use cable modems. Among the other countries with substantial broadband penetration, Canada, the United States, the Netherlands and Austria have a large share connected to cable modems. Countries with little cable television, such as France, Italy, New Zealand, and Spain tend to have much lower broadband penetration[10].

Some economists contend that incumbent telephone companies have been reluctant to deploy DSL because this service 'cannibalizes' the incumbents' highly profitable high-speed business services, such as DS-1. Although there is no empirical evidence to support this claim as yet, it is a plausible argument because of regulators' attempts to keep basic local residential telephone service rates low by allowing regulated incumbent carriers to charge supra-competitive prices in

[10] Germany and the United Kingdom have substantial cable television penetration, but little cable modem subscription. In the U.K., the cable companies have performed badly and have recently been reorganized through merger into two large carriers. In Germany, Deutsche Telekom, the incumbent telephone company, owns most of the cable network and has been unwilling to pursue cable modem development.

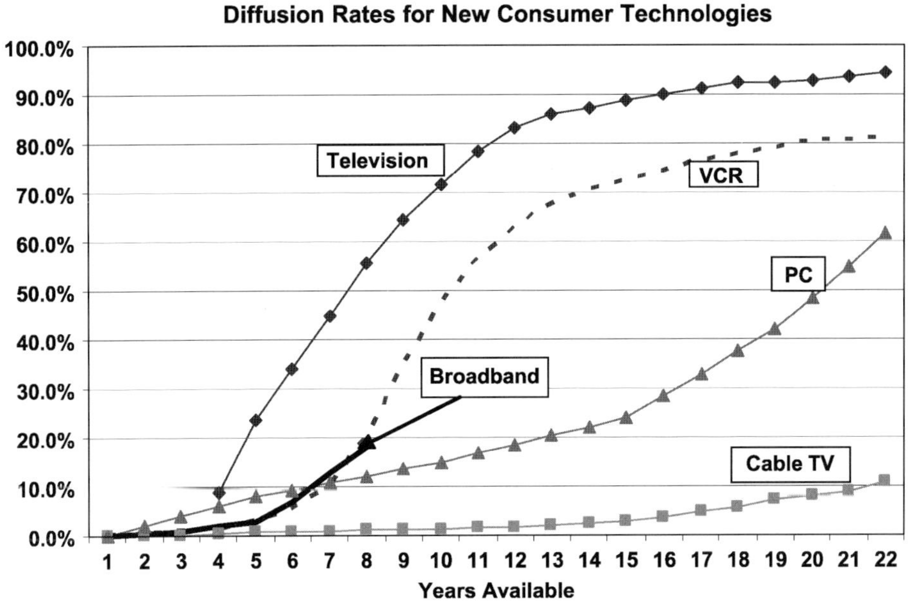

Fig. 5. Diffusion rates for new consumer technologies. Source: Adapted from Owen (2002).

business markets[11]. This theory underlies much of the regulatory rationale for requiring 'local loop unbundling' that is described below.

2.2. Comparing broadband diffusion with other 'breakthrough' technologies

Given that the Internet has existed only since 1990[12] and that household Internet connections did not approach 25 percent in even the most developed countries until 1995 or later, the growth of broadband has been quite rapid. But how does its diffusion compare with the spread of earlier technological breakthroughs in consumer durables or services, such as radio, television, the VCR, or the home computer?

Economists and technologists often measure the speed at which new inventions or media spread throughout the population by calculating 'diffusion curves.' Owen (2002) measures the rate of diffusion for major new consumer technologies. An updated version of his results is reproduced in Fig. 5. The diffusion of consumer broadband services, shown with a bold, solid line, occurred about as rapidly in its first few years as the diffusion of the VCR, and it spread more rapidly

[11] See Crandall and Waverman (2000).

[12] Formally, the Internet was created on 1 January 1983 when the arpanet split into the Arpanet and milnet.

than cable television or personal computers. Only television gained consumer favor more rapidly than broadband in Owen's analysis.

Faulhaber (2002) makes the same point in analyzing the barriers to broadband deployment. He finds that broadband diffusion has been as rapid as that of the VCR in its first few years and far outstrips the diffusion rate for wireless telephony. As a result, he doubts that regulatory policies had been a major impediment to widespread broadband penetration through 2002.

These comparisons of historical diffusion rates are somewhat misleading. When television sets were first introduced, they were quite expensive, as were the first wireless telephones. A new television set in 1948 would have represented a much larger share of a household's budget in 1948 than the purchase of a household computer or broadband service today[13]. The diffusion issue will arise again in a later discussion of network and 'bandwagon' effects.

3. The economics of broadband supply

The provision of broadband communications services is subject to many of the economies of scale, scope, and density generally observed in the telecommunications sector, particularly if the service is delivered over wires. Given these economies, what is the likely market structure of the broadband industry?

3.1. Cable modems, DSL, and fixed wireless

Faulhaber and Hogendorn (2000) have undertaken an analysis of the cost of building a hybrid fiber-coaxial cable system *de novo* and the likelihood of competitive entry into the delivery of broadband services over such networks. They conclude that such entry is feasible under certain conditions. Using a Cournot model of spatial competition, they analyze the likely deployment of new hybrid fiber-coaxial cable networks to deliver broadband services to residences. (They assume that DSL is not a viable competitor.) They find that a second firm enters the market when the demand for broadband rises by 50 percent from its initial level of 15 percent at a $50 per month price. The third firm enters when demand rises to 85 percent above this level. Once broadband subscriptions rise to the level of current cable television subscription, their model predicts that 70 percent of households will have a choice of three facilities-based providers.

Jackson (2002) also examines the economics of offering broadband over DSL on an incumbent telephone company's network and via cable modem over a cable television network. He finds that the costs are relatively similar. However, Jackson does not analyze the effects of the necessary investment in modifying existing telephone networks or cable systems to deliver these new broadband services.

[13] Faulhaber acknowledges this point *vis a vis* wireless telephony in his article Faulhaber (2002).

Table 2
Annual cost per line for broadband over three platforms ($/(line/year))

Density (lines/sq. mi.)	Fixed wireless*	DSL	Cable modem
0–5	250–336	707	646
5–100	248–308	364	292
100–200	230–304	274	189
200–650	233–287	228	136
650–850	227–302	212	121
850–2250	217–270	202	113
2250–5000	212–259	195	109
5000–10,000	207–258	199	114
>10,000	214–241	181	110
Average	225–286	236	151

*Excludes spectrum costs that may add as much as $22.60/(line/year). Wireless may also offer bandwidth (speed) that is different from typical cable modem or DSL bandwidth.
Source: Wanichkorn and Sirbu (2002).

Wanichkorn and Sirbu (2002) have provided a detailed engineering analysis of the capital costs of delivering broadband over three platforms: fixed wireless, DSL, and cable television, using the results provided by Fryxell (2002) for cable modems and DSL. They conclude that cable modem service enjoys a cost advantage over DSL in every population density range, but that fixed wireless has an advantage over cable at extremely low-population densities (i.e., less than 100 lines per square mile) (see Table 2). Their costs exclude all operating costs, marketing costs and SG&A, and they also exclude the spectrum costs of the fixed wireless option. When spectrum costs are included, the cost advantage of fixed wireless diminishes and perhaps disappears for all but the areas with fewer than 5 lines per square mile.

Given the large customer-acquisition (marketing) costs involved in selling a new service, none of these studies provides an analysis of the full costs of delivering broadband services. In addition, they do not take into account the joint costs of marketing and operating a communications company, such as a cable television company or an incumbent telephone company that offers a variety of video, wireless telecommunications, and wireline telecommunications. If there are substantial joint economies in offering and marketing such services, the market for broadband services may evolve into one involving only platform competition from large, diversified carriers.

3.2. *Fiber to the home*

The costs of broadband delivered on a FTTH system are substantial because of the need to extend fiber farther into the network and to install a variety of equipment in the subscriber's premises to translate the optical signal into an

Table 3
The capital cost per subscriber of alternative broadband technologies

Technology	Data speed	Approximate cost per home served
DSL	640 Kbp(s-8Mbps)	\$500
Cable	500 Kbps +	\$400
FTTH	20 Mbps +	\$2000–\$6000
MMDS	2 Mbps +	\$650
LMDS	2 Mbps +	\$1300
Satellite	400 Kbps +	\$1000 + Earth station

Source: Hopkins (2001) as reported in Carnegie Mellon University (2002).

electrical signal that existing household appliances can receive. Several U.S. communities have considered the construction of FTTH networks. Palo Alto, CA, has been investigating the economics of such a system for some time and has estimated that a system designed to serve the entire city would cost about \$1000 per home passed plus another \$1500 per home connected to the network. Thus, if 25 percent of homes subscribed, the network could cost as much as \$5500 per subscriber, or more than three times the cost of current cable or telephone networks[14].

A more recent analysis of current and prospective FTTH technologies has been undertaken by a consortium led by Carnegie Mellon University (2002). It focuses principally on FTTH technologies. It provides a useful comparison of the costs of FTTH vs. other technologies based upon a study performed by Hopkins (2001) and summarized in Table 3.

However, according to the CMU analysis, the capital cost of FTTH systems could be much greater than the \$2000–6000 shown in Table 3. The capital cost per home served using the PON architecture in an urban area rises from about \$2000 with 80 percent subscriber penetration to \$5000 with only 20 percent subscriber penetration. In small towns, such a system would cost about \$3500 per subscriber if 80 percent subscribed, but more than \$10,000 if only 20 percent subscribed.

These costs have made the widespread deployment of new FTTH networks uneconomic unless a large share of households subscribe and the cost of the subscriber equipment and installation declines substantially. The extension of current fiber-coaxial cable networks, owned by cable television companies, or fiber-copper systems, owned by telephone companies, appear to offer better economics. However, as the cost of the electronic equipment required declines, FTTH could evolve into an attractive option for subscribers desiring extremely fast network connections.

[14] See 'FTTH Business Case' at http://www.pafiber.net/references/20021002-UAC-packet.htm.

3.3. The weak dominance of cable

The analyses summarized above suggest cable modem service currently enjoys a substantial advantage over DSL, particularly in areas where the cable infrastructure is already deployed and where the copper telephone plant requires substantial upgrading. This advantage may be mitigated by the degradation of speed that occurs when large numbers of subscribers are connected to the same cable node. Such performance degradation can be overcome, however, if cable operators devote more than 6 MHz of their 750 MHz of capacity to cable modem service or if fiber is extended farther out toward the subscriber. Why they do not do so at present is discussed below.

Fixed wireless and satellite systems are generally higher-cost technologies that dominate cable and DSL only in rural areas. Finally, new FTTH systems are being deployed in the United States, often with municipal government participation, and in Japan, a country with high population density. These systems appear to have rather high costs at the present and will only be economical as the cost of the required electronics declines and household demand for high-speed outruns the speeds available via cable modems and DSL.

4. The demand for broadband

As with all new services or technologies, particularly those that feature network externalities, it is very difficult to predict the demand for mass-market broadband service. The nature of any new service may be evolving rapidly at first, or in the case of broadband, the potential use of the service may be changing rapidly. Some households may not be interested until others show them the value of adopting the service or provide them the opportunity to reap the benefits of consumption externalities generated by the service.

The first empirical studies to examine broadband adoption are Madden and Savage (1996) and Madden et al. (1996). They utilize survey data from a sample of Australian households that asked participants to indicate their preferences for a broadband service that had not yet been deployed. They find that the interest in broadband is directly related to education, at least one member of the household originating from Europe or Asia, and age. Households were less interested in broadband if one or more people in the household was 65 years or older. They interpret their results as portending a danger of the creation of a class of 'information poor.' This danger seems to have receded since then as the real cost of personal computers and Internet subscriptions has fallen dramatically.

One of the earliest studies of the demand for high-speed connections in the United States was undertaken by Hal Varian in an experimental setting at the University of California, Berkeley. In 1998–1999, 70 members of the Berkeley community were provided with various access speeds up to 128 Kbps at different prices, admittedly a low ceiling by today's standards but not by the standards of 1998–1999. Varian found that his sample of users exhibited a rather high price elasticity of demand for bandwidth. The own-price elasticities of demand for the higher speeds were in the range of –2.0 to –3.1. Moreover, the implied value of the time exhibited by these 70 participants was astoundingly low, generally in the range of 1–5 cents per minute. Varian attributes this low demand for bandwidth to the absence of applications requiring high speed and the ability of users to occupy their time productively in other pursuits while waiting for file downloads.

Obviously, Varian's results predate the development of Napster and the increased use of the Internet for downloading video or engaging in real-time electronic games. More current estimates of the demand for broadband are reflected in the work of Rappoport et al. Their work addresses two related issues: (1) what are the own and cross price elasticities of demand for broadband and narrowband services?; and (2) how different are the users and uses of broadband and narrowband Internet services?

Using a sample of 20,000 United States households that were surveyed in 2000, Rappoport et al. (2001a) estimated a nested logit model of Internet subscription and choice of narrowband or broadband access. Because cable modem service is available to some households in the United States where DSL is unavailable, DSL is available to others where cable modem service is unavailable, and both are available to yet others, the authors are afforded a rich set of possible choices to examine. They find that the own-price elasticity of demand for broadband service is much greater than the demand elasticity for narrowband. Rappoport et al., find that the own-price elasticity is in the range of –0.6 to –0.8 for cable modem service and –1.4 to –1.5 for ADSL service. In general, they find that dial-up service and broadband services are substitutes for each other, a result that conflicts with the results in Hausman (2002a). Hausman (2002a) points out that narrowband and broadband services are not likely in the same market given that residential dial-up telephone rates vary substantially across states while broadband rates typically do not. As broadband diffuses throughout the economy, the degree of substitution between broadband and narrowband connections will surely decline even farther.

Using data from 2000–2001, Crandall et al. (2002) estimate a model similar to that used by Rappoport et al., and obtain own-price elasticities for both services equal to –1.2. In a later paper that uses August 2001 Internet user data, Rappoport et al. (2002) examine the intensity and nature of Internet use of households with narrowband or broadband connections. They show that the decision to use a broadband connection depends on the opportunity cost of time for the user and intensity of Internet use. Higher income households who use the Internet intensively are most likely to subscribe to broadband. Broadband subscribers tend to visit more Internet sites, particularly 'entertainment' and 'Internet services' sites and spend more time online. However, in this later paper, Rappoport et al. find that the demand for broadband is substantially less price elastic (–0.47) than it was a year earlier. This suggests that broadband is moving from the 'luxury' category to one closer to a 'necessity', particularly as consumers learn to use it and increasingly understand its potential benefits.

More recent work by Rappoport et al. (2003) employs 2002 survey data on households' willingness to pay for DSL and cable modem service. Therefore, these results reflect the most recent reaction of households to the choice of Internet service. Rappoport et al. (2003) find that willingness to pay is inversely related to age and not very strongly related to household income. More importantly, they find that the own-price elasticity of demand for both broadband technologies declines from about –3.0 to approximately –1.0 as the price falls from $50 per month to $20 per month and that the own-price elasticity of demand is higher for DSL than for cable modem service.

5. Network and bandwagon effects

Broadband access to the Internet opens up vast new opportunities for subscribers to use their connections. What began as a popular medium for exchanging e-mails may now evolve into one that allows users to manipulate large video files, watch a variety of filmed content, or participate in real-time video games with others around the world. It is likely that these and totally unforeseen new uses will develop in the next few years, but it is not clear that the existing market organization provides sufficient incentives for the efficient development of this content.

5.1. Internalizing network externalities

As each new mass medium developed, the owners of the distribution facilities or networks were initially integrated into the supply of content. Such integration allowed them to exploit the network effects created by their investments. The value of any increment of content is a direct function of the deployment of distribution assets, and the value of the distribution assets is also directly related to the amount of content available. Therefore, vertical integration between content production and distribution creates incentives for production and investment that would not exist if the two stages of production were independent and unable to negotiate efficient, enforceable long term contracts (Katz and Shapiro, 1985, 1994).

For instance, motion picture companies integrated production, distribution, and even exhibition into their operations in the early twentieth century (Crandall (1975, 2001). Shortly after World War II, the new television networks initially produced a substantial share of their own programs. As a vibrant program production sector developed, generally among the motion picture companies in Hollywood, the networks receded to financing new programs that would be contracted out to independent producers (Crandall, 1972; Fisher, 1985). Three decades later, when the cable television industry was freed of regulation that had been designed to protect the television broadcasters, cable companies expanded their channel capacity dramatically and integrated backward into programming to fill these channels (Chipty, 2001).

Broadband Internet access presents opportunities and problems similar to those experienced by motion picture distributors, television networks, and cable networks. The investment in distribution capacity is obviously driven by perceptions of consumer demand, which in turn requires the development of consumer applications for use over these high-speed networks. One only has to look back as far as 1998–1999 to Varian's INDEX experiment to find evidence that a set of informed consumers placed little value on increased access speed for using the Internet.

Who will provide new broadband content? It is unlikely that regulated telephone companies will have a comparative advantage in developing new, innovative uses of the Internet. The owners of some broadband content, the large motion picture companies, appear to be afraid of the lack of intellectual property protection if they make their products widely available on the Internet. Some integration between entities with experience in creating content and broadband networks distributing it would seem to be required for efficient exploitation of network economies.

This type of integration was what was anticipated when AOL, the leading U.S. Internet service provider, announced a merger with Time Warner, a major cable, media, and motion picture company. Unfortunately, this combination did not succeed for reasons that are not entirely clear. Nevertheless, both the U.S. Federal Trade Commission and Federal Communications Commission delayed the merger's completion through extensive investigations and the imposition of several conditions before they would approve the merger.

5.2. Network effects and first-mover advantages

The existence of network effects can be shown to lead to a 'tipping' of the market towards the successful first mover under certain conditions[15]. Even if the first-mover chooses an inferior technology, he may dominate the market because subsequent producers cannot overcome the advantages that the first mover obtains through early adoptions. The most cited examples of this tipping towards inefficient technologies are the QWERTY typewriter keyboard and the VHS videotape format. These examples of market failure have been challenged by Liebowitz and Margolis[16].

Are there any possibilities of market failures due to tipping in the case of broadband? Specifically, is it possible that vertical integration into content by a large broadband distributor could allow that distributor to dominate broadband distribution and even to direct resources to an inefficient distribution technology?

[15] See Katz and Shapiro (1994).

[16] Liebowitz and Margolis (1994, 2001).

Faulhaber (2002) demonstrates that the likelihood of such tipping in the case of broadband Internet services is remote. The key question in Faulhaber's analysis is whether the integrated supplier chooses *interoperability* with other broadband content.

If the integrated firm does not have an overwhelming share of broadband distribution and cannot use its current position in distribution or content to deny rivals in either sector the ability to survive, it has the incentive to offer its software to competing platforms and to distribute other firms' content on its platform. It must weigh the benefits from wider distribution of its own content against the possible benefits to its distribution operations from denying rival distributors its content. Similarly, it must weigh the benefits that it can reap from greater subscriptions to its distribution services from allowing other content suppliers to distribute over its platform vs. the possible benefits in the content market from denying rival content suppliers access to its distribution network.

In many of the earlier examples related to consumer markets with network effects, such as motion pictures, television, and cable television, the integrated owners of distribution facilities decided to interoperate with content provided by rival content suppliers. The notable example of noninteroperability was the early AT&T Company, which denied independent distributors (local telephone companies) interconnection with its local and long distance facilities (Mueller, 1997).

In the case of broadband Internet services, no distributor or owner of content has or is likely to have sufficient first-mover advantages to tip the market. In the U.S., the large cable companies perhaps come closest to such a position, but none has exhibited any ability to leapfrog ahead of rival distributors through investments in content. Indeed, one of the early cable-company ISPs, @Home, declared bankruptcy and disappeared after a few years. Even if they were to achieve a dominant position, however, one would have to weigh the benefits from more rapid development of content facilitated by vertical integration against any possible loss of output to monopoly in a 'tipped' market.

5.3. The broadband 'bandwagon'

Rohlfs (2001) has provided an intriguing analysis of a form of network effects that he calls 'bandwagon effects.' As more and more people adopt a service or buy a good, others are attracted to it. Even if there are no network externalities, *per se*, these bandwagon effects can be important to suppliers of the service. Rohlfs examines how bandwagon effects developed for several new products or technologies, including telephones, VCRs, personal computers, and the Internet. His analysis of the development of bandwagon effects through the Internet is quite simple. Interlinking hundreds or even thousands of networks reduced transactions costs and brought a flood of free information to subscribers. As the Internet became virtually ubiquitous, new uses for it developed, such as the handling of commercial transactions, searching for recent news stories, looking for weather predictions in distant locations, or finding alternative entertainment options.

Broadband simply extends the scope of the Internet's bandwagon effects because it permits new and potentially more attractive uses of the Internet to be developed. As more subscribers sign up for high-speed access, more and more Internet content or applications will develop, leading still others to subscribe. As this chapter is being written, only about 25 percent of households in OECD countries have broadband access. With penetration this low, the returns to developing new content are attenuated. In addition, the owners of content that is currently exploited commercially through other media, such as motion pictures, audio recordings, and books, may be reluctant to try to adapt their products for broadband distribution because of the absence of intellectual property protection in many countries.

5.4. Consumer value after the bandwagon

Crandall and Jackson (2003) have attempted to estimate the value of broadband to consumers in a hypothetical future world in which broadband services become virtually ubiquitous among households, much as basic telephone service has become. They conclude that broadband could confer as much as \$300 billion per year in consumers' surplus on U.S. households. Their analysis is based on the likely price elasticity of demand for broadband when and if U.S. broadband subscriptions are as ubiquitous as ordinary telephone service (i.e., when broadband subscriptions spread to 94 percent of U.S. households). As Rappoport et al. have shown, the price elasticity of demand for broadband has fallen as subscriptions have risen. Crandall and Jackson (2003) assume a linear demand curve and an own-price elasticity of -0.12 at 94 percent penetration, which is substantially greater in absolute terms than the elasticity of demand for ordinary telephone service. If, however, the demand curve is loglinear with a 'choke point', the estimates are likely to be substantially smaller (Crandall et al., 2002). Crandall and Jackson also support their estimate by adding up the potential consumer-welfare gains from broadband service applications that are likely to exist once a large share of households subscribe to high-speed services.

6. Regulation and competition

In its first few years, broadband deployment has been driven by incumbent telephone companies and cable television companies. Neither of these two participants' networks was initially able to deliver high-speed, two way Internet connections without substantial additional investment in network capacity and electronics. This process has inevitably been a slow one; indeed, many countries even lack a cable television infrastructure through which cable modem service could be delivered. In others, regulatory disputes have slowed the deployment of network facilities or at least clouded the prospects for successful exploitation of such investment. Nevertheless, *platform* competition between DSL and cable modem service remains a very important component of most countries' strategies for obtaining rapid diffusion of broadband.

Because broadband services evolved from a regulated network industry—telecommunications—that was being liberalized in most advanced countries, much of its early development has occurred in an environment of contentious regulatory disputes, particularly in the United States and Europe. These disputes generally involve two major issues:

- The need for incumbent telecommunications carriers to allow entrants to use their facilities in offering competitive broadband services; and
- Whether carriers, particularly cable television companies, should be required to provide 'open access' to Internet service providers or other suppliers of broadband content.

6.1. Platform competition

Telecommunications networks exhibit obvious economies of density, scale, and scope. Entrants into the provision of broadband services may find it difficult to build their own networks to compete with incumbents, particularly if the incumbent is a 'natural monopoly.' For this reason, there is a strong argument that entrants should not be required to engage in wasteful duplication of network assets to compete with incumbents in the delivery of traditional 'narrowband' voice-data services. Incumbents might therefore be required to share the last-mile copper loop network with entrants if the loop is a bottleneck 'essential facility' because there are no other economically viable technological options for reaching the subscriber.

In the case of broadband, however, the copper loop is not the only means of reaching the consumer. Indeed, as shown above, there are other technologies available, and one of these alternatives—cable television—appears to enjoy a cost advantage over copper loop telephone systems. In addition, in many jurisdictions the copper loop plant requires substantial additional investment to allow it to deliver a DSL service to large numbers of subscribers. In areas of low-population density, this investment may involve the extension of fiber optics far out from the telephone company's central office and the location of the electronics (DSLAMs) in remote terminals. If the incumbent still has to build this fiber, in the form of 'digital loop carrier,' it is not an essential facility in the usual sense. Anyone could build it. Indeed, Faulhaber and Hogendorn (2000) have shown that competition among new fiber-coaxial cable networks is possible in the delivery of broadband services.

The importance of platform competition has been noted in numerous studies of broadband and its regulation. For instance, in its Eighth Report on implementation of telecommunications reform in Europe, The European Commission (2002) noted that:

> Cable modem access had notable success in the Netherlands, Belgium, and the United Kingdom, while it is also quite strong in Spain, Austria, and France. Cable network operators are faced with a series of regulatory and financial obstacles concerning the upgrading of their networks, which means that, outside of the above countries, they are not in a position to offer serious competition for the moment, nor to develop their broadband facilities at a pace sufficient to keep up with the speed of development of competing DSL providers.

For these reasons, the EU has stressed the need for local loop unbundling for competitive DSL providers to supplement platform competition.

The OECD (2001) has also stressed the importance of platform competition in its analysis of the adoption of broadband technology in the 30 OECD member countries. The OECD report found that "[o]ne of the key ingredients in why some countries are forging ahead, is whether there is competition between different networks and networks with different technologies," and that "[t]here is a significant correlation between the growth of cable modems and DSL services."

An early study of broadband deployment in the United States by Gabel and Kwan (2000) used a logit regression to estimate the determinants of the availability of DSL and cable modems in February 2000 across 286 wire centers. They find that household income, the age of the head of household, population density, and the age of the housing stock have an effect on the availability of high-speed service (DSL and/or cable modem service). Similar but less pronounced results are also obtained for the availability of DSL service alone. They test for the effects of asymmetric regulation of Regional Bell Operating Companies (RBOCs) by simply inserting dummy variables for the RBOCs and find no significant effect. However, their test does not capture the effects of differential regulation of RBOC broadband services across states and is therefore not very powerful.

More recently, Burnstein and Aron (2002) have studied the effect of platform competition on broadband penetration in the United States. Using state data for 46 states in 2000, they estimate a 'reduced form' logit model of broadband penetration that includes both demand and supply variables, such as household income, average length of the telephone local loop, the number of telephone lines per square mile, the wholesale price of an unbundled local loop, household Internet penetration, dummy variables for each RBOC, and a measure of the availability of cable modem service and DSL service in the state. They find that the length of loop, the density of telephone lines, education, and Internet penetration are all strongly related to broadband penetration. More importantly, they find that the presence of both cable modem service and DSL service is associated with greatly increased household subscription to broadband services. Their result suggests that in 2002, when about 8 percent of households subscribed to broadband, the presence of both cable modem service and DSL was associated with an increased penetration of 6.5 percentage points. However, because this is a single-equation analysis, there remains the question of whether the supply of service by both networks responds to greater demand or whether greater penetration is the result of increased competition.

6.2. Interconnection and network unbundling

Despite the fact that the copper loop may not be an essential facility for the delivery of broadband, because there are competing platforms, there is still an argument for requiring incumbent telephone companies to unbundle the last-mile copper loop. First, because of 'universal service' regulation, telephone companies are often allowed to charge very high rates on high-speed business lines to defray the losses on underpriced residential lines. These companies may therefore be reluctant to deploy DSL services for fear that such services will compete with their own high-speed business services. Second, given decades of regulation, incumbent telephone companies may not be sufficiently entrepreneurial to develop new services, such as DSL. Allowing entrants to use their copper loops at wholesale prices may therefore accelerate the deployment of the new service. Finally, cable television may not exist in many jurisdictions, often because of past government policy. Without the external competition of cable modem service and while fixed wireless or satellite services are still in their infancy, it is sometimes argued that allowing entrants access to the incumbents' copper loops may be the most effective for stimulating the deployment of DSL.

6.2.1. Broadband unbundling in the United States

The United States led the way among developed countries in opening its telecommunications markets to competition, capping its efforts with the 1996 Telecommunications Act. This statute embraced a concept of 'unbundling' network facilities that are needed by new entrants to compete and offering them to entrants at cost-based rates, but it did so without the specific use of the phrase 'essential facilities'[17].

The U.S. Federal Communications Commission was embroiled in controversy over the required extent of unbundling of incumbents' network facilities for broadband DSL services for more than eight years.

[17] The United States was not the first country to mandate unbundling. Hong Kong required its incumbent telephone company to unbundle copper loops in 1985 despite the fact that Hong Kong had the highest density of telephone lines was major political jurisdiction in the world.

Table 4
High-speed subscriber lines in the United States (lines offering 200 Kbps or more in at least one direction)

Technology	12/31/99	12/31/00	12/31/01	12/31/02	12/31/03	12/31/2004
ADSL total	369,792	1,977,101	3,947,808	6,471,716	9,509,442	13,817,280
Incumbents		*1,814,776*	*3,839,666*	*6,156,854*	*9,030,574*	*13,220,203*
Entrants		*162,225*	*108,142*	*314,862*	*478,868*	*597,077*
Other wireline	609,909	1,021,291	1,078,597	1,216,208	1,305,070	1,486,566
Coaxial cable	1,411,977	3,582,874	7,059,598	11,369,087	16,446,322	21,357,400
Fiber	312,204	376,203	494,199	548,471	602,197	697,779
Satellite or fixed wireless	50,404	112,405	212,610	276,067	367,118	541,621
Total	2,754,286	7,069,874	12,792,812	19,881,549	28,230,149	37,810,646

Source: U.S. Federal Communications Commission (2005).

Initially, the FCC ruled that the electronics equipment required to deliver broadband services, the DSLAMs and ATM switches, did not have to be unbundled and provided to entrants at cost-based rates[18]. However, the Commission required that the copper loop be unbundled for both narrowband and broadband uses. Because some entrants wanted to offer DSL alone, they asked the Commission to require that incumbents *share* the loop with them, a request that the FCC eventually granted.

Because much of the U.S. has low-population density, incumbents have deployed substantial amounts of digital loop carrier and need to continue to extend fiber farther out from many central offices. This, in turn, has led to controversy over the effectiveness of the FCC's unbundling and line sharing mandates in promoting competition. Entrants have not wanted to deploy their own electronic equipment in hundreds or even thousands of remote terminals, asking instead that regulators require the incumbents offer them a wholesale product that extends from the central office to the subscriber at regulated, cost-based rates. These demands were being arbitrated in various states when the FCC ruled in February 2003, in response to an appellate court remand of its rules, that line sharing would no longer be required and that the electronics portion of the incumbent's broadband network need not be unbundled[19].

Despite its very aggressive unbundling and line sharing policy before 1993, the FCC was not successful in stimulating sustainable entry from independent providers of DSL services. As Table 4 shows, competition for the incumbents' DSL broadband services is provided by cable television companies, who enjoy a lead in

[18] I do not attempt to cite to each of the FCC decisions or the court appeals that followed. The interested reader may find the entire history extensively recorded at http://www.fcc.gov.

[19] See 'FCC Adopts New Rules for Network Unbundling Obligations of Incumbent Local Phone Carriers,' February 24, 2003, available at http://www.fcc.gov/wcb/cpd/.

network development and subscriber enrollments, but not from independent suppliers of DSL services. Most entrants who have tried to offer DSL by leasing a portion of the incumbents' lines have failed and have abandoned the business. This undoubtedly explains Burnstein and Aron's (2002) result that overall broadband penetration is not related to the regulated wholesale price of an unbundled loop in the 46 U.S. states that they studied.

Several studies have criticized the effects of the pre-2003 U.S. broadband unbundling policy. Hausman (2002a) has been particularly critical of U.S. unbundling policies. Utilizing Hausman and Sidak's (1999) consumer-welfare test for determining whether network facilities should be unbundled, he concludes that there is no compelling case for requiring incumbent telephone companies in the United States to supply network facilities to entrants at regulated rates because the incumbents do not have monopoly power in the broadband market. In addition, entrants can feasibly construct their own broadband facilities in many urban areas. He concludes that such regulation has impeded the incumbents' deployment of the network facilities required for DSL, conveying market power on the cable operators who control two-thirds of the U.S. broadband market.

Faulhaber (2002a) is less critical of unbundling in general, but he agrees that network unbundling should not be required for new facilities because of the effect of such a policy on investment incentives.

Bittlingmayer and Hazlett (2002) show that proposed U.S. legislation to curb wholesale unbundling for DSL services is associated with a significant increase in the value of the common equities of facilities-based competitive carriers, but has no statistically significant effect on incumbent telephone companies' stock prices or the stock prices of competitive carriers using unbundled elements. The latter result may reflect the fact that, as Table 4 shows, entrants have simply not been able to succeed in offering DSL through unbundled elements. Hazlett (2002) also suggests that the mere threat of regulation could explain why U.S. cable television operators allocate only one 6 MHz channel of their 750 MHZ of capacity to broadband Internet services despite problems with consumer congestion on this single channel[20]. Were they to devote a second or third 6 MHZ channel to this service, regulators might opportunistically seek to impose 'common carrier' regulation on such services according to Hazlett.

Others, such as Glassman and Lehr (2002), claim that any reduction of network unbundling for broadband places downward pressure on the competitive carriers' equity prices, thereby reducing investment by entrants in network facilities. However, the failure of most of these entrants even before the FCC's 2003 decision ending line sharing and the unbundling of the electronics required for broadband delivery suggests that these companies were not viable anyway.

[20] It is likely, however, that many cable companies already must allocate more than one channel for their own cable modem service as usage increases and congestion becomes a problem.

6.2.2. Wholesale regulation and unbundling in Europe

The European Commission did not require national authorities to mandate unbundling of network facilities until 1 January 2001, and even then it only required the unbundling of the local loop. Unlike the United States, the EU did not mandate unbundling to promote narrowband voice/data services competition but rather to accelerate the deployment of broadband services. Germany had begun local loop unbundling (LLU) in 1998 when the EU first began to liberalize telecommunications, but a large of share of unbundled lines were devoted to the competitive supply of ISDN, not broadband.

Local loop unbundling was not a notable success in the EU in its first 2 years. Very few DSL lines were supplied by entrants over unbundled loops until recently. As of early 2004, most of the DSL competitors in Europe were still simply reselling the incumbents' services, such as 'bitstream access,' which they acquire, at regulated wholesale rates. This resale accounted for 20 percent of DSL lines at the end of March 2004 while unbundled local loops and line sharing accounted for less than 10 percent of DSL lines and only about 7 percent of all broadband lines. As in the United States, competition in European broadband was largely between the incumbent telephone companies (or their ISPs) and cable companies. In countries with very weak cable television systems, such as France, Germany, and Italy, such competition was not very strong and broadband subscriptions languished in the range of 10–14 percent of telephone access lines. By contrast, broadband subscriptions in countries with well-developed cable television systems, such as Belgium, Denmark, and the Netherlands, subscriptions were equal to between 26 and 33 percent of access lines.

Kosmides (2002) suggests that the EU decision to require network unbundling was driven by the European Commission's view that the slow start to broadband in the United States was due to 'incentive problems' among incumbent telephone companies. But her analysis concludes that platform competition has worked to stimulate broadband in Europe while local loop unbundling has been a failure thus far. She favors a technology-neutral system of broadband regulation for Europe perhaps supplemented with regional or state subsidies for infrastructure or broadband applications.

One puzzle that emerges from the data in Table 5 is the United Kingdom's relatively low rate of broadband penetration despite the country's extensive cable television network and emphasis on platform competition. The former U.K. regulator, OFTEL, favored a facilities-based competition policy for nearly two decades, and it created an environment that was conducive to the growth of cable as an alternative to the incumbent telephone company, British Telecom[21]. This policy was a success in inducing competition for local telephone service. The two major cable companies have more local telephone subscribers than cable

[21] For a summary of OFTEL Policy, see OFTEL (2003).

Table 5
Broadband competition in Europe as of 31 March 2004 (subscribers)

Country	Incumbent retail DSL lines	Entrant DSL lines resale	Entrant DSL lines using LLU or line sharing	Cable modem	Other technologies	Total broadband subscribers (as share of access lines)
Austria	227,700	64,100	27,000	350,000	0	668,800 (0.22)
Belgium	718,605	143,207	7064	482,400	0	1,351,276 (0.26)
Denmark	401,350	47,589	78,561	252,528	41,753	821,781 (0.33)
Finland	235,000	18,600	83,000	81,000	3200	420,800 (0.13)
France	2,122,300	1,413,300	477,000	416,800	0	4,429,400 (0.14)
Germany	4,400,000	0	550,015	65,000	54,000	5,069,015 (0.10)
Greece	6757	7620	779	0	1416	16,572 (0.00)
Ireland	30,780	7220	1380	4900	1350	45,630 (0.03)
Italy	2,081,000	484,000	290,092	0	266,000	3,121,092 (0.13)
Luxembourg	12,180	795	350	2030	220	15,575 (0.05)
Netherlands	950,464	2160	281,824	1,035,000	0	2,269,448 (0.30)
Portugal	206,109	29,897	2341	327,583	0	565,930 (0.14)
Spain	1,473,601	373,312	23,303	404,473	0	2,275,089 (0.13)
Sweden	417,000	144,200	72,842	205,000	190,000	1,029,042 (0.19)
United Kingdom	1,035,000	1,270,000	10,500	1,491,000	8500	3,815,000 (0.13)
Total EU	14,317,846	4,006,400	1,906,051	5,117,714	566,439	25,914,450 (0.14)

Source: ECTA (2004).

television subscribers, and BT has lost one-third of the local calling market to cable companies and other rivals (European Commission, 2002). Why, then, did cable not move aggressively to exploit broadband Internet connections? One possible explanation is that the policy of licensing large numbers of cable companies throughout the U.K. denied them the economies of scale and scope enjoyed by U.S. cable companies, but the U.S. cable industry began in much the same way with thousands of locally-franchised companies. Therefore, the failure of platform competition to develop in the U.K. broadband market through 2004 remains a mystery.

6.2.3. Success without unbundling—Korea

Korea provides a fascinating case study in the deployment of broadband because it has moved so far ahead of all other countries in broadband penetration without any formal wholesale regulation of the incumbent telecommunications carrier, Korea Telecom. As Figure 3 shows, Korea has about 55 percent more broadband subscribers per 100 persons than Canada, which is in second place, more than twice as many as the U.S., and about three times as many as the European Union. Surely, this stunning level of adoption of a new technology cannot be explained on the basis of household disposable income or even the relative price of broadband. Korea has a lower average household income than Japan, North America, or the European Union, and its broadband prices are no lower than those of Japan.

Korea has very densely populated residential areas with large, relatively new high-rise buildings to which the government has directed the construction of an advanced fiber optic network by the electric utility company, Korea Electric Power Corporation, which is not in the business of supplying retail broadband services (see Hausman, 2002a). With this network in place, competitive suppliers of broadband services can gain ready access to transmission facilities. Thrunet, Hanaro Telecom, and a variety of other carriers compete aggressively with Korea Telecom, the incumbent telephone company. Because so many of the residential facilities are new high-rise apartments, these buildings have been apparently constructed in a manner that permits easy access to rival carriers. Therefore, competition in Korea does not depend on mandated unbundling or the provision of wholesale broadband services by the incumbent carrier. Despite the absence of such regulation, DSL commands nearly two-thirds of this very large and growing broadband market.

Among the reasons for the popularity of broadband in Korea are the high speeds available and the considerable interest in 'network gaming,' OECD's (2001) comparison of DSL across 30 OECD countries showed that Korea's DSL had the highest speed, up to 8.0 Mbps compared with less than 1.0 Mbps in all EU and North American countries. Survey data for the same period show that 54 percent of Korean households used the Internet for networked games in a survey month compared to only 6 percent in the United States (Lee, 2002). Korea is far ahead of the U.S. in broadband penetration despite the fact that the U.S.

leads Korea in Internet domains per capita, e-commerce, and secure servers. As Lee (2002) explains, Koreans use the Internet much more for content than for e-commerce.

Despite its success in achieving a huge early lead in broadband penetration, Korea implemented local loop unbundling in 2003. It appears that its motivation in pursuing this regulatory strategy is to increase competition in narrowband services, not in broadband.

6.2.4. Unbundling in Japan

If Korea provides the best example of the benefits of regulatory forbearance, perhaps Japan provides the best counterexample. Local loop unbundling was introduced in Japan in September 2000 and the wholesale rate for a loop was set in December 2000 (Fuke, 2002). At that time, there were fewer than 10,000 DSL subscribers and 625,000 cable modems in Japan. After local loop unbundling was authorized, Softbank established Yahoo! Japan, 'Yahoo! BB,' to begin offering ADSL service in September 2001. Yahoo has established a very low-monthly rate for its service, roughly US$ 20 per month, for a service with speeds up to 8.0 Mbps and launched a vigorous promotional program that included a free modem. The result was an enormous surge in DSL subscriptions, which totaled nearly 4 million by August 2002, a 10-fold increase in just 1 year. At the same time, cable modem subscriptions increased to 1.8 million (Tsuji, 2002).

The recent surge of broadband in Japan is clearly due to one company's aggressive pricing and promotion policies in offering a service that relies entirely on NTT's local loops. Unfortunately, Yahoo! BB is incurring huge losses that may not be sustainable in the long run. It hopes to grow rapidly and then deploy a variety of new services, such as video on demand that it can exploit over its large subscriber base. In essence, it is attempting to create its own 'bandwagon' and to exploit the benefits from it through much higher revenues per subscriber than it currently obtains. This may prove to be a great success that can be attributed to loop unbundling. On the other hand, it may be just one more failure to add to the impressive list of failed new telecom enterprises in the U.S. and the EU, particularly those relying on incumbents' facilities.

6.2.5. Hong Kong

Hong Kong was the first to pass political jurisdiction mandating local loop unbundling. It began to require unbundling for narrowband services in 1995 despite its very dense population. Hong Kong has an average of 15,000 lines per square mile, surely an attractive environment for facilities-based competition. Hong Kong also has a well-developed cable television company, but despite the availability of platform competition, it decided in 2000 that it would require PCCW, the incumbent telephone company, to unbundle and/or share its local loops with broadband competitors (Telecom Authority of Hong Kong, 2000). In

2003, this decision was implemented. A major competitor and lessee of PCCW's loops is the cable television company, which has apparently decided not to invest in telephony and advanced services over its own facilities. Hong Kong's experience would appear to support the theory that network unbundling creates disincentives for network investment. Apparently the Hong Kong regulator agreed with this analysis, for in 2004 it announced that it would phase out network unbundling.

6.3. Cable open access

The regulation of incumbent broadband service companies also may include a requirement that Internet service providers (ISPs) or suppliers of content be granted access to the broadband network at regulated rates or even for a zero price. Such proposals often emanate from those who attribute the success of the Internet to the 'end to end' architecture that places much of the intelligence in the network at the top of a layered system. Under this Internet model, the lower levels of the network infrastructure—the broadband 'pipes' provided by telephone companies or cable companies—would be designed to be flexible and simple, but open to all potential suppliers of content (Lemley and Lessig, 2000). To end-to-end advocates, the benefits of 'open access' far exceed the internalization of network effects through vertical integration.

In the United States, telephone companies have provided open access to competing ISPs and content providers because they are subject to common-carrier regulation of their services, including broadband services. Cable companies, on the other hand, are not regulated common carriers under U.S. law, but are subject to state and local franchising regulation as well as FCC video regulation. As a result, cable companies began to offer cable modem service through their own ISPs, At Home and Road Runner. At first, this vertical integration was uncontroversial; however, once cable modem service began to grow rapidly, the ISPs who thrived under the common carriage dial-up regime felt threatened. Eventually, they persuaded state and local regulators to begin to require that cable systems provide open access to rival ISPs. After a series of court battles, the open access issue was sent to the Federal Communications Commission to resolve, and it decided not to require such access[22].

Ultimately, the open access 'end-to-end' issue may be distilled into one which the benefits of vertical integration must be traded off against the dangers of monopoly leveraging. Because U.S. cable companies account for nearly 60 percent of broadband subscriptions, some economists fear that the cable companies might be able to leverage their monopoly power into downstream content markets. This was the essence of the AOL-Time Warner merger policy deliberations described by Faulhaber (2002b and 2003) above.

[22] Details may be found at the FCC Website, http://www.fcc.gov.

If the cable companies obtain sufficient market power in downstream content, they may be able to discriminate against rival distribution systems (i.e., deny these systems access to their content), and thereby foreclose entry into or even monopolize the broadband service itself. Rubinfeld and Singer (2001) describe this as 'content discrimination.' Alternatively, owners of distribution services with downstream content investments may be able to discriminate against rival content (i.e., deny rival suppliers of content access to their distribution system), thereby discouraging entry into content. Rubinfeld and Singer call this 'conduit discrimination.' The cable open access issue in the United States involves the latter form of discrimination.

Rubinfeld and Singer show that incentive to engage in content discrimination is likely to decline as a cable company's footprint (i.e., the share of the nation that its cable systems cover), grows because the gain from content discrimination is primarily from increased sales outside its distribution areas. On the other hand, the likelihood of conduit discrimination increases with the size of the cable system's footprint. Given the national concentration levels in cable television distribution when they were writing, Rubinfeld and Singer find that such discrimination would be unlikely. Therefore, the case for mandated open access would appear to be weak.

The other side of the open-access debate is the effect of vertical integration in internalizing network effects (Crandall et al., 2002; Hazlett, 2002). Without the ability to internalize the network externalities inherent in this network industry, much as motion picture companies, television networks, and cable companies did in earlier examples of the deployment of new consumer services, broadband may grow much less rapidly. Owen and Rosston (1998) argue that imposing open-access on U.S. cable firms would reduce investment incentives required for the deployment of the cable infrastructure necessary for the delivery of broadband services, particularly in the early stages of broadband development. Noll (2002) summarizes the trade-offs involved but does not reach a definitive policy conclusion on the issue. He shows that the benefits from mandating equal access are directly related to cable's market power and the likelihood of a 'lock-in' effect on consumers who buy the bundled cable modem/ISP service, but that these benefits are offset by the probability that cable companies will delay broadband deployment if regulated.

Bittlingmayer and Hazlett (2002) have analyzed the effect of open access proposals on U.S. carriers' common equities. They find that events that reflect setbacks to proposals for open access are associated with increases in the value of Internet stocks. They surmise that the financial markets see open access as so delaying broadband infrastructure deployment that it reduces the prospects for independent ISPs even if they do not enjoy the benefits of equal access to cable platforms.

7. Subsidies, universal service, and the 'digital divide'

The development of any new technology that is likely to be valuable to consumers, and that is expensive when first purchased by early adopters, inevitably creates concerns for consumers who cannot or will not purchase it. This has been especially true for the personal computer, the Internet, and broadband. Concerns over the development of a societal 'digital divide' have been expressed quite widely and given particular stress by the U.S. Department of Commerce (1999, 2000, 2002). The initial concern was that the cost of personal computers and Internet service, coupled with low-educational attainment, could create a divide between wealthier citizens and poorer citizens, including minorities in the United States. Similar concerns were expressed much earlier about Australia by Madden et al. (1996). More recently, the stress in the United States has been on an urban–rural divide in broadband subscriptions, created in part by low-deployment rates in rural areas.

Given the likely network externalities from broadband subscriptions and use, it is possible to make an economic case for subsidizing populations or areas with low-broadband penetration. But subsidy programs inevitably suffer from two problems: they confer substantial benefits on households already subscribing, and they enjoy political support long after the need for them has diminished or expired. For this reason, Goolsbee (2002) suggests that a subsidy program target the *fixed* costs of network deployment in areas in which broadband networks have not already been built. Unfortunately, he can find only a few areas where such a subsidy would be beneficial on net. His ideal subsidy program for the United States would total only $14 million and generate benefits of $210 million in net present value. This is not a program that has captured the imagination of U.S. politicians because of its limited dimensions.

8. Conclusions

It is far too early to draw any definitive conclusions about the economics of broadband Internet service. With only a decade of experience with this new medium to draw upon and the inevitable information and publishing lags, we do not yet have a robust literature on any of the important issues or phenomena relating to broadband.

Because broadband has been developed and deployed by firms in a heavily regulated sector of the economy that is now being liberalized, much of the literature that has developed has grown out of the parochial battles for regulatory favor in North America, Europe, and Asia. Some of this literature is cited in this chapter because, frankly, there is nothing better available at present.

The rapid changes in communications technologies also create problems for those studying this new medium. For example, econometric analyses of the demand for 600 kbps DSL services in the United States may neither tell us much about the demand for 20 or 100 megabits per second services provided over fiber optics in the future, nor will the demand for broadband given the content available currently be a good predictor of demand when video streaming of motion pictures or other content becomes routinely available.

Similarly, the analysis of the effects of the current market structure on the rate of deployment of new facilities may become irrelevant once broadband becomes ubiquitous. Similarly, if wireless, satellite, and wire-based systems compete aggressively for subscribers, the current concerns over potential vertical market foreclosure evaporate.

At this early stage of the development of broadband, there is surprisingly little attention given to the obstacles to developing the content that is necessary to keep the 'bandwagon' effects flowing. Are digital copyright laws too broad or are they unenforceable across the global domain of the Internet? Does the failure of the AOL-Time Warner merger signify that vertical integration will not solve the coordination problem and internalize the network externalities inherent in broadband deployment? These and many other questions are not answered by the literature now available.

References

Bittlingmayer, G. and T. W. Hazlett, 2002, The financial effects of broadband regulation, in Robert W. Crandall and James H. Alleman, eds., Broadband: Should We Regulate High-Speed Internet Access?, Washington, DC: The Brookings Institution.

Burnstein, D. E. and D. J. Aron, 2003, Broadband adoption in the United States: An empirical analysis, in Allan L. Shampine, ed., Down to the Wire: Studies in the Diffusion and Regulation of Telecommunications Technologies, Haupaugge, NY: Nova Science Press.

Carnegie Mellon University and 3 Rivers Connect, 2002, Digital Rivers Final Report, April 11. available at *http://www.3rc.org/downloads/Digitalrivers.pdf.*

Chipty, T., 2001, Vertical integration, market foreclosure, and consumer welfare in the cable television industry, American Economic Review, 91, 428–453, June.

Church, J. and N. Gandal, 2000, Systems competition, vertical merger and foreclosure, Journal of Economics and Management Strategy, 9, 25–51.

Computer Science and Telecommunications Board, 2002, National Research Council, Broadband: Bringing Home the Bits, National Academy Press.

Crandall, R. W., 1972, FCC regulation, monopsony, and network television program costs, Bell Journal of Economics and Management Science, 3, 483–508.

Crandall, R. W., 1975, The postwar performance of the motion picture industry, The Antitrust Bulletin, Spring.

Crandall, R. W., 2001, The failure of structural remedies in sherman act monopolization cases, Oregon Law Review, Spring, 109–198.

Crandall, R. W. and L. Waverman, 2000, Who pays for universal service?, When Telephone Subsidies Become Transparent, Washington, DC: The Brookings Institution.

Crandall, R. W., J. Gregory Sidak and H. J. Singer, 2002, The empirical case against the regulation of broadband access, Berkeley Technology Law Journal, 17, 953–987.

Crandall, R. W., R. W. Hahn and T. J. Tardiff, 2002, The benefits of broadband and the effect of regulation, in R. W. Crandall and J. H. Alleman, eds., Broadband: Should We Regulate High-Speed Internet Access?, Washington, DC: The Brookings Institution.

Crandall, R. W. and C. L. Jackson, 2003, The $500 billion opportunity: The potential economic benefit of widespread diffusion of broadband Internet access, in A. L. Shampine, ed., Down to the Wire: Studies in the Diffusion and Regulation of Telecommunications Technologies, Haupaugge, NY: Nova Science Press.

European Commission, 2002, Telecommunications Regulatory Package, Eighth Implementation Report, October.

European Competitive Telecommunications Association, 2003, ECTA DSL Scorecard, Accessed at *http://www.ectaportal.com/html/index.php.*

Faulhaber, G. R., 2002a,Broadband deployment: Is policy in the way?, in Robert W. Crandall and James H. Alleman, eds., Broadband: Should We Regulate High-Speed Internet Access?, Washington, DC: The Brookings Institution.

Faulhaber, G. R., 2002b, Network effects and merger analysis: Instant messaging and the AOL-Time Warner case, Telecommunications Policy, 26, June–July, 311–333.

Faulhaber, G. R., 2003, Access and network effects in a 'New Economy' merger: AOL-Time Warner, in J. Kwoka and L. White, eds., The Antitrust Revolution, Oxford University Press.

Faulhaber, G. R. and C. Hogendorn, 2000, The market structure of broadband telecommunications, Journal of Industrial Economics, XLVIII(3), 305–329, September.

Fisher, F. M., 1985, The financial interest and syndication rules in network television: Regulatory fantasy and reality, in F. M. Fisher, ed., Antitrust and Regulation: Essays in Memory of John J. McGowan, 263–298.

Fryxell, D., 2002, Broadband local access networks: An economic and public policy analysis of cable modems and DSL, Ph.D. Dissertation, Department of Engineering and Public Policy, Pittsburgh, PA: Carnegie Mellon University.

Fuke, H., 2002, Facilities Based Competition and Broadband Access, Paper presented at International Conference on Convergence in Communications Industries, Warwick Business School, November.

Gabel, D. and F. Kwan, 2000, Accessibility of Broadband Telecommunications Services by Various Segments of the American Population, Paper Prepared for the Telecommunications Policy Research Conference, August.

Gillett, S. and W. Lehr, 1999, Availability of Broadband Internet Access: Empirical Evidence, 27th Annual Telecommunications Policy Research Conference, September 25–27.

Goolsbee, A., 2002, Subsidies, the value of broadband, and the importance of fixed cost, in R. W. Crandall and James H. Alleman, eds., Broadband: Should We Regulate High-Speed Internet Access?, Washington, DC: The Brookings Institution.

Hausman, J. A., 2002a, Internet related services: The results of asymmetric regulation, in R. W. Crandall and J. H. Alleman, eds., Broadband: Should We Regulate High-Speed Internet Access?, Washington, DC: The Brookings Institution.

Hausman, J. A., 2002b, From 2G to 3G: Wireless competition for Internet-related services, in Robert W. Crandall and James H. Alleman, eds., Broadband: Should We Regulate High-Speed Internet access?, Washington, DC: The Brookings Institution.

Hausman, J. A., J. G. Sidak and H. J. Singer, 2001, Residential demand for broadband telecommunications access to unaffiliated Internet content providers, Yale Journal on Regulation, 18.

Hausman, J. A., J. G. Sidak and H. J. Singer, 2001a, Cable modems and DSL: Broadband Internet access for residential customers, American Economic Review, Vol. 91(2), 302–307, May.

Hausman, J. A. and J. G. Sidak, 1999, A consumer welfare approach to the mandatory unbundling of telecommunications networks, Yale Law Journal, 109, 417–505.

Hazlett, T. W., 2002, Regulation and vertical integration in broadband access supply, in Robert W. Crandall and James H. Alleman, eds., Broadband: Should We Regulate High-Speed Internet Access?, Washington, DC: The Brookings Institution.

Hopkins, M., 2001, IP Local Loop: Key Facts, London: Analysys Research.

Howell, B., 2002, Broadband Uptake and Infrastructure Regulation: Evidence from the OECD Countries, New Zealand Institute for the Study of Competition and Regulation, Working Paper BH02/01, February, available at *http://www.iscr.org.nz/navigation/research.html.*

Jackson, C., 2002, Wired high-speed access, in R. W. Crandall and J. H. Alleman, eds., Broadband: Should We Regulate High-Speed Internet Access?, Washington, DC: The Brookings Institution.

Katz, M. and C. Shapiro, 1985, Network externalities, competition, and compatibility, American Economic Review, 75, 424–440, June.

Katz, M. and C. Shapiro, 1994, System competition and network effects, Journal of Economic Perspectives, 8, spring.

Kosmides, M. S., 2002, Deployment of broadband infrastructure in the EU: Is state intervention necessary? Paper presented at the 2002 Telecommunications Policy Research Conference, September.

Lee, Choongok, 2002, Competitive advantage of the broadband Internet: A case study between the South Korea and the United States, M.A. Thesis, Department of Mass Communication, University of Florida.

Lehr, Williamand and L. W. McKnight, 2003, Wireless Internet Access: 3G vs. WiFi? MIT Research Program on Internet and Telecoms Convergence.

Lemley, Mark, A. and L. Lessig, 2000, The End of End-to-End: Preserving the Architecture of the Internet in the Broadband Era, John M. Olin Program in Law and Economics, Stanford University.

Liebowitz, S. J. and S. E. Margolis, 1994, Network externality: An uncommon tragedy, Journal of Economic Perspectives, 8, 133–50, Spring.

Liebowitz, S. J. and S. E. Margolis, 2001, Network effects, in G. Martin Cave, Sumit K. Majumdar and Ingo Vogelsang, eds., Handbook of Telecommunications Economics, Volume 1, Amsterdam: Elsevier Science, BV.

Madden, G. and S. J. Savage, 1996, A probit model of household broadband service subscription intentions: A regional analysis, Information Economics and Policy, 8, April, 249–67.

Madden, G., S. J. Savage, G. Coble-Neal and P. Bloxham, 2000, Advanced communications policy and adoption in rural western Australia, Telecommunications Policy, 24, May.

Madden, G., S. J. Savage and M. Simpson, 1996, Information inequality and broadband network access: An analysis of Australian Household Survey Data, Industrial and Corporate Change, Volume 5, 1049–1066.

Mueller, M. L., 1997, Universal Service, Interconnection and Monopoly in the Making of the American Telephone System, MIT Press and AEI Press.

Noll, R., 2002, Resolving Policy Chaos in High-Speed Internet Access, Stanford Institute for Economic Policy Research.

OFTEL, 2003, OFTEL's Internet and Broadband Brief, London.

Organization for Economic Cooperation and Development, 2001, The Development of Broadband Access in OECD Countries, Paris, October.

Owen, B. M., 1999, The Internet Challenge to Television, Cambridge: Harvard University Press.

Owen, B. M., 2002, Broadband mysteries, in Robert W. Crandall and James H. Alleman, eds., Broadband: Should We Regulate High-Speed Internet Access?, Washington, DC: The Brookings Institution.

Owen, B. M. and G. L. Rosston, Cable modems, access and investment incentives. Report prepared for National Cable Television Association.

Rappoport, P. N., D. J. Kridel and L. D. Taylor, 2001b, An econometric model of the demand for access to the Internet by cable modem, Forecasting the Internet, in D. G. Loomis and L. D. Taylor, eds., Understanding the Explosive Growth of Data Communications, Kluwer Academic Publishers.

Rappoport, P. N., D. J. Kridel, L. D. Taylor and K. Duffy-Deno, 2003, Residential demand for access to the Internet, in G. Madden, ed., Handbook of Telecommunications Economics, Volume 2, Elgar.

Rappoport, P. N., D. J. Kridel and L. D. Taylor, 2002, The demand for broadband: Access, content, and the value of time, in Robert W. Crandall and J. H. Alleman, eds., Broadband: Should We Regulate High-Speed Internet Access?, Washington, DC: The Brookings Institution.

Rappoport, P. N., D. J. Kridel and L. D. Taylor, 2003, Willingness-to-pay and the demand for broadband service, in Allan L. Shampine, ed., Down to the Wire: Studies in the Diffusion and Regulation of Telecommunications Technologies, Haupaugge, NY: Nova Science Press.

Rohfls, J., 2003, Bandwagon Effects in High Technology Industries, Cambridge, MA: MIT Press.

Rubinfeld, D. and H. J. Singer, 2001, Vertical foreclosure in broadband access, The Journal of Industrial Economics, 69, 298–318, September.

Spulber, D. F. and C. S. Yoo, 2003, Access to networks: Economic and constitutional connections, The Cornell Law Review, 88, May.

Telecommunications Authority of Hong Kong, 2000, Broadband Interconnection, November.

Tsuji, M., 2002, Network Competition in the Japanese Broadband Infrastructure: From the Last Mile to the Last Quarter Mile, Paper presented at International Conference on Convergence in Communications Industries, Warwick Business School, November.

U.S. Department of Commerce, National Telecommunications & Information Administration, 1999, Falling through the Net: Defining the Digital Divide, July.

U.S. Department of Commerce, National Telecommunications & Information Administration, 2000, Falling through the Net: Toward Digital Inclusion, October.

U.S. Department of Commerce, National Telecommunications & Information Administration, 2002, A Nation Online: How Americans Are Expanding Their Use of the Internet, February.

U.S. Federal Communications Commission, 2000, The Availability of High-Speed and Advanced Telecommunications Services, August 3.

U.S. Federal Communications Commission, 2005, High-Speed Services for Internet Access: Status as of December 31, 2004, July.

Varian, H. R., 2005, The demand for bandwidth: Evidence from the INDEX Project, in R. W. Crandall and J. H. Alleman, eds., Broadband: Should We Regulate High-Speed Internet Access?, Washington, DC: The Brookings Institution.

Wanichkorn, K. and S. Marvin, 2002, The Role of Wireless Access Networks in the Deployment of Broadband Services and Competition in Local Telecommunications Market, Paper presented at TPRC 2002, Alexandria, VA, September, 28–29.

Woroch, G. A., 1998, Facilities competition and local network investment, Industrial and Corporate Change, Volume 7, 601–614.

Chapter 6

CABLE TELEVISION

THOMAS W. HAZLETT*

George Mason University

Contents

* Professor of Law & Economics, George Mason University. The author served as Chief Economist of the Federal Communications Commission (1991–1992). He wishes to thank Roberto Muñoz and Bruno Viani for their outstanding research assistance.

Handbook of Telecommunications Economics, Volume 2, Edited by S. Majumdar et al.

DOI: 10.1016/S1569-4054(05)02006-3

1. Spectrum in a tube

MIT's Nicolas Negroponte augured some years ago that, while we were born into a world in which we received our telephone calls over wires and our television programs over-the-air, we would live to see a world just reversed. The prediction came to be famously known as the 'Negroponte Switch' (Negroponte, 1996, p. 24). It is now well along, as some nations experience near-universal cable subscription. Alas, the decline of traditional broadcast TV has also been boosted by the emergence of direct broadcast satellite (DBS) systems—wireless, in defiance of Negroponte's technical parameters, but similarly transformative.

While this overview of the cable television industry will focus on the United States market, a broader perspective helps to introduce essential aspects of the analysis. Across OECD member states, cable penetration rates (the percent of households subscribing to cable TV) range from 1 percent to 96 percent. Cable television systems are now providing 'last-mile' broadband links, and in some countries (e.g., the United States and Canada) cable modem users outnumber DSL (the primary broadband alternative) households (Table 1). Cable telephony is now gaining ground as well, inverting the Negroponte Switch by delivering plain old telephone service via fiber optics and coaxial cable.

The historical development of the cable television industry is interesting in its economic and political dimensions. Take the alternative paths traveled in Canada and Italy. In the former, regulators boast: "Today, Canada is the most cabled country in the world, with a penetration rate of roughly 80 percent"[1]. Italian policy makers, ironically, embrace just the opposite result: "The Italian radio-television market is typified by the absence of cable TV, which was compensated with the total liberalization of the radio-television sector in 1976"[2].

In fact, open entry into broadcasting gave Italy, by 1985, the densest concentration of TV stations in the developed world (Noam, 1992, p. 88). Demand for cable TV service was thusly undercut by abundant over-the-air video programming[3]. In Canada, meanwhile, regulatory policy tightly constrained broadcast TV, both with respect to the number of channels and with respect to 'domestic content'[4].

[1] Michael Koch 1998–1999. The claim is slightly overstated. See Table 1.

[2] Autorità per le Garanzie Nelle Comunicazioni (AGCOM), The regulation of the media in Italy, 31 August 2001, www.agcom.it/eng/resp_reg.htm.

[3] Ironically, the open entry policy was first adopted in cable, where an unlicensed cable TV system won a 1973 court case establishing the right to provide cable service without a franchise. Similar rights were soon asserted, and won, by unlicensed entrants into over-the-air television broadcast markets. Ibid.

[4] The definition of 'domestic content' television programming is curious. Under Canadian rules, a TV show shot in a U.S. city and appearing to be an 'American program' to American audiences qualifies as Canadian programming if certain members of the cast or production crew are Canadian citizens. See Stanbury, 1998.

Table 1
Cable TV and cable modem penetration in OECD countries

Country	HH passed by cable TV (%) (1999)	HH subscribing to cable TV (%) (1999)	HH passed by cable modem (%) (2000)	Cable modem subscribers per 100 inhabitants (2000)
Australia	34	16	34	0.34
Austria	53	30	n.a.	1.22
Belgium	100	96	n.a.	1.00
Canada	93	69	60	3.01
Czech Republic	n.a.	24	n.a.	0.10
Denmark	70	56	30	0.70
Finland	63	42	n.a.	0.29
France	32	12	n.a.	0.21
Germany	86	53	n.a.	0.08
Greece	0	0	n.a.	0.00
Hungary	66	48	n.a.	0.03
Iceland	n.a.	n.a.	n.a.	0.00
Ireland	50	49	n.a.	0.00
Italy	5	<1	n.a.	0.00
Japan	n.a.	17	n.a.	0.49
Korea	48	12	n.a.	3.32
Luxemburg	100	90	n.a.	0.00
Mexico	32	10	n.a.	0.02
Netherlands	94	89	n.a.	1.58
New Zealand	n.a.	n.a.	n.a.	0.02
Norway	47	40	n.a.	0.34
Poland	n.a.	31	n.a.	0.00
Portugal	47	16	n.a.	0.25
Slovak Republic	n.a.	n.a.	n.a.	0.00
Spain	8	3	n.a.	0.03
Sweden	65	50	n.a.	0.71
Switzerland	72	n.a.	n.a.	0.38
Turkey	n.a.	7	n.a.	0.00
United Kingdom	51	13	n.a.	0.03
United States	96	67	59	1.36

Source: OECD, 2001.

When Canadian cable TV operators won the right to carry the major American TV networks (CBS, NBC, and ABC), cable penetration rose rapidly.

Much of the history of cable TV is embedded in this intermodal rivalry. In allocating licenses for over-the-air (OTA) TV, governments have typically been parsimonious. Cable television has been a way to bypass the regulatory bottleneck, bringing additional programming to viewers by creating new bandwidth, or 'spectrum in a tube'. Yet, the same political forces that fueled restrictions in broadcasting have impeded the development of cable television. In the United

States, the history of the sector can be seen as a Long March from the wilderness to, ultimately, economic success. Today, cable TV operators and cable TV programmers dominate the video marketplace while operating under relatively benign regulatory constraints.

1.1. Emergence of cable television

In its 1952 TV Station Allocation Table, the US allowed for hundreds of local stations but effectively precluded the financial viability of more than three national broadcasting networks[5]. Cable television operators had already begun wiring outlying areas where broadcast reception was poor and could be improved via an antenna service with long wires[6].

Federal regulators took little interest in cable's initial build-outs, which simply extended the reach of broadcast signals to additional viewers. While the case for regulating cable as a common carrier was made twice by FCC staff in the 1950s, both proposals were voted down by the Commission (Powe, 1987, Chapter 12). The regulatory equilibrium changed dramatically in the early 1960s, as cable operators launched their first competitive forays by bringing alternative programming to households that could receive OTA television (Huber et al., 1999, p. 1161).

Firms began cabling larger markets where distant programming could be obtained via microwave or long-distance lines and retransmitted to subscribers. For instance, Cox Cable offered households in San Diego, California the broadcasts of Los Angeles stations located 120 miles to the north. Viewers proved eager for additional program choices, especially those featuring popular sporting events.

In a series of decisions between 1962 and 1966, regulations to limit the advance of cable operators were enacted to preempt the possibility that they would 'siphon' viewers from over-the-air television. "It is claimed that audience splintering through CATV is especially harmful to UHF stations, which for technical and set conversion reasons reach smaller audiences than VHF stations..." (Federal Communications Commission, 1965, p. 690). This violated the regulatory purpose of cable: "CATV serves the public interest when it provides program choices not locally available off the air and acts as a supplement rather than a substitute for off-the-air television service..." (Federal Communications Commission, 1966,

[5] Four networks were actually operating in 1952, with programs distributed via the 108 stations licensed prior to a national plan being adopted. The Dumont network argued that the FCC licensing scheme would prevent it from reaching audiences sufficient to attain critical mass. The prediction proved correct; the network ceased operations in 1955. A fourth network would not again be seen until the formation of Fox in 1986.

[6] The initial 'cable TV' systems simply consisted of a shared antenna. Indeed, the industry was called 'community antenna television' (CATV). The first U.S. system was either in Mahoney City, Pennsylvania (1948) or Astoria, Oregon (1949), (Crandall & Furchtgott-Roth, 1996, p. 1).

p. 735). The market structure was altered by cable competition, threatening the system of cross-subsidies[7].

Citing concerns that "the burgeoning CATV industry would have a future adverse impact on television broadcast service" (Ibid., p. 736), the Commission implemented a series of complex requirements. "The pay-TV or 'antisiphoning' rules adopted in the 1960s seem laughable today in their baldly anticonsumer, protectionist tone. Cable operators could not offer pay-TV (per-channel or per-program) service that included (a) sports programming that had been on free television within the past 4 years, (b) series programs, or (c) movies more than two or less than 6 years old. The idea was to keep cable operators from competing. The rationale was to protect viewers' supposed interest in having their free programs (with commercials) saved from siphoning to pay-TV..." (Owen, 1999, pp. 117–118).

These rules thwarted cable industry expansion. By 1974, however, a thaw was in evidence (Besen, 1974), and a 'deregulation wave' 1976–1980 melted most of the anticable regulations[8]. The race to wire America was on. As seen in Table 2, fewer than 11 million homes subscribed to cable in 1976; in 1987, nearly 45 million did, more than one-half of television households. As of year-end 2003, cable subscribership stood at about 66 million[9]. This was down from a 2001 peak of about 72 million subscribers, a trend attributed to the emergence of satellite TV. By November 2003, some 24.38 million households subscribed to noncable multichannel video services (primarily satellite)[10]. When combined with other subscription services, such as DBS, over 85 percent of US households were multichannel video subscribers (Federal Communications Commission, 2002c, p. 75).

1.1.1. UHF and cable television

An intriguing aspect of the early problems encountered by cable TV is that the government's rationale for limiting industry expansion was the specific threat it posed to ultra-high frequency (UHF) television broadcasting. Then allocated channels 14–83, UHF TV was an important part of the FCC's plan for broadcasting.

[7] "If the New York independents sought translators [microwave links] to place their signals over the Philadelphia area, it could not seriously be argued that we should grant such applications on the ground that...events in the market place should be allowed to give the answer" (Federal Communications Commission, 1966, p. 776).

[8] Changing financial interests undoubtedly played a role in the regulation/deregulation cycle. See the discussion in Besen (1974); Besen and Crandall (1981); Crandall and Furchtgott-Roth (1996). See, generally, Peltzman, (1989).

[9] The National Cable Telecommunications Association counted 73.6 million subscribers in July 2002, over six million beyond the FCC's total shown in Table 2. See: http://www.ncta.com/industry_overview/indStat.cfm?indOverviewID = 2 (visited January 20, 2003). FCC subscriber counts have conflicted both internally and with external sources. They are displayed in Table 2 with this caveat.

[10] National Cable & Telecommunications Association, http://www.ncta.com/Docs/PageContent.cfm?pageID = 86 (visited April 15, 2004).

Table 2
Cable TV in the United States, 1976–2003

Year	U.S. households (million)	Cable TV subscribers (million)	Cable TV subs/HHs (%)	Availability, HP/HH (%)	Total industry revenue (million $)
1976	72.9	10.79	15.1	31.69	932
1977	74.1	12.17	16.6	32.66	1200
1978	76.0	13.39	17.9	35.26	1476
1979	77.3	14.81	19.4	37.90	1875
1980	80.8	17.67	22.6	43.19	2549
1981	82.4	23.22	28.3	50.73	3656
1982	83.5	29.34	35.0	59.28	4984
1983	83.9	34.11	40.5	66.63	6425
1984	85.3	37.29	43.7	70.93	7774
1985	86.8	39.87	46.2	74.54	8938
1986	88.5	42.24	48.1	78.42	10144
1987	89.5	44.97	50.5	81.68	11765
1988	91.1	48.64	53.8	84.74	13595
1989	92.8	52.56	57.1	89.22	15678
1990	93.3	54.87	59.0	92.18	17855
1991	94.3	55.79	60.6	93.74	19463
1992	95.7	57.21	61.5	93.73	21044
1993	96.4	58.83	62.5	93.98	22755
1994	97.1	60.49	63.4	94.34	23010
1995	99.0	62.96	65.7	93.65	25558
1996	99.6	64.65	66.7	94.05	27950
1997	101.0	65.93	67.3	93.65	30388
1998	102.5	67.01	67.4	93.24	32162
1999	103.9	68.54	68.0	93.00	34642
2000	104.7	69.30	67.8	94.36	37967
2001	104.9	72.96	69.2	97.24	43994
2002	106.7	66.10	61.9	96.30	46862
2003	106.7	65.90	61.8	97.00	51295

Sources: Hazlett and Spitzer 1997, pp. 115–116; FCC, 2001, p. B-7; FCC, 2001, p. 87; FCC, 2002c, pp. 11, 15; FCC, 2004, Table 4, and App. B; NCTA, Cable Television Developments Fall, 1994; NCTA, Industry Statistics: Television and Cable Households 1979–2001 (3 February 2003). http://www.ncta.com/industry_overview/indStats.cfm?statID=1. United States Bureau of the Census, Household by type: 1940 to present, http://www.census.gov/population/www/socdemo/hh-fam.html

The TV station allocation table had resulted in just three national networks even as it 'consumed' the very-high frequency (VHF) TV band, channels 2–13. To offer some competition to this triopoly, the FCC focused on UHF stations. With the 1962 All-Channel Receiver Act, which mandated that TV sets sold in the United States after 1 January 1964 include UHF tuning capability, the Commission determined that,

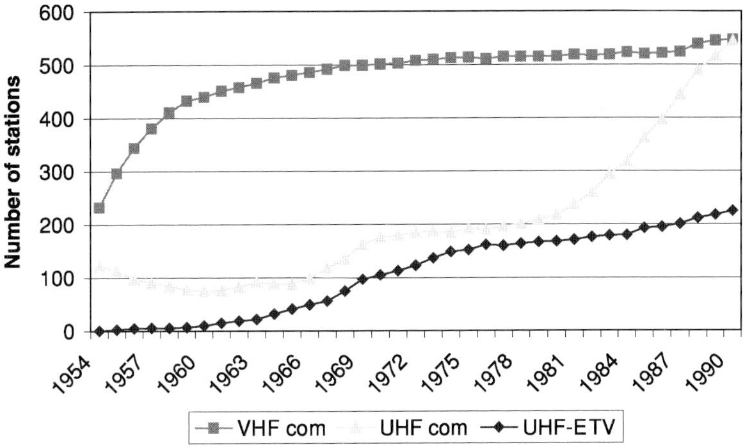

Fig. 1. U.S. television stations, 1954–1991. Note: com = commercial TV station; ETV = public (educational) TV station.

"Congress and the American public have staked a great deal on the development of UHF" (Federal Communications Commission, 1966, p. 771)[11].

But that investment was threatened by cable operators. "Both CATV and UHF broadcasting... are entering the larger markets, most often in an effort to bring programming that is not now available to these markets.... The most critical question is how these two trends mesh in the ensuing years" (Ibid., p. 772). Regulators decided that UHF and CATV were substitutes rather than complements, and "the problem posed is whether, if CATV succeeds greatly—for example, to the 50–85-percent figure predicted by its optimistic proponents—there is correspondingly a grave danger to UHF broadcasting" (Federal Communications Commission, 1966, p. 775). The commission decided to preempt this threat by imposing substantial entry barriers[12].

The historical record allows us to evaluate the FCC's strategy. The combined policy approach, blocking cable TV deployment and mandating all-channel TV receivers, was associated with roughly a doubling of commercial UHF stations 1964–1970 (Figure 1)[13]. Douglas Webbink examined these data and found that the marginal impact on UHF viewing did not justify the outlays consumers were required to make under the all-channel receiver law (Webbink, 1969). Moreover,

[11] For details of television policy during this period, see Noll et al., 1973 and Levin, 1980.

[12] In concluding that cable deployment was deleterious to UHF broadcasting, the Commission noted its reliance on a paper written by Franklin Fisher, then an associate professor of economics at MIT. The "Fisher report" was submitted by the National Association of Broadcasters (Federal Communications Commission, 1965, pp. 691–93; Fisher and Ferrall, 1966).

[13] Noncommercial stations rely on government funding and private donations, and are not subject to the same competitive constraints as commercial stations.

UHF station growth was virtually flat throughout the following decade, just when the effect of the mandated UHF tuners was presumably marked[14]. Then, from 1978 to 1990, the commercial UHF station total rose from about 200 to well over 500. This increase followed federal cable deregulation, coinciding with the period in which cable TV subscribership was growing most rapidly.

Regression analysis reveals that, while UHF station growth did not increase around the time of the all-channel receiver act, the number of commercial UHF stations was highly and positively correlated (throughout the period) with cable television subscribership (Hazlett, 2001). This relationship holds true for neither commercial VHF nor public UHF stations, suggesting that cable carriage resulted in improved financial viability for fledgling, ad-supported UHF TV broadcasters. This result was predicted by Rolla Park, whose 1972 study examined the FCC's assumption that cable was "a potential threat to UHF stations"[15]. Park concluded that cable deployment "will help, not hurt, UHF independents through the 1970s"[16]. By 1979, the FCC concurred[17].

1.2. *Cable television's rise to dominance*

What is known as the 'deregulation of cable television' can be categorized in two sets of reforms. The first involved a series of federal decisions to eliminate rules blocking cable to favor broadcasters. In 1976, the FCC dropped its 'antileapfrogging rules,' which had prevented cable systems from carrying any but the closest broadcast TV stations, and in 1977 relaxed technical requirements (for instance, dropping the requirement that cable systems use 9-m dishes to bring in satellite signals, permitting a 4.5-m diameter), lowering costs. Also in 1977 a federal court struck down regulations on premium networks offering recent movies for a per-channel fee[18]. The FCC largely deregulated cable television programming in September 1980, after embracing competition between cable and broadcast TV stations in an April 1979 report[19].

The second round of 'deregulation' occurred in the Cable Communications Policy Act of 1984. The legislation ended most rate controls, preempting local governments. This is discussed in Section 3.

[14] Consumers were purchasing about nine million TV sets per year when the mandate was imposed. See FCC, 1966, p. 771.

[15] Park, 1972, p. 208. Park noted that "FCC policy restricting cable growth ... is to shield these stations from the (supposed) harmful impact of cable and to maintain a climate favorable to the establishment of new UHF stations."

[16] Ibid. 'Independents' are commercial TV stations not affiliated with national networks, a category specifically identified by the FCC as at-risk.

[17] 71 FCC 2d 632.

[18] Home Box Office vs. FCC, 567 F. 2d 9 (D.C. Circ.), cert. den. 434 U.S. 829 (1977).

[19] 71 FCC 2d 632. For a summary of the key regulatory and deregulatory events between 1965 and 1987, see Fournier and Campbell, 1993, appendix.

Economic trends both shaped the changing policy landscape and were reinforced by it. In particular, satellite distribution of video programming was made possible with the 'open skies' policy adopted in 1972 (Owen, 1999, p. 146). Home Box Office had begun as a cable-only movie service in 1972, but its cable-only programs were 'bicycled' from system to system, and distribution was limited. With its launch via satellite in 1975 it became cheaply accessible to cable systems nationwide. Economies of scale were important to creating cable-only TV programs, and the rising quantity and quality of such shows increased demand for subscriptions to cable—positive feedback in network development.

In 1976, with satellite distribution and the end of antileapfrogging rules, Atlanta station WTBS began beaming its signal to cable systems nationwide. Home Box Office (HBO), the first national premium cable network, was joined by competitor Showtime in 1976. Cable systems were no longer just community antenna service for broadcasting, and new programming networks formed as operators sought to augment demand for subscriptions.

The most important early entrants were ESPN (1979), Nickelodeon (1979), C-SPAN (1979), and CNN (1980). Each tapped into a niche program market: ESPN in sports, Nickelodeon in children's shows, CNN with around-the-clock news, 'Cable Satellite Public Affairs Network' (C-SPAN) with live sessions of Congress, political events, and public policy forums[20].

Basic cable networks are included in the customer's monthly fee, along with retransmission of local TV signals. Cable networks charge cable operators for carriage (license fees) and sell commercial time (advertising spots), achieving a dual revenue stream. Premium networks, such as HBO and Showtime do not sell ad time, but charge considerably higher license fees and are typically sold to viewers on a per-channel basis[21]. Compared to the mass audience programming produced by broadcasting networks, cable menus have come to feature a diverse array of programs, including many less expensive, and some more expensive, programs.

By the 1970s, 12-channel cable systems were giving way to systems with 35 channels; by the late 1970s, 64 channel capacity was state of the art. This abundant capacity was both driving, and driven by, the development of new program networks. Systems were able to serve much more diverse viewing interests and, in delivering a package of channels for a fee, financially motivated to offer a broad range of narrowly-targeted services. The fact that cable services are sold as a package by an operator who enjoys monopoly market share (in the multichannel video program distribution market) reinforces this tendency toward product

[20] The network that has produced two clones, C-SPAN2 and C-SPAN3, is a nonprofit venture owned by major cable television operators. It charges cable systems carriage fees but does not accept advertising.

[21] Both cable and satellite TV systems have begun offering premium channels in bundles, or 'premium tiers.'

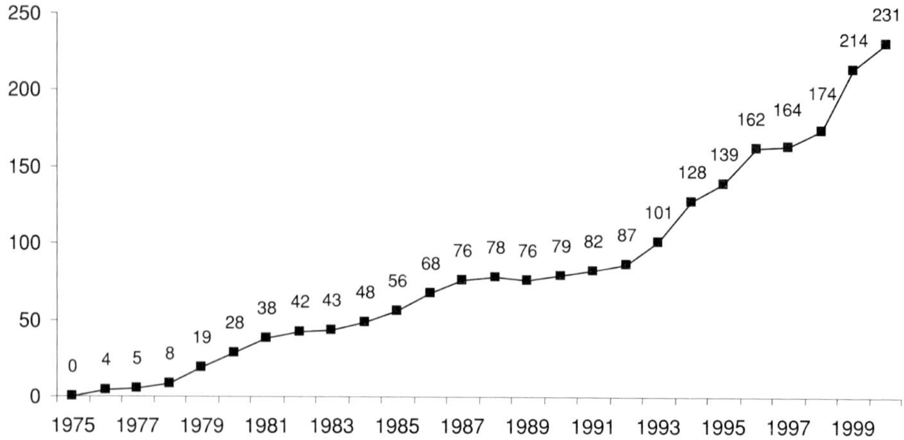

Fig. 2. National Video Cable Program Networks, 1975–2000. Sources: For 1975–1994: Hazlett and Spitzer, 1997, p. 96; for 1995–2000: FCC, 1997, p. B-5; FCC, 1999c, p. B-6; FCC, 2001, p. 89.

diversification. Hotelling (1929), Steiner (1952), Spence and Owen (1977), Wildman and Owen (1985), Hamilton (2004).

Cable subscribership began growing, as viewers were attracted to the new cable networks, and new cable networks formed in response to growing cable audiences, further fueling subscriber growth. A virtuous circle formed. The successful launch of DBS systems in 1994[22] and 1996[23] further stimulated demand for program networks. By 2002, well over 200 national program networks were available (Figure 2). This excludes dozens of regional sports networks and at least 37 state and local news or public affairs channels (Hazlett, 2000, p. 193).

In 1976, fewer than one third of U.S. homes were passed by a cable system, and just one-sixth of U.S. households subscribed (Table 3). With restrictions loosened, however, incentives for both operators and programmers improved. By 1996 well over 90 percent of households were able to subscribe to cable and over two-thirds did. The average subscriber, who was delivered just 21 channels of basic service in 1984 was treated to 46 in 1996, and to over 59 in

[22] Hughes Network Systems, owner of DirecTV, began offering service in 1994 with United States Satellite Broadcasting (USSB). Customers could buy programming from either, or both, firms and receive it with the same (HNS) satellite receiver system. DirecTV acquired USSB in a merger in April 1999. See: http://www.fcc.gov/Speeches/Kennard/Statements/stwek919.html (visited 20 January 2003).

[23] EchoStar began offering DBS service in 1996. It announced plans to acquire DirecTV in October 2001, but the proposal was opposed by both the U.S. Department of Justice Antitrust Division and the Federal Communications Commission. Rather than challenge the regulatory agencies in court, EchoStar abandoned the attempted merger in December 2002. See: EchoStar break-up fee boosts Hughes profit, Lycos News, http://news.lycos.com/news/story.asp?section=Technology& storyId=625022&topic=echostar (visited 20 January 2003).

Table 3
Key averages for the U.S. cable television industry, 1976–2001

Year	Monthly total revenue per subscriber (US$)	Weighted avg. no. of basic channels	Price of most popular basic cable tier (US$/month)	Mean sales price of cable systems per subscriber (US$)
1976	6.58		6.45	
1977	7.94		6.86	
1978	8.66		7.13	
1979	9.89		7.4	
1980	11.06		7.69	
1981	13.25		7.99	
1982	15.10		8.3	922
1983	17.05		8.61	1026
1984	18.94	21	8.98	948
1985	20.30		9.73	1008
1986	21.29		10.67	1339
1987	23.01		12.18	1723
1988	24.79		13.86	2003
1989	26.50		15.21	2291
1990	28.78		16.78	2049
1991	30.37		18.1	1795
1992	31.77		19.08	1765
1993	33.15	38.5	19.39	2160
1994	32.12	39.6	21.62	1869
1995	34.30	40.2	23.07	1829
1996	36.68	46.1	24.41	2155
1997	39.08	47.9	26.48	2044
1998	40.55	50.1	27.81	2875
1999	42.89	53.6	28.92	3995
2000	46.19	56.2	30.37	5755
2001	53.44	59.3	31.58	4872

Sources: Hazlett and Spitzer, 1997, pp. 115–116; FCC, 1998, p. B-8; FCC, 1999c, pp. B-1, B-7, B-9; FCC, 2001, pp. 87, 90, 92; FCC, 2002c, pp. 11, 15–16; NCTA, Industry statistics, average monthly price for expanded basic programming packages: 1983–2001. http://www.ncta.com/industry_overview/indStats.cfm?ststID=27 (visited 15 January 2003); FCC, 2002a, p. 21; FCC, 1999a, p. 23; FCC, 1996, Table 2, paragraph 25. Note for No. of channels: Data is from noncompetitive cable systems in annual FCC surveys. Values weighted by system share of total cable subscribers. Weights for 1993–1995 different than for later years. Average channels in 1984 is from: NCTA Cable TV Handbook 2001, page 1-C-3.

2001. When combined with DBS, multichannel video distribution providers (MVDPs) delivered service to over 85 percent of households and accounted for 2001 revenues of about $53 billion[24]. By comparison, broadcast TV stations

[24] U.S. Census Bureau, Table 3.0.1: Information sector services (NAICS 51)-Estimated revenue for employer and nonemployer firms: 1998 through 2001.

Table 4
All day television viewing shares for broadcast and cable TV, 1983–1998

	1983	1986	1990	1994	1998
Broadcast					
Network affiliates	70	66	55	51	39
Independent	18	18	20	22	20
Public	3	3	3	4	3
Total broadcast	91	87	78	77	62
Cable					
Basic	7	11	21	26	40
Premium	5	5	7	6	7
Total cable	12	16	28	32	47

Source: Kagan, 2000, p. 55.
Notes: Shares aggregate to over 100 due to multiple television sets being viewed simultaneously in some households. Survey universe is 'all TV homes'. Shares are averages for the calendar year. Superstation shares split equally between basic cable and independent TV stations. Fox, UPN, and WP affiliates counted as independent stations.

produced $33 billion in 2001 revenues[25]. Basic and premium networks (in aggregate) surpassed the viewing shares of broadcast television in 2002[26]. In 1983, broadcast TV audiences were *over seven times as large* as the aggregate viewership of basic and pay cable networks (Table 4).

1.3. Basic structure of the cable television industry

The basic structure of the cable television business is pictured in Figure 3. Programmers produce cable networks featuring video content for distribution to subscribers. These networks can be categorized as basic, premium, and pay-per-view (including video-on-demand) services (see Section 4).

Basic channels are bundled together and sold as packages (rather than on a per-channel or per-view basis). The 'limited basic' tier carries broadcast TV signals, PEG channels[27], and a small number of basic cable networks. It is the entry-level service upon which 'expanded basic', pay channels, pay-per-view, and other services are added. Cable systems rarely market the service on a stand-alone basis, and nearly all customers subscribe to upper tiers.

[25] Ibid.
[26] "2002 was the year cable outperformed broadcast, grabbing a higher aggregate share of the audience for the first time ever." Andrew Wallenstein, Cable claims first TV title, The Hollywood Reporter, 18 December 2002, http://story.news.yahoo.com/news?tmpl=story2&cid=91&u=/bpihw/20021218/en_bpihw/cable_claims_first_tv_title&printer =1 (visited 18 December 2002).
[27] The acronym for public access, educational and local government channels routinely provided for in municipal cable franchises.

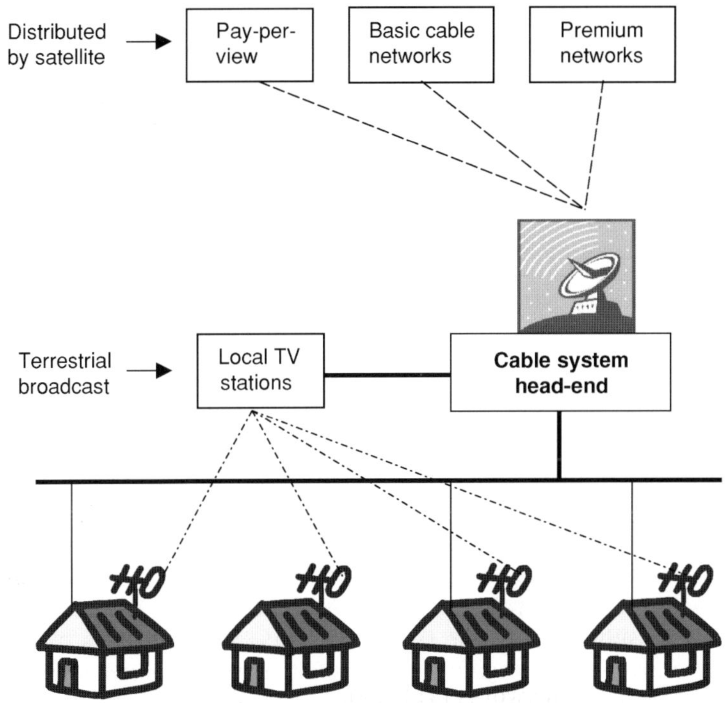

Fig. 3. The basic structure of the cable-television industry.

The structure of the programming menu is at least partly driven by regulation. Local governments have often maintained authority to regulate the lowest-priced tier, and the revenues generated by this tier are used to calculate copyright liability for broadcast TV content procured under a compulsory license. Hence, a limited basic tier creates an accounting advantage for the cable operator, blunting the impact of rate regulations and royalty obligations.

As systems upgrade to digital transmission technology, they offer a large package of higher quality signals to augment analog tiers. In 2001, there were 16.7 million subscribers to digital cable, or 24.4 percent of total subscribers. By 2010 this is estimated to increase to 62.5 million households, or 84.2 percent of subscribers (Kagan, 2002, p. 10). Cable operators see digital cable not only as an incremental revenue stream, but as a defense against DBS offering 150-channel digital tiers.

Local TV station broadcasts can be received by households either via cable TV carriage or (in most cases) rooftop antennae. Cable carriage enhances signal quality and, by integrating cable programming controls, improves viewer accessibility. Prior to the 'must carry' rules enacted in the 1992 Cable Act, cable systems provided 'A/B switch' functionality, which allowed subscribers to click to broadcast signals

Table 5
Services offered by top 10 U.S. cable system operators

Service	Homes passed	Subscribers	Availability (%)	Penetration (%)	Net 2003 additions	Net 2002 additions
Basic cable	105,900,000	58,870,000		55.6	(103,000)	(754,000)
Digital Cable	104,900,000	21,375,000	99	20.4	2,973,000	4,319,000
Broadband	96,750,000	15,338,000	91	15.9	4,486,000	4,217,000
Telephone	16,400,000	2,375,000	15	14.5	156,000	685,000

Source: Leichtman Research Notes (1Q 2004), p. 5.

received by antenna while watching cable. Broadcasters opposed this system in favor of cable carriage, and argued for rules granting them preferential numerical assignments on the cable channel line-up. These were granted in the act[28].

Cable TV systems increasingly use their distribution networks to provide telecommunications and information services as well as video programs[29]. These include telephony and high-speed Internet access (Table 5)[30].

Industry structure is interesting in both its horizontal and vertical dimensions. The horizontal aspect is conceptualized by looking from the cable system to consumers (i.e., subscribers). The evidence is that cable operators typically price above competitive levels, forging a consensus that market power is exercised by cable television operators lacking a direct (wireline) competitor. In this context, however, a merger between a cable system in New Jersey and a cable system in Florida does not alter market concentration, as the local market defines the scope of competitive rivalry.

In the vertical dimension, the situation is somewhat different. Cable operators purchase specialized inputs upstream, including video programs retransmitted to television viewers. These drive demand for subscriptions and, hence, subscriber fees. Audiences of such programming are also 'sold' to advertisers. The existence of horizontal market power is crucial here, as well, although it is measured differently. The cable operator enjoying monopoly power over subscribers in a

[28] The 1992 Cable Act also granted TV stations 'retransmission consent,' giving them the option to forego 'must carry' in favor of licensing arrangements negotiated with cable operators. See discussion in Section 4.

[29] In the United States, regulators define 'cable,' 'telecommunications,' and 'information' services with great care. Cable operators have largely avoided common carrier requirements for cable modem service by gaining classification as an 'information' service. Had the FCC instead labeled cable broadband access a 'telecommunications' service, it would presumably have borne the extensive access obligations levied on local exchange operators. See Speta (2000); George A. Chidi, Jr., FCC declares cable Internet an information service, Network World Fusion, 14 March 2002, http://www.nwfusion.com/news/2002/0314fccis.html (visited 21 January 2003).

[30] The data in Table 5 are from different sources than those in Table 3, and are limited to the top 10 multiple cable system operators (MSOs), excluding smaller systems.

given market also enjoys monopsony power over programmers wishing to reach viewers there. A bilateral negotiation takes place to set license fees and other terms. In this, the extent of the geographic market served by the operator becomes a potentially important factor. Now a merger of the New Jersey and Florida cable systems may alter the balance of power. Monopsony power of cable operators may also serve to offset market power exerted by cable program networks, lowering prices toward competitive levels. Vertical integration may also reduce prices, ending double marginalization.

1.4. Basic structure of a cable television system

Cable systems operators receive satellite signals via earth stations linked to 'head-ends'. With the assistance of a rich assortment of electronic processing equipment, these services are then distributed to households connected, via wires, to the system head-end. Sometimes wireless links are used to connect various hubs in the distribution system, but the end user is connected via coaxial cable[31]. This connection features a relatively generous allotment of bandwidth; standard cable TV architecture now features 750 MHz bandwidth. In contrast, a standard twisted pair copper telephone wire features about 1 MHz of bandwidth.

With a standard U.S. television channel allocated 6 MHz, state of the art cable systems transmit 78 analog channels in the 50–550 MHz band (which includes a gap to avoid interference with FM radio broadcasts at 88–108 MHz). The band stretching from 0 to 5 MHz is little used. Frequencies between 5 and 42 MHz are susceptible to 'ingress noise,' but with techniques that dynamically select utilizable bandwidth, provide upstream (user return path) communications. The adjacent bandwidth (42–50 MHz) is then blocked by filters separating upstream and downstream transmissions (Maxwell, 1999, pp. 180, 250).

This leaves 200 MHz (550–750 MHz) to provide downstream digital cable channels (where five to twenty video programs can be sent via one analog channel) video-on-demand, cable telephony, and cable modem service (Maxwell, 1999, pp. 148, 179–180). Routinely, just 6 MHz of bandwidth (delivering about 30 Mbps) is devoted to the data-intensive downstream path in the latter service (Figure 4)[32].

Traditionally, cable systems were one-way, analog, 'buses'—all the data (or video signals) boarded at the head-end and traveled in the same direction. The way in which customers received customized services (e.g., premium channels) was via 'traps,' physical devices cable operators would attach to cables to block certain

[31] Some operators used fixed wireless technologies to connect end users; these are generally referred to as 'wireless cable' systems. The oxymoron is testament to the degree to which a given technology is viewed as synonymous with the application it delivers. The point is made more striking by the fact that 'wireless cable' was the original distribution technology—also known as over-the-air television broadcasting.

[32] Diagram found in Maxwell (1999, p. 179).

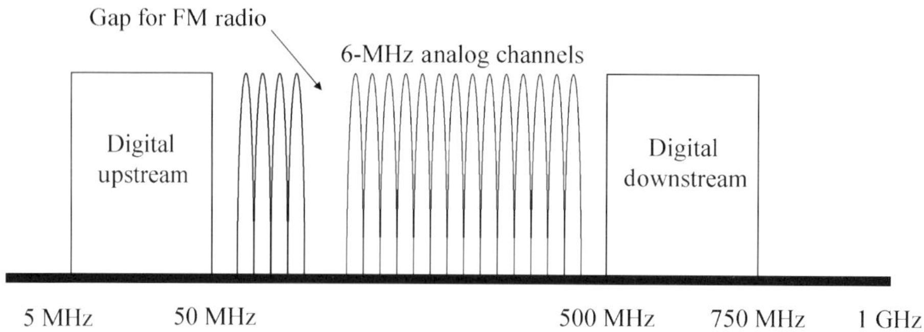

Fig. 4. Spectrum allocation in cable television systems: standard architecture[33].

signals from traveling from the feeder line to the 'drop,' the last cable strand connecting to the customer's premises. Technicians would remove traps for customers who paid for additional services. (Often basic customers would mount telephone poles, or open in-ground cable pedestals, and remove traps themselves, leading to piracy losses.)

Over time, systems have been upgraded to give operators control of services flowing from the head-end (a feature known as addressability). The current generation of cable technology has brought digitization, and a new standard architecture: hybrid fiber-coax (HFC). This upgrades the old one-way systems by replacing copper trunks with fiber optics. These very high capacity lines link to neighborhood nodes of about 500 homes each (in the usual configuration). Optical interfaces (translating data from the fiber transmission to the electronics used by customer premises equipment) are still relatively expensive, so fiber trunks are connected to coaxial cables that provide 'last mile' links to end users. HFC systems provide 'interactive broadband services,' both analog and digital (Maxwell, 1999, p. 151). Subscribers can be sold basic and premium cable, pay-per-view, video-on-demand, voice (telephony), and data (cable modems). The set-top box is increasingly functional in this multiproduct, interactive environment, acting as a digital traffic controller for a bundle of services.

Cable's continuing migration through product space is captured in Table 6. Traditional cable services—subscription fees for basic, premium, equipment, and installation—are projected to decline from over 95 percent of total cable system revenues in 1990 to under 55 percent in 2010.

2. Market power in local cable television service

Regulators became interested in cable when systems threatened to compete with broadcast TV stations. This triggered rules to limit cable deployment and so protect broadcast revenues, important in promoting the 'public interest' obligations,

Table 6
U.S. cable TV operator revenues (U.S. million $)

	1990	1995	2000	2001	2010(e)
Cable operator revenue					
Basic/expanded basic	10,674	16,860	24,729	27,031	41,890
Premium revenue (Pay)	5,105	4,306	5,115	5,617	6,961
Pay-per-view/VOD	253	498	751	993	5,623
Advanced analog/digital	0.0	30	1,088	2,365	11,407
Home Shopping Commissions	96	129	239	260	530
Advertising revenue	476	1,075	2,430	2,430	6,216
Install and Equipment revenue	1,068	1,888	2,451	2,463	2,698
High-speed Internet access, cable telephony, video games	0.0	0.0	1,164	2,835	19,316
Total revenue	17,672	24,786	37,967	43,994	94,641
Percent of total revenue					
Basic/expanded basic	60.4%	68.0%	65.1%	61.4%	44.3%
Premium revenue (Pay)	29.0%	17.4%	13.5%	12.8%	7.4%
Pay-per-view/VOD	1.4%	2.0%	2.0%	2.3%	5.9%
Advanced analog/digital	0.0%	0.1%	2.9%	5.4%	12.1%
Home Shopping Commissions	0.5%	0.5%	0.6%	0.6%	0.6%
Advertising revenue	2.7%	4.4%	6.4%	5.5%	6.6%
Install and Equipment revenue	6.0%	7.6%	6.4%	5.6%	2.9%
High-speed Internet access, cable telephony, video games	0.0%	0.0%	3.1%	6.4%	20.4%

Source: Levy, Ford-Livene, and Levine (2002).

which broadcast licensees assumed[33]. Yet, the FCC also instituted rules to protect cable operators from telephone companies, local utilities, which supplied important cable system inputs: telephone poles.

The average cable system is about 70 percent aerial and 30 percent underground. All else equal, underground construction costs substantially more. The poles owned by telephone companies are available for use on a common carrier basis. But, because telephone companies had some interest in providing cable TV service, and because rate regulated utilities might enjoy economic incentives to be uncooperative with suppliers in adjacent, less regulated markets, the issue of discrimination arose. To preempt telephone companies from using their incumbency to thwart cable entry, the Federal Communications Commission instituted the so-called cable/telco cross-ownership ban in 1970 (Federal Communications Commission, 1988).

[33] Such obligations included coverage of local issues and involvement of local citizens in ownership and management, educational programming for children, and coverage of controversial issues from balanced perspectives. That such obligations were rarely fulfilled did not appear to undermine the arrangement, which included barriers for competitors. See Noll et al. (1973), Levin (1980).

The ban on telephone company ownership of cable systems helps calibrate industry development. Having earlier instituted rules protecting broadcasters from cable, in 1970 the FCC sought to protect cable from telephone companies. In 1987, the FCC reversed course, arguing that the ban had served its purpose, promoting nationwide cable deployment. It then identified cable's new-found market power as the extant policy concern, and recommended elimination of the cross-ownership ban on the grounds that telephone companies were the most likely entrants into local cable TV markets. From infant industry to incumbent monopolist in 17 years[34].

2.1. *Defining cable's market power*

Evidence of the market power cable TV systems has surfaced in market share, pricing, and system valuation data. In addition, entry barriers and various indicia of strategic behavior support the conclusion that market power is often exercised by cable operators.

Perhaps the least conclusive is evidence provided by market shares. Local cable markets have exhibited similar levels of concentration when operators struggled for financial solvency and when they have been capitalized at levels several times the cost of tangible capital. Cable systems in 1982 typically accounted for 100 percent of multichannel video sales in their franchise areas but exercised less market power than in 2003, when cable's average share of multichannel subscriptions had declined to 75 percent (Federal Communications Commission, 1994, p. 3).

Still, it is important that major competitors have made inroads in recent years, particularly DBS. U.S. cable TV operators are now losing video subscribers, year-on-year. Direct broadcast satellite rivalry has prompted cable systems to invest in digital upgrades and to provide abundant new programming options, a response to the broader packages offered by satellite TV. Cable systems are also driven to promote nonvideo services, such as telephony and broadband access—services in which they enjoy distinct advantages over satellite rivals.

It is instructive to observe how video rivalry has cascaded into adjacent service markets. When satellite TV operators began accumulating market share in the mid-1990s, cable TV incumbents—who fearfully referenced the technological entrant as the 'death star'—undertook serious network expansion. The standard architecture had been a 450 MHz one-way analog system; this was quickly displaced by 750 MHz, two-way digital systems. Capital investment was

[34] In 1984, however, Congress codified the cross-ownership ban and would not repeal it until the Telecommunications Act of 1996. In the interim, the Commission created "video dial tone," wherein local telcos would be allowed to build video infrastructure but could not own or select programs. This is discussed below.

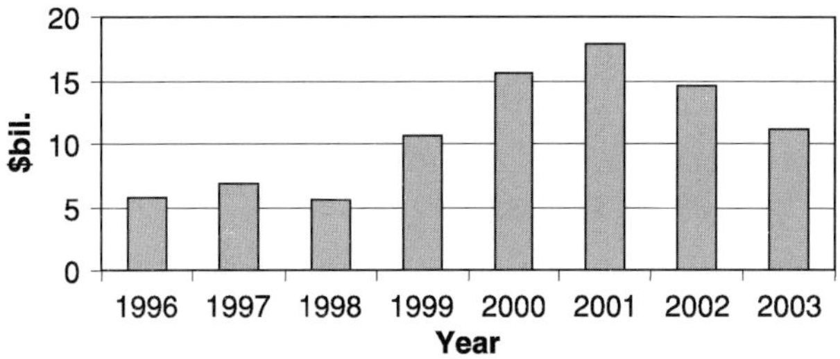

Fig. 5. U.S. cable TV infrastructure expenditures: 1996–2003. Source: NCTA web page (*http://www.ncta.com*) based on Kagan World Media data.

substantial (Figure 5). Industry outlays for fixed assets, 1999–2003, amounted to $70 billion—approximately $1000 per subscriber.

These investments were highly complementary to the provision of broadband Internet access. By 2002, 80 percent of U.S. households were able to subscribe to cable modem service; cable operators have captured over 60 percent of residential broadband subscribers[35]. Local telephone companies, which had been slow to deploy a competing broadband technology, digital subscriber lines (DSL), were forced to play catch-up. The race between broadband providers now encompasses local exchange service rivalry, given the emergence of voice-over-Internet (VoIP) applications.

The market share of cable operators remains high whether the market is defined as 'cable TV' or 'multichannel video.' In 1998, the U.S. Department of Justice (DOJ) blocked Primestar, a DBS company, from acquiring 'satellite assets' owned by MCI and NewsCorp, finding it "would effectively foreclose the use of this scarce and valuable asset to challenge defendants' monopoly power"[36]. Primestar, the third largest satellite TV company, was owned by a group of large MSOs. The DOJ alleged that ownership of Primestar was, since 1990, a way for cable operators to limit competition. By buying satellite assets and marketing only a modestly attractive package of video services, it would make some sales to outlying homes not passed by cable, and would prevent more aggressive rivalry by denying the use of frequencies and orbital slots to others.

When the Primestar purchase was stymied by regulators, the MCI/NewsCorp assets were purchased by EchoStar, and DirecTV bought Primestar. This reduced the number of DBS rivals from three to two, but the Justice Department

[35] "The top cable broadband providers now have a 63% share of the overall market vs. DSL and account for over 15.5 million high-speed Internet subscribers compared to 9.1 million for DSL." Leichtman Research Group, Broadband Internet Grows to 25 Million in the U.S. (March 8, 2004), http://www.leichtmanresearch.com/press/030804 release.html.

[36] United States vs. Primestar et al., Complaint filed by the United States Department of Justice (12 May 1998).

applauded the outcome as pro-competitive. It believed that the surviving (independent) DBS players would offer more protean competition to the dominant incumbent, cable[37].

2.2. Pricing power

Where cable operators compete head-to-head (offering service to identical households), quality-adjusted prices are substantially below levels obtaining when such competition is absent. Hazlett and Spitzer (1997, p. 30) summarize 11 different surveys or studies during the period 1984–1992. All indicate competitive (by which is meant, duopolistic) prices below those of noncompetitive cable systems. Price differentials range from 8 percent to 34 percent.

An important set of estimations is provided by Emmons and Prager (1997), who found that, "it seems reasonable to infer that approximately 20 percent of the price of basic service provided by private monopoly cable operators, in both 1983 and 1989, can be attributed to monopoly rents." (Ibid., p. 746) Not only do the authors adjust for standard cost and demand variables in estimating the effect of market structure on rates, the consistency of their results yields important information. In 1983, local and state governments could regulate basic cable TV rates, and most did. In 1989, these price caps were preempted by the Cable Communications Policy Act of 1984 (freeing basic cable rates as of 29 December 1986). That the monopoly price premium did not change between regimes leads Emmons and Prager to conclude that, "regulation was not very effective in limiting basic cable rates ..." (Ibid., p. 741) (see Section 3).

The Federal Communications Commission has also investigated the issue of monopoly pricing and found that competitive entry matters. In 1994 the Commission estimated a price regression with a dummy for overbuilt systems, finding prices averaging about 16 percent below monopoly systems (Federal Communications Commission, 1994, App. E). Municipally-owned cable system rates were even lower than for private competitive systems, but appear to be of lower quality (Emmons and Prager, 1997) and may be advantaged by public subsidies. Beard et al. (2002) found that the degree of overlap—meaning the percentage of the market actually overbuilt by rivals' cable—has a large bearing on the degree of price competition. The average overbuild involves 58 percent of the subscribers of the overbuilt cable system, and produces an average price reduction of 18 percent. The predicted price reduction in a complete overbuild rises to 42 percent.

Two recent reports from the U.S. General Accounting Office extend these findings, and shed light on the effects of DBS entry. In an October 2003 report, the agency found that cable competition from another wire-based company is

[37] The rejection of the proposed merger of EchoStar (acquiring) DirecTV in 2002 has not changed this analysis. With the DBS market whittled to two rivals, and multichannel at three (two DBS plus one cable), further concentration was opposed by regulators.

associated with a decrease in rates of 15 percent (GAO, 2003, p. 4). This study measures a competitive margin beyond the competitive infusion provided by two DBS operators, using a key variation in DBS service. In approximately the 40 largest television markets (of 210 total), satellite providers offered local TV stations for about $5 per month (incremental to other subscription fees). The GAO found that markets having such local satellite service exhibited DBS subscription levels about 40 percent higher than elsewhere. This was seen to produce a quality adjustment by incumbent cable operators, as basic cable menus included 5 percent more channels than elsewhere (GAO, 2003, p. 3)[38].

The General Accounting Office also releases an analysis of 'overbuild' competition involving head-to-head wireline rivals called 'broadband service providers' (BSPs) (GAO, 2004). The analysis relied on the case study method, selecting six overbuilt markets, which were matched with six similar markets lacking such competition. Rates were found to be 15–41 percent lower in five of the duopoly markets; 3 percent higher in the sixth. The GAO concluded that, "Incumbent cable operators often responded to BSP entry by lowering prices, enhancing the services that they provide, and improving customer service (GAO, 2004, p. 4)."

These empirical findings regarding the magnitude of price reductions associated with enhanced competition are broadly consistent with own-price demand elasticity estimates found in various investigations. In 1994, the FCC examined five such estimates ranging from 1.31 to 3.38, with a mean value of 2.03 (Federal Communications Commission, 1994b, p. H-13). The Lerner Index implies that, for profit-maximization:

$$\frac{(P - \text{MC})}{P} = \frac{1}{0},$$

where P = price, MC = marginal cost, and 0 = elasticity of demand. Assuming that average variable cost is a reasonable proxy for marginal cost, and arbitrarily setting $P = 1$, we can solve for 0. The mean 'cash flow margin' among U.S. cable television operators (revenue minus operating expenses) was 37.4 percent in 2001 (Kagan, 2002, p. 37). This implies average variable cost = 62.4 percent, and elasticity of demand = 2.7[39].

[38] The emerging cable-DBS competition is the subject of two new research papers, either of which finds that DBS entry lowers quality-adjusted cable rates. Austan Goolsbee and Amil Petrin (2004, p. 351) find that "without DBS entry cable prices would be about 15 percent higher and cable quality would fall." Reiffen, Ward and Wiegand (2004, p. 3) focus on the availability of local TV signals via satellite, concluding that "cable prices are about 2% lower in markets in which local programming is available . . . [while] cable companies have increased their quality in response . . ."

[39] Using 1993 cable data for U.S. systems, Beard et al. (2002, p. 19) estimate elasticity = 2.27. This implies a cash flow ratio = 44 percent. In 1993, the average cable industry cash flow ratio was 45.5 percent (Kagan, 1995, p. 37).

Recent survey data from the FCC shed additional light on the magnitude of price reductions and the margins along which competition occurs (see Table 7). Rates are lower, and channel packages are larger. The price of basic service in monopoly systems is 6.6 percent higher (including expanded tiers and equipment charge), and more channels are included on competitive menus. The resulting per-channel monopoly premium is about 19 percent.

Quality competition is seen most vividly in the entry strategy of DBS operators who charge substantially more than cable system incumbents, as seen in Table 8. Satellite operators offer digital signal quality (higher resolution than standard over-the-air analog signals) and a far broader range of programming. Either DBS system features hundreds of channels of capacity. Even given the large allocation

Table 7
Mean U.S. cable TV rates in competitive and noncompetitive systems (Federal Communications Commission survey data, 1 July 2001)

	(a) Noncompetitive ($/month)	(b) Competitive ($/month)	(a) – (b) Difference ($/month)	[(a) – (b)]/(b) Noncompetitive premium (%)
Basic programming and equipment rental ($)	37.13	34.82	2.31	6.64
No. of channels	59.27	63.35	−4.08	−6.44
Rate per channel per month ($)	0.60	0.51	0.09	18.31
No. of satellite channels	44.79	48.19	−3.40	−7.05
Rate per satellite channel per month ($)	0.805	0.68	0.13	18.65

Note: Competitive system sample is weighted average of overbuilds involving local exchange carriers (60.35%) and privately-owned wireline entrants (16.14%). Direct broadcast satellite, municipal, and low penetration 'overbuilds' are excluded. "Basic programming" includes limited and expanded basic. Source: FCC, 2002a, pp. 20–22.

Table 8
Cable vs. DBS snapshot comparison, U.S. 2001 data

	Total subscribers	Video rev/sub/mo.	Projected 2002 subscriber growth rate
Cable	68.6 million	$43.46	0.5%
DBS	19.5 million	$58.19	14.1%

Sources: Kagan, 2000, p. 15; Kagan, 2002, pp. 10–11.

made for delivering local TV stations[40], operators deliver much more diverse content than is available via analog cable. As noted, cable incumbents have responded by upgrading systems to offer digital cable tiers.

2.3. Cable asset valuation

The q ratio measures pricing power as anticipated by investors, where $q =$ market value of assets/replacement cost of tangible capital. In a highly competitive market, it is likely to observe $q \approx 1$ where $q > 1$ obtains it suggests that supra-competitive returns are being capitalized[41]. Whether these constitute monopoly rents or returns to productive economic activity depends on the dynamic process wherein the supracompetitive returns are generated. A high q ratio among drug companies may reflect survivorship bias, for instance.

Cable TV franchises in the United States have exhibited q ratios considerably in excess of unity over a substantial period of time, in the absence of efficiency explanations. Rents are observed industry-wide, and there has not been a pattern of exit which would suggest that monopoly franchises are atypically risky investments[42]. Nor is the magnitude of customer acquisition costs sufficient to explain the valuation/cost ratio. Indeed, the accounting firm Deloitte, Haskins and Sells recommended a pro forma financial structure in 1987 that attributed just 40 percent of system value to physical capital, implying a generic $q = 2.5$ (Hazlett and Spitzer, 1997, p. 24). This implies competition for franchises, but this is characterized as rent seeking (discussed below).

In analyzing the cable industry q, one must consider both asset valuation and capital costs. On the cost side, standard cable architecture was fairly stable in terms of dollars per subscriber outlays through the 1980s and most of the 1990s, with the industry benchmark at about \$600 per subscriber. Smiley (1989) yields an estimate of \$619 per subscriber in capital cost for a prototypical cable build-out, and the Federal Communications Commission (1994, p. I-4) cited 1993 data provided by four publicly listed cable companies to produce a weighted average capital cost per subscriber of \$583[43].

This changed in the late 1990s, however, when cable systems began upgrading from analog to digital technology. This permitted the delivery of new services, including broadband access and digital video channels, and substantially raised per subscriber capital costs. According to the Federal Commnications Commission, (1999b, Chart 2), the new benchmark was '\$800 to \$1000' per subscriber. Figure 6 uses the upper bound, and assumes a discrete transition to the higher

[40] Due to satellite 'must carry' rules, DBS systems must deliver all local TV broadcasts to subscribers in any market where they deliver any local TV signals.

[41] See Lindenberg and Ross (1981), Gilligan and Marshall (1984).

[42] Cable TV bankruptcies have essentially been confined to overbuilders or highly leveraged systems. Market power can, of course, be capitalized as either debt or equity.

[43] This includes "replacement cost per subscriber" and "other tangible assets" per subscriber.

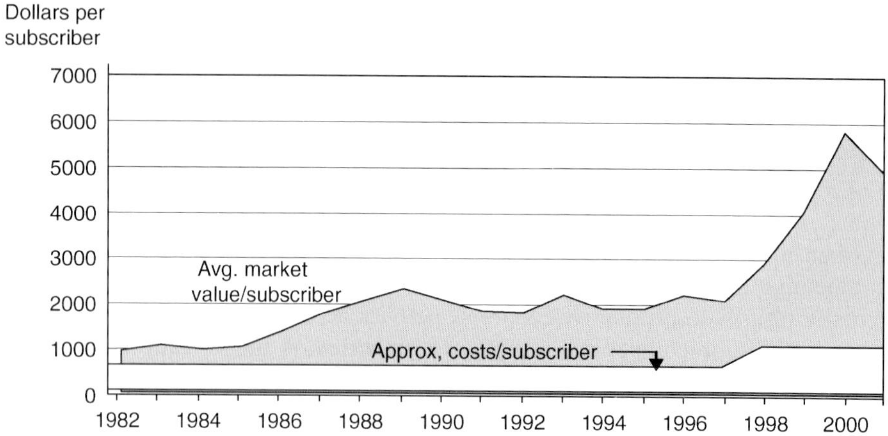

Fig. 6. Economic rents in cable TV systems. Sources: See Table 3 and text.

Table 9
Cable industry *q* ratios, 1985, 1993, and 2001

	Market value per subscriber	Capital replacement cost per subscriber	*q*	Source
1985	$1008	$619	1.6	Text (above)
1993	$2326	$583	4.0	FCC (1994, I-4)
2001	$4872	$1000	4.9	Text (above)

Note: The FCC miscalculated the ratio as 4:5.

replacement cost plateau in 1998, just as system values are increasing, presumably due to the additional network services which cable systems are anticipated to deliver. By 2001, 90 percent of U.S. cable subscribers were served by upgraded systems (Kagan, 2002, p. 10) (Table 9).

As the FCC concluded in 1994, "the current *q* ratios suggest that, overall, cable television operators possess substantial market power" (Federal Communications commission, 1994, p. 101). Despite the emergence of DBS operators who have captured substantial market share of multichannel video subscriptions, the cable industry looks to have substantially maintained its economic position. This is likely due to expanded opportunities to provide new services.

2.4. *Entry barriers*

2.4.1. *Franchising*

In the 'gold rush' to wire major residential markets for cable television in the late 1970s and 1980s, cities issued franchises under their regulatory authority over public rights-of-way. Municipal governments discovered that they could extract substantial rents by awarding licenses on favorable terms to the applicant. In the 1960s, New York Mayor John Lindsay proclaimed cable franchises "urban oil wells beneath our city streets"[44]. This produced a decided bias in favor of monopoly, which would improve expected returns and so raise the 'bid' from prospective applicants.

This suggests the solution to the natural monopoly problem offered by Demsetz (1968), which pointed out that competition *for* the market, could fully substitute for competition *within* the market (Baumol et al., 1982). The theoretical conclusion was that, by moving rivalry to the correct margin, consumers (aided by their bargaining agents) could realize the best of both worlds, economies of scale for cost savings and competitive pricing. In the Demsetz auction, a bargaining agent for consumers invites prospective (monopoly) suppliers to offer a long-term contract. It then selects the lowest price schedule bid. This process reveals the efficient solution and, for good measure, obviates the need for regulation—the upfront auction sets a market price[45].

When actually applied to cable markets, however, complications arose. Williamson (1976) documented the complexity of the cable franchise, the numerous margins on which opportunistic behavior could occur, and the demand (by both authorities and cable company) to change terms over time. Moreover, the pro-consumer agent necessary to arrange a long-term service contract was elusive. Cable franchise proposals were multidimensional; the superior offer was not easy to identify. The optimal system for consumers was unknown and, without investing heavily in the process, voters would not know if their agents had been loyal to their preferences.

Hence, principal-agent problems arose wherein franchisors used the bidding process to arrange cross-subsidies at the expense of lower rates. Zupan (1989, p. 406) surveyed a group of 66 cable systems in Massachusetts and found that franchising agents tended to impose costly requirements on cable TV systems. These amounted to about 26 percent of capital costs and 11 percent of operating costs. "The pursuit of nonprice concessions by franchisors has two negative consequences. First, many of the nonprice concessions appear to provide only limited economic benefits. Second, by raising costs, nonprice concessions lead to higher cable prices" (Ibid., p. 440).

Beutel (1990) discovered, through probit analysis of successful and unsuccessful franchise proposals by prospective cable systems, that firms offering lower prices enjoyed a lower probability of winning the award. The result is consistent with the theory that franchising agents internalize gains by creating, and then directing, monopoly rents. Hazlett (1986) reviewed the terms on which franchising competitions developed, finding that competing bidders often distributed equity shares,

[44] Albin Krebs, Cities reassured on cable-TV rights, New York Times, 6 February 1973, p. 73.

[45] A standard textbook in industrial organization logically offers the cable franchising process as an example of the Demsetz auction solution. See Viscusi et al., 1995.

on highly favorable terms, to influential local residents. The practice was widely known as 'rent-a-citizen.'

The evidence suggests the model of regulatory behavior outlined in Posner (1971). In lieu of imposing direct taxes and spending public funds through standard budgetary appropriations, regulators impose barriers to entry to generate rents. This forms a funding source that can be apportioned through nonbudgetary channels, creating political benefits for franchise regulators. In general, interest groups well informed about the process will be remunerated at the expense of the less informed. While monopoly barriers impose an inefficiently high tax rate, advantages are afforded key policy makers. Even if cities sought to use franchise bidding purely to extract franchise rents for public coffers, the 1984 Cable Act capped municipal franchise fees at 5 percent of gross receipts. The 'price control' created excess demand for monopoly licenses, steering bids to alternative margins.

Cities sometimes invite local competition. Yet, they often refuse to license second entrants, impose costly requirements on new competitors, and even litigate to protect monopoly franchises (Hazlett, 1990; Hazlett and Ford, 2001). Because the franchise competition results in a distribution of rents that includes both the monopoly cable operator and politically influential constituencies, it offers a bulwark against entry.

2.4.2. *Predation*

Cable television markets have also proven fertile hunting grounds in the search for strategic behavior. Bolton et al. (2000, p. 2293) find that a competitive skirmish in the Sacramento, California cable market in the late 1980s "provides a vivid context in which to illustrate application of the strategic approach to financial predation." The authors stress that, by dramatically cutting prices and aggressively remarketing subscribers in the specific neighborhood where an entrant first places its cables, financial performance of the investment can be discredited in capital markets. Where access to outside financing is important, the cost of capital may rise, impeding entry.

In the aforementioned case study, a cable operator passing 400,000 homes succeeded in ending two competitive build-outs. Entry deterrence began in 1983, when the franchise authority denied a request for a competitive license. A lawsuit filed by a prospective entrant was won in 1987, collapsing that barrier. Another cable firm serving an adjacent community was first to compete, extending its wires past about 700 homes. The incumbent responded with free service for just these households. After seven months the entrant was bought out by the incumbent. The free offers stopped.

The second entrant invested five million dollars, building a head-end and beginning cable construction in a 5000-home area. By the time marketing commenced, the incumbent had already blanketed the submarket with additional promotions, including free cable service and even free TV sets for customers who would sign extended service contracts. The established system opened a special customer service office in the overbuilt area, and internally formed a 'S.C. A.T.' task force (as with a S.W.A.T.—special weapons and tactics—team assembled by law enforcement) to attack the competitor.

Executives of the incumbent firm quantified the revenue it stood to lose if the entrant survived, reducing the incumbent's prices—$3.45 per subscriber per month—and the company CEO noted that, "Taking Pac West [the entrant] out of the picture early has considerable value" (Hazlett, 1995, pp. 614–622)[46]. The entrant abandoned its construction project (and considerable sunk costs) to purchase the local 'wireless cable' firm[47]. While an inferior delivery system in virtually all dimensions, wireless had at least one strategic advantage: the incumbent could not pinpoint where, among 400,000 homes, its competitor's next sale would be made.

The episode illustrates some strategic advantages incumbent suppliers may be unable to exploit elsewhere[48]. First, franchise barriers can be used to block entrants without substantial expenditure by the incumbent. Even when possible to surmount, franchise barriers offer a useful long-run defense, providing an early warning system (system architecture and build-out plans must typically be detailed in franchise applications) and offering numerous dimensions in which rivals' costs may be raised, including burdensome regulatory requirements and simple administrative delay[49]. After a costly campaign of predation, an incumbent benefits to the degree that the exit of one entrant is not quickly succeeded with entry by another.

Second, a cable incumbent need not lower prices across an entire market to deter an entrant, only where the rival has physically placed a wire. Indeed, the nature of a cable overbuild requires the entrant to notify the incumbent *prior* to

[46] One internal memo argued for eliminating the rival thusly: "[W]e won't face a revenue hit of 1/3 ($30 mm in 1990!) by being forced to reduce prices. I don't need to tell you what our expense lines will look like if we have to provide 0% call abandonment, instant service response, local service centers, high profile advertising, and on and on." Hazlett, 1995, p. 619.

[47] "Wireless cable" systems use microwave frequencies to deliver multichannel video programming. In the United States, they have used the MMDS band (2.5 GHz), with up to thirty-three 6 MHz channels of capacity. In 2002, there were about 500,000 wireless cable subscribers. Subscribership is actually declining as MMDS systems transition from video services to providing high-speed Internet access. FCC, 2002b, p. 75.

[48] Similar postentry responses by cable operators are described in Barrett, 1996. Antitrust litigation has resulted from the strategic behavior described, although it does not appear that the courts have entirely succeeded in promoting competitive outcomes (Hazlett, 1995; Bolton et al., 2000; Edlin, 2002).

[49] Cable operators have pursued both state statutes and municipal franchises guaranteeing a "level playing field" (Hazlett and Ford, 2001). This provides that, should a competitor secure a franchise, its terms will not be "less burdensome" than that held by the incumbent. Nominal parity, however, can lead to asymmetric economic results. For instance, a common recommendation by incumbents is that new franchisees be required to provide "universal service". The economic burden associated with such an obligation is lower for a monopolist than it is for a firm charging competitive rates and splitting the market with an established rival. Predatory intent is revealed by incumbents who lobby for such provisions. It is inimical to the financial interests of the incumbent to expand the competitive frontier; the motive is to deter entry altogether.

placing its wire on utility poles or burying them in underground conduits. This is done both through the franchising process, where system architecture must be detailed, and in 'make ready,' which occurs (as the term implies) before cables are attached to poles or buried underground. In this process, existing wired networks, including that owned by the incumbent cable operator, are given the opportunity to rearrange their aerial wiring to accommodate a new cable, billing the entrant for the expense. In areas of underground wiring, the incumbent networks are given time to locate, for the benefit of the entrant (who would otherwise be liable for the lines it would inadvertently cut), its underground plant. This notification process is the devil's playground for strategic behavior, as cable incumbents able to preemptively respond to entry narrowly targeted to those homes, which will be offered competitive service first. This makes price competition much more affordable for the incumbent and potentially devastating for even an equally efficient entrant.

Third, an incumbent operator that establishes exclusive contracts with a small number of key program services may create a substantial hurdle for the entrant. If the incumbent is able to match the services offered by the entrant, and match (or undercut) prices, exclusive control of a few popular channels may tilt the outcome of a competitive struggle predictably in favor of the incumbent. This, of course, acts to preempt the competitive challenge to begin with. See the program access discussion in Section 4.

2.5. *Intermodal competition*

The strategic difficulties encountered by wireline overbuilders enhanced the case for relaxation of the cable/telco cross-ownership ban. Since the object of targeted price cutting and marketing campaigns is to disrupt the entrant's ability to obtain financing, a well-funded video competitor committed to constructing large, regional infrastructure presents a more viable threat. The rule prohibiting local phone companies from providing video service was repealed in the 1996 Telecommunications Act.

Paradoxically, more competition ensued from the assistance rendered cable overbuilders than traditional local exchange carriers. Ameritech, one of the seven postdivestiture Regional Bell Operating Companies until purchased by fellow RBOC SBC in 1999, did make a foray into cable markets. But SBC sold these cable systems to an independent broadband service provider, WideOpenWest, in 2001.

But the repeal of the cross-ownership band importantly allowed overbuilders to offer bundled video, voice, and high-speed data. Companies like RCN and Knology have integrated into telephony and are identified as CLECs—competitive local exchange carriers. Residential phone access is not considered a lucrative market; as the incumbent exchange carrier's regulated rates are generally at or below cost, with losses offset with subsidies generated in business and long-distance service. But the ability to bundle a range of services

Table 10
Broadband service providers (overbuilders), 2002

	Video subscribers (June 2002)	Homes passed
Three Largest		
RCN	506,700	1.5 million
WideOpen West (WOW)	310,000	
Knology	124,700	
Total BSPs	1,000,000	>4 million

Sources: FCC, 2002c, pp. 48–49; Broadband Service Providers Association, 2002, p. 5.
Notes: Total is for members of the Broadband Service Providers Association.

allows transaction efficiencies and is a key to gaining, and retaining, subscribers. By 2002, overbuilders had managed to gain about 1 million video customers out of just over four million homes passed (Table 10). Both subscribers to the entrant and subscribers to the incumbent (which reacts with lower prices and better service) benefit from entry.

As with DBS, a 'compete large' strategy is seen in the investment decisions of broadband service providers. The overbuilding business model embraces higher capacity and enhanced functionality compared to the incumbent's system. This includes building more bandwidth by constructing many more nodes (with fewer subscribers per node to share the physical network)[50]. Entering a market with greater capacity differentiates the entrant's product, countering the effectiveness of price wars (Shaked and Sutton, 1982).

Satellite TV has enjoyed substantially more competitive success, which may be a product of strategic advantages. First, discriminatory marketing offers are difficult for the incumbent to target when confronted by nationwide satellite competition. Direct broadcast satellite operators sell subscriptions across a broad market; identifying the competitive fringe is problematic[51].

Second, many local impediments to competition are swept away—municipal franchise barriers, universal service (or other costly) requirements for entrants, and tactics delaying access to rights-of-way.[52] Third, program networks are easier to deal with. A national DBS operator, even with modest market share, looms as a

[50] RCN's standard architecture, for instance, delivers 860 MHz, and features 125 homes per cable modem node. This delivers much higher channel capacity and access speed when compared to the industry standard of 750 MHz and 500 homes, respectively.

[51] Efforts to do so, however, persist, as seen in the marketing campaign of the then-largest MSO: "AT&T looks for dishes in neighborhoods they service and leaves a leaflet which includes an offer to buy back the satellite for $200 credit, plus a free installation. They leave the dish but take the satellite receiver. Time Warner pays up to $200 and offers two free months of its Digital Supreme package." J. Ewoldt, Buyer's edge: satellite or digital cable TV?, Star Tribune, Dec. 6, 2001; http://www.startribune.com/stories/1229/870861.html.

[52] State taxes on satellite TV service may be imposed, however, and have generated much controversy.

more substantial player in video distribution than a new competitor in just a few local markets. For a programmer to discriminate against this video retailer augurs to be more expensive than an arrangement accommodating an incumbent cable system desiring exclusivity in a local market[53].

Finally, the DBS operator enjoys scale economies. These are important in purchasing programming, as volume discounts are commonly part of cable network licensing fee schedules[54]. In utilizing national media outlets for advertising campaigns, DBS can achieve efficiencies unavailable to any individual cable operator, the largest of which serves less than one-third of the national market. In marketing services, the ability to use national retailers has also proven an important competitive tool. And scale economies extend to equipment manufacturing, where customized receivers and set-top boxes are mass produced by major electronics suppliers, opportunities not readily available to a local overbuilder. While the wireline broadband service provider may readily access equipment mass produced for other cable TV systems, it is relatively expensive to build-in extra functionality. Direct broadcast satellite entrants, conversely, have offered digital boxes with added features (like parental controls) to differentiate from cable rivals.

While satellite TV systems are disadvantaged in bundling telecom services, with the latency of satellite transmissions affecting voice and data service, joint marketing arrangements can potentially overcome such obstacles. Local telephone companies are an obvious match[55]. By 2002, the two primary satellite TV competitors had attained such economic importance that regulators determined that a merger of the companies would substantially reduce competition (Federal Communications Commission, 2002b). Having surmounted many of the entry barriers that cable overbuilders have confronted, DBS has brought additional programming, and retail competition, to video markets in the United States (GAO, 2003; Goolsbee and Petrin, 2004; Reiffen et al., 2004). This success has not been greeted enthusiastically either by incumbent providers of cable TV service or television station owners[56].

[53] Some program access rules in the 1992 Cable Act appear to have helped both DBS and overbuilders. See discussion below.

[54] Ford and Jackson (1997) and Chipty and Snyder (1999) have found that larger system size lowers programming costs for both operators and subscribers.

[55] Vince Vittore, Telco Video Plans, Take Two: Bell Atlantic, SBC sign marketing alliance with DirecTV, Telephony (March 9, 1998).

[56] "Like the US cable television industry, television and radio broadcasters have opposed DBS [direct broadcast satellite]. In general, the National Association of Broadcasters [U.S. trade association for commercial stations] has been anxious about more competition. One of the core arguments used by the NAB against DBS is its potential threat to the economic viability of local stations and network affiliates. Over-the-air broadcasters believe that because of the low costs involved in reaching each consumer through DBS, direct broadcasters pose a direct threat to their audience share" (Comor, 1998, p. 169, emphasis in original).

3. Regulation of rates

Given market power, cable TV systems would seem likely candidates for rate regulation. They have been—and for deregulation, as well. In a recurring pattern, the cable TV market has cycled through at least two full generations of regulation-and-deregulation in the United States. The evidence reveals that, despite market power, price caps have not effectively lowered prices for subscribers.

3.1. Deregulation in the 1984 Cable Act

The 1984 Cable Communications Policy Act delivered cable incumbents generous benefits: monopoly franchises at unregulated prices. While preempting local government regulation of rates (triggered on 29 December 1986), the measure mandated local cable franchises[57] and barred telephone companies from providing video services in their local exchange territories. While the measure is commonly referred to as 'deregulatory,' it substantially raised the value of incumbency in cable markets (Zupan, 1989; Jaffe and Kantor, 1990; Prager, 1992).

Basic cable rates rose substantially in the postderegulation period. Reports issued by the federal government's General Accounting Office documented these rates increases (GAO, 1989, 1990, 1991). Rubinovitz (1993) deconstructed the price changes, finding that about one-half could not be explained by cost increases, and were attributable to the market power of cable operators.

This evidence, however, is consistent with the finding that subscriber growth in cable markets was at least as robust following deregulation as previously under rate controls. The implication is that regulation did not suppress real, quality-adjusted cable TV rates. Rates do appear to rise more rapidly in the deregulated period (1987–1992) than previously (Figure 7), but the observation of rising prices over time does not suggest the release of binding price controls, which would prompt discrete rate increases following deregulation. Rather, it suggests systematic changes were altering the demand for cable services over several years.

The most plausible explanation is that demand was increasing for cable television more rapidly during the deregulated period, and that these shifts were prompted by investments in programming and service by cable operators and networks. Price increases were offset by quality increases (Beard et al., 2001). Quality appeared highly responsive to regulatory constraints, undermining the ability of price caps to improve consumer welfare. Cable subscriber growth markedly increased during deregulation, output gains that were statistically significant and not explained by alternative factors, such as income growth (Hazlett, 1996). Cable programming expenditures and audience shares for basic cable

[57] Some jurisdictions had issued generic licenses to cable TV systems, regulating the use of public rights-of-way but not otherwise regulating cable systems. The federal franchise requirement effectively ended this, raising barriers to entry.

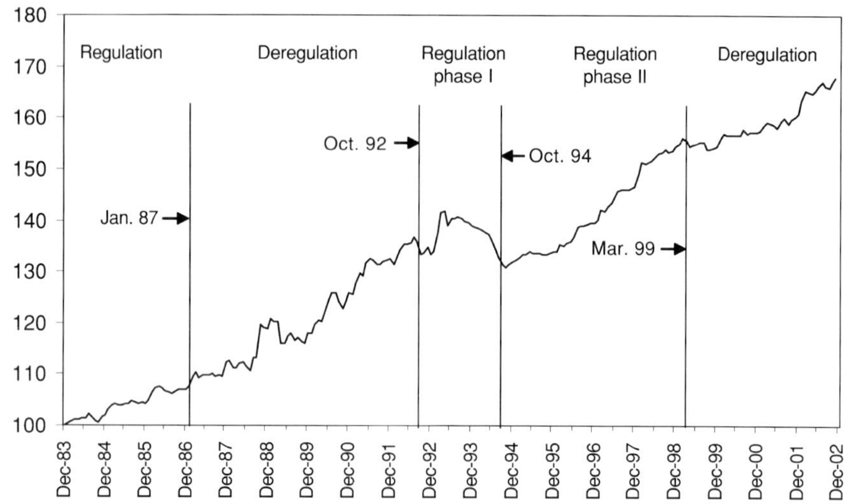

Fig. 7. Real U.S. cable TV rates, 1984–2002. Note: Cable rates for urban consumers deflated using CPI for urban consumers. Source: Bureau of Labor Statistics. http://data.bls.gov/cgi-bin/srgate.

networks also grew rapidly during deregulation, consistent with quality improvements. While supracompetitive pricing was evident, rate regulation proved ineffective (Hazlett and Spitzer, 1997, pp. 69–101)[58].

3.2. Reregulation in the 1992 Cable Act

Rate regulation was nonetheless reintroduced in the Cable Act of 1992, which directed regulators to impose rate caps on cable TV systems wherever 'effective competition' was found to be lacking[59]. This excluded few systems, leading to a regime under which the FCC set rate benchmarks and municipal governments handled enforcement.

The FCC imposed price controls in three discrete steps:

[58] Studies comparing rates charged by locally regulated vs. locally deregulated systems prior to the federal deregulation produce conflicting results (Mayo and Utsuki, 1991; Prager, 1990). Cross-sectional studies of local rate regulation, however, encounter at least two substantial problems: a lack of marginal cost data, and the endogeneity of the local government's decision to regulate rates. Observing system prices and quantities adjusted in response to an exogenous shock (such as state or federal rate deregulation) yields a clearer view of the impact of rate regulation. These studies reveal regulatory ineffectiveness, as does Emmons and Prager (1997) which finds that local competition under local rate controls (1983) and under federal deregulation (1989) had virtually identical price-lowering effects.

[59] The statute defined competition in three ways, the most important of which determined that "effective competition" obtained where a second multichannel video distributor could serve at least 50% of local households and did, in fact, serve at least 15%. This recognizes competition from wireline rivals, "wireless cable" systems, and satellite operators.

- A 'rate freeze,' announced in April 1993, capped rates at levels prevailing 30 September 1992, just before the Cable Act became law.
- A 10 percent rate rollback was imposed in September 1993.
- An additional 7 percent rate rollback was instituted by July 1994.

The FCC implemented rollbacks based on its estimate of the difference between monopoly and duopoly prices observed in local cable markets. Rate caps were adjusted for the number of channels offered customers, and applied to all basic programming tiers (limited basic and expanded basic) but excluded premium channels and pay-per-view. The second rollback was prompted by a new FCC estimate that the observed competitive pricing differential was not 10 percent but 17 percent, a reconsideration that was itself prompted by a public outcry, following the first round of price caps. News stories alleged that many customers' bills had actually increased under regulation.

The price data reported by the Bureau of Labor Statistics revealed that there were sharp reductions in cable prices during reregulation. Real basic cable charges in October 1994 were about 9 percent below where they would have been had the preregulation trend continued (Hazlett, 1997, p. 179). Yet, cable TV subscriber "growth did not rebound after the economy recovered in 1992–1994. In fact, subscribers numbered more than 2 million less than a prediction based on a time trend, a post-1985 time trend, and real per capita income. No more than 300,000 of the 2 million can be attributed to DBS. Given the evidence that basic subscriber rates may have been depressed by as much as 10 percent, the failure to even match the trend should pose a warning to regulators that something may be amiss" (Crandall and Furchtgott-Roth, 1996, p. 81).

The introduction of price regulation triggered a complex set of reactions by cable television operators, competitors, input suppliers, and customers. In general, cable systems adjusted to the new constraints by shifting capital investments and marketing efforts to less regulated activities. Multiple tiers of service were introduced; pricing and service menus became more complex. Basic revenues shifted towards higher tiers, where price constraints were less binding (Figure 8). Premium services, not subject to regulation, were promoted more heavily. Firms also invested in regulatory process, overwhelming federal officials in administrative issues including 'cost of service' proceedings which would lift price caps for individual cable systems. Eventually, major operators negotiated 'social contracts' with the FCC, allowing rate increases in exchange for commitments to invest in cable infrastructure, commitments widely viewed as nonbinding by industry analysts. Given the collapse of subscriber growth during the period of rate reductions, the regulators were anxious for face-saving agreements that would provide political cover for their retreat on price controls.

The negative deviation from long-run trend was particularly stark in the subscriber numbers reported by basic cable networks, a constituency that lobbied vigorously for the FCC to allow higher rates. Programmers supply the regulated

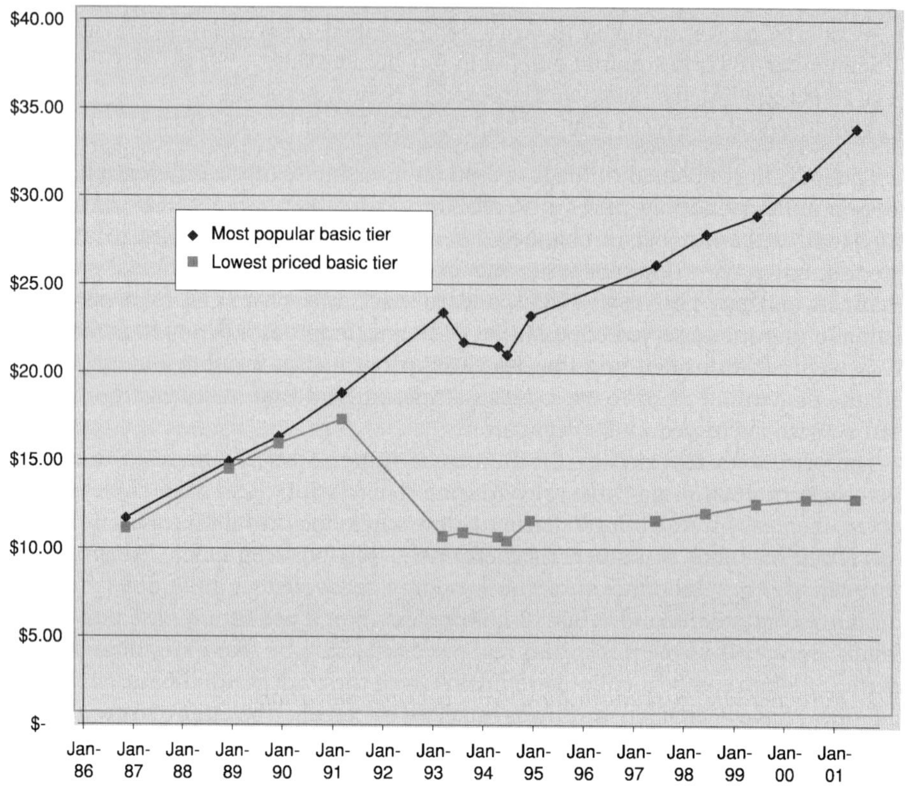

Fig. 8. Monthly rates for limited basic and expanded basic services (national means, 1986–2001).

industry; while having no direct interest in rate regulation, reduced demand for programming had a depressing effect on network sales. In particular, network start-ups (often spin-offs from existing services) had been encouraged by the emergence of satellite TV (DirecTV began serving subscribers in mid-1994), but found cable carriage agreements difficult to procure. In a reversal of industry practice, new program services began paying cable systems to gain 'shelf space'.

In response to pressure from operators and programmers, the FCC relented, allowing substantial price increases under 'going forward' rules issued in Nov. 1994. Both real cable rates and subscriber growth rebounded. Again, quality responses of cable operators had undermined the effectiveness of retail price regulation. Crandall and Furchtgott-Roth found that "real prices for basic service and the average channel capacity of system increased almost in lockstep; the correlation of these annual series is more than 0.99" between 1982 and 1993. The price-quality relationship (with channel capacity a proxy for quality) would appear to hold through the reregulation period, as well.

Mirroring previous research, economists found that cable operators were financially harmed by rate regulation (Carroll and Lamdin, 1993). But other sectors gained. Havenner et al. (2001) show that broadcasters were rewarded for advancing the 1992 Cable Act, as television station equities rose in value as the legislation was enacted. In addition to the other pro-broadcaster measures in the act, broadcasters revealed strong support for binding rate controls on basic cable service. As a staunch cable rival for audience share, the interest group support for rate regulation is consistent with the observed results: quality diminished sufficiently to more than offset consumer gains from price reductions.

Crawford (2000) investigated changes in household welfare between 1992 and 1995 to determine the effect of rate regulations imposed in 1993–1994. He concluded that, "while regulations mandated price reductions of 10–17 percent for cable services, observed system responses yielded at best no change in household welfare." Operators proved worthy adversaries in the regulatory game: "Cable systems control many aspects of their services: what programming to offer, how to bundle that programming into services, and how to price those services. The results suggest that one should carefully consider the product and price responses of cable systems to further regulations, and that alternative policies promoting competition in multichannel video programming markets may prove more effective at increasing household welfare in cable markets" (Ibid., p. 446).

3.3. Deregulation pursuant to the 1996 Telecommunications Act

Regulators were frustrated: "What indeed was the point of the regulation," asked the FCC Chairman, "if the beneficiaries were neither thankful nor economically better off?" (Hundt, 2000, p. 56). The FCC implicitly deregulated cable rates in late 1994 by allowing benchmark rate ceilings to rise, and formal deregulation (with federal preemption of local rate controls) was then included in the Telecommunications Act of 1996, which effectively ended cable rate controls as of 31 March 1999.

As in the previous episode, the end of rate controls did not prompt prices to fly up. Following deregulation, average real rate increases had actually been substantially less than during 1994–1999, when rate controls were in effect, and marginally below those seen throughout the reregulation imposed by the 1992 Cable Act (2.4 percent annualized) (Figure 9)[60].

The historic pattern is provocative. Deregulated cable prices increased somewhat faster in the 1980s deregulation than they did previously, but these price increases were correlated with increases in quality and subscriber penetration. With reregulation in the 1990s, prices were suppressed for a short period, but cable growth slowed while program networks were hurt, leading regulators to

[60] Note: Cable CPI for urban consumers deflated using monthly overall CPI for urban consumers. Source: Bureau of Labor Statistics, http://data.bls.gov/cgi-bin/srgate.

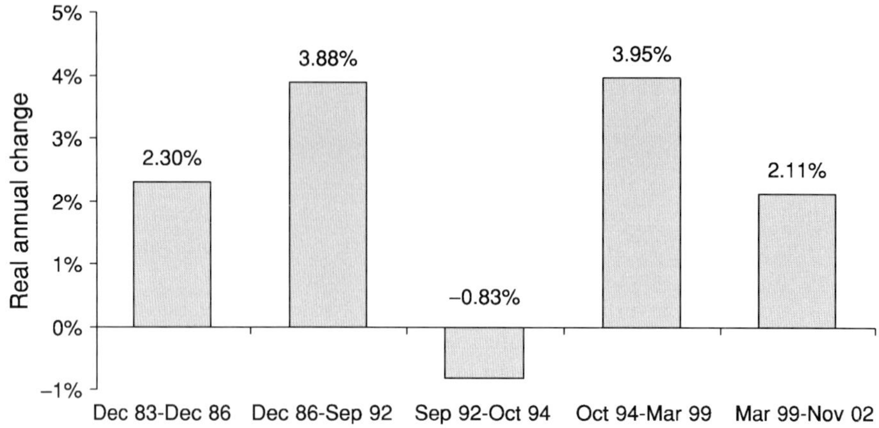

Fig. 9. Real cable rate increases during regulatory windows.

reverse course. They allowed price increases almost identical in magnitude to those seen during deregulation a decade previous. With the relaxation of controls in 1999 we observe a reduction in rate increases; it appears that market conditions, including perhaps competition from DBS, is constraining cable operator behavior. Cable rate regulation has faded as an effective option, although public calls for reregulation are a seemingly permanent feature of the political landscape.

These pronouncements, however, appear to have become informed by our long experience. Referring to that year's legislative effort, a United States Senator noted in 2002, "The only reason we didn't reregulate cable ... is because that's been the only competition to the Bell companies on getting us the broadband"[61]. Even when threatening to again impose controls, policy makers evince knowledge that cable rate regulation serves to curtail investment and diminish network development.

4. Cable television programming

A recurring theme in the history of cable television is that the sector would be more efficient with separation of operators and programmers. David Waterman and Andrew A. Weiss begin their book on vertical integration in cable by recalling that, "President Nixon's [1974] Cabinet Committee on Cable Communications proposed a complete vertical divestiture of cable systems from program suppliers ... the committee wanted to eliminate any incentives that cable systems might have to exclude or otherwise disadvantage program suppliers with which they had no vertical affiliation" (Waterman and Weiss, 1997, p. 1).

[61] Ted Hearn, Hollings: Cable reregulation possible next year, Multichannel News (30 July 2002).

Table 11
Top 10 cable program networks by subscriber size, 1998

Network	Launch date	Subs (mil.)	License rev ($mil.)	Gross ad rev ($mil.)	Average license fee ($/sub/mo)	Average prime time audience (000)
TBS	12/76	76.5	165	575	0.18	1393
Discovery	6/85	75.9	158	280	0.18	847
ESPN	9/79	75.7	537	664	0.65	1031
CNN/HN	6/80	75.6	290	402	0.17	1814
USA	4/80	75.2	328	458	0.37	1726
C-SPAN	3/79	75.2	n.a.	0	n.a.	n.a.
Nickelodean	4/79	74.9	205	636	0.24	1401
Fox Family	4/77	74.2	132	220	0.15	970
A&E	2/84	74.0	126	298	0.15	920
TNN	3/83	73.7	125	235	0.14	626

Sources: Kagan, 2000, pp. 39, 127, 172, 203, 223, 253, 335, 340, 376, 410, 427–28, 431–32.
Note: CNN/HN entries include CNN and Headline News.

The vertical separation rules were never imposed, yet the cable television industry developed a vibrant programming segment. In 2001 there were over 230 national programming networks. In 1998, an industry analyst calculated that the top 59 networks were worth, in aggregate, $61 billion. These networks then accounted for $11.2 billion in revenue, of which $6.9 billion was attributed to advertising, the residual to license fees (Table 11)[62].

4.1. Monopsony power

The issue of whether cable operators exercise market power when purchasing programming inputs became a regulatory matter when, in the 1992 Cable Television Consumer Competition and Protection Act, the FCC was directed to set a limit for national cable market share. The rule adopted was that no one cable operator could control over 30 percent of total subscribers, which did not necessitate any divestiture[63]. The analysis was based on a traditional approach to market concentration.

[62] Kagan, 2000, pp. 4, 7.

[63] "Commission rules ... impose a 30% limit on the number of multichannel video subscribers that may be served by a multiple cable system operator (the 'horizontal limit') and a 40% limit on the channel capacity that an operator may use to attributable programming, with such limit applicable only up to 75 channels (the 'vertical limit')" (Eisenach and May, 2002, p. 4, footnote omitted). The Commission's rationale was that a new programming service needed at least a 20 percent share of 80 million MVPD households, and had a 50–50 chance of gaining carriage when offering its program service to nonintegrated operators. Hence, a 40 percent 'open field' requirement was established under a 30% cap on each operator, blocking the two largest MSOs from controlling more than 60% of the market. These rules were invalidated in Time Warner Entertainment Co., L.P. vs. United States, 240-F. 3d 1126 (D.C. Cir. 2001).

According to the Commission, the national cable programming market is relatively unconcentrated because the Herfindahl-Hirschman Index is below 1000[64].

Waterman and Weiss argue, however, that traditional measures of concentration may not be appropriate. Cable systems do not (except in the special case of overbuilds) directly compete for customers, meaning that each firm enjoys far higher concentration *in its service area* than would be indicated by the national HHI. While program distributors look to the big picture, desiring to market their product through various media to a large national market, to reach a specific set of customers—(i.e., the households in a cable operator's franchise area)—the programmer deals with one dominant firm. This gives the cable operator influence over programming locally which is, after all, the full extent of the operator's scope. Should the programmer fail to sell to a given operator, the 'deconcentration' of cable does not imply that a competitor will soon steal customers from the recalcitrant cable operator by procuring and offering the missing content.

Able to reach a large, national audience, a cable program service can amortize expensive productions, gaining audience share. Absent such scale, however, costly investments may prove uneconomic, and viewers will be more difficult to attract. Audience size and program quality, in other words, tend to be mutually reinforcing. A large MSO, then, may tip the scales between two (or more) rival services with its individual carriage decisions. The result: power over the price of programming.

This is the interpretation given to a famous industry episode in 1990, where Lifetime, an independent cable network, bid $40 million for The Learning Channel. "TCI [then the largest MSO] reportedly threatened to remove TLC from virtually all its cable systems, which served approximately 20 percent of all U.S. cable subscribers, if the sale was consummated. Lifetime then reportedly withdrew its bid, and TCI's subsidiary purchased TLC a few months later for $32 million" (Waterman and Weiss, 1997, p. 65).

Such bargaining by cable operators may reduce investments in welfare enhancing content, or may reduce the exercise of market power in the programming market. Studies have found that larger firms obtain lower prices from cable program networks, but that these input discounts result in reduced prices for cable subscribers. Nonprice dimensions may be improved from the consumer's viewpoint, as well. A 1985 start-up by broadcasting network NBC (owned by General Electric), was deterred by cable systems because it was considered a close substitute for the all-news channel, CNN. NBC refocused its content, launching CNBC as a financial news channel. It gained widespread carriage. Cable operators then had no ownership interest in CNN, which was later purchased by cable operators TCI and Time Warner, but revealed a preference for program diversity. "Cable operators don't need another all-news channel," said the CEO of

[64] In 2001, the cable industry HHI was 905; in 2002, it was 884. FCC, 2002c, pp. 59–60.

Group W, a large MSO. "Basically, it would be taking up extra channel capacity with a duplicated service ..." (quoted in Waterman and Weiss, 1997, p. 64).

The success of CNBC led to the December 1995 creation of CNN/fn, CNN's entry into the financial news space. In July 1996, NBC/GE enlisted Microsoft to help counter the move by forming MSNBC, a direct CNN rival. Fox News then entered the 24/7 general news segment in October 1996 (Kagan, 2000, pp. 32–33). The latter is owned by News Corp., which does not own cable television systems[65]. In 2002, Fox News established a clear lead over CNN and MSNBC in audience share.

Yet, "bargaining power is not the same as market power" (Owen and Wildman, 1992, p. 244). The former transfers wealth, whereas the latter reduces output. Cable programming networks are not sold as commodities, and heterogeneous prices obtain. Sharp bargaining will not necessarily discourage efficient investments in programming. Still, if bargaining leaves total costs unremunerated—a possibility given the public good (high first copy costs) nature of programs—inefficiency may result. When a large MSO uses its muscle to avoid contributing to fixed programming overhead, it essentially free rides. One solution to this problem is vertical integration.

4.2. Vertical integration

Efficiencies possible from cable operator ownership of programming include elimination of the operator's incentive to suppress payments to the programmer via the exercise of monopsony power, as well as the reverse, elimination of monopoly pricing by the cable programmer (i.e., double marginalization). In addition, transaction cost savings, tighter coordination of long-run investment decisions (matching hardware capacity with software availability), risk sharing (as cable carriage helps build future demand for programs), and precluding opportunistic behavior are potentially achievable through vertical integration.

Warren-Boulton (2002) notes an essential tension between input manufacturers and downstream retail distributors when the former exhibit high fixed costs and low marginal costs, a situation obtaining in cable programming. The tension springs from the fact that the retailer sees input price as marginal cost, whereas the input supplier actually realizes a rent or quasi-rent on each additional unit. "Thus all producers of differentiated products would like retailers to devote more shelf space or other inputs to their product than the retailer would choose to supply absent special inducements ..." (Ibid.).

This conflict is mitigated with some cable networks via agreements that operators will include the channel on the basic programming tiers, guaranteeing shelf space and tying it to the basic cable marketing efforts undertaken by the system. When cable operators attempt to split certain basic networks into specialized program

[65] In 2003, however, News Corp. bought 35% of satellite provider DirecTV.

tiers, programmers typically resist—fiercely. With respect to pay cable networks, a quasi-integration has developed wherein per-channel fees paid by subscribers are shared by the programmer and operator, thus incenting operator sales effort. Again, vertical integration is a means by which incentives may also be aligned.

Vertical foreclosure theories give rise to the possibility that cable operator ownership of content is not entirely benign. In particular, integration may raise entry barriers either in the retail segment (as when cable system competitors are discouraged by an inability to obtain popular programming) or the upstream input segment (where independent cable networks have difficulty forming because of an inability to gain cable carriage), yielding integrated cable firms potentially higher profits.

Various studies have attempted to analytically separate (or identify) efficient and monopolistic integration. Klein (1989) found that cable systems do favor programming that they own, but that the behavior was consistent with efficiency because integrated operators were not less likely to carry unaffiliated programming than nonintegrated cable systems. Salinger (1988) did uncover evidence that cable operators vertically integrated into pay networks not only favored their own programming but offered fewer total choices. This finding is consistent with foreclosure.

In an attempt to examine the consumer welfare effects of vertical integration in cable, Chipty (2001) estimated a structural model. While finding that integration is associated with operators' favoring programming in which they maintain ownership, Chipty's results indicate that local markets served by cable systems with programming interests demonstrate higher levels of consumer surplus. Hence, vertical integration in cable TV may actually benefit consumers.

Cable television's emerging dominance in video distribution appears to have benefited from investments in cable programming made both by cable operators and, ironically, by cable's intraindustry rival, broadcasting (Table 12). A review of the equity shares in the top 33 cable television networks (ranked by cash flows) reveals that many of the most profitable cable channels are not owned by cable operators. ESPN, the leading generator of cash flows, and Nickelodeon, the second most profitable network, are examples. Moreover, broadcasters' ownership stake is more than twice the level of cable operator ownership, using equity shares weighted by cash flows (64 percent vs. 31 percent). The strong involvement in programming by either subsector suggests that scale efficiencies, including those from portfolio diversification, are important in producing television content.

The historical development of cable networks also suggests that ownership ties to operators can be valuable but are not necessary for success. The first cable network was HBO, undertaken in 1972 by a cable system attempting to differentiate its product from traditional broadcasting[66]. This was an important example

[66] "Home Box Office [HBO] forever divided cable from broadcast TV in viewers' minds. It also fed demand for cable in the cities, which typically had good broadcast reception and therefore little need for a duplicative television service." Keating, 1999, p. 46.

Table 12
Ownership of top 33 cable networks ranked by cash flow, 1998

Cable network	Cable MSO shares	Broadcast shares	Cash flow (1998) ($mil.)	Cable share weighted by percent of total CF	Broadcast share weighted by percent of total CF
ESPN	0.0%	100.0%	459.7	0.0	11.7
Nickelodeon	0.0%	100.0%	407.8	0.0	10.4
TNT	100.0%	0.0%	285.1	7.3	0.0
CNN + HN	100.0%	0.0%	276.1	7.0	0.0
TBS	100.0%	0.0%	246.4	6.3	0.0
MTV	0.0%	100.0%	238.9	0.0	6.1
Lifetime	0.0%	100.0%	186.8	0.0	4.8
Discovery	73.5%	24.5%	186.4	3.5	1.2
USA	0.0%	100.0%	184.5	0.0	4.7
Disney	0.0%	100.0%	175.0	0.0	4.5
Top 10			2647	24.1	43.4
Top 33			3919	30.7	64.3

Source: Kagan, 2000, pp. 66–67.

of a cable operator going upstream to create valuable inputs. The first basic cable 'network,' on the other hand, was actually a broadcasting station beamed nationally via satellite, WTBS. This programming was offered to cable systems by the Turner Broadcasting System, which was to launch CNN, Turner Network Television, and Turner Classic Movies, becoming a vital input supplier without vertical integration[67]. Another of the most successful cable programmers has been Viacom, owner of MTV, Nickelodeon, and Showtime. While Viacom was an MSO, it chose to sell its systems in 1995, buying a TV broadcast network, CBS, in 1999.

4.3. Program access rules

In the 1992 Cable Act, the FCC was mandated to promulgate rules giving cable competitors the opportunity to purchase cable networks on reasonable terms and conditions. The experience of many entrants was that exclusive agreements made it difficult for them to purchase popular programs, such as National Football League games shown on ESPN, and that without such content their efforts were doomed. In the rules adopted pursuant to the Act, "cable operators generally are prohibited from entering into exclusive distribution arrangements with vertically integrated-programming vendors" (Federal Communications Commission, 2002a, p. 61)[68].

[67] Turner was purchased by Time Warner, the second largest MSO, in 1996, following creation of the listed networks.

[68] "While nonvertically integrated cable networks are prohibited from certain 'unfair' behavior, the program-access provisions fall basically on vertically integrated programming suppliers" (Waterman and Weiss, 1997, p. 147, footnote omitted).

These rules "apply only to satellite-delivered and not to terrestrially delivered programming" (Ibid.).

The regulations have been credited with helping to open the market for overbuilders and satellite TV operators, but the loopholes are large. Vertical integration is essentially irrelevant to the question of discrimination: an exclusive agreement between unaffiliated firms creates a barrier equal in magnitude to one imposed by an integrated firm. When American Cable Co. overbuilt an incumbent cable system in Columbus, OH, it objected to exclusive contracts its rival had entered into, denying it access to the Sci-Fi Network (owned by USA Networks and NBC) and ESPN (Disney). The complaint filed by American was dismissed, however, which "puts into relief the cable act's illogical distinction between vertically affiliated and unaffiliated programming suppliers" (Waterman and Weiss, 1997, p. 150). As cable programmers have begun distributing signals via fiber optic transmission lines in lieu of satellites, the access rules can also be sidestepped by simply switching transport modes.

4.4. Carriage of broadcast TV signals

The complexities of the complement/substitute relationship between cable system operators and TV broadcasters have led to controversy in cable's carriage of off-air TV signals. In the early days of cable, operators retransmitted local signals to households lacking off-air reception capability, and serious conflicts did not arise. As operators began importing distant signals, however, viewers were being 'siphoned' from local stations. Hence, broadcasters collectively objected to what they saw as an appropriation of their property for the beneficial use of a market competitor.

To some degree, the pro-broadcaster rulings issued by the FCC beginning in 1962 were motivated by a desire to protect the intellectual property rights of television broadcasters. Yet, when the U.S. Supreme Court heard the issue, it twice ruled that the retransmission of broadcast TV shows to cable subscribers was not an appropriation. Just as a television set, a tuner, or a rooftop antenna assists members of the audience to receive signals, a cable TV system carrying TV broadcasts to subscribers unaltered (i.e., leaving commercial messages intact) constituted fair use[69].

The rulings were followed by a legislative resolution. In the Copyright Act of 1976, a complicated compensation scheme for program owners was devised. In simple terms it allowed cable TV systems to use a 'compulsory license' under which they had to carry the signal of all local TV stations and could also carry some distant stations, paying royalty rates set by the government. Local stations got free carriage but were essentially uncompensated; out-of-market signals (and

[69] Fortnightly Corp. vs. United Artists, 392 U.S. 390 (1968); Teleprompter Corp. vs. Columbia Broadcasting System, 415 U.S. 395 (1974).

the owners of programming on those signals) were paid according to how valuable their signals were to the cable systems that carried them (as determined by a royalty tribunal). The terms were not considered generous by broadcasters, particularly when the 'must carry' provisions were struck down by federal courts as unconstitutional in the 1980s. The total revenue generated by the compulsory license was a modest fraction of industry receipts; in 1992 it totaled about $180 million, most of it apportioned to the owners of movies, syndicated TV shows, and professional sports events.

Broadcasters sought to regain legislative advantage in the 1992 Cable Act. In addition to reregulating cable TV rates, the law granted TV stations 'must carry' and 'retransmission consent' rights. The dual award worked as follows. A station would determine whether to exercise its must carry rights. If it chose to do so, it would be given free carriage on each cable TV system serving its market—without monetary compensation. Alternatively, it could pass on must carry and negotiate a carriage agreement with cable systems, similar to license fees paid to cable networks.

When the 1992 Cable Act provisions went into effect the following year, "Ninety percent of network affiliates chose retransmission consent over must-carry . . . [and] 80 percent of independent stations chose must-carry" (Lubinsky, 1996, p. 153). Based on the license fees charged by cable TV networks, broadcasters had anticipated that they would receive substantial payments for retransmission consent. It was advantageous for a cable operator to feature local TV signals as part of its bundled offerings, but it was also valuable for the broadcaster to be included among the signals received by cable viewers. Cable operators bargained sharply, with the largest, TCI, announcing it would not pay for carriage. When the emergent fourth broadcast network, Fox, responded by offering its network-owned stations to local cable systems with the only compensation being a fairly standard carriage agreement for its new F/X cable network, broadcaster demands collapsed. Other companion network and cross-promotional deals were announced, but no substantial license fees were generated for broadcast retransmission.

4.5. Common carrier video

While his 'switch' is well underway, Nicolas Negroponte's claim to prescience is likely to be weaker in this prognostication: "A major step forward was made on 20 October 1994, when the FCC approved so-called 'video dial tone" (Negroponte, 1996, p. 26). The opinion reflects the theme, noted in the Nixon Cabinet Report and elsewhere (Noam, 1982), that public policy should separate content from conduit in multichannel video. Exploiting scale economies in distribution, while leaving content open to multiple rivals, tidily solves the apparent conflict between productive and allocative efficiency.

Yet common carrier video has not succeeded. The FCC's video dial tone (VDT) rulemaking spanned 1987–1996. Local telephone companies, banned from providing video content in their service territories, could create platforms to distribute programming supplied by third parties. Telephone companies were to build video systems and offer tariffed services to independent programmers and cable operators, but not be allowed to own more than 5 percent of the programming offered. But the rules proved contentious and ultimately unproductive. The VDT proceeding succeeded in licensing just one provider—New Jersey Bell, a subsidiary of Bell Atlantic. The sole system built, in Dover, New Jersey, signed up 1250 subscribers before the 1996 Telecommunications Act ended VDT, replacing it with a new common carrier model, Open Video Systems (OVS).

Open Video Systems allows for greater participation in content by the provider of the physical network: at least 35 percent of channel capacity, more if third parties (independent cable operators leasing OVS transport) are not displaced. But no independent programmers have yet materialized to utilize OVS facilities. The OVS model has allowed some prospective cable TV entrants to gain leverage over municipal franchising agents; broadband service providers have obtained federal OVS licenses before going on to procure standard cable franchises, shedding the regulatory obligations of OVS. Total subscribership stalled at 66,000 in 1998, fell to 60,000 in 2002 (Federal Communications Commission, 2002c, p. 76), a small fraction of overbuilds and a tiny increment of the MVPD market.

Economists anticipated this outcome. Leland Johnson noted that "the imposition of common-carriage video dial tone requirements on the LECs will make competition [with cable TV operators] even more difficult" (Johnson, 1994, p. 60). Robert Crandall and Harold Furchtgott-Roth predicted in 1996 that, "Were the telephone companies permitted to offer only common-carrier video dial tone service, it is unlikely that they would become full-fledged competitors of cable MSOs. The riskiness of investments in new services or programming requires that there be some probability of very large returns. As common carriers, telephone companies would be unable to participate in the successes of those services" (Crandall and Furchtgott-Roth, 1996, p. 91). Structural separation alters investment incentives. When vertical efficiencies are present, the modular architecture of common carriage may eliminate market power by simply eliminating the market[70].

A scaled down version of common carrier video obtains with respect to 'leased access'. Since 1972 cable operators have had the obligation to make individual channels available to unaffiliated programmers or other independent content

[70] This logic has driven regulatory decisions in broadband markets, where the FCC has been asked to impose "open access" rules such that independent Internet Service Providers may use cable modem connections to provide high-speed Internet connections at regulated wholesale rates. Denying such requests in 1999, then-FCC Chairman William Kennard stated: "Today ... we don't have a duopoly, we don't have a monopoly, we have a no-opoly." Federal Communications Commission Press Release, "Chairman Kennard Calls on Cable Franchising Authorities to Promote National Broadband Policy; Vows Continued Consumer Protection" (15 July 1999); http://ftp.fcc.gov/Bureaus/Miscellaneous/News_Releases/1999/nrmc9041.txt.

providers on reasonable terms. The rules have also proven contentious, and the use of channels virtually nonexistent—'least access' in one account (Lampert, 1992).

5. The evolution of cable

When viewed over a half-century, the emergence of cable television has been spectacular. The industry was recently a fledgling upstart attempting to gain a market toehold, economically weak and politically dominated by television broadcasting. Today, the near-universal penetration of subscription video service in the U.S., as well as in Canada, the Netherlands, and elsewhere, begs the question: should TV broadcasting be allowed to die a natural death?

Alternatively, the market position of cable is itself changing. In 2002, U.S. cable subscribership began declining, quarter-on-quarter, for the first time in history. The emergence of a serious multichannel rival, DBS, is primarily responsible. In countries where cable is less developed, DSL may preempt development of cable TV in the broadband space and, with help from satellite and TV-over-Internet technologies, provide alternative ways to bring 'spectrum in a tube' to viewers eager for video content. Like Italy, where abundant broadcasting outlets undercut the market opportunity for multichannel systems, these alternative video delivery systems may prove so effective that demand for cable is thwarted.

Quite by happenstance, however, cable TV infrastructure now proves valuable in providing a vector of telecommunications services. With technological convergence driven by digitization, coaxial cable grids use their capacious bandwidth to deliver not only video but voice, high-speed data, and other interactive services. Cable's video dominance may be nearing an end. But the bundles it delivers may compete effectively in an even larger market (Legg Mason, 2002). By 2010, just one-half of cable system revenue is projected to result from Plain Old Cable Service. Future research within the sector will no doubt focus on the competitive challenges of a dynamically evolving product package. Cable operators will be modeled as both incumbents and entrants.

References

Barrett, M., 1996, Strategic behavior and competition in cable television: Evidence from two overbuilt markets, Journal of Media Economics, 9, 43–62.

Baumol, W. J., J. C. Panzar and R. D. Willig, 1982, Contestable markets and the theory of industry structure, Jovanovich, New York: Harcourt, Brace.

Beard, T. R., R. B. Ekelund, Jr., G. S. Ford and R. S. Saba, 2001, Price-quality tradeoffs and welfare effects in the cable television market, Journal of Regulatory Economics, 20, 107–123.

Beard, T. R., G. S. Ford and R. P. Saba, 2002, Fragmented duopoly: A conceptual and empirical investigation, Auburn University manuscript, www.telepolicy.com.

Besen, S. M., 1974, The economics of the cable television 'consensus,' Journal of Law and Economics, 17, 39–51.
Besen, S. M. and R. W. Crandall, 1981, The deregulation of cable television, in Law and Contemporary Problems 1981, 79–124.
Beutel, P., 1990, City objectives in monopoly franchising: The case of cable television, Applied Economics, 22, 1237–1247.
Bittlingmayer, G. and Hazlett, T. W., 2002, Open access: The 'ideal' and the real, Telecommunications Policy, 26, 295–310.
Broadband Service Providers Association, 2002, Comments of Broadband Service Providers Association, Federal Communications Commission, MB docket no. 02–145, July 29.
Bolton, P., J. F. Brodley and M. H. Riordan, 2000, Predatory pricing: Strategic theory and legal policy, Georgetown Law Journal, 88, 2239–2330.
Carroll, K. A. and D. J. Lamdin, 1993, Measuring market response to regulation of the cable TV industry, Journal of Regulatory Economics, 5, 385–399.
Chipty, T. and C. M. Snyder, 1999, The role of firm size in bilateral bargaining: A study of the cable television industry, Review of Economics and Statistics, 81, 326–340.
Chipty, T., 2001, Vertical integration, market foreclosure, and consumer welfare in the cable television industry, American Economic Review, 91, 428–453.
Comor, E. A., 1998, Communication, commerce and power: The political economy of America and the direct broadcast satellite, 1960–2000, New York: St. Martin's Press.
Crandall, R. W. and H. Furchtgott-Roth, 1996, Cable TV: Regulation or competition?, Washington, DC: Brookings Institution.
Crawford, G. S., 2000, The impact of the 1992 Cable Act on household demand and welfare, Rand Journal of Economics, 31, 422–449.
Demsetz, H., 1968, Why regulate utilities?, Journal of Law and Economics, 11, 55–65.
Edlin, A. S., 2002, Stopping above-cost predatory pricing, Yale Law Journal, 111, 941–991.
Eisenach, J., and R. May, 2002, Comments of the Progress and Freedom Foundation, Federal Communications Commission, In the matter of implementation of Section 11 of the Cable Television Consumer Protection and Competition Act of 1992, CS Docket No. 98–82, January 4.
Emmons, W. M, III and R. A. Prager, 1997, The effects of market structure and ownership on prices and service offerings in the U.S. cable television industry, Rand Journal of Economics, 28, 732–750.
Federal Communications Commission, 1965, Amendment of Subpart L, part 11,* to adopt rules and regulations to govern the grant of authorizations in the business radio service for microwave stations to relay television signals to community antenna systems, second report and order, Docket No. 14895, adopted April 22.
Federal Communications Commission, 1966, Amendment of Subpart L, Part 91, to adopt rules and regulations to govern the grant of authorizations in the business radio service for microwave stations to relay television signals to community antenna systems, second report and rder, Docket No. 14895. adopted March 4.
Federal Communications Commission, 1988, Telephone company-cable television cross-ownership rules, Sections 63.54–63.58, further notice of inquiry and notice of proposed rulemaking, CC Docket No. 87–266, 3 F.C.C. Rcd. 5849.
Federal Communications Commission, 1994, Annual assessment of the status of competition in the market for the delivery of video programming: First annual report, CS Docket No. 94–48, September 28.
Federal Communications Commission, 1996, Report on cable industry prices in the matter of implementation of Section 3 of the Cable Television Consumer Protection and Competition Act of 1992, MM Docket No. 92–266, December 31.
Federal Communications Commission, 1997, Annual assessment of the status of competition in the market for the delivery of video programming, fourth annual report, CS Docket No. 97–141, December 31.

Federal Communications Commission, 1998, Annual assessment of the status of competition in the market for the delivery of video programming, fifth annual report, CS Docket No. 98–102, December 17.

Federal Communications Commission, 1999a, Inquiry concerning the deployment of advanced telecommunications capability to all Americans in a reasonable and timely fashion, and possible steps to accelerate such deployment pursuant to Section 706 of the Telecommunications Act of 1996, CC Docket No. 98–146, Report, 14 FCC Rcd. 2398.

Federal Communications Commission, 1999b, Report on cable industry prices in the matter of implementation of Section 3 of the Cable Television Consumer Protection and Competition Act of 1992, MM Docket No. 92–266, May 5.

Federal Communications Commission, 1999c, Annual assessment of the status of competition in the market for the delivery of video programming, sixth annual report, CS Docket No. 99–230, December 30.

Federal Communications Commission, 2001, Annual assessment of the status of competition in the market for the delivery of video programming, eighth annual report, CS Docket No. 01–129, December 27.

Federal Communications Commission, 2002a, Report on cable industry prices in the matter of implementation of Section 3 of the Cable Television Consumer Protection and Competition Act of 1992, MM Docket No. 92–266, April 1.

Federal Communications Commission, 2002b, In the matter of Application of EchoStar Communications Corporation, (a Nevada Corporation), General Motors Corporation, and Hughes Electronics Corporation (Delaware Corporations) (Transferors) and EchoStar Communications Corporation (a Delaware Corporation) (Transferee), hearing designation order, CS Docket No. 01–348, October 18.

Federal Communications Commission, 2002c, Annual assessment of the status of competition in the market for the delivery of video programming, ninth annual report, MB Docket No. 02–145, December 31.

Federal Communications Commission, 2004, Annual assessment of the status of competition in the market for the delivery of video programming, tenth annual report, MB Docket No. 03–172, January 28.

Fisher, F. M. and V. E. Ferrall, Jr., 1966, Community antenna television systems and local television station audience, Quarterly Journal of Economics, 60, 227–251.

Fournier, Gary, M. and Ellen S. Campbell, 1993, Shifts in broadcast policy and the value of television licenses, Information Economics and Policy, 5, 87–104.

Ford, G. S. and J. D. Jackson, 1997, Horizontal concentration and vertical integration in the cable television industry, Review of Industrial Organization, 12, 501–518.

General Accounting Office (GAO), 1989, National survey of cable television rates and services, GAO/RCED-89-183, August 3.

General Accounting Office (GAO), 1990, Follow-up national survey of cable television rates and services, GAO/RCED-90-199, June 13.

General Accounting Office (GAO), 1991, 1991 survey of cable television rates and services, GAO/RCED-91-195, July 17.

General Accounting Office (GAO), 2003, Issues related to competition and subscribers rates in the cable television industry, GAO-04-8, October.

General Accounting Office (GAO), 2004, Wire-based competition benefited consumers in selected markets, GAO-04-241, February.

Goolsbee, A. and A. Petrin, 2004, The consumer gains from direct broadcast satellites and the competition with cable TV, Econometrica, 72, 351–381.

Hamilton, J., 2004, All the News That's Fit to Sell, Princeton, NJ: Princeton University Press.

Havenner, A. M., T. W. Hazlett and Z. Leng, 2001, The effects of rate regulation on mean returns and non-diversifiable risk: The case of cable television, Review of Industrial Organization, 19, 149–164.

Hazlett, T. W., 1986, Private monopoly and the public interest: an economic analysis of the cable television franchise, University of Pennsylvania Law Review, 134, 1335–1409.
Hazlett, T. W., 1990, Duopolistic competition in cable television: Implications for public policy, Yale Journal on Regulation, 7, 65–148.
Hazlett, T. W., 1991, The demand to regulate franchise monopoly: Evidence from CATV rate deregulation in California, Economic Inquiry, 29, 275–296.
Hazlett, T. W., 1995, Predation in local cable TV markets, Antitrust Bulletin, 40, 609–644.
Hazlett, T. W., 1997, Prices and outputs under cable TV reregulation, Journal of Regulatory Economics, 12, 173–195.
Hazlett, T. W., 1998, Assigning property rights to radio spectrum users: Why did FCC license auctions take 67 years? Journal of Law and Economics, 41, 529–575.
Hazlett, T. W., 1996, Cable television rate deregulation, International Journal of the Economics of Business, 3, 145–163.
Hazlett, T. W., 2000, Digitizing "must-carry" under Turner Broadcasting v. FCC (1997), The Supreme Court Economic Review, 8, 141–207.
Hazlett, T. W., 2001, The U.S. digital TV transition: Time to toss the Negroponte Switch, in AEI-Brookings Joint Center for Regulatory Studies Working Paper, 1–15.
Hazlett, T. W. and M. L. Spitzer, 1997, Public Policy Toward Cable Television: The Economics of Rate Controls, Cambridge, MA: MIT Press.
Hazlett, T. W. and G. S. Ford, 2001, The fallacy of regulatory symmetry: An economic analysis of the 'Level Playing Field, in cable TV franchising, statutes, Business & Politics, 3, 21–46.
Hotelling, H., 1929, Stability in competition, Economic Journal, 34, 41–57.
Huber, P. W., M. K. Kellogg and J. Thorne, 1999, Federal telecommunications law, second edn., New York: Aspen Law & Business.
Hundt, R. E., 2000, You say you want a revolution: A story of information age politics, New Haven: Yale University Press.
Jaffe, A. B. and D. M. Kanter, 1990, Market power of local cable television franchises: Evidence from the effects of deregulation, Rand Journal of Economics, 21, 226–234.
Johnson, L. L., 1994, Toward competition in cable, Cambridge, MA: MIT Press.
Kagan, P., 1995, The cable TV financial databook, California, Carmel: Paul Kagan Associates.
Kagan, P., 2000, The state of DBS 2001, California, Carmel: Paul Kagan Associates.
Kagan, P., 2002, Broadband cable financial databook 2002, California, Carmel: Paul Kagan Associates.
Keating, S., 1999, Cutthroat: High stakes and killer moves on the electronic frontier, Boulder, CO: Johnson Books.
Klein, B., 1989, The competitive consequences of vertical integration in the cable industry, Report on behalf of the National Cable Television Association, Washington, DC.
Koch, M., 1998–1999, Two solitudes: Canadian communications regulation applied to the Internet, International Journal of Communications Law and Policy, 2.
Lampert, D. M., 1992, Does leased access mean least access? Federal Communications Law Journal, 44, 245–284.
Lampert, D. N., F. H. Cate and F. W. Lloyd, Cable television leased access, Annenberg Washington Program, Northwestern University.
Legg, Mason, 2002, Battle of the bundles, Baltimore: Legg Mason Equity Research.
Levin, H., 1980, Fact and fancy in television regulation: An economic study of policy alternatives, New York: Russell Sage Foundation.
Levy, J., M. Ford-Livene and A. Levine, 2002, Broadcast television: Survivor in a sea of competition Federal Communications Commission Office of Plans and Policy Working Paper No. 37, September.
Lindenberg, E. and S. Ross, 1981, Tobin's q ratio and industrial organization, Journal of Business, 54, 1–32.

Lubinsky, C., 1996, Reconsidering retransmission consent: An examination of the retransmission consent provision (47 U.S.C. § 325(b)) of the 1992 Cable Act, Federal Communications Law Journal, 49, 99–165.
Maxwell, K., 1999, Residential broadband: An insider's guide to the battle for the last mile, New York: John Wiley & Sons.
Mayo, J. and Y. Otsuka, 1991, Demand, pricing, and regulation: Evidence from the cable TV industry, Rand Journal of Economics, 22, 396–410.
Negroponte, N., 1996, Being digital, second edn., New York: Vintage.
Noam, E. M., 1982, Towards an integrated communications market: Overcoming the local monopoly of cable television, Federal Communications Law Journal, 34, 209–257.
Noam, E. M., 1992, Television in Europe, New York: Oxford University Press.
Noll, R. G., M. J. Peck and J. J. McGowan, 1973, Economic aspects of television regulation, Washington, DC: Brookings.
OECD, 2001, The development of broadband access in OECD countries, Paris: OECD.
Owen, B. M. and S. Wildman, 1992, Video economics, Cambridge, Massachusetts: Harvard University Press.
Owen, B. M., 1999, The Internet challenge to television, Cambridge, MA: Harvard University Press.
Parkman, A., 1981, The growth of CATV and the value of TV stations, Quarterly Review of Economics and Business, 21, 64–73.
Peltzman, S., 1989, The economic theory of regulation after a decade of deregulation, Brookings Papers on Economic Activity, Microeconomics, 1989, 1–41.
Posner, R. A., 1971, Taxation by regulation, Bell Journal of Economics and Management Science, 2, 22–50.
Powe, L. A., Jr., 1987, American broadcasting and the First Amendment, Berkeley: University of California Press.
Prager, R., 1990, Firm behavior in franchise monopoly markets, Rand Journal of Economics, 21, 211–225.
Prager, R., 1992, The effects of deregulating cable television: Evidence from the financial markets, Journal of Regulatory Economics, 4, 347–463.
Reiffen, D., M. R. Ward and J. Wiegand, 2004, Duplication of public goods: Some evidence on the potential efficiencies from the proposed Echostar/DirecTV merger, manuscript, March, http://www.uta.edu/faculty/mikeward/dbspaper.pdf.
Rubinovitz, R. N., 1993, Market power and price increases for basic cable service since deregulation, Rand Journal of Economics, 24, 1–18.
Salinger, M., 1988, A test of successive monopoly and foreclosure effects: Vertical integration between cable systems and pay services, Graduate School of Business, Columbia University, manuscript.
Shaked, A. and J. Sutton, 1982, Relaxing price competition through product differentiation, Review of Economic Studies, 49, 3–13.
Smiley, A. K., 1989, Direct competition among cable television systems, U.S. Department of Justice Antitrust Division, Economic Analysis Group Working Paper No. 86–89.
Smiley, A. K., 1990, Regulation and competition in cable television, Yale Journal on Regulation, 7, 121–139.
Smirlock, M., T. Gilligan and W. Marshal, 1984, Tobin's q and the structure performance relationship, American Economic Review, 74, 1051–1060.
Spence, A. M. and B. M. Owen, 1977, Television programming, monopolistic competition, and welfare, Quarterly Journal of Economics, 91, 103–126.
Speta, J. B., 2000, Handicapping the race for the last mile? A critique of open access rules for broadband platforms, Yale Journal on Regulation, 17, 39–91.
Stanbury, W. T., 1998, Canadian content regulations: The intrusive state at work, Vancouver, B.C.: Fraser Institute.

Steiner, P. O., 1952, Program Patterns and Preferences and the Workability of competition in radio broadcasting, Quarterly Journal of Economics, 66, 194–223.

Viscusi, W. K., J. M. Vernon and J. E. Harrington, 1995, Economics of regulation and antitrust, Cambridge, MA: MIT Press.

Warren-Boulton, F. R., 2002, The contribution of the merger guidelines to the analysis of non-horizontal mergers, U.S. Department of Justice Antitrust Division, 20th Anniversary of the 1982 Merger Guidelines: The Contribution of the Merger Guidelines to the Evolution of Antitrust Doctrine, June 10, http://www.usdoj.gov/atr/hmerger.htm.

Waterman, D. and A. A. Weiss, 1997, The effect of vertical integration between cable television systems and pay cable networks, Journal of Econometrics, 72, 357–395.

Webbink, D. W., 1969, The impact of UHF promotion: The all-channel television receiver law, Law and Contemporary Problems, 34, 535–561.

Williamson, O. E., 1976, Franchise bidding for natural monopoly–In general and with respect to CATV, Bell Journal of Economics, 7, 73–104.

Wildman, S. S., 2003, Assessing Quality-Adjusted Changes in the Real Price of Basic Cable Service, attached to Comments of the National Cable & Telecommunications Association submitted to the Federal Communications Commission, In the Matter of Annual Assessment of the Status of Competition in the Market for the Delivery of Video Programming, MB Docket No. 03–172, September 11.

Wildman, S. S. and B. M. Owen, 1985, Program competition, diversity, and multichannel bundling in the new video industry, in E. Noam, ed., Video Media Competition: Regulation, Economics, and Technology, New York: Columbia University Press, 244–279.

Zupan, M. A., 1989, The effect of franchise bidding schemes in the case of cable television: Some systematic evidence, Journal of Law and Economics, 32, 401–456.

Chapter 7

WIRELESS COMMUNICATIONS

JOSHUA S. GANS

University of Melbourne

STEPHEN P. KING

University of Melbourne

JULIAN WRIGHT

National University of Singapore

Contents

Handbook of Telecommunications Economics, Volume 2, Edited by S. Majumdar et al.

DOI: 10.1016/S1569-4054(05)02007-5

1. Introduction

In 1895, Guglielmo Marconi opened the way for modern wireless communications by transmitting the 3-dot Morse code for the letter 'S' over a distance of 3 kilometers using electromagnetic waves. From this beginning, wireless communications has developed into a key element of modern society. From satellite transmission, radio and television broadcasting to the now ubiquitous mobile telephone, wireless communications has revolutionized the way societies function.

This chapter surveys the economics literature on wireless communications. Wireless communications and the economic goods and services that utilize it have some special characteristics that have motivated specialized studies. First, wireless communications relies on a scarce resource—radio spectrum—the property rights, which were traditionally vested with the state. In order to foster the development of wireless communications (including telephony and broadcasting), those assets were privatized. Second, use of spectrum for wireless communications required the development of key complementary technologies, especially those that allowed higher frequencies to be utilized more efficiently. Finally, because of its special nature, the efficient use of spectrum required the coordinated development of standards. Those standards in turn played a critical role in the diffusion of technologies that relied on spectrum use.

In large part our chapter focuses on wireless telephony rather than broadcasting and other uses of spectrum (e.g., telemetry and biomedical services). Specifically, the economics literature on that industry has focused on factors driving the diffusion of wireless telecommunication technologies and on the nature of network pricing regulation and competition in the industry. By focusing on the economic literature, this chapter complements other surveys in this Handbook. Hausman (2002) focuses on technological and policy developments in mobile telephony rather than economic research per se. Cramton (2002) provides a survey of the theory and practice of spectrum auctions used for privatization. Armstrong (2002a) and Noam (2002) consider general issues regarding network interconnection and access pricing, while Woroch (2002) investigates the potential for wireless technologies as a substitute for local fixed-line telephony. Finally, Liebowitz and Margolis (2002) provide a general survey of the economics literature on network effects. In contrast, we focus here solely on the economic literature on the mobile telephony industry.

The outline for this chapter is as follows. The next section provides background information regarding the adoption of wireless communication technologies. Section 3 then considers the economic issues associated with mobile telephony, including spectrum allocation and standards. Section 4 surveys recent economic studies of the diffusion of mobile telephony. Finally, Section 5 reviews issues of regulation and competition, in particular, the need for and principles behind access pricing for mobile phone networks.

2. Background

Marconi's pioneering work quickly led to variety of commercial and government (particularly military) developments and innovations. In the early 1900s, voice and then music was transmitted and modern radio was born. By 1920, commercial radio had been established with Detroit station WWJ and KDKA in Pittsburgh. Wireless telegraphy was first used by the British military in South Africa in 1900 during the Anglo-Boer war. The British navy used equipment supplied by Marconi to communicate between ships in Delagoa Bay. Shipping was a major early client for wireless telegraphy and wireless was standard for shipping by the time the Titanic issued its radio distress calls in 1912[1].

Early on, it was quickly recognized that international coordination was required for wireless communication to be effective. This coordination involved two features. First, the potential for interference in radio transmissions meant that at least local coordination was needed to avoid the transmission of conflicting signals. Secondly, with spectrum to be used for international communications and areas such as maritime safety and navigation, coordination was necessary between countries to guarantee consistency in approach to these services. This drove government intervention to ensure the coordinated allocation of radio spectrum.

2.1. Spectrum allocation

Radio transmission involves the use of part of the electromagnetic spectrum. Electromagnetic energy is transmitted in different frequencies and the properties of the energy depend on the frequency. For example, visible light has a frequency between 4×10^{14} and 7.5×10^{14} Hz[2]. Ultraviolet radiation, X-rays, and gamma rays have higher frequencies (or equivalently a shorter wavelength), while infrared radiation, microwaves, and radio waves have lower frequencies (longer wavelengths). The radio frequency spectrum involves electromagnetic radiation with frequencies between 3000 Hz and 300 GHz[3].

Even within the radio spectrum, different frequencies have different properties. As Cave (2001) notes, the higher the frequency, the shorter the distance the signal will travel, but the greater the capacity of the signal to carry data. The tasks of internationally coordinating the use of radio spectrum, managing interference and setting global standards are undertaken by the International Telecommunication Union (ITU). The ITU was created by the International Telecommunications Convention in 1947, but has predecessors dating back to

[1] Numerous references provide a history of wireless communications. For a succinct overview, see Schiller (2000).

[2] One Hertz (Hz) equals one cycle per second.

[3] One GHz equals one billion cycles per second.

approximately 1865[4]. It is a specialist agency of the United Nations with over 180 members.

The Radiocommunication Sector of the ITU coordinates global spectrum use through the Radio Regulations. These regulations were first put in place at the 1906 Berlin International Radiotelegraph Conference. Allocation of the radio spectrum occurs along three dimensions—the frequency, the geographic location, and the priority of the user with regards to interference. The radio spectrum is broken into 8 frequency bands, ranging from Very Low Frequency (3–30 kHz) up to Extremely High Frequency (30–300 GHz). Geographically, the world is also divided into three regions. The ITU then allocates certain frequencies for specific uses on either a worldwide or a regional basis. Individual countries may then further allocate frequencies within the ITU international allocation. For example, in the U.S., the Federal Communications Commission's (FCC's) table of frequency allocations is derived from both the international table of allocations and U.S. allocations. Users are broken into primary and secondary services, with primary users protected from interference from secondary users but not vice versa.

As an example, in 2003, the band below 9 kHz was not allocated in the international or the U.S. table. The 9–14 kHz was allocated to radio navigation in both tables and all international regions while 14–70 kHz is allocated with both maritime communications and fixed wireless communications as primary users. There is also an international time signal at 20 kHz. But the U.S. table also adds an additional time frequency at 60 kHz. International regional distinctions begin to appear in the 70–90 kHz range with differences in use and priority between radio navigation, fixed, radiolocation, and maritime mobile uses. These allocations continue right up to 300 GHz, with frequencies above 300 GHz not allocated in the U.S. and those above 275 GHz not allocated in the international table[5].

The ITU deals with interference by requiring member countries to follow notification and registration procedures whenever they plan to assign frequency to a particular use, such as a radio station or a new satellite.

2.2. *The range of wireless services*

Radio spectrum is used for a wide range of services. These can be broken into the following broad classes:

- *Broadcasting services*: including short wave, AM, and FM radio as well as terrestrial television;
- *Mobile communications of voice and data*: including maritime and aeronautical mobile for communications between ships, airplanes, and land; land mobile for

[4] See Productivity Commission (2002) and the ITU website at www.itu.int

[5] For full details see The FFC's on-line table of frequency allocations, Federal Communications Commission Office of Engineering and Technology, revised as of January 17, 2003, http://www.fcc.gov/oet/spectrum/table/fcctable.pdf

communications between a fixed base station and moving sites such as a taxi fleet and paging services; and mobile communications either between mobile users and a fixed network, or between mobile users, such as mobile telephone services;

- *Fixed Services*: either point to point or point to multipoint services;
- *Satellite*: used for broadcasting, telecommunications, and Internet, particularly over long distances;
- *Amateur radio*; and
- *Other Uses*: including military, radio astronomy, meteorological, and scientific uses[6].

The amount of spectrum allocated to these different uses differs by country and frequency band. For example, in the U.K., 40 percent of the 88 MHz to 1 GHz band of frequencies are used for TV broadcasting, 22 percent for defense, 10 percent for GSM mobile, and 1 percent for maritime communications. In contrast, none of the 1 GHz to 3 GHz frequency range is used for television, 19 percent is allocated to GSM and third-generation mobile phones, 17 percent to defense, and 23 percent for aeronautical radar[7].

The number of different devices using wireless communications is rising rapidly. Sensors and embedded wireless controllers are increasingly used in a variety of appliances and applications. Personal digital assistants (PDAs) and mobile computers are regularly connected to e-mail and Internet services through wireless communications, and wireless local area networks (LANs) for computers are becoming common in public areas like airport lounges. However, by far the most important and dramatic change in the use of wireless communications in the past 20 years has been the rise of the mobile telephone.

2.3. *The rise of mobile telephony*

The history of mobile telephones can be broken into four periods. The first (precellular) period involved mobile telephones that exclusively used a frequency band in a particular area. These telephones had severe problems with congestion and call completion. If one customer was using a particular frequency in a geographic area, no other customer could make a call on that same frequency. Further, the number of frequencies allocated by the FCC in the U.S., to mobile telephone services, was small, limiting the number of simultaneous calls. Similar systems, known as A-Netz and B-Netz were developed in Germany.

The introduction of cellular technology greatly expanded the efficiency of frequency use of mobile phones. Rather than exclusively allocating a band of frequency to one telephone call in a large geographic area, a cell telephone breaks down

[6] Adapted from Productivity Commission (2002)

[7] See Annex A from Cave (2001)

a geographic area into small areas or cells. Different users in different (nonadjacent) cells are able to use the same frequency for a call without interference.

First generation cellular mobile telephones developed around the world using different, incompatible analogue technologies. For example, in the 1980s in the U.S., there was the Advanced Mobile Phone System (AMPS), the U.K. had the Total Access Communications System (TACS), Germany developed C-Netz, while Scandinavia developed the Nordic Mobile Telephone (NMT) system. The result was a wide range of largely incompatible systems, particularly in Europe, although the single AMPS system was used throughout the U.S.

Second-generation (2G) mobile telephones used digital technology. The adoption of second-generation technology differed substantially between the U.S. and Europe, and reverses the earlier analogue mobile experience. In Europe, a common standard was adopted, partly due to government intervention.

Groupe Speciale Mobile (GSM) was first developed in the 1980s and was the first 2G system. But it was only in 1990 that GSM was standardized (with the new name of Global System for Mobile communication) under the auspices of the European Technical Standards Institute. The standardized GSM could allow full international roaming, automatic location services, common encryption, and relatively high quality audio. GSM is now the most widely used 2G system worldwide, in more than 130 countries, using the 900 MHz frequency range.

In contrast, a variety of incompatible 2G standards developed in the U.S. These include TDMA, a close relative of GSM, and CDMA, referring to Time and Code Division Multiple Access, respectively. These technologies differ in how they break down calls to allow for more efficient use of spectrum within a single cell. While there is some argument as to the 'better' system, the failure of the U.S. to adopt a common 2G standard, with the associated benefits in terms of roaming and switching of handsets, meant the first-generation AMPS system remained the most popular mobile technology in the U.S. throughout the 1990s.

The final stage in the development of mobile telephones is the move to third-generation (3G) technology. These systems will allow for significantly increased speeds of transmission and are particularly useful for data services. For example, 3G phones can more efficiently be used for e-mail services, and downloading content (such as music and videos) from the Internet. They can also allow more rapid transmission of images, for example from camera phones.

An attempt to establish an international standard for 3G mobile was being moderated through the ITU, under the auspices of its IMT-2000 program. The IMT-2000 determined that 3G technology should be based on CDMA systems, but there are (at least two) alternative competing systems and IMT-2000 did not choose a single system but rather a suite of approaches. At the ITU's World Radiocommunication Conference in 2000, frequencies for IMT-2000 systems were allocated on a worldwide basis. By 2002, the only 3G system in operation was in Japan, although numerous companies have plans to roll out 3G systems in the next few years.

The growth in use of mobile telephones has been spectacular. From almost a zero base in the early 1980s, mobile penetration worldwide in 2002 was estimated at 15.57 mobile phones per 100 people worldwide. Of course, the level of penetration differed greatly between countries. In the U.S., there were 44.2 mobile telephones per 100 inhabitants, with penetration rates of 60.53 in France, 68.29 in Germany, 77.84 in Finland, and 78.28 in the U.K. Thus, in general, mobile penetration is lower in the U.S. than in the wealthier European countries. Outside Europe and the U.S., the penetration rate in Australia is 57.75, 62.13 in New Zealand, and 58.76 in Japan. Unsurprisingly, penetration rates depend on the level of economic development, so that India had only 0.63 mobile telephones per 100 inhabitants in 2002, with 1.60 for Kenya, 11.17 for China, and 29.95 for Malaysia. The number of mobile phones now exceeds the number of fixed-wire telephone lines in a variety of countries including Germany, France, the U.K., Greece, Italy, and Belgium. However, the reverse holds with fixed-lines outnumbering mobiles in the U.S., Canada, and Argentina. Penetration rates were close to equal in Japan in 2001, but in all countries, mobile penetration is rising much faster than fixed lines[8].

The prices for mobile phone services are difficult to compare between countries. In part this reflects exchange rate variations, but more importantly, pricing packages and the form of pricing differs significantly between countries. Most obviously, different countries have different charging mechanisms, with 'calling party pays' dominating outside the U.S. But in the U.S. and Canada 'receiving party pays' pricing often applies for calls to mobile telephones. Different packages and bundling of equipment and call charges also make comparisons difficult. A major innovation in mobile telephone pricing in the late 1990s was the use of prepaid cards. This system, where customers pay in advance for mobile calls rather than being billed at a later date, has proved popular in many countries. For example, in Sweden, prepaid cards gained 25 percent of the mobile market within 2 years of their introduction (OECD, 2000, p. 11).

Despite the changing patterns of pricing, the OECD estimated that there was a 25 percent fall in the cost of a representative 'bundle' of mobile services over its member countries between 1992 and 1998 (OECD, 2000, p. 22).

3. Economic issues in wireless communications

3.1. Spectrum as a scarce resource

Radio spectrum is a natural resource, but one with rather unusual properties. As noted above, it is nonhomogeneous, with different parts of the spectrum being best used for different purposes. It is finite in the sense that only part of the

[8] All statistics from the ITU-D Country Database.

electromagnetic spectrum is suitable for wireless communications, although both the available frequencies and the carrying capacity of any transmission system depend on technology. The radio spectrum is nondepletable; using spectrum today does not reduce the amount available for use in the future. But it is nonstorable.

Under ITU guidance, spectrum has been allocated to specific uses and then assigned to particular users given the relevant use. Traditionally, user assignment was by government fiat. Not infrequently, the user was government owned.

Privatizations in the 1980s and 1990s, and the success of (at least limited) mobile telephone competition in some countries, resulted in a more arms-length process of spectrum allocation developing in the 1990s. Users of radio spectrum, and particularly users of 2G and 3G mobile telephone spectrum, have generally been chosen by one of two broad approaches since the early 1990s—a 'beauty contest' or an auction.

A 'beauty contest' involves potential users submitting business plans to the government (or its appointed committee). The winners are then chosen from those firms submitting plans. There may be some payment to the government by the winners, although the potential user most willing to pay for the spectrum need not be among the winners. For example, the U.K. used a beauty contest approach to assign 2G mobile telephone licenses in the 1990s. Sweden and Spain have used beauty contests to assign 3G licenses.

France used a beauty contest to assign four 3G licenses. The national telecommunications regulator required firms to submit applications by the end of January 2001. These applications were then evaluated according to preset criteria and given a mark out of 500. Criteria included employment (worth up to 25 points), service offerings (up to 50 points), and speed of deployment (up to 100 points). Winning applicants faced a relatively high license fee set by the government. As a result, there were only two applicants. These firms received their licenses in June 2001, with the remaining two licenses unallocated (Penard, 2002).

The concept of using a market mechanism to assign property rights over spectrum and to deal with issues such as interference goes back to at least the 1950s when it was canvassed by Herzel (1951) and then by Coase (1959). But it was more than 30 years before spectrum auctions became common. New Zealand altered its laws to allow spectrum auctions in 1989 and in the early 1990s auctions were used to assign blocks of spectrum relating to mobile telephones, television, radio broadcasting, and other smaller services to private management (Crandall, 1998). In August 1993, U.S. law was modified to allow the FCC to use auctions to assign radio spectrum licenses and by July 1996, the FCC had conducted seven auctions and assigned over 2100 licenses (Moreton and Spiller, 1998). This included the assignment of two new 2G mobile telephone licenses in each region of the

U.S. through two auctions[9]. In 2000, the U.K. auctioned off five 3G licenses for a total payment of approximately $34 billion[10].

Auctions have involved a variety of formats including 'second price sealed bid' in New Zealand, modified ascending bid in the U.S., and a mixed ascending bid and Dutch auction format in the U.K.[11] Bidders may have to satisfy certain criteria, such as service guarantees and participation deposits, before they can participate in the auctions. Limits may also be placed on the number of licenses a single firm can win in a particular geographic area, so that the auction does not create a monopoly supplier. From an economic perspective, using an auction to assign spectrum helps ensure that the spectrum goes to the highest value user.

While auctions have been used to assign spectrum to different users, they still involve a prior centralized allocation of bands of spectrum to particular uses. Economically, this can lead to an inefficient use of spectrum. A user of a particular frequency band (e.g. for 3G services) might have a much higher willingness-to-pay for neighboring spectrum than the current user of that neighboring spectrum (e.g. a broadcaster or the military). But the prior allocation of frequency bands means that these parties are unable to benefit from mutually advantageous trade. It would violate the existing license conditions to move spectrum allocated to one use into another use even if this is mutually advantageous. Building on the work of Coase (1959), Valletti (2001) proposes a system of tradable spectrum rights, using the market to both allocate spectrum to uses and simultaneously assign it to users. Interference can be dealt with through the assignment of property rights and negotiation between owners of neighboring spectrum. Valletti notes that both competition issues and issues of mandated standards would need to be addressed in a market for spectrum rights. We deal with the issue of standards later in this section while competition issues are considered in Section 5 below.

Noam (1997) takes the concept of tradable spectrum assignment one stage further. Technological advancements, such as the ability for a signal to be broken into numerous separate digital packets for the purposes of transmission and then reassembled on reception, means that the concept of permanent spectrum assignment may become redundant in the near future. As technology advances, Noam argues, spot and forward markets can be used to assign use within designated bands of spectrum. The price of spectrum use would then alter to reflect

[9] Formally, the FCC spectrum auctions were for personal communications services (PCS). In addition to 2G mobile telephones, relevant services included two-way paging, portable fax machines, and wireless computer networks. See McAfee and McMillan (1996) for an early appraisal of these auctions.

[10] See Binmore and Klemperer (2002) for a review of the U.K.'s 3G auction. For a survey of auctions in telecommunications see Cramton (2002).

[11] McMillan (1994) notes problems that arose in early spectrum auctions including the 1990 spectrum auctions in New Zealand and the 1993 satellite television auctions in Australia.

congestion of use. DeVany (1998) also discusses market-based spectrum policies, including the potential for a future 'open, commoditized, unbundled spectrum market system' (p. 641).

Conflicts in the allocation of spectrum allocation arose in the FCC auctions in the U.S. The 1850–1910 MHz and 1930–1990 MHz bands to be allocated by these auctions already had private fixed point-to-point users. The FCC ruled that existing users had a period of up to 3 years to negotiate alternative spectrum location and compensation with new users. If negotiations failed, the existing user could be involuntarily relocated. Cramton, Kwerel, and Williams (1998) examine a variety of alternative 'property rights' regimes for negotiated reallocation of existing spectrum and conclude that the experience of the U.S. reallocations is roughly consistent with simple bargaining theory.

While economists have generally advocated the assignment of spectrum by auction, auctions are not without their critics. Binmore and Klemperer (2002) argue that a number of the arguments against auctions are misguided. But both Noam (1997) and Gruber (2001b) make the criticism that spectrum auctions automatically create a noncompetitive oligopoly environment. Gruber argues that technological change has generally increased the efficiency of spectrum use and increased the viability of competition in wireless services. For example, in terms of spectral efficiency, GSM mobile telephone services are approximately 4 to 30 times more efficient than earlier analogue systems (Gruber, 2001b, Table 1). An auction of spectrum rights, however, is preceded by an allocation of spectrum. The government usually allocates a fixed band of spectrum to the relevant services. Further, the government usually decides on the number of licenses that it will auction within this band. So the price paid at the auction and the level of *ex post* competition in the relevant wireless services are determined by the amount of spectrum and the number of licenses the government initially allocates to the service. While the auction creates competition for the scarce spectrum, it does not allow the market to determine the optimal form of competition. Noam argues that flexibility of entry needs to be provided by the assignment system in order to overcome the artificial creation of a noncompetitive market structure.

3.2. Complementarities in spectrum use

Using spectrum to produce wireless communications services can lead to synergies between services and between geographic regions. In the U.K., 3G spectrum auction, the potential synergies between 2G and 3G mobile telephone infrastructure was noted by Binmore and Klemperer:

> [T]he incumbents who are already operating in the 2G telecom industry enjoy a major advantage over potential new entrants .... Not only are the incumbents' 2G businesses complementary to 3G, but the costs of rolling out the infrastructure (radio masts and the like) necessary to operate a 3G industry are very substantially less than those of a new entrant, because they can piggyback on the 2G infrastructure. (2002, p. C80)

Thus, there are synergies in terms of being able to supply new products to an existing customer base using existing brands, and economies of scope between 2G and 3G services.

Geographic synergies are evident from the FCC 2G auctions. Moreton and Spiller (1998) examine the two 1995–96 mobile phone auctions in the U.S. They run a reduced-form regression on the winning bid for each license and a number of factors designed to capture the demographics of the relevant license area, the competitive and regulatory environment, and the effects of any synergies. These were ascending bid auctions so that the winning price is approximately equal to the second-to-last bidder's valuation for the license. As such, the relevant synergies relate to the network of the second-to-last bidder, to capture any effect of this network on the value of that bidder.

To capture the effect of geographic synergies, Moreton and Spiller assume that the expected network associated with any bidder is the same as the actual post-auction network. They categorize geographic synergies as either 'local' or 'global'. Local synergies consider the relationship between value of a license in one area and ownership of 2G licenses in neighboring geographic areas. Global synergies look at the total extent of the second-to-last bidder's national network.

Moreton and Spiller find strong evidence of local synergies. "At the local level, our results indicate that groups of two or more adjacent licenses were worth more to a single bidder than to separate bidders" (p. 711)[12]. These local synergies appear to fall rapidly as the geographic area covered by adjacent licenses increases and evidence of global synergies is weak.

Local coverage by existing cellular services tended to reduce the price paid for 2G licenses in the Moreton and Spiller study. This appears to run counter to the Binmore and Klemperer argument for economies of scope between different mobile telephone services. Moreton and Spiller argue that the negative relationship may reflect a reduction in competition. Firms are reluctant to bid strongly against existing analogue mobile telephone incumbents and prefer to use their limited resources elsewhere. This argument, however, is weak. In an ascending bid auction, participants will bid up to their own valuations and if there are positive synergies between existing analogue mobile services and 2G services, this should raise the value of the license to the second-to-last bidder regardless of any other parties bids.

As expected, Moreton and Spiller find that the value of a 2G license increases with market population and population growth rate and decreases with the size of the area served. These results are broadly consistent with Ausubel et al. (1997) and are intuitive. Population and demand are likely to be positively correlated so that for any given level of competition, increased population will tend to increase

[12] Ausubel et al. (1997) also find evidence that local synergies were a significant determinant of price in the U.S. mobile telephone spectrum auctions.

expected profits. But increased geographic region tends to raise the roll-out cost of the 2G cellular network for any population size, lowering expected profits.

The Moreton and Spiller study find some evidence that those jurisdictions, where regulators require tariff filing for the existing analog mobile phone networks tend to have higher values for the 2G licenses. This suggests that tariff filing on existing services may have an anticompetitive effect leading to higher prices overall. The potential anticompetitive effects associated with regulatory notification and publication of price information has been shown in other industries (e.g., Albaek et al., 1997).

3.3. Standards

The adoption of standards has been a long-standing issue in wireless communications. The international spectrum allocation system is a form of standardization—ensuring that certain frequencies are used for certain purposes on a regional, or worldwide basis. Standardization, often with government involvement, has also been a key factor in the success of new wireless technology.

For example, when television was first introduced in the U.S., there were a variety of competing potential technologies. In 1939, NBC began experimental television broadcasts in New York, but the FCC ordered it to cease broadcasting until the FCC had approved a standard. As a result, in 1940, the National Television Standards Committee (NTSC) was formed and their recommended standard was adopted by the FCC. The FCC mandated that television broadcasters must use this standard[13]. Governments in other countries also had a strong role in determining television standards, often to be broadcast by state-owned companies. For example, in the 1950s, Western European countries adopted alternative technologies that were incompatible with NTSC, known as PAL and SECAM.

In contrast, AM radio in the U.S. did not involve government-mandated standards. In fact, radio competition evolved before both the 1934 Communications Act and the formation of the FCC. A significant difference in mobile telephony between the U.S. and Europe relates to standards. For analogue cellular systems, Europe had a variety of incompatible systems while the AMPS system was ubiquitous in the U.S. But the reverse holds true for 2G mobile, with GSM being the government-imposed standard in Europe but incompatible alternative systems being offered in the U.S.

There has been extensive analysis of the incentives for private firms to adopt standardized products or components. Economides (1989) considers both pricing and industry profits when firms can produce either compatible or incompatible component products. If components are compatible, then customers can assemble a system using different sellers' individual components. If the components are

[13] See Rohlfs (2001, Ch. 12) for background on the adoption of standards in U.S. television.

incompatible, however, a consumer can only buy a system by assembling components from the same seller. For a given number of firms, Economides shows that both prices and profits are higher when firms adopt compatible technology even in the absence of any direct consumer network externalities. This is due to the effect of compatibility on the intensity of competition. Given the prices set by other firms, a reduction in a firm's own price of one component tends to raise the demand for just this component when systems are compatible. But the same price decrease raises the demand for the firm's entire system of components when they are incompatible across firms. This leads to more intense price competition when components are incompatible.

For single products, in the absence of network externalities, differentiated consumer tastes mean that product variety is likely to be preferred. If there are direct network externalities, so that each consumer benefits when more consumers buy a product that is compatible with their own choice, standardization can create consumer benefits. Again, incompatibility tends to result in more intense competition, so that prices and profits are higher when a desirable standard is chosen.

As Shy (2001) notes, both the components and the network-externality models of compatibility show that choosing a standard tends to raise prices, lower consumer welfare and to raise social welfare. This does not, however, mean that private and social incentives for standardization are aligned. As Katz and Shapiro (1985) note, firms may have too little or too great an incentive for standardization from a social perspective. Further, even if private firms adopt a standard through market interaction, this need not be the most desirable standard from a social perspective (Economides, 1996)[14].

If competitive markets do not necessarily result in optimal standard choice, then there is a potential role for government. However, it cannot be assumed that government standard setting is necessarily superior to market-based processes. Governments may not set appropriate standards. This, in part, reflects the information limitations government faces, particularly when dealing with rapidly evolving technology. Standards are also often associated with proprietary intellectual property. This means that governments must deal with problems relating to 'essential' intellectual property when choosing a standard. For example, Bekkers, Verspagen, and Smits (2002) consider the intellectual property conflicts that arose in the choice of the GSM standard in Europe. Further, being chosen as the appropriate standard can be profitable for the firm that owns and controls that standard, leading to the potential for corruption in the process of standard choice.

History shows that government choice of standard is not a panacea for market failure. As Rohlfs notes,

[14] See Economides (1996) for an extensive survey of the economics of compatibility choice in network industries. For more recent surveys, see Shy (2001) and Liebowitz and Margolis (2002).

> [T]he history of television illustrates the wide range of possible outcomes from governmental standard setting—from effective action to utter failure.... The history of television illustrates many of the substantial defects that inhere in government decision-making; for example, the role of political influence/corruption and the pursuit of unworthy protectionist goals. Nevertheless, in bandwagon markets, government standard setting is sometimes superior to the alternative inefficient competitive process. (2001, pp. 164–165)

While standards have been important in wireless communication, there has been little formal economic analysis focused specifically on wireless. Some recent papers, including Lehenkari and Miettinen (2002) and Haug (2002) focus on the history of technological adoption and the role of government for the NMTS and GSM standards.

Standardization is most important when there are strong network externalities. It can be argued that mobile telephony, unlike say television broadcasting, has few network externalities. Direct network externalities appear limited. Historically, mobile phone calls have tended to be to or from the fixed line network. Even for mobile-to-mobile calls, there is little need for compatibility as calls from say a CDMA phone can terminate on an AMPS phone through the fixed-line network. The main network benefit for mobile phones relates to roaming—where the out-of-area coverage of a particular technological standard becomes important. Thus, while incompatible standards may lead to customer lock-in as moving service providers require purchase of a new telephone, there are few network externalities that justify a government imposed standard. In such circumstances, it is most likely better to allow competing standards to 'fight it out in the market' (Rohlfs, 2001, p. 141. See also Shapiro and Varian, 1999, pp. 264–267).

Koski and Kretscher (2002) empirically estimate the effects of standardization through two alternative approaches. For a series of 32 industrialized countries, they consider how a variety of factors, including standardization influence (a) the timing of the initial introduction of digital mobile telephony in a country, and (b) the diffusion of digital mobile technology in the country. The initial adoption decision is modeled by considering the probability of adoption by a country after 1991—the year before retail 2G services were first offered in Finland. In other words, the authors model the probability that a country will adopt digital mobile technology in any year t given that it has not previously introduced 2G technology. Diffusion is modeled through a standard S-shaped (logistic growth curve) process.

Koski and Kretscher find that standardization has a positive but insignificant effect on the timing of initial entry of 2G services. Thus, the adoption of a specific national standard for 2G is statistically unrelated to the timing of adoption of 2G by a country. This is not unexpected, even if there are network effects. While Koski and Kretscher argue that standardization should encourage early entry, by reducing technological uncertainty, when the standard is chosen by a government or regulator there is no reason to expect that the choice of standard will be timely.

Network effects are more likely to be important for the diffusion of mobile technology. Koski and Kretscher find that standardization has significantly facilitated the diffusion of 2G mobile technology. This result is broadly in line with Gruber and Harold (2001a), who analyze the diffusion of both analog and digital mobile technology over 118 countries. Countries that had competing analog systems had significantly slower rates of diffusion than single standard countries. They also find that standardization is associated with faster diffusion for digital technology, although this effect is both smaller and less precise than for analog technology. Gruber and Harold (2001a) conclude that "the disadvantages of competing systems (network effects and scale economies) were dominant during the analogue era. During the digital era, the disadvantages may have been partly balanced by the advantages from technological systems competition." (p. 1210)

A strong empirical prediction from the theoretical literature on standards is that standardization can lead to higher prices as it dampens competition. This prediction is consistent with Koski and Kretschers' findings. Their regressions show a significant positive relationship between standardization and price, and they conclude that 'firms implement less aggressive pricing strategies when competition takes place within a single standard.' (2002, p. 26)

4. Diffusion and demand for mobile telephony

4.1. *Diffusion*

While standardization appears to increase the rate of diffusion of mobile technology within a country, what else affects the rate of take up of mobile phones? Gruber and Verboven (2001b) estimate the diffusion of mobile technology over the 15 states in the European Union using an S-shaped diffusion path[15]. They address a number of specific issues:

a. *Did the switch from analogue to digital systems increase the rate of diffusion?* Gruber and Verbovens' analysis suggests that the move to 2G systems in Europe led to a rapid increase in the diffusion of mobile technology. This diffusion effect was significantly greater than the acceleration effect built into an S-shaped diffusion process. However, Gruber and Verboven do not isolate the cause of this acceleration. One explanation is standardization. The 2G technology was introduced using the GSM standard across the EU, replacing a variety of incompatible analog systems. Thus, the network effects associated with roaming may underlie the increased diffusion. Alternatively, as noted above, 2G mobile systems are significantly more efficient in terms of

[15] For a nontechnical overview of the diffusion of mobile telephones in the EU, see Gruber (1999).

spectrum use than analog systems. The capacity expansion associated with the introduction of 2G systems could lead to a drop in prices that encourages mobile take-up, regardless of the degree of competition.

b. *Does increased competition increase the rate of diffusion?* Intuitively, increased competition (say from monopoly to duopoly) should lead to lower prices and increase the rate of diffusion of mobile technology. This intuition is verified by Gruber and Verboven. Moving from monopoly to duopoly increased the rate of diffusion although this effect was more important for analog technology and was smaller than the effect on diffusion of the move to digital technology.

c. *Do 'late' adopters catch up with 'early' adopters?* Gruber and Verboven show that convergence is occurring within the EU. Late adopting countries tend to have a rate of diffusion that is faster than early adopters, although the lead by early adopters is predicted to last for more than a decade on their analysis.

These empirical findings are broadly supported by other work. Koski and Kretschmer (2002) find that a wider diffusion of analog mobile technology has a significant effect on the early take-up of 2G technology. This may reflect that spectrum capacity was being reached in those countries with greater analog mobile phone use, and is consistent with the reduced capacity constraints of digital systems accelerating the rate of 2G diffusion.

Koski and Kretschmer (2002), Gruber and Verboven (2001a), and Gruber (2001a) all find that increased competition increases the rate of diffusion of 2G technology. Ahn and Lee (1999) estimate the demand for mobile connections using data from 64 countries. Consistent with the 'competition' hypothesis, they find that mobile demand is decreasing in price, although only the effect of monthly charges is statistically significant.

Koski and Kretschmer (2002) do not consider 'catch up' explicitly but do find that countries with a higher GDP per head adopted 2G technology earlier. This is consistent with Ahn and Lee (1999). Gruber and Verboven (2001a) also find that 'countries with higher income per capita tend to be more advanced in adopting mobile phones' (p. 1204). Further, they consider convergence in terms of the level of a countries development and show that convergence in diffusion of mobile technology is much faster in rich countries than in poor countries. Thus, a 'late starter' among poorer nations will take a lot longer to catch up with an early starter than would a comparable 'late starter' among rich countries.

4.2. The relationship between fixed and mobile telephony

Are fixed-line telephones and mobiles complement or substitute in demand? Theoretically, the answer is ambiguous. To the extent that mobile telephones offer similar call functions to fixed-line telephones, we would expect there to be substitution in demand (Woroch, 2002). But mobile telephones are often used for short calls that would not be possible on a fixed-line telephone and such calls are often made to or from fixed-line telephones. Thus, the diffusion of mobile technology increases the benefits accruing to a fixed-line subscriber, potentially increasing demand for fixed-line services.

In the U.K., Office of Telecommunications (OFTEL) concluded on the basis of qualitative survey evidence that fixed-line and mobile telephones, to a significant degree, are complements. "[T]he advent of the mobile has, to a significant degree, expanded the market for making calls, rather than substituting for fixed calls, implying that a large majority of mobile calls are complementary to fixed calls." (OFTEL, 2001, paragraph A1.14)

This conclusion is backed up by Gruber and Verboven (2001a). "[C]ountries with a large fixed network tend to be more advanced in adopting mobile phones" (pp. 1024–1025). While this effect is diminishing over time, Gruber and Verboven conclude that "the fixed network is largely viewed as a complement to mobile phones." Similarly, Ahn and Lee (1999) find that "[t]he number of fixed lines per person ... has a positive influence on the probability of mobile telephone subscription." (p. 304)

A number of recent studies, however, suggest that mobile and fixed-line telephony are substitutes in demand. Horvath and Maldoom (2002) criticize the OFTEL conclusion as potentially confusing complementarity with individual tastes. If individuals who have a greater propensity to make telephone calls tend to have both mobile and fixed-line telephones, then simply noting the correlation between ownership does not give any information about the degree of substitution between mobile and fixed-line services. They use survey data to try and correct for this taste effect, concluding that an individual's spending on fixed-line telephony decreases significantly when the individual also has a mobile telephone.

While Horvath and Maldoom consider call spending, other studies relate mobile and fixed-line penetration rates. Cadima and Barros (2000) find that the ability to access mobile telephony reduces demand for fixed-line services. The availability of mobile services leads to approximately a 10 percent decrease in the fixed-wire telephony penetration rate in their study. Sung and Lee (2002) use Korean data and estimate that a 1 percent increase in the number of mobile telephones results in a reduction of 0.10–0.18 percent in new fixed-line connections and a 0.14–0.22 percent increase in fixed-line disconnections.

The conflict between these empirical results may partially be explained by differences in countries and the life cycle of mobile technology. As Gruber (2001a) and Gruber and Verboven (2001a) find increased waiting lists for fixed-line telephones has a positive effect on the diffusion of mobile telephones. Thus, in lesser developed countries we would expect access to a mobile phone to substitute for rationed access to fixed-line connections. In developed countries, initial use of mobile telephones is likely to involve mainly mobile-to-fixed or fixed-to-mobile calls. In such circumstances, call benefits can dominate leading to complementarity between fixed and mobile services. But as mobile penetration rises and mobile-to-mobile calls increase in importance, mobile phones may become substitutes for fixed-line services. This latter effect may be exacerbated as technology advances, both reducing the cost of mobile services and improving mobile functionality.

While there has been significant empirical work on the demand-side relationship between fixed and mobile services, there has been little work considering the supply-side. Koski and Kretschmer (2002) find that those countries with more competition in fixed-line telephone services introduced 2G mobile services earlier. They argue that this reflects substitution in supply between fixed and mobile telephones. Liberalization of fixed-line services tends to raise competition and lower profits in these services, making early entry into mobile look relatively more attractive for telecommunications firms.

4.3. Costs

A small number of papers have attempted to determine the presence of scale economies in mobile telephony. The results from these studies tend to be contradictory. While McKenzie and Small (1997) find that mobile telephony generally exhibits diseconomies of scale, and at best has constant returns to scale, Foreman and Beauvais (1999) find that mobile telephony exhibits modest increasing returns to scale.

The disagreement between such cost studies is unsurprising. In fixed-line local telephony, the presence or absence of increasing returns to scale has been subject to heated debate for much of the last 30 years (e.g. see Shin and Ying, 1992). In local telephony, the debate has immediate regulatory implications regarding the presence or absence of a natural monopoly in the local loop. In contrast, for 2G mobile, robust competition has been shown to be viable in many countries. Even if it is possible to find some degree of increasing returns to scale in mobile telephony, both the practicality and the benefits from mobile competition seem undeniable.

5. Regulation and competition

This section surveys the literature dealing with possible anticompetitive behavior in wireless markets and the associated regulation of wireless services. The focus is on issues that are of potential concern to policy makers—including potential barriers to wireless competition (arising from incomplete coverage, roaming, standards, and number portability), the possibility of tacit collusion, discriminatory on- and off-net pricing, and the regulation of access prices.

5.1. Limitations of wireless competition

Given the success of wireless services, at least in terms of the spectacular growth in cellular subscriptions (see Section 2.3 for details), it is perhaps surprising that a main focus for the economics literature on wireless has been on the limitations of wireless competition. To some extent this probably reflects the historical structure of the industry, which given the allocation of scarce spectrum often only allowed two networks to exist, for example in the U.S. and U.K. A natural question to ask is whether two networks are enough to ensure competitive conditions, a question we will address in Section 5.1.1, which reviews the literature on tacit collusion in wireless markets.

Section 2 introduced several other properties of the wireless market that might cause policymakers to be concerned about the likely levels of competition and prices. These included sunk costs, high switching costs (lock-in occurs through long-term contracts, a lack of phone portability, and a lack of number portability), differentiation in coverage, and inconsistent standards. In Section 5.1.2, we review the competition and regulatory aspects of these issues. We also review the literature relating to pricing in wireless markets, dealing with the possibility of discriminatory on- and off-net pricing and the high price of international roaming.

5.1.1. Tacit collusion

A number of studies have suggested tacit collusion was present during the early period of wireless competition, when many wireless markets contained just two operators. For the U.K., Valletti and Cave (1998, pp. 115–116) point to tacit collusion to explain the 7 years of stable and similar prices (1985–1991), when the market was a duopoly, followed by the sudden decrease in prices, the emergence of multiple tariff options, and the divergence of prices between operators following the entrance of two new competitors in 1993.

Stoetzer and Tewes (1996, pp. 305–307) suggest tacit collusion characterized the German market during the duopoly period up to 1994. In addition to the fact that there were only two operators during this period, entry was not possible (due to spectrum not being available to competitors), tariffs were relatively simple and easily observable by competitors, and the operators appeared to be quite similar. These factors all tend to make tacit collusion easier to sustain. For Germany, Stoetzer, and Tewes point to stable prices but strong competition in the service dimension (such as geographic coverage and network-specific value added features) as further evidence the two operators were colluding over prices.

Parker and Röller (1997) and Busse (2000) conduct formal econometric analyses of the tacit collusion hypothesis using U.S. data. In each subnational geographic market, the U.S. Federal Communications Commission (FCC) granted licences to two cellular operators, with one going to the existing local wireline operator. A duopolistic structure was thus established. Both Parker and Röller and Busse use panel data on the different geographic markets over the period 1984–1988 to test the degree to which duopolistic competition is consistent with the observed market outcomes. Both find evidence of noncompetitive prices consistent with tacit collusion.

Parker and Röller estimate a standard structural model of competition (see Bresnahan, 1989 for a survey of this type of approach), making use of conjectural variations to imbed different types of competitive behavior. The relevant conduct parameter θ equals one for a monopoly, zero for perfect competition, one-half for noncooperative duopoly (Cournot), and is greater than one-half for collusive duopoly. They allow this parameter to depend on a variety of market characteristics that might increase the likelihood of tacit collusion, such as the extent to which operators compete with each other in multiple markets (multimarket contact), and the extent to which competitors have cross-ownership in some other market. They also allow the parameter to depend on the extent of state regulation (regulators sometimes requested or required operators to file tariffs for informational purposes).

An appealing feature of the data on wireless competition for this kind of estimation is that regulators fixed the structure of the industry. After the licenses were awarded, there was an initial monopoly period, followed by duopoly until around 1996 when further entry occurred. The length of this initial monopoly period depended on technical constraints. The first operator to provide a cellular service knew they would face a competitor (given the duopoly policy) and so there was presumably no point trying to price low initially to deter entry. Once a duopoly was established, there was also no point for the firms to try to deter further entry, since such entry was restricted by the lack of additional licenses. Additional licenses were not awarded until 1996, or later, and according to Parker and Röller, there was anecdotal evidence that wireless operators did not expect any entry over the period they study. This allows market behavior under duopoly to be compared to behavior under the initial monopoly without worrying about the possibility of entry deterrence behavior in either case. It also allows the initial monopoly period to be used to test the appropriateness of the empirical specification.

Parker and Röller allow their conduct parameter θ to vary across the monopoly period and the duopoly period. They estimate θ to be 1.079 with a standard error of 0.17 in the former period, which means their model cannot reject monopoly behavior during the period the market was known to be one of monopoly. In comparison, for the duopoly period, they estimate θ to be 0.857 with a standard error of 0.042, meaning they can reject perfect competition, noncooperative behavior, and cartel (monopoly) behavior. The estimate is consistent with collusive behavior somewhere between that of Cournot and a monopoly.

The conduct parameter was found to depend positively and significantly on variables that are thought to make tacit collusion easier; namely, the extent of multimarket contact, the extent of cross-ownership in other markets, and the ability of operators to voluntarily report prices to a regulatory authority[16]. Overall, their empirical evidence is consistent with the view that over the period 1984–1988, wireless operators in the U.S. were engaging in tacit collusion to sustain high prices.

[16] Mandatory disclosure of prices did not have a significant effect on the conduct parameter.

Busse (2000) is interested in how firms tacitly collude. She hypothesizes that U.S. wireless operators tacitly colluded over the period 1984–1988 by setting identical prices in multiple markets in which they serve. Identical pricing arises when a firm sets the same price schedule across different markets—rather than different firms setting the same price schedules within the same market. In her view, multimarket contact facilitated tacit collusion not only by enhancing the ability of firms to punish deviators, but also by increasing firms' scope for price signalling and coordination. Thus, multimarket contact alone may not be sufficient to enable wireless operators to tacitly collude. They may need a way to communicate their behavior, which could be established by setting identical price schedules in multiple markets.

Using a probit regression, she finds multimarket contact makes operators significantly more likely to set identical price schedules in the different markets they engage in, controlling for the similarity of the characteristics of the markets, the geographic proximity of the markets, and an indicator variable for each operator (to capture any idiosyncratic tendency of firms to price identically across the markets they serve). Second, using a panel regression of a firm's price on various demand characteristics, firm and time effects, and dummies for the use of identical pricing, she finds that when a wireless operator sets identical prices across markets, its price increases by 6.9 percent (with a t-statistic of 2.69)[17]. Combined, this evidence is suggestive of identical pricing across markets being used as a way to facilitate tacit collusion.

The above analysis suggests it is the combination of multimarket contact and identical pricing that is responsible for higher prices. Given this, it is surprising that Busse does not interact these variables directly. Instead, she adds multimarket contact as an additional explanatory variable in the price regressions and finds it is not important after controlling for the identical pricing dummy variable. The finding that multimarket contact does not lead to higher prices even when the identical pricing dummy variable is dropped, is inconsistent with the finding from Parker and Röller that multimarket contact leads to more collusive behavior. Given both studies consider essentially the same sample, this suggests one of the specifications is misspecified. Interestingly, Busse also finds higher prices are associated with price-matching, in which multiple competitors set the same price schedules in a given market, suggesting an additional avenue for tacit collusion.

Put together, the above studies are certainly indicative of tacit collusion during the early period of duopoly in wireless markets. The extent to which this estimated collusion has persisted in the face of more recent entry is something worthy of further empirical investigation. The anecdotal evidence from Valletti and Cave (1998) and Stoetzer and Tewes (1996) is that competition has indeed become

[17] The endogenous variable, price, is averaged over usage levels for each firm's price schedule. Similar results are obtained using a single point along each firm's price schedule, say, corresponding to usage of 500 minutes per month.

stronger following entry in the U.K. and Germany. It would be interesting to use U.S. data to see whether any breakdown of tacit collusion could be linked to specific market features, such as the process of entry, the number of new entrants, deregulation, or a breakdown in identical pricing. It would also be interesting to compare the estimated conduct parameter (a'la Parker and Röller) in the post-entry period with the preentry period.

The finding of tacit collusion raises issues of whether regulation can be used to limit tacit collusion, or whether regulation is in fact a source of tacit collusion. The result of Parker and Röller that tacit collusion is stronger in markets where operators can voluntarily report prices to a regulatory authority, but not in markets where operators must report prices is consistent with the view that regulation that allows for voluntary price reporting promotes tacit collusion and higher prices. Hausman (1995) and Hausman (2002) make the case that cellular prices are higher because of regulation[18]. Hausman uses data from the 26 states in the U.S. that had voluntary or mandatory disclosure of mobile prices (regulated states), and compares it to data from the other 25 states, which did not. Based on instrumental variables estimation in which 'regulation' is treated as a jointly endogenous variable (along with price), he finds that regulated states have prices that are 15 percent higher, holding other variables constant (population, average commuting time, average income). In 1996, after such regulation was eliminated due to an act of Congress, prices in the previously regulated states were not significantly different from the nonregulated states.

Duso (2003) takes a different approach, explicitly accounting for the endogeneity of regulation. He estimates an equation for the determinants of regulation. This allows for reverse causality in which regulation can be less likely when it would be most effective in reducing prices, reflecting the effect of lobbying activity on the regulatory choice. Controlling for this endogeneity of regulation, Duso finds that prices in regulated markets are lower than the prices cellular operators would have set had these markets not been regulated, although this effect is not statistically significant. Duso also finds that regulation would have resulted in a significant decrease in prices had the nonregulated markets been regulated. The latter result reflects the success of lobbying activity against regulation that would have reduced prices a lot.

Another paper that accounts for the endogeneity of regulation (in fact of deregulation) is that of Duso and Röller (2003). They consider the impact that deregulation (that is, entry) has had on productivity in the mobile phone sector across OECD countries. They again test for reverse causality, in this case whether countries that are more productive are more likely to deregulate. They estimate two equations—a policy equation and a market equation. The first is the effect of

[18] Shew (1994) also argues that regulation is responsible for higher cellular prices. Ruiz (1995) found that the regulatory variables did not significantly explain prices, and concluded that the analysis did not imply any policy suggestions.

deregulation on productivity. The second is the effect of productivity on the decision to deregulate. They find that deregulation does lead to higher productivity, but the impact is about 40 percent smaller once they control for reverse causality (the second equation).

The unique features of the regulatory environment for wireless markets make it ideal for testing theories of industrial organization and regulation. We just briefly mention some other recent work that utilizes these unique features. Duso and Jung (2003) provide an empirical investigation of the relationship between firms' lobbying expenditures and product market collusion using data from the U.S. cellular industry between 1984 and 1988. They find a significant negative two-way relationship between the strength of collusion in the product market and firms' lobbying expenditures. Collusive conduct decreases political activities, while higher contributions increase competition in the product market. Reiffen et al. (2000) use data from the cellular industry in the U.S. to study the possibility that wireless operators that are also local exchange carriers supplying interconnection to rival cellular operators may discriminate against their rivals[19]. The exogenous geographic differences in the carriers' ability and incentive to discriminate in the U.S. make the industry ideal for such a study. Miravete and Röller (2002) also use historical data on wireless markets in the U.S. to estimate an equilibrium oligopoly model of horizontal product differentiation when firms compete in nonlinear tariffs.

5.1.2. Strategic instruments to limit competition

Aside from the empirical studies of tacit collusion and regulation, most of the literature on competition in wireless markets is based on descriptive accounts of competition, with a focus on the barriers to effective competition. For instance, Valletti and Cave (1998) provide a detailed discussion of competition in the U.K. wireless market up to 1997. In addition to the possibility of tacit collusion in the preentry period, Valletti and Cave also discuss the various strategic instruments that wireless firms may have used to raise switching costs and enhance network effects, so as bind consumers to their networks. These include a lack of number portability, the use of SIM card locking, incompatible standards, and network-based price discrimination. The possibility of limiting competition through differential coverage is raised by Valletti (1999, 2002), while roaming policy has been the subject of recent regulatory interest. These themes are discussed here in relation to the available literature.

5.1.2.1. Number portability. Number portability ensures wireless subscribers can maintain the same phone number when switching to a rival's network. Since consumers generally value maintaining the same phone number, number

[19] Reiffen and Ward (2002) provide a survey of this line of work.

portability reduces switching costs. What are the implications of reduced switching costs for wireless competition and prices?

Farrell and Klemperer (2002) provide a survey of the switching cost literature, arguing in general that the lock-in associated with switching costs has an ambiguous effect on competition and prices. Generally, it leads to less aggressive *ex post* competition (once customers are locked-in), but it also leads to more aggressive *ex ante* competition (to attract market share in the first place). On balance, the present value of current and future prices and profits can be higher or lower as a result. Switching costs suggest a particular pattern of prices, a 'bargain-then-rip off' structure. This approach predicts that as wireless ownership matures, the percentage of locked-in consumers will increase and prices should rise. However, this does not necessarily imply wireless firms are earning supranormal profits. Through their *ex ante* competition for customers (for the market), they may have already competed away these future *ex post* rents[20]. This is less likely to be true if for the period over which operators were building market share they engaged in tacit collusion.

While there is no general and unambiguous result on the effects of switching costs, there seems to be some consensus that an increase in switching costs increases industry profits, especially where firms try to sustain or artificially create switching costs (Farrell and Klemperer, 2002). If this is true, then the wireless industry should be opposed to the introduction of number portability, which indeed seems to be the case[21]. There is an offsetting efficiency effect that could also explain why operators are opposed to number portability. If number portability is costly, but the operators bearing the cost are unable to fully pass on this cost to the consumers who wish to port their number, then there may be excessive use of number portability (with switching), resulting in higher costs to operators. Number portability could also have differential effects on new entrants vs. established incumbents, also explaining why it might be difficult to get industry agreement in support of it.

Aoki and Small (2001) construct a model of number portability in telecommunications between two suppliers of differentiated products, each of which sets a two-part tariff. Number portability is modeled as involving a reduction in switching costs and an increase in the marginal cost of making calls (as additional routing-related tasks must be performed to establish connections). The model of competition is one in which lower switching costs result in lower fixed fees set by

[20] Although policy intervention to control *ex post* rents of any incumbent that exploits locked-in consumers amounts to expropriation of the incumbents' *ex ante* investments in attracting customers, if incumbents expect such a policy the result could be a more efficient time path of prices, avoiding the usual 'bargains-then-rip off' pattern.

[21] Perhaps because of this, number portability has been mandated in the telecommunications sector in Australia, Denmark, Hong Kong, the U.K., and the U.S., among other countries. In the EU, there is a directive that states that all EU countries have to supply mobile number portability.

carriers, but in which the increased marginal cost of call production raises usage fees. The implications of number portability for consumer or social surplus in this model are ambiguous, a finding that contrasts with the normal regulatory presumption in favor of number portability. Their modeling suggests empirical work is needed to measure the two opposing effects, and to quantify the consumer and welfare effects of number portability.

In Aoki and Small, the consumer and social benefits depend on the market structure (symmetric or asymmetric) and who (incumbent or entrant) bears the cost of porting consumers. They consider several different market configurations with the most relevant to wireless being a structure in which not all consumers buy initially from the firms (infant industry). In this model there are two periods. In the first period there is a monopoly. In the second period some new consumers enter, and an entrant competes with the incumbent. Unlike the first period customers, the new customers can choose between two firms without incurring a switching cost[22]. Aoki and Small conclude, "In this case, under plausible assumptions, reductions in the cost of switching benefit consumers and the entrant, and have no effect on the incumbent. The consumer gains come entirely from expansion of the market."

Gans and King (2001b) also model number portability as a switching cost-reducing device. They focus on the case in which an established incumbent, which already has all of the potential customers, faces a new entrant. Unlike Aoki and Small, they allow the consumer to be charged for porting, which reduces any excessive usage of porting by consumers. However, this (along with the case in which entrant pays for porting) raises the possibility that the incumbent will then inflate the cost of providing number portability, so as to reduce switching. In the face of this effect, a superior regulatory solution may be to provide the customer with the opportunity to own their own phone number, and the carriers the option of buying this back from customers. Provided Coase-type bargains can be reached, this should ensure number portability is chosen only if it is efficient, and that the incumbent provides number portability in the most cost effective way[23].

The above models are orientated toward addressing number portability for the wired-line sector, between an established incumbent and a new entrant, in a mature market. A more interesting benchmark for wireless is one in which multiple (possibly symmetric) carriers are already established, and in which there are new customers to be attracted. In this case, switching costs may actually

[22] However, as with the Gans and King paper discussed below, this is still a static model of competition in that firms only compete in one period. As such, there is no *ex ante* competition for the market, and so the normal trade-off between *ex ante* competition for the market and *ex post* competition in the market does not apply. This explains why both papers have the property that higher switching costs imply higher prices.

[23] In practice, as Gans and King (2001b) point out, asymmetric information will likely prevent this first best outcome being achieved. See also the discussion in Gans, King, and Woodbridge (2001).

encourage entry, as a large customer base encourages incumbents to harvest its base relative to winning new customers, thus letting smaller firms catch up (Farrell and Klemperer, 2002). On the other hand, switching costs will tend to deter entry if entrants have to attract existing customers to cover their fixed costs, which could be the case when an expensive block of spectrum has to be purchased to enter in the first place.

Finally, it is worth noting that even if number portability is mandated, this does not mean it will be effective. Networks may find ways to limit number porting, through poor advice about the possibility of number porting, or by making porting troublesome for consumers (OFTEL, 2001, Section 2.2.3).

5.1.2.2. SIM card locking. SIM-locking stops mobile handsets obtained through one operator from being used to get a rival operator's service. For a fee (currently around 20 pounds in the U.K.), handsets can be unlocked to be used on a rival's network. Such fees are a way to tie together the ownership of a handset to the subscription to a wireless service. As such, their main impact exists when consumers do not already have contracts involving the subsidized purchase of a handset and an ongoing subscription service. Even in such cases, SIM-locking is only likely to have a limited impact, both since the fee is not huge, and since many consumers will wish to update their handsets anyway.

Unlike number portability, the decision to maintain SIM-locking rests with each individual carrier. In the U.K., fees for unlocking SIM cards have remained despite increased competition and despite regulatory investigations into the fees. The OFTEL (2002a) summarizes the arguments that have been put forward for and against the fees.

Justifications for the fees include that without SIM-locking there is a risk of a 'chicken and egg' situation, whereby providers would not develop services without being sure of having enough available handsets with the functionality to deliver them, but manufacturers would not add this functionality without more certainty that it would be used, and that SIM-locking enables operates to provide a handset subsidy, which encourages more participation in the market (to the benefit of existing users). On the other hand, OFTEL argues SIM-locking raises switching costs for consumers[24], prevents competition for services over the same handset (where consumers have multiple SIM cards), limits consumers choosing separately their preferred handsets and preferred service provider, and prevents suppliers providing just handsets or just service provision without providing the other.

The OFTEL's current response to these fees is to make consumers more aware of them in the first place. Although this will not reduce the switching costs implied by these fees, it may make consumers more likely to take the fees into account in

[24] This increase in switching costs may be offset by handset subsidies. The OFTEL (2001, Section 2.3.1) notes, "... it is possible that the current levels of switching are partly supported by handset subsidies because these subsidies lower the cost of switching when handsets are SIM-locked."

deciding which network to join in the first place, thereby increasing the demand for operators that set lower fees for unlocking SIM cards.

5.1.2.3. Standards and competition. The adoption of incompatible standards in wireless telephony increases switching costs, and has an effect similar to SIM-locking. For instance, in New Zealand, the two wireless networks (Telecom and Vodafone) operate on different technologies so that to switch between them, consumers need to purchase a different handset. To the extent that consumers expect to update their handsets fairly regularly, this may not be a large barrier to switching operators. Moreover, with a well-developed secondhand market for handsets, the switching cost implied by incompatible standards is further reduced (to the transaction costs of trading in the secondhand market).

5.1.2.4. Network-based price discrimination. Historically, the number of mobile-to-mobile calls has not been very significant, at least relative to the number of mobile-to-fixed calls[25]. As the penetration of wireless subscribers has increased, so has the proportion of mobile-to-mobile calls. This makes the use of discriminatory tariffs between on- and off-net calls more relevant[26].

In a model with two symmetric but horizontally differentiated telecommunications firms, Laffont et al. (1998) show that if firms are allowed to price discriminate, they will set their on- and off-net usage prices at cost and recover all profits through their fixed charges (monthly rentals). This implies that the incentive for network-based price discrimination (along the lines above) only arises to the extent the cost of terminating calls on the rival wireless network is higher than the cost of terminating calls on one's own wireless network. The main reason for any difference in the on-net vs. off-net cost of terminating calls is if the interconnection charge agreed for mobile-to-mobile calls is set above or below cost. Using the Laffont et al. (1998) framework, Gans and King (2001a) show that the operators will want to set this interconnection price below cost to soften competition[27]. This implies that wireless firms should charge consumers *less* for calls made to rival networks than to their own network. Clearly, this is not the case in practice. If there is any price discrimination, it tends to be to price off-net calls more than on-net calls[28]. This suggests there is some other reason for network-based price discrimination.

[25] According to Valletti (1999), in the U.K. in mid 1997, around 95 percent of wireless calls terminate on wired-line networks.

[26] By 2000/2001, off-net mobile-to-mobile calls in the U.K. accounted for 13 percent of all revenue generated (OFTEL, 2001, Section 1.6).

[27] Mobile-to-mobile interconnection is also discussed in Section 5.2.

[28] In OFTEL's review of the U.K. wireless sector, it noted the relatively high price of off-net mobile-to-mobile calls as an area of concern (OFTEL, 2001, Section 2.6.1).

Berger (2002) provides a model that can explain why operators may want to set off-net retail prices higher than on-net prices. His main idea is that when consumers care about receiving calls (this would seem to be quite important for wireless), firms will want to make it expensive to call the rival network, so as to lower the demand for subscription to the rival network. This effect makes networks set higher off-net prices than otherwise would be the case, since by doing so they obtain the added benefit of making their rival less attractive to subscribe to. They then compete by setting lower on-net prices (in his model firms set only usage fees). Like Gans and King, in Berger's model, firms may still want to set the interconnection price below cost, although in his model this helps offset the inflated off-net prices and so is generally a good thing[29].

5.1.2.5. Differential coverage. An important attribute of wireless is its coverage. Over what geographic area (or percentage of the population) does the mobile phone work? A network that allows consumers to use their phones over a wider area is, other things equal, more valuable. Full coverage can be viewed as the wireless equivalent of universal service, in that it requires more expensive locations to be provided for, although not necessarily at the same price. This could serve some social (or political) objectives, such as ensuring people living in remote areas can contact and be contacted by others on the network, and ensuring that people can always remain in contact even when they travel into more remote areas.

Valletti (1999) emphasizes the role of differential coverage in limiting wireless competition. He argues that operators may vertically differentiate themselves, opting either for full coverage or for minimal coverage (where the minimum allowable coverage is determined by regulation).

In Valletti's model, consumers differ in their willingness to pay for calls (income) but are otherwise homogenous with respect to their need for coverage. Firms play a two-stage competition game, choosing coverage in the first period and price in the second period. As a result of the vertical differentiation structure, in equilibrium firms differentiate themselves in terms of coverage, with one firm opting for maximal coverage and the rival firm opting for less coverage (minimum coverage if the regulated minimum coverage level is binding). The analysis suggests if regulators ensure a high minimum coverage of all operators, they will reduce vertical differentiation, resulting in more intense price competition[30].

A classic result in the literature on vertical differentiation is that even if incumbent operators earn positive economic profits, no matter how low are investment costs, there may be no further entry (Sutton, 1991). Valletti (2003) applies this

[29] Section 5.2 provides a different reason why mobile firms may set high off-net prices, which reflects inflated mobile-to-mobile termination charges.

[30] Minimum coverage conditions were required by OFTEL in the U.K. Note, however, minimum coverage conditions may be easily met if a few cells in large population centers enable carriers to say they cover a large proportion of the population.

result in the wireless context, showing in a model in which consumers move randomly between city and rural areas, and in which wireless operators choose their coverage in the rural area (as well as whether to enter the city), that there will be a 'natural oligopoly' equilibrium. Despite the existence of positive economic profits, there will be no further entry. Operators have a strong incentive to use differing levels of coverage in order to vertically differentiate and relax price competition.

Both of Valletti's models assume away horizontal differentiation. With both horizontal and vertical differentiation, firms may no longer want to differentiate vertically, and instead may tend to set similar coverage levels[31]. This seems to be the norm in practice, with most providers attempting to provide full coverage. Valletti argues that as more spectrums are released, price competition will become stronger, and the incentives to vertically differentiate will become greater. In this regard, evidence from the different geographic regions of the U.S. should be telling.

5.1.2.6. Roaming. Roaming arises when a network has incomplete coverage. If a geographic area is not covered by a subscriber's own network then the wireless subscriber may still be able to make or receive calls in the area if it is able to 'roam' on another network's infrastructure. It is useful to distinguish 'domestic' roaming from 'foreign' roaming. Domestic roaming happens when an operator wants roaming rights on a *rival's* network. Foreign roaming happens when an operator wants roaming rights on another network that does not (normally) compete for the same customers. In most countries, wireless firms operate at a national level so domestic roaming is equivalent to national roaming and foreign roaming is equivalent to international roaming. In the U.S., these need not be the same, as some firms operate across many regions and others in only specific regions. Likewise, with the development of some multimarket mobile operators, international roaming may be achieved within the same network (or alliance of networks).

Where firms do not compete against one another, as is frequently the case when operators are in different countries, the networks are complements to each other, and they both can benefit from roaming agreements. Consistent with this, Valletti (2003) notes that international roaming is far more prevalent than national roaming. However, international roaming is often alleged to be excessively expensive (Sutherland, 2001). Surveys conducted in 1999 indicate roaming charges are between 2 and 10 times the price of the same or a similar domestic call. The difference in price between roamed and nonroamed international mobile calls to the same destination within the EU can be up to 500 percent. The OFTEL (2001, Section 2.70) and Sutherland suggest a lack of customer information

[31] See for instance, Dos Santos Ferreira and Thisse (1996). Similarly, if firms tacitly collude, then they may not want to vertically differentiate.

may explain why retail prices for international roaming have remained so high. While plausible, this does not explain high wholesale charges. Typically, the operator providing the roaming service charges a wholesale fee based on its normal retail prices plus a margin. The home operator than adds a mark-up to this, often between 10 percent and 25 percent. The result appears to represent a case of double marginalization, which would be neither in the operators' or the consumers' interests.

One reason such high roaming rates could be in the operators' interests is if the demand for roaming is very price inelastic (compared to regular calls). Given the common costs of a mobile network, Ramsey-style pricing could be used to justify setting high rates for providing international roaming services (both wholesale and retail), although not necessarily to the levels chosen by the operators. With high wholesale charges for roaming, one might also expect that each foreign operator will undercut its rival (at the wholesale level) to be the network of choice for the home networks. Perhaps collusion at the wholesale level (through the negotiation of roaming agreements) can prevent such undercutting. High retail mark-ups are then sustained by the lack of customer information.

When operators compete for the same customer base, networks are substitutes and denying competitors domestic roaming is sometimes argued by policymakers as a way to make it more difficult for small entrants to compete. Entrants have to provide full coverage to compete on equal terms with incumbent operators, and it may be difficult to attract customers until they have comprehensive networks. This tends to raise the stakes of entry, although it is not so clear how it can prevent entry. In fact, it seems quite possible that if two partial coverage networks agree on high-roaming fees, retail price competition could be weakened.

Roaming is also likely to affect investment incentives. On the face of it, (cheap) roaming would seem to weaken the incentive of firms to invest in infrastructure (full coverage), since using a rival's network may be more profitable. A free-riding problem could result in insufficient investment. However, limiting their own incentive to invest could be good for wireless operators, if one models them as competing in a Cournot fashion. This is the insight Foros et al. (2001) develop in a model of investment and roaming in a wireless market.

In particular, they explore the implications of roaming policy for the upgrade of networks from 2G to 3G. The framework used is that of a three-stage game with two firms, where in the first stage roaming quality is chosen, in the second stage firms choose how much to invest, and in the final stage they compete in quantities. An extended version of the model also incorporates a third firm, a virtual operator that competes in the third stage of the game.

If wireless firms are not allowed to cooperate at the investment stage, they will choose too much roaming quality in the first stage. To understand this result, note that under roaming, each firm's investment increases the demand for the other firms' networks. By choosing a high-roaming quality, the effect of each firm's investment on the demand faced by rival firms is also high. It is assumed this spillover cannot be internalized through a pricing mechanism. With large spillovers under noncooperative investment, each firm will invest too little. This has the effect of lowering output and raising prices in the third stage.

When a virtual operator is allowed to enter in stage 3, the incentives for roaming change. The two physical operators may now want to choose a high level of roaming between themselves and a low level with the virtual operator, so as to block the third firm from sharing the market. Such blocking of the third firm may be socially desirable, since with a high level of roaming the virtual firm may induce the incumbent firms to invest less at the second stage.

Some countries, such as the U.K., have mandated roaming during the period over which competitors are rolling out their network. The above analysis suggests it could be important to take into account the investment incentives such a policy may have, as well as any direct effects roaming has on retail competition. Mandated roaming may result in entrant networks choosing limited coverage, which tends to perpetuate the regulation of roaming and delay full-blown infrastructure competition. Hausman (2002) also emphasizes the negative implications for investment of giving competitors a 'free option' to roam existing network.

5.2. Access pricing and caller vs. receiver pays

5.2.1. Varieties of access pricing

Broadly speaking, there are three different access prices that are of interest in wireless markets:

1. A call from a mobile operator to a fixed-line network (a mobile-to-fixed call) involves the mobile operator collecting all the revenue for the call, but the fixed-line network providing the termination of the call. As a result, the fixed-line network usually charges for the termination of mobile-to-fixed calls. Typically, the termination of mobile-to-fixed calls is treated in the same way as the termination of other calls to fixed-line networks, being regulated on a cost-basis.
2. A call from a fixed-line network to a mobile operator (a fixed-to-mobile call) often leads to an access price being set by the mobile operator. In the majority of markets, the termination of fixed-to-mobile calls has been priced many times higher than the termination of mobile-to-fixed calls, sometimes resulting in the regulation of fixed-to-mobile termination prices[32]. An exception is the U.S., where this access price is the same as the mobile-to-fixed access price due to the symmetry requirements under the 1996 *Telecommunications Act*. Other

[32] Mobile termination charges are regulated in the U.K using a price-cap approach, with the cap reducing toward estimates of long-run incremental cost over 4 years. In Australia, the regulator ties year-on-year changes in mobile termination charges to changes in retail prices of mobile services.

exceptions are Canada, Hong Kong, and Singapore, where there are no fixed-to-mobile termination charges. Instead, like the U.S., mobile operators recover the cost of terminating fixed-to-mobile calls from the mobile callers directly (receiver pays).

3. A call from a mobile operator to another mobile operator (a mobile-to-mobile call) may involve mobile operators charging each other for termination. Such access prices are often negotiated privately between mobile operators.

In this section, we will review the recent literature that has developed on fixed-to-mobile termination, dealing with the related issues of caller vs. receiver pays and mobile-to-mobile interconnection as we go.

5.2.2. Fixed to mobile termination

The first paper to model the optimal pricing of fixed-to-mobile termination was that of Armstrong (1997). Armstrong noted the high price of fixed-to-mobile calls in the U.K. In July 1996, the average per minute retail charge for fixed-to-mobile calls was around 25 pence. In his view, this high price reflected two distortions—an inflated fixed-to-mobile termination charge (16.5 pence per minute), and a margin that fixed-line operators put on fixed-to-mobile calls over and above the high termination charge (8.5 pence per minute)[33]. Armstrong provides a simple model of fixed-to-mobile calls, which deals with the optimal pricing of the two components.

Armstrong assumes mobile operators are perfectly competitive and symmetric. There are a fixed number of subscribers that the mobile operators compete for. Each mobile operator is assumed to charge the same amount for terminating fixed-to-mobile calls, and earns a margin equal to this termination charge less its costs of terminating calls[34]. Competition drives mobile operators to return this margin to its subscribers, which takes the form of a subsidy to consumers to connect to the network. Armstrong argues that high fixed-to-mobile termination charges can explain the handset subsidies so often observed in the wireless market.

Assuming fixed-line operators obtain no margin on fixed-to-mobile calls, Armstrong shows the socially optimal fixed-to-mobile termination charge is equal to cost. This ensures fixed-to-mobile calls are priced at cost, which is the only source of a possible inefficiency in the model. Such cost-based termination of fixed-to-mobile calls eliminates any subsidy to mobile subscribers.

Armstrong considers how optimal termination charges change when three important assumptions are relaxed. The optimal termination charge is lower if

[33] Even back in 1997, these issues had already been dealt with by regulation in the U.S. The 1996 Telecommunications Act already required that fixed-to-mobile termination be priced at the same level as mobile-to-fixed termination, and fixed-to-mobile retail prices be regulated on the same basis as other outgoing fixed-line calls.

[34] In his model, mobile-to-mobile calls and mobile-to-fixed calls do not play any role in the analysis.

mobile subscribers get utility from receiving calls, since then mobile subscribers benefit from lower fixed-to-mobile prices[35]. The optimal termination charge is also lower if the price of fixed-to-mobile calls involves a margin above the termination cost, as might be expected with a dominant fixed-line network that is free to set the fixed-to-mobile price.

A lower termination charge can help offset the mark-up put on fixed-to-mobile calls by a dominant fixed-line operator. Note this 'subsidizing a monopolist' approach is also applicable to mobile carriers to the extent competition between them is imperfect, suggesting it could also be used to justify higher termination charges. Finally, Armstrong notes that the optimal termination charge is higher to the extent a subsidy to mobile subscribers encourages more consumers to subscribe. An increase in mobile subscribers provides a benefit to fixed-line consumers who now can make more fixed-to-mobile calls, which increases the social benefits of above-cost termination charges.

An implicit assumption in Armstrong's model is that fixed-line callers alone determine the number and length of calls to a mobile subscriber. If in practice the number and length of fixed-to-mobile calls is jointly determined by both caller and receiver, then above-cost termination charges should lead competing mobile carriers to set *negative* reception fees. By paying consumers to receive calls, mobile carriers will more directly promote increased fixed-to-mobile termination revenue than an equivalent subscription subsidy. This can explain why consumers are not charged for receiving calls in countries with high (above-cost) termination charges. However, negative reception fees are not observed in practice. This could reflect the fact that with high fixed-to-mobile prices and no reception fees, the number and length of calls will be predominately determined by the fixed-line caller and not the mobile receiver. That is, fixed-line callers will usually be the first to want to hang up for price reasons. In this case, there will be little or no benefits to mobile carriers of further subsidizing reception fees rather than subscription fees[36].

Doyle and Smith (1998) develop a model of fixed-to-mobile calls that allows caller-pays and receiver-pays to be compared. Their model assumes two competing mobile operators and a single unregulated fixed-line network. They assume the demand for outgoing mobile calls determines the demand for fixed-to-mobile calls. They show that each mobile network will set its fixed-to-mobile termination charge above cost, and on top of this, the fixed-line network will impose its own bottleneck mark-up, resulting in a double marginalization problem. They then modify their model to consider the receiver pays principle. They consider a set-up

[35] It is worth noting that the same logic could also be applied to other types of calls, such as mobile-to-fixed calls, where the fixed-line subscriber obtains benefits from receiving calls from mobile subscribers. Traditionally the benefits of call receivers have been ignored.

[36] There may be other problems with negative reception fees. In this regard, Tommaso Valletti pointed out in private correspondence that some time ago in Italy, mobile operators set negative reception fees only to find that people were calling their mobile phones from office lines so as to obtain these rebates.

in which the fixed-to-mobile caller pays a price equal to the normal price of a fixed-line call. The mobile subscriber receiving the call makes up the difference between the posted price of a fixed-to-mobile call, set by the mobile operator, and the lower charge paid by the fixed-line caller. They argue under this receiver-pays principle, double marginalization is eliminated and prices are lower. In some cases, termination charges set by the mobile network will be lower than under the caller-pays principle.

Their analysis can be interpreted in an alternative way. Their set-up is equivalent to regulating the price of a fixed-to-mobile call at the price of a regular fixed-line call, and imposing zero termination charges for mobile carriers. With no termination revenues from the fixed-line network, mobile operators will be forced to charge mobile callers directly to recover their costs of terminating calls. With positive (and potentially high) reception charges, call receivers will have a strong interest in the number and length of calls to their phones. A lower reception charge will encourage mobile subscribers to accept more calls (or allow longer conversations). Both reception and subscription charges will be uniquely determined. In this case, competition between mobile carriers will drive their reception charges toward the cost of terminating calls. Under this interpretation, it is not that the receiver-pays principle helps solve the problem of inflated termination charges, but rather that the regulation of low (or zero) termination charges leads carriers to charge receivers for incoming calls.

5.2.3. *Operators' incentives in fixed-to-mobile termination*

Gans and King (2000) and Wright (1999) model the incentives of competing mobile operators in setting termination charges. They assume two mobile operators that compete in a Hotelling framework, and that fixed-line subscribers choose the number of fixed-to-mobile calls[37]. Gans and King consider the case in which the fixed line network can set differential fixed-to-mobile prices depending on which mobile operator the calls terminate on, but where consumers are ignorant about which mobile network they are calling. Wright makes a formally equivalent assumption, that a uniform fixed-to-mobile price is set. In both cases, with a fixed number of mobile subscribers in the market, competing mobile operators set their respective termination charges to the point where they completely choke off all demand for fixed-to-mobile calls.

Using a more general model of competition, Wright (2002) clarifies the source of this market failure. When a mobile carrier increases its termination charge, there will be two types of effects on its profits. First, the increase in its termination charge will increase the fixed-to-mobile price, and so decrease the number of

[37] As noted in the discussion of reception charges, the fact that fixed and mobile subscribers do not jointly determine the number of fixed-to-mobile calls is broadly consistent with the fact that in equilibrium it is the fixed-line caller that pays for the call.

fixed-to-mobile calls. This will change the carrier's termination profit obtained per subscriber, and so have a direct impact on its profits. In addition, any change in termination profit per subscriber will influence the strength of competition for mobile subscribers, having an indirect impact on the carrier's profits (and other carriers' profits).

When the fixed-line networks set differential fixed-to-mobile prices depending on which mobile network calls terminate on (and assuming callers know which operator they are calling), competition will lead mobile operators to set termination charges to the same level as would be set by a single monopoly mobile operator. Competing cellular firms will want to maximize termination profits so as to subsidize their mobile subscribers as much as possible.

With uniform fixed-to-mobile pricing (or customer ignorance), any increase in termination charges and decrease in fixed-to-mobile demand is shared across all mobile carriers. Essentially, fixed-to-mobile demand becomes more elastic. This effect implies each mobile operator will want to set a higher (monopoly) termination charge. However, as termination charges are increased above this monopoly level, there are two additional effects. On the one hand, termination profits per subscriber will decrease, which will decrease the mobile operator's profits. On the other hand, the increase in termination charges will increase the uniform fixed-to-mobile price, which will decrease the termination profits per subscriber that rivals can capture. This makes rivals compete less aggressively for mobile subscribers. When both operators set their termination charges at the monopoly level, the first effect will vanish while the second effect remains. This implies that each operator will want to set a termination charge above the monopoly level. In the special case of the Hotelling model considered by Gans and King (2000) and Wright (1999), the second effect continues to dominate no matter how high the two carriers termination charges are set, and equilibrium termination charges in fact eliminate fixed-to-mobile calls altogether. Each operator will want to set termination charges a few cents above its rival.

5.2.4. Outcomes from regulation of fixed-to-mobile termination

Relative to this rather extreme outcome, regulation of lower termination charges generally benefits all parties, including mobile operators and mobile subscribers[38]. It is unambiguously desirable. Relative to the monopoly termination charges that result when discriminatory fixed-to-mobile pricing is effective, regulation of lower termination charges benefits the fixed-line network and fixed-to-mobile callers, but generally makes mobile operators and mobile subscribers worse off. However,

[38] In the Hotelling model of competition used by Gans and King (2000) in which there are a fixed number of mobile subscribers, mobile operators are indifferent over the common fixed-to-mobile termination charge. However, in a more general model of competition, including when demand by mobile subscribers is elastic, mobile operators will be better off avoiding termination charges above the monopoly level. See Wright (2002).

starting from the monopoly termination charge, a small decrease in the termination charge will have only a second-order effect on mobile operators and subscribers, but will have a first-order effect on fixed-to-mobile prices. Welfare will be higher when termination charges are regulated below those set by competing firms.

Wright (1999) calculates welfare maximizing termination charges when consumers vary in their subscription benefits of a mobile phone. When consumers already have access to the fixed-line network, but many do not have access to the mobile network, a subsidy to mobile subscribers from high termination charges has two beneficial effects. First, both fixed-line subscribers and existing cellular subscribers benefit from being able to call (and be called by) new mobile subscribers who join because of subsidized subscription charges. Efficiency is also enhanced by above-cost termination charges when retail competition between mobile carriers is limited.

This 'monopoly subsidy' justification for a high-termination charge is neutralized to the extent the fixed-to-mobile price also needs to be subsidized to offset monopoly mark-ups. Wright takes into account these various effects and finds using a calibrated version of the model that the socially optimal termination charge can be several times costly when half of all consumers have mobile subscriptions (in equilibrium). As mobile subscription tends toward saturation point, the need to tax fixed-to-mobile calls to subsidize mobile subscription diminishes. Given an increased tax-base (an increased number of fixed-to-mobile calls), the efficient mark-up of termination charges above cost decreases.

Armstrong (2002a) analyses socially optimal termination charges, assuming away complications arising from mobile-to-mobile calls and other nonlinear network externalities. In addition to the effects discussed previously, he shows that when mobile subscribers internalize the benefits of those calling them (e.g., they might care about the price friends and family pay to call them), mobile carriers will have less incentive to set high-termination charges. Mobile subscribers directly care about how much fixed-to-mobile callers pay and, so are less inclined to join a network that offers cheap subscription only because it is expensive for others to call the network[39].

The models of fixed-to-mobile termination of Armstrong, Gans, and King, and Wright all assume away any role for the fixed and common costs of operating fixed-line and mobile networks. Implicitly, it is assumed that retail mark-ups on these services are sufficient for fixed and common costs to be fully recovered. In the face of fixed and common costs, solving for the Ramsey optimal termination charges taking into account the various externalities described above would be

[39] Such subscribers may still join a cheaper network funded by high fixed-to-mobile termination charges since they can always take advantage of cost-based mobile-to-fixed prices to call back fixed-to-mobile callers. This tends to make fixed-to-mobile demand more elastic, and the monopoly fixed-to-mobile termination charge lower.

required to determine the optimal solution[40]. However, given Ramsey considerations are at least as important for the fixed-line operator as they are for mobile operators, the Ramsey argument alone does not provide an automatic justification for high fixed-to-mobile termination charges. If fixed-to-mobile demand is inelastic, then it may be socially optimal for fixed-line networks to exploit this by setting high fixed-to-mobile retail mark-ups, thereby enabling other fixed-line prices to be lowered. High fixed-to-mobile mark-ups would then imply socially optimal fixed-to-mobile termination charges are lower.

5.2.5. *Competitive bottlenecks and market power*

Armstrong calls the situation in which firms compete for customers but in which access to these customers is required by third parties one of 'competitive bottlenecks'. Mobile operators have a bottleneck because when someone wants to call a subscriber to their network, the calling party has no choice but to call their network. However, unlike a typical bottleneck situation, mobile firms compete to attract the subscribers in the first place. Competition transfers the rents from the access bottleneck to the mobile subscribers.

Despite the bottleneck for mobile termination, mobile operators do not necessarily have market power. The market is a two-sided one, with services being provided to both fixed-to-mobile callers and mobile subscribers[41]. It is wrong to isolate the effects on just one side of the market, and ignore the effects on the other. After all, without mobile subscribers, there can be no fixed-to-mobile calls. The competitive bottleneck does not necessarily lead to higher overall prices of fixed-to-mobile calls and mobile services. Rather, with a competitive mobile sector, it leads to a distortion in the *structure* of prices between fixed-to-mobile calls and mobile services.

Regulation may be required to achieve a more efficient structure of prices in this market. However, this does not indicate a problem of market power. As Gans (1999) argues, "the *less concentrated* the mobile network market, the higher will be the level of fixed to mobile call charges." With perfect competition between mobile

[40] The Competition Commission in the U.K. has argued against the use of Ramsey pricing to set fixed-to-mobile termination charges, arguments summarized and rebutted in OFTEL (2002b). See also Competition Commission (2003) for a further lengthy discussion of why it rejects Ramsey pricing.

[41] A recent literature on two-sided markets has emerged that can be applied to these issues. Two-sided markets have the property that there are two types of users that wish to make use of a common platform, each of which obtains benefits that depend on the number of users of the opposite type. As Armstrong (2002b) notes, fixed and mobile telephony is a type of two-sided market. One type of user is the mobile subscriber who values receiving calls, and the other is the fixed-to-mobile caller who values being able to call mobile subscribers. The two-sided markets literature suggests similar issues to those discussed here arise in many different industries including shopping malls, yellow pages, software/hardware, payment systems, and matchmaking services to name a few. See also Caillaud and Jullien (2001), Guthrie and Wright (2003), Parker and Van Alstyne (2000), Rochet and Tirole (2001), and Schiff (2003).

operators, the distortion in prices between the two sides of the market may be at its most severe (for instance, if there is uniformity of fixed-to-mobile prices, or if there is some customer ignorance over which network they are calling). Moreover, the need to regulate termination charges of small carriers may be even stronger, since small carriers will tend to set even higher termination charges[42].

A more general framework of fixed and mobile networks would have the number of fixed-to-mobile calls be jointly determined by both fixed and mobile callers. Taking into account that both callers and receivers obtain some utility from calls, and allowing for the possibility that networks can charge consumers directly for making and receiving calls; in general the socially optimal termination charge could be negative, zero, below cost, at cost, or above cost[43]. Seen in this light, the termination charge is just the instrument that the fixed-line and mobile networks can use to get closer to their desired structure of retail fees (caller and reception fees). Bill and keep (zero termination charges) might have some appeal in this context, as suggested by DeGraba (2002), but depending on various asymmetries in the two sides of the market, so might various other approaches.

It would be especially interesting to consider the case in which users get random utility from calls, and so can choose to be callers, or wait to be receivers. In this case, the question is whether there are any particular advantages of having callers pay (given originators of calls are more likely to be the ones with higher willingness to pay), and therefore whether there is a general argument for positive fixed-to-mobile termination charges. Empirically, it would also be fascinating to try and address whether receiver pays, as opposed to caller pays, has restrained growth in mobile services[44].

5.2.6. *The receiver-pays principle*

Finally, a word on mobile-to-mobile calls. Traditionally, mobile-to-mobile termination has been thought about in the same way as other two-way interconnection problems. In this case, the literature predicts that operators will agree to below cost termination (see Gans, 1999; Gans and King, 2000; and Berger, 2002). However, with high mobile penetration, mobile-to-mobile calls are an important substitute for fixed-to-mobile calls. If consumers are charged too much for making fixed-to-mobile calls, they will simply make mobile-to-mobile calls instead. This substitution possibility should help constrain the ability of the fixed-line operator to mark up fixed-to-mobile calls.

[42] To the extent that fixed-to-mobile callers act as though there is a uniform price of calling mobile subscribers, a small mobile carrier will face very inelastic fixed-to-mobile demand.

[43] Hermalin and Katz (2001) and Jeon et al. (2001) provide frameworks in which callers and receivers may obtain benefits from calls, and be charged for calls.

[44] The faster growth of mobile subscriptions in (caller-pays) Europe compared to (receiver-pays) North America, the rapid growth of subscriptions in Mexico and Peru following their shift to caller-pays from receiver-pays, and the lackluster performance of mobile subscription growth in India under receiver pays (Srivastava and Sinha, 2001) are all noteworthy in this regard.

To the extent that mobile-to-mobile calls are cost based, this will also undermine the ability of mobile carriers to raise their fixed-to-mobile termination charges much above cost. Fixed-to-mobile demand will become highly elastic. To overcome this customer arbitrage, mobile carriers may want to set mobile-to-mobile termination charges at similar (high) levels as fixed-to-mobile termination charges. This could explain the emergence of off-net price discrimination involving higher retail prices for mobile-to-mobile calls that terminate on a rival's network vs. calls that terminate on the same mobile network[45]. This also suggests the analysis of mobile-to-mobile termination charges may be quite closely tied to that of fixed-to-mobile termination. Future research would be usefully directed toward a deeper analysis of mobile-to-mobile calls.

6. Conclusions

The unique resource of wireless communications is the radio spectrum. The evolution of mobile telephony is therefore a story of how regulation and technology development interact to drive the diffusion of ever more radio spectrum use. While auctions have been used to assign spectrum to different users, they still involve a prior centralized allocation of bands of spectrum to particular uses. But the prior allocation of frequency bands means that these parties are unable to benefit from mutually advantageous trade. While the auction creates competition for the scarce spectrum, it does not allow the market to determine the optimal form of competition.

The diffusion of wireless communications follows an S-shaped path. The speed of adoption has been accelerated by the switch from analogue to 2G technology and by increased competition. Late adopters in the EU have been able to make up some of the lead of the early adopters. Mobile telephony seems to be moving toward becoming a substitute for fixed-line telephony both on-the-demand and on-the-supply side.

While mobile telephony is characterized by some amount of competition, this competition can be highly imperfect. Evidence exists for tacit collusion through multimarket contacts (U.S.) and soft end-user price regulation (U.S.). Switching costs through lack of number portability could lessen competition, but introducing number portability may be more costly than justified by improved

[45] Off-net calls averaged over four times as much as on-net calls in the U.K. in 2000/01. See Competition Commission (2003, 2.122). More surprisingly, they were significantly higher than on-net calls, even allowing for differences in termination charges. See Competition Commission (2003, footnote 1). As discussed in Section 5.1.2(d) above, Berger (2002) provides one explanation for this phenomenon when consumers value incoming calls. An alternative explanation for why a large mobile network may want to set high charges for calls to rival mobile networks is to generate an artificial network effect, so that customers will be drawn to its own (large) network to avoid higher off-net prices.

competition. Similar ambiguities arise with respect to SIM-locking. Network-based price discrimination is common among mobile networks. It runs counter to theoretical models predicting lower off-net than on-net charges. This could be explained by the utility gained by called parties from receiving calls. Regional coverage of networks seems to have become similar over time so that competitive advantages and disadvantages from differentiated coverage have largely disappeared. This, however, does not yet hold for international roaming.

High rates for roaming may simply reflect inelastic demands from users while abroad; but they may also suggest inefficiencies from double marginalization. Such high-roaming rates would disadvantage carriers with limited own international network coverage. High fixed-to-mobile termination charges may be compatible with intense overall competition between mobile carriers. In this case, such termination charges would not signal exploitation of consumers but rather allocative and competitive distortions between fixed and mobile networks, and between different mobile services, such as receiving calls and receiving handset subsidies. Mobile operators may even set termination charges above the monopoly level.

High fixed-to-mobile termination charges may be allocatively justifiable at low mobile penetration levels. At higher penetration levels they cannot even be justified by Ramsey principles, as such would necessitate high mark-ups for outgoing calls for the fixed networks as well. Mobile-to-mobile termination charges should ordinarily be set low but, with increased substitution between fixed-to-mobile and mobile-to-mobile calls; they are likely to be set equal to fixed-to-mobile termination charges.

The results of the economic literature on wireless communications are still in quite some flux, which is commensurate with the dynamic development of the sector.

References

Ahn, H. and M.-H. Lee, 1999, An econometric analysis of the demand for access to mobile telephone networks, Information Economics and Policy, 11, 297–305.

Albaek, S., P. Mollgaard and P. Overgaard, 1997, Government-assisted oligopoly coordination? A concrete case, Journal of Industrial Economics, 45, 429–443.

Armstrong, M., 1997, Mobile telephony in the U.K., Regulation Initiative Discussion Paper No. 15, London Business School.

Armstrong, M., 2002a, The theory of access pricing and interconnection, in M. Cave, S. Majumdar and I. Vogelsang, eds., Handbook of Telecommunications Economics, North-Holland, 295–384.

Armstrong, M., 2002b, Competition in Two-sided Markets, mimeo, Nuffield College, University of Oxford.

Ausubel, L., P. Cramton, P. McAfee and J. McMillan, 1997, Synergies in wireless telephony: Evidence from the MTA auction, Journal of Economics and Management Strategy, 6, 497–527.

Bekkers, R., B. Verspagen and J. Smits, 2002, Intellectual property rights and standardization: The case of GSM, Telecommunications Policy, 26, 171–188.

Berger, U., 2002, Network competition and the collusive role of the access charge, mimeo, Vienna University.
Binmore, K. and P. Klemperer, 2002, The biggest auction ever: The sale of the British 3G telecom licences, The Economic Journal, 112, C74–C96.
Bresnahan, T., 1989, Empirical studies in industries with market power, in R. Schmalensee and R. Willig, eds., Handbook of Industrial Organization, North-Holland.
Busse, M., 2000, Multimarket contact and price coordination in the cellular telephone industry, Journal of Economics and Management Strategy, 9(3), 287–320.
Cadima, N. and P. Barros, 2000, The impact of mobile phone diffusion on the fixed-link network, CEPR Discussion Paper 2598, London.
Caillaud, B. and B. Jullien, 2001, Competing cybermediaries, European Economic Review, 45, 797–808.
Cave, M., 2001, Independent Review of Radio Spectrum Management, Consultation Paper, HM Treasury and Department of Trade and Industry, London.
Coase, R., 1959, The Federal Communciations Commission, The Journal of Law and Economics, 2, 1–40.
Competition Commission, 2003, Vodafone, O_2, Orange, and T-Mobile, U.K, 18 February 2003.
Cramton, P., 2002, Spectrum auctions, in M. Cave, S. Majumdar and I. Vogelsang, eds., Handbook of Telecommunications Economics, North-Holland, 605–640.
Cramton, P., E. Kwerel and J. Williams, 1998, Efficient relocation of spectrum incumbents, Journal of Law and Economics, 41, 647–675.
Crandall, R., 1998, New Zealand spectrum policy: A model for the United States? Journal of Law and Economics, 41, 821–840.
DeGraba, P., 2002, Bill and keep as the efficient interconnection regime? A Reply, Review of Network Economics, 1, 61–65.
DeVany, A., 1998, Implementing a market-based spectrum policy, Journal of Law and Economics, 41, 627–646.
Doyle, C. and J. Smith, 1998, Market structure in mobile telecoms: Qualified indirect access and the receiver pays principle, Information Economics and Policy, 10, 471–488.
Duso, T., 2003, Lobbying and regulation in a political economy: Evidence from the U.S. cellular industry, Public Choice, forthcoming.
Duso, T. and A. Jung, 2003, Lobbying and collusion in a regulated industry: The U.S. mobile telecommunications market, mimeo, Berlin: Humboldt University.
Duso, T. and L.-H. Röller, 2003, Endogenous Deregulation: Evidence from OECD Countries, mimeo, Wissenschaftszentrum Berlin (WZB).
Economides, N., 1996, The Economics of networks, International Journal of Industrial Organization, 14, 673–700.
Economides, N., 1989, Desirability of compatibility in the absence of network externalities, American Economic Review, 79, 1165–1181.
Farrell, J. and P. Klemperer, 2002, Coordination and lock-in: Competition with switching costs and network effects (forthcoming chapter), Handbook of Industrial Organization, Volume 3.
Dos Santos Ferreira, R. and J.-F. Thisse, 1996, Horizontal and vertical differentiation: The Launhardt model, International Journal of Industrial Organization, 14, 485–506.
Foreman, R. D. and E. Beauvais, 1999, Scale economies in cellular telephony: Size matters, Journal of Regulatory Economics, 16, 297–306.
Foros, O., B. Hansen and J.-Y. Sand, 2001, Demand-side spillovers and semicollusion in the mobile communications market, NHH Department of Economics Discussion Paper 32, December.
Gans, J. S., 1999, An evaluation of regulatory pricing options for mobile termination services, mimeo, University of Melbourne, December.

Gans, J. S. and S. P. King, 2000, Mobile network competition, customer ignorance and fixed-to-mobile call prices, Information Economics and Policy, 12, 301–327.

Gans, J. S. and S. P. King, 2001a, Using 'Bill and Keep' interconnect arrangements to soften network competition, Economic Letters, 71(3), 413–420.

Gans, J. S. and S. P. King, 2001b, Regulating endogenous customer switching costs, Contributions to Theoretical Economics, 1, 1–29.

Gans, J. S., S. P. King and G. Woodbridge, 2001, Numbers to the people: Regulation, ownership and local number portability, Information Economics and Policy, 13(2), 167–180.

Gruber, H., 2001a, Competition and innovation: The diffusion of mobile telecommunications in central and eastern Europe, Information Economics and Policy, 13, 19–34.

Gruber, H., 2001b, Spectrum limits and competition in mobile markets: The role of license fees, Telecommunications Policy, 25, 59–70.

Gruber, H., 1999, An investment view of mobile telecommunications in the European Union, Telecommunications Policy, 23, 521–538.

Gruber, H. and F. Verboven, 2001a, The evolution of markets under entry standards and regulation—The case of global mobile telecommunications, International Journal of Industrial Organization, 19, 1189–1212.

Gruber, H. and F. Verboven, 2001b, The diffusion of mobile telecommunications services in the European Union, European Economic Review, 45, 577–588.

Guthrie, G. and J. Wright, 2003, Competing Payment Schemes, Working Paper 245, University of Auckland.

Haug, T., 2002, A commentary on standardization practices: Lessons from the NMT and GSM mobile telephone histories, Telecommunications Policy, 26, 101–107.

Hausman, J., 1995, The cost of cellular telephone regulation, mimeo, MIT Press.

Hausman, J., 2002, Mobile telephony, in M. Cave, S. Majumdar and I. Vogelsang, eds., Handbook of Telecommunications Economics, North-Holland, 563–604.

Herzel, L., 1951, Public interest and the market in color television regulation, University of Chicago Law review, 18, 802–816.

Horvath, R. and D. Maldoom, 2002, Fixed-mobile substitution: A simultaneous equation model with qualitative and limited dependent variables, DotEcon Discussion Paper, DP 02/02, www.dotecon.com, London.

Jeon, D.-S., J.-J. Laffont and J. Tirole, 2001, On the receiver pays principle, mimeo, IDEI: University of Toulouse I.

Katz, M. and C. Shapiro, 1985, Network externalities, competition and compatibility, American Economic Review, 73, 424–440.

Koski, H. and T. Kretschmer, 2002, Entry, standards and competition: Firm strategies and the diffusion of mobile technology, mimeo, London: LSE.

Laffont, J.-J., P. Rey and J. Tirole, 1998a, Network competition I: Overview and nondiscriminatory pricing, RAND Journal of Economics, 29(1), 1–37.

Laffont, J.-J., P. Rey and J. Tirole, 1998b, Network competition II: Price discrimination, RAND Journal of Economics, 29(1), 38–56.

Lehenkari, J. and R. Miettinen, 2002, Standardization in the construction of a large technological system—The case of the Nordic mobile telephone system, Telecommunications Policy, 26, 109–127.

Liebowitz, S. J. and S. E. Margolis, 2002, Network effects, in M. Cave, S. Majumdar and I. Vogelsang, eds., Handbook of Telecommunications Economics, North-Holland, 75–96.

McAfee, R. P. and J. McMillan, 1996, Analyzing the airwaves auction, Journal of Economic Perspectives, 10, 159–175.

McKenzie, D. and J. Small, 1997, Econometric cost structure estimates for cellular telephony in the United States, Journal of Regulatory Economics, 12, 147–157.

McMillan, J., 1994, Selling spectrum rights, Journal of Economic Perspectives, 8, 145–162.
Miravete, E. and L.-H. Röller, 2002, Competitive nonlinear pricing in duopoly equilibrium: The early U.S. cellular telephone industry, mimeo, Department of Economics, University of Pennsylvania, August 4.
Moreton, P. and P. Spiller, 1998, What's in the air: Interlicense synergies in the Federal Communication Commission's broadband personal communications service spectrum auctions, Journal of Law and Economics, 41, 677–716.
Noam, E. M., 1997, Beyond spectrum auctions—Taking the next step to open spectrum access, Telecommunciations Policy, 21, 461–475.
Noam, E. M., 2002, Interconnection practices, in M. Cave, S. Majumdar and I. Vogelsang, eds., Handbook of Telecommunications Economics, North-Holland, 385–422.
OECD, 2000, Cellular mobile pricing structures and trends, May, Paris.
OFTEL, 2001, Effective competition review: Mobile, Office of Telecommunications, U.K., 26 September.
OFTEL, 2002a, Review of SIM-locking policy, Office of Telecommunications, U.K., 26 November.
OFTEL, 2002b, Ramsey pricing—Oftel's response to letter of 4 July, U.K., 12 July.
Parker, G. and M. Van Alstyne, 2000, Information Complements, Substitutes, and Strategic Product Design, Tulane University and University of Michigan, Working Paper No. 299, March.
Parker, P. and L.-H. Röller, 1997, Collusive conduct in duopolies: Multimarket contact and cross-ownership in the mobile telephone industry, RAND Journal of Economics, 28(2), 304–322.
Penard, T., 2002, Competition and strategy on the mobile telephony market: A look at the GSM business model in France, Communications and Strategies, 45, 49–79.
Productivity Commission, 2002, Radiocommunications, Report no. 22, Canberra: AusInfo.
Reiffen, D., L. Schumann and M. Ward, 2000, Discriminatory dealing with downstream competitors: Evidence from the cellular industry, Journal of Industrial Economics, XLVIII, 253–286.
Reiffen, D. and M. Ward, 2002, Recent empirical evidence on discrimination by regulated firms, Review of Network Economics, 1, 39–53.
Rochet, J. and J. Tirole, 2001, Platform Competition in Two-sided Markets, mimeo, Toulouse University.
Ruiz, L. K., 1995, Pricing strategies and regulatory effects in the U.S. cellular telecommunications duopolies, in G. W. Brock, ed., Toward a Competitive Telecommunications Industry, Mahwah, NJ: Lawrence Erlbaum Associated, 13–54.
Schiff, A., 2003, Open and closed systems of two-sided networks, Information, Economics and Policy-forthcoming.
Schiller, J., 2000, Mobile Communications, Addison-Wesley.
Shapiro, C. and H. Varian, 1999, Information Rules: A Strategic Guide to the Network Economy, Boston: Harvard Business School Press.
Shew, W. B., 1994, Regulation, competition and prices in the U.S. cellular telephone industry, Working Paper, American Enterprise Institute.
Shin, R. and J. Ying, 1992, Unnatural monopolies in local telephone, RAND Journal of Economics, 23, 171–183.
Shy, O., 2001, The Economics of Network Industries, Cambridge, U.K.: CUP.
Srivastava, L. and S. Sinhab, 2001, TP case study: Fixed-mobile interconnection in India, Telecommunications Policy, 25, 21–38.
Stoetzer, M. and D. Tewes, 1996, Competition in the German cellular market? Telecommunications Policy, 20, 303–310.
Sung, N. and Y-H. Lee, 2002, Substitution between mobile and fixed telephones in Korea, Review of Industrial Organization, 20, 367–374.
Sutherland, E., 2001, International roaming charges: Overcharging and competition law, Telecommunications Policy, 25, 5–20.

Sutton, J., 1991, Sunk costs and market structure, Cambridge, MA: MIT Press.

Valletti, T., 1999, A model of competition in mobile communications, Information Economics and Policy, 11, 61–72.

Valletti, T., 2001, Spectrum trading, Telecommunications Policy, 25, 655–670.

Valletti, T., 2003, Is mobile telephony a natural oligopoly? forthcoming, Review of Industrial Organization.

Valletti, T. and M. Cave, 1998, Competition in U.K. mobile communications, Telecommunications Policy, 22, 109–131.

Woroch, G. A., 2002, Local network competition, in M. E. Cave, S. Majumdar and I. Vogelsang, eds., Handbook of Telecommunications Economics, North-Holland, 641–716.

Wright, J., 1999, Competition and Termination in Cellular Networks, Working Paper, Centre for Research in Network Economics and Communications, University of Auckland.

Wright, J., 2002, Access pricing under competition: An application to cellular networks, Journal of Industrial Economics, L, 289–315.

Chapter 8

THE ECONOMIC GEOGRAPHY OF INTERNET INFRASTRUCTURE IN THE UNITED STATES

SHANE M. GREENSTEIN

Northwestern University

Contents

Handbook of Telecommunications Economics, Volume 2, Edited by S. Majumdar et al.

DOI: 10.1016/S1569-4054(05)02008-7

1. Introduction

At some broad level, it is no great surprise that there are plenty of Internet infrastructures in Manhattan and not in the Mojave Desert. After all, the strongest factors that shape the location of demand for Internet services are the density of human settlement and the location of industry. Yet, that simple piece of economic reasoning only goes so far in addressing important questions about the location of Internet infrastructure.

In other areas of communications a combination of high sunk costs during installation and economies of density in operation make many areas very costly to serve. Remarkably, Internet infrastructure seems to defy that history. Less than a decade into commercialization, the supply of the most basic (i.e., dial-up) Internet infrastructure has become available nearly everywhere in the United States—except in the poorest rural locations. Was this achievement the intended or unintended consequence of government policy? Does that experience offer lessons about the strengths or weaknesses of relying on market forces to build-out information infrastructure? These concerns are intertwined with a related event, the diffusion of the next generation of high-speed Internet infrastructure, generically called *broadband.* Unlike dial-up infrastructure, the earliest build-out of broadband has resulted in uneven geographic coverage that favors urban areas. Is this coverage an artifact of slow build-out or a permanent feature of relying on commercial markets for this service?

Addressing these questions presents quite a challenge. The Internet is unlike any communications network that came before it. It is a complex technology embedded in a multilayered network, and many different participants operate its pieces. When the National Science Foundation (NSF) began to commercialize the Internet around 1992 there were no obvious precedents from which to forecast the Internet's geographic reach, nor were there simple lessons to borrow from other commercially supported communications networks.

In this chapter, the industrial economics is placed at the center of the explanation for the geographic properties of Internet infrastructure, applying this established mode of analysis to the unique cost structures and demand characteristics of these markets. It is challenging to do because Internet infrastructure involves an unprecedented mix of new economic actors and incumbent participants, something that one should expect of a new and large economic opportunity. However, unique and unprecedented market activity does not imply the presence of new economic precepts. Some basic (and often not mysterious) economic factors shaped the location of Internet infrastructure.

This analysis makes reference to the same familiar economic concepts that have been seen with other communication networks. Especially central are economies of density and scale in operation, the economies of entry into services with high sunk costs, and the economics of retrofitting technical upgrades on existing infrastructure. The analysis also highlights the economics of competitive behavior for growing markets, behavior not unusual for many new technology markets, but somewhat novel for parts of communications. These themes are often neglected in common narratives about the Internet's geographic features. Because the review also seeks to synthesize its themes with the already rich cross-disciplinary dialogue about the Internet's geography, the review presents its analysis at an intuitive level, eschewing formal economic theory, referring the reader to the writing of others when it is available.

The narrative is organized around six different topics. It begins with an overview of the origins of the Internet and why the economic geography of Internet infrastructure was unclear prior to commercialization. This first section highlights why the setting contributed to the Internet being both a surprise to most commercial participants, and, yet, also contributed to the Internet's fast diffusion. This analysis gives rise to themes that arise repeatedly later in the review, so it also serves as a foundation for analysis of the determinants of economic activity during the era of rapid geographic expansion.

The next sections analyze four overlapping markets of Internet infrastructure: dial-up service, backbone, broadband, and business provision of Internet infrastructure services. The section on dial-up focuses on why this mode of access became geographically ubiquitous so quickly. The discussion about broadband highlights why commercial activity in the backbone contributed to geographic ubiquity, and was not a hindrance to expansion, as some feared. The section on broadband highlights why broadband is comparatively more available in dense areas than areas of low density. The discussion about business infrastructure highlights progress and open questions in this comparatively newer area of research.

The concluding section summarizes the answers to several key questions motivating this broad research area. Why did near geographic ubiquity arise after commercialization? Why did market forces encourage extensive growth after commercialization? Has the Internet diffused disproportionately to urban areas? Is this Internet a substitute or a complement for urban agglomeration? Which policies mattered most and were these the intended or unintended consequences? What are the lessons for other countries?

2. Dispersion and concentration of Internet infrastructure

Before the Internet diffused there was no consensus about the shape it would take. To be sure, there was an active discussion among the telecommunications research and policy community about the economic geography of communications networks[1]. Even so, this only helped frame general issues, without providing sufficient precision to forecast specific outcomes. The presence of uncertainty is

[1] For reviews, see, for example, Greenstein et al. (2003), Shampine (2001), Greenstein and Lizardo (1999), National Telecommunications and Information Administration (1995), and Kahin (1997).

understandable, in retrospect. To oversimplify, new sets of economic actors developed new services, seeking demand for which there was only limited precedent. As with many new markets, participants had to experience a variety of activities to learn about key drivers of underlying costs, demand and operational efficiencies. It simply took a while to distinguish between long-term lessons and short-lived phenomena.

National and local policies during this era were also a mix of foresighted programs, which were then followed by reactive corrections to unanticipated issues. Some national and local policies anticipated some of the salient questions about the Internet's geography, because those questions arose as a by-product of concerns about regional comparative advantages, or about universal access to public communications networks. Even prior to the diffusion of the Internet, definitions for universal service had come under stress, as the diffusion of digital technologies and wireless communications highlighted regional disparities in access to frontier technology[2]. This experience helped to frame the first questions about the Internet. Only as the Internet grew much larger than anyone had forecast did it become apparent that policy makers needed to ask a new set of questions.

This topic provides useful background to the all the later sections, so it is natural to discuss first. The discussion focuses on three broad areas:

1. How could the Internet be a surprise and, yet, diffuse ubiquitously in only a few years after it became commercially available?
2. What were the origins of the commercial Internet and what aspects from this early era persisted into the commercial form? What changes caught most observers by surprise?
3. What pattern of geographic diffusion was first forecast and why? What issues did suppliers anticipate they would have to overcome to make the commercial Internet available outside a narrow set of big cities?

2.1. What is Internet infrastructure?

In its most common usage, *Internet infrastructure* is synonymous with durable investments in software, communication and computing equipment, and related activities associated with operating information technology. This common and broad definition of Internet infrastructure encompasses quite a lot: capital equipment, such as mainframes, minicomputers, PCs (personal computers), local area

[2] There are many places to enter the discussion about universal service issues in an era of technical change (Cherry et al., 1999; Crandall and Waverman, 2000; Compaine, 2001; Noll and colleagues, 2001), as well as many of the articles cited in this review. For new approaches more consistent with the concerns of urban geography, see Kotkin (2000), Castells (2002), Malecki (2002c), or Gorman and McIntee (forthcoming).

networks (LANs), wide area networks (WANs), local and long-distance telephone equipment, private and quasi-public switching equipment, wireless networks for data transmission—and software—both packaged and customized. Notice that it also incorporates human capital, a key (and often local) input along any value chain for Internet services.

In addition, a proper economic accounting for the stock of information infrastructure encompasses the results of activities at user enterprises, such as training, maintenance, and operational experience. A proper economic accounting should also account for the stocks of knowledge and the flows of tacit and explicit knowledge about the technological frontier, which also often has local dimensions[3].

During the early years of the commercialization of the Internet, the location of human capital was important. Universities had been the spawning ground for the Internet in two ways. First, the NSF subsidized the connection of many universities, which effectively trained many students in the basic operations. Second, these same subsidies encouraged many users to become familiar with the benefits of the technology's application (e.g., e-mail and file-transfers at one point followed later by the browser). This experience helped seed demand for commercial versions of the same service, particularly among recent university graduates.

It was often observed that the early commercial firms and users tended to be located near universities. Casually speaking, the high-speed 'pipe' was nearby, as were potential demanders and potential entrepreneurs to operate new businesses. As a result, there was an open research question at the outset of the Internet: as the commercial market for Internet services acquired its own market-oriented incentives, would this technology take the path of many new high-technology industries and concentrate within the location from which it grew? Or would the Internet divorce economic activity from the location of the major research universities? The answer turned out to be divorce with an exclamation point. The divorce occurred quickly.

2.2. *The origins of Internet infrastructure*

Nobody set out to design a commercial Internet with specific geographic features. These features arose endogenously, both borrowing from the geographic features of the network's origins and adopting new features to suit market forces. The following overview of the early Internet helps explain why it was hard to forecast the geography of the commercial Internet from the features of the noncommercial Internet. Table 1 highlights some key events in the historical progression to the commercial Internet.

[3] For a mildly different conceptual model and an attempt to trace the dollar flows of the Internet value chain, see O'Donnell (2001). Another related model can be found in Gorman and Malecki (2002), who distinguish between the layers of the Internet affiliated with application and data transport, arguing that the geography of data transport affects the performance and value created in application.

Table 1
Abridged chronology of the first decade of the commercial Internet

1988	NSF bids out contracts to build the NSFNet as a backbone for researchers
1989	Internet host computers reaches 100K
1990	Arpanet decommissioned, leaving a network with over 300K host computers
1991	NSF begins commercialization by lifting restrictions on commercial use
1992	Internet host computers reach 1 million
1992	First commercial ISPs begin to interconnect legally with Internet
1993	NSCA at University of Illinois releases the browser named Mosaic
1994	April, Netscape is founded
1994	May, first World Wide Web Consortium (W3C) conference held
1994	August, Compuserve offer Internet access to their users
1994	Internet host computers pass 4 million
1994	November, MCI introduces an Internet connection service, AOL buys ANS
1995	February, Yahoo founded as a commercial service
1995	March, BBN rolls out first nationwide Internet service—with over 500 access points
1995	May, UUNET and PSInet go public
1995	NSFNet ends contract with ANS, MCI becomes largest Internet backbone provider
1995	August, Netscape goes public, valued at 2.2 Billion on first day of trading
1995	September, AT&T launches business Internet services
1995	October, InternNIC begins charging for domain name registrations
1996	Telecom Act passes U.S. Congress
1996	January, 100K web sites online
1996	February, AT&T begins to sell $20 home Internet access
1996	April, Yahoo goes public, reaching market value of $848M on first day of trading
1996	May, MFS communications buys Uunet technologies for 2 Billion
1996	Internet host computers passes 15M
1996	August, WorldCom offers $14.4B to buy MFS
1996	November, AOL adopts flat rate fee of $19.95 per month for unlimited access
1997	March, Network Solutions (NSI) registers the 1 millionth Internet domain name
1997	May, GTE buys BBN and Genuity
1997	May, UUNET and PSnetannounces it will charge small ISPs for interconnection
1997	September, WorldCom buys Compuserve's infrastructure, AOL buys its customer base
1997	October, WorldCom bids for MCI. WorldCom wins in November
1997	Internet host computers passes 25M
1998	January, AT&T buys Teleport
1998	February, ITU announces 56K modem standards, ending war over specification
1998	May, Network Solutions registers 2 millionth domain name
1998	June, AT&T buys TCI, a cable company
1998	Over 26 percent of U.S. households have Internet access
1999	March, 5 million web sites on line
1999	May, AT&T announces agreement to buy Media One, a cable company
1999	Peak of the dot-com boom, over 490 IPOs take place during the year
2000	January, AOL announces merger with Time Warner
2000	March, Nasdaq peaks above 5000, height of 'dot-com bubble'
2000	April, 'Dot-com crash'. Many Internet IPOs withdrawn or delayed indefinitely
2000	September, 25M web sites online
2000	Over 41 percent of U.S. households have Internet access
2001	Internet host computers passes 100 M hosts

(continued)

Table 1 (*continued*)

2001	July, AT&T receives bid from ComCast for cable assets
2001	Estimates for retail e-commerce sales at $25 Billion in U.S.
2001	Over 50 percent of U.S. households have Internet access, less than a fifth of those are broadband
2002	July, MCI/WorldCom declares bankruptcy

Sources: Juliussen and Petska-Juliussen, 1998, Chapter 7, NTIA, 2002, E-Stats. www.census.gov/eos/www/ebusiness614.htm.

Even before its commercialization, Internet access was ultimately a local activity. In its dial-up form, the Internet employed a series of components, supplied by vendors and bought by users, retrofitted onto the U.S. voice network. That is, the Internet first grew on top of 'plain old telephone service.' Just as the user did not need the permission of the local telephone company to hook up a facsimile machine, the user of the Internet could hook his or her PC into the telephone system in order to download messages. This involved a local phone call to a local phone number, where a modem bank and server automatically handled the call.

Part of this was not an accident. Many advanced features of local telephony—such as clean lines free of static that made it feasible to send a digital signal—had been the focus of purposeful policies aimed at upgrading the U.S. telephone system. To be sure, many of these upgrades were done to enable services other than the Internet and the advance to digital technology in local telephone switches was not necessary for the diffusion of the first generation of the Internet, but their presence made later transitions to advanced technology almost incremental. Similarly, during the 1980s the U.S. telephone switch network made a technical leap to digital equipment from legacy analog equipment (Shampine, 2001). This leap spawned many new applications and services, everything from data-traffic to better faxes. Generally, there were many potential uses for digital switches, and the Internet was not the primary motivation at first.

Prior to 1992, the Internet consisted of the operations found at an academic modem pool or research center, with network interconnection handled by NSF-sponsored 'regional' operations (Kahin, 1992). National Science Foundation's acceptable use policy prevented large scale commercial activity from occurring on the network. University centers were small-scale operations, typically serving no more than several hundred users, involving a mix of frontier and routine hardware and software. A small operation required: (1) a server to monitor traffic and act as a gatekeeper, (2) a router to direct traffic between the Internet and users at PCs within a LAN, and (3) an arrangement with a regional network, connecting the University to the Internet backbone or data-exchange point. Revenues were not regularly collected in these arrangements and budgetary constraints were not representative of what might arise with commercial operations and competitive pressures.

The array of services matched the needs of academic or research computing, which had only a partial overlap with the needs of commercial users[4]. Many small colleges had opened their Internet connections with NSF subsidies. A small staff of students or information technology professionals operated these centers. The organizational arrangement within research computing centers also was idiosyncratic, usually with only loose ties, if any, to the professionally run administrative computing centers of a university or research organization.

In all but the wealthiest universities, academic access centers operated in the cheapest way possible. They borrowed equipment from already existing facilities (i.e., PC and network servers) and charged little to students and staff. They borrowed practices common to bulletin board operations, and took advantage of local telephone pricing (i.e., by anticipating free local phone calls at no expense to users *and* by making no provision for long-distance calls or any other services). Points of presence (or POPs in industry parlance) began sprouting up in every university to support a geographically concentrated local user base. Although the primitive operations would eventually lend some aspects to the commercial operations, these university operations provided little basis for forecasting the form eventually taken by the commercial Internet.

2.3. Scale and urban location

In the beginning the Internet was a network for use by only researchers at laboratories, universities, and military research centers. The Internet of 1992 was a world of insiders who treated it as a labor of love. The collection of insiders consisted of academic engineers who loved operating a network with frontier features, software gurus who loved experimenting across a wide array of multilocation applications, and policy experts who loved discussing policy issues for communications markets. Representative views from this group can be found in policy-oriented publications of the time (e.g., Kahin, 1992). These views display a mix of pride about accomplishments and competing visions for the future. One might also say that these insiders were focused on turning the Internet into a

[4] See Kahin (1992), Frazer (1995), Mowery and Simcoe (2002) or Kahn (1995) or Waldrop (2002) for a history of the development of the Internet under academic auspices and the policies supporting it, as well as how this translated into commercial development. Kende (2000) provides an excellent description of the commercial activities that aided the transition into the commercial era, such as the establishment of the Commercial Internet Exchange (CIX) and the privatization of NSF's Network Access Points (NAPS) and their eventual expansion. Note, too, university access providers did not operate in a complete vacuum. The technology and operations in ISPs and bulletin boards had much in common. For this viewpoint, see editions of Boardwatch Magazine from the early 1990s, stored on www.boardwatch.com.

self-sustaining network (i.e., one that covered its own costs), but not necessarily a commercially successful network with appeal to the mass market.

Up until this point, the assets for operating the Internet were mostly in the public domain. The exception arose at private laboratories with government funding. In such a setting ownership of some Internet assets was private, but they operated in conjunction with the networks in the public sphere, such as the regional networks operated under the ultimate guidance of the NSF. After 1992, the Internet was headed toward an unknown future, where private firms could make all the investments in Internet infrastructure. The innocent past gave little hint about what would occur after the Internet commercialized. Nobody was openly predicting the type of wild growth that soon would take place.

Did anyone in 1992 forecast whether the Internet would agglomerate within major cities or not? The question is almost unfair in retrospect. As with many other government-subsidized technologies, it was unclear *whether* the Internet would grow at all outside of the quasi-public sphere that spawned it. The question of *where* it would grow was less paramount in 1992. To be sure, by 1992 investment decisions for the network backbone had been devolved to regional decentralized decision makers (Mandelbaum and Mandelbaum, 1992), but these arrangements came about as a result of NSF's guiding hands mixed with the cooperative nature of academic networking among insiders. A self-organized network operated by privately motivated participants was just starting at this point and—with NSF's blessing—could have gone in any number of directions.

Three considerations informed the discussion at the time about geographic diffusion. First, Internet infrastructure displayed economies of scale, but it was unclear how this would show up in a commercial setting. Second, the Internet was an advanced technology, and hard to support. Third, the policy environment at the time was in flux. This review addresses each of these in turn.

The technology of the Internet might have led to economies of scale, as found in the voice network. In voice networks there are economies of density in the provision of services linked to telephone switches and lines. This arises for two reasons. First, there are economies of scale in switching equipment. A large switch has a higher capacity per dollar than a smaller switch. Second, the lines for serving a denser area require less material and operational expense per customer[5]. It was possible that the building blocks of local Internet networks in the early 1990s—e.g., hubs and routers, Ethernet cards, and T-1 lines hooked into local area networks—might give rise to some form of density economies. Third, there were well-known forecasts that economies of scale in transmission technology would continue to increase (i.e., as fiber optic technology advanced and data backbone lines became capable of carrying more traffic).

[5]See, for example, the discussion in Rosston and Wimmer (2001).

Yet, in the early 1990s it was not at all clear that these economies would continue to extend beyond the floor of an office or a large office building, as scale increased to city blocks, neighborhoods and regions. The NSFnet at the time was too special to give any hints about where or how it would deploy when the scale of use increased dramatically (see the discussion in Frazer [1995] or, earlier, Kahin [1992]). If anything, it was mildly misleading to see it used as a network designed to facilitate sharing supercomputing.

Moreover, Internet infrastructure initially appeared to be sensitive to a different geographic factor—the location of personnel to solve technically difficult problems. As with other advanced technologies, this one might favor urban areas where there are thicker labor markets for specialized engineering talent or supporting services. Similarly, close proximity to thick technical labor markets facilitates the development of complementary service markets for maintenance and engineering services[6]. Indeed, as commercialization got well underway, a pessimistic view emerged in the mid-1990s, one that became affiliated with the notion labeled 'the digital divide.' It forecast that the Internet was diffusing disproportionately to urban areas with their complementary technical and knowledge resources for supporting PCs and other computing, leaving rural areas behind[7].

There was only limited experience from which to forecast the importance of urban location. The NSF had deployed the Internet in a configuration that favored the needs of universities and researchers, with some emphasis on the needs of supercomputer users in the late 1980s (Smarr and Catlett, 1992; Frazer, 1995; Kahn, 1997). This focus provided few general lessons. The self-contained operations found at university campuses or large laboratory settings provided no market model for the structure of profit-oriented operations, sales, and support to a large number of businesses and households. The commercial firms who first provided TCP/IP interconnection for a fee also faced large unknowns about the basic features of demand, such as its location, time-of-day properties, and so on.

Moreover, the NSF (following DARPA) had adopted an 'end-to-end architecture,' a feature that served academic needs by encouraging development of applications at the 'ends of the system,' where users could customize applications to their needs. That is, the applications ran in the PC or workstation client after data traveled across the Internet. Applications were disembodied from the specific location, the identity of the data carrier, or even the network configuration for moving data (Blumenthal and Clark, 2001). If location were not supposed to play a role in application development, how would a commercial deployment vary by location? It was not at all clear how a new set of customers would alter the deployment of Internet infrastructure across geographic space.

[6] These are closely related to the major reasons given for industrial agglomeration (Krugman, 1991).

[7] Articulation of this view is particular prominent in the National Telecommunications and Information Administration's 'Falling Through The Net' series (See National Telecommunications and Information Administration 1995, 1997).

The commercial uncertainty was irreducible. It was unclear at the outset which of several potential maturation processes would occur after commercialization (e.g., for a standard discussion, see Rogers [1995]). If advancing Internet infrastructure stayed exotic and difficult to use, then its geographic distribution would depend on the location of the users most willing to pay for infrastructure. If advancing Internet infrastructure became less exotic to a greater number of users and vendors, then commercial maturation would produce geographic dispersion over time, away from the areas of early experimentation. Then, the diffusion would depend on the cost of building, supporting, and operating the network. Similarly, as advanced technology became more standardized, it would be also more easily serviced in outlying areas, again contributing to its geographic dispersion.

The other major source of uncertainty concerned the policy environment. It was not enough for the NSF to simply declare the Internet private, sit back and watch firms interconnect. It was well understood that several cooperation principles—such as voluntary interoperability among the components of infrastructure—kept the precommercial Internet operating. For example, all operators of components of the noncommercial infrastructure voluntarily interoperated with each other. What was to prevent private parties in the new era from choosing to not interconnect if it served a commercial purpose? Would the Internet balkanize from such action?

There was simply no experience in the precommercialization period with uncoordinated commercial forces developing a communication network such as this. As a result, it was unclear what commercial governance structure was most appropriate for privately motivated investment between interconnecting firms[8]. The Domain Name System also had to be transferred into the public domain, a process that turned out not to be smooth (Weinberg, 2001). The appropriate pricing structure for use of assets by private parties was also unknown[9]. There were open questions: Would the Internet require close regulation, as in the phone industry? The NSF put control over data-exchange points into regional hands, but would this cooperative structure be sustainable with infrequent guidance by government oversight?

[8] Kende (2000) provides an excellent discussion about the emergence of several private institutions, such as private peering and privately operated public exchange points, which did not interfere with the operation of the commercial Internet.

[9] This situation gave rise to uncertainty in many aspects of the operations of the Internet. For a skeptical early discussion, see Mackie-Mason and Varian (1995). See Meeker and Dupuy (1996) for an early and comparatively sober assessment (that is, in comparison to later analyses) of the commercial prospects of many firms in the Internet infrastructure market. In some respects, the Internet's geographic reach seemed almost secondary to many contemporary policy observers, who were concerned about many seemingly more fundamental operational issues, such as covering costs and the scope of use. For further discussions of many of the policy and operational issues (Kahin, 1992, 1997; Kahin and Keller, 1997; Kahin and Nesson, 1999; Kenney, 2003).

In addition, there were national political pressures to reduce regional inequities in the deployment of the voice network, especially between urban and rural areas. It was not at all clear how, or whether, these policies applied to Internet infrastructure (for a full history, see Kahin, 1997). The uncertainty extended to the highest levels of the Federal government. Vice President Gore had supported federal subsidies for an 'Information Superhighway,' especially while it was primarily used for research. It was less obvious how this broad policy translated into specific laws after commercialization. For example, several federal agencies had historically pursued policies associated with mandates against regional inequality of access to advanced technology (e.g., they applied pressure for the adoption of digital switches in rural areas). It was unclear how to apply such principles to Internet access or broadband, if at all.

As it turned out, early into the commercial development of the Internet, the U.S. Congress passed the 1996 Telecom Act. This Act included provisions for the 'E-rate program,' which was aimed at alleviating geographic inequities in the provision of the Internet, among other goals[10]. It eventually raised over two billion dollars a year from long-distance telephone bills. This money was administered by the FCC, and primarily given to disadvantaged schools and libraries to develop Internet connections. In 1996, it appeared as if this might help those in isolated or rural locations. At its passage, however, neither it was clear whether this government program would survive court challenges, nor how it would shape the geographic reach of the Internet. This would take several years to figure out.

Moreover, the 1996 Act altered the legal relationship between local telephone companies, who were usually local monopolies, and the competitive firms who interconnected with them. These provisions were quickly reviewed by the FCC and then challenged in court. These regulations had consequences for how broadband Internet reached businesses and homes, but any informed insider could easily understand that these regulations would not settle into a permanent set of rules for several years. It was another source of uncertainty.

2.4. Scale and standardization

Internet technologies associated with textual information had incubated for 20 years prior to commercialization. Many of these building blocks of the mass-market Internet were well past the degree of technical maturity necessary for mainstream use. Many were 'standardized' by the early 1990s—in the sense of having refined interfaces that permitted interoperability with other pieces commonly found within homes and offices (Kahin and Abbate, 1995). Several of these standards will be referred to throughout this narrative because they shaped the

[10] Due to its close identification with Vice President Gore, this provision of the Act was labeled the 'Gore Tax' by opponents. To understand how this provision arose out of a longer history of policy initiatives about information technology in the US, see Kahin (1997).

economic geography of the Internet—through bringing scale to new application development and scale to Internet users. That is, the technical problem faced by potential users of TCP/IP applications in different locations was *similar*.

Some of this was a propitious result of the fact that the most common computing configurations in the U.S. were alike. Client server architecture in most business computing in the early 1990s involved a PC client, a network connection, and a server. Clients were largely IBM-compatible PCs with Intel-compatible chip sets. Apple and work station clients were in the minority. Network connections were Ethernet-compatible, with a few proprietary technologies—such as IBM's token ring—in minority use[11]. There was variety in servers. There were minicomputers, mainframes, or workstations. The latter were mostly Unix boxes and most of these were compatible with the Internet and TCP/IP because such compatibility had been a procurement requirement for many years at DARPA, which bought many workstations throughout the 1980s. Finally, FCC mandates for physical connection resulted in similar plugs and sockets everywhere in the stock of communications customer premise equipment most commonly found at residences and businesses[12].

Yet, many proprietary technologies seemingly stood in the way of the use of a network with nonproprietary standards, such as the Internet. Novell's Netware was widely used, but could be made compatible with TCP/IP messages only with great effort. Only later generations of Netware made the transition easier[13]. DOS users and generations of Windows prior to Windows 95 also needed effort to download text messages (though, to be sure, any sophisticated user could do it). Once Windows 95 diffused, all TCP/IP programs became easier to use[14].

Another key event was the end of IBM's opposition to converting its mainframes to server purposes using nonproprietary technologies in the mid-1990s[15].

[11] The standardization on the descendents of the IBM PC by the early 1990s is widely documented, for an analysis (Bresnahan and Greenstein, 2001), for an account of the standardization to Ethernet (Von Burg, 2001).

[12] See Kahin (1997), Oxman (1999), and Cannon (2001), for descriptions of the origins of these policies at the FCC and how these led to mandates over the equipment design of anything plugging into the US telephone system.

[13] For more on this, see, e.g., Forman (2002) for an account of business establishments' conversion to the Internet, or Kenney (2003) for an analysis of factors shaping the development of Internet applications.

[14] Microsoft built TCP/IP compatibility into Windows in order to (1) foster diffusion of Windows NT as a server operating system, and (2) facilitate the use of TCP/IP programs on the client. This activity facilitated 'plumbing,' not applications. In this sense it helped the diffusion of the Internet after 1995. This should be distinguished from making a browser, which Microsoft did later. Netscape/Mosaic showed it could be very popular with users and profitable to sell (Cusumano and Yoffie, 1998).

[15] This was a by-product of the strategy of IBM's new CEO, Lou Gerstner, to turn around the firm's performance by abandoning many proprietary legacy practices, instead focusing on developing integrated services for large scale enterprises, IBM's traditional customer base.

With this decision one of the last major barriers to nonproprietary technology at large enterprises was gone. The remaining major barrier at large enterprises was Electronic Data Interchange (EDI)—an electronic standard for communicating transactions between firms. It had started to diffuse in the early 1990s and was famously complicated and difficult to use (and so, accordingly, only wealthy firms could afford to invest in it). Despite its drawbacks, it had high value to enterprises that had large numbers of regularized transactions, such as auto parts suppliers. The diffusion of EDI slowed considerably after the commercialization of the Internet. Yet, EDI applications were not immediately retired, because these investments were expensive, complex and essential to many on-going operations. More effort in the late 1990s went into making TCP/IP applications and existing EDI applications compatible.

Resistance to networking technology in the early 1990s was partly, but not primarily, associated with technical incompatibilities. Studies of adoption behavior (Bresnahan and Greenstein, 1997) associated resistance to networking with the costs of altering organizational functions and procedures. Organizational procedures only changed slowly. Any utopian forecast about getting a new network technology rapidly adopted at businesses would have had to face this rather sobering prospect. That said, there was no particular geographic pattern to this resistance.

As it turned out, by the mid-1990s, at many business establishments it was possible to use a nonproprietary standard, such as TCP/IP, as long as the ultimate purpose was simple interorganizational communications. For simple applications, such as e-mail or browsing, converting Netware was the largest barrier to overcome in the mid-1990s. Otherwise, most businesses could quite easily convert. For complex applications, however, many technical and organizational issues also mattered (Forman, 2002). At homes, the ownership of a PC and a modem also enabled conversion to use of the Internet for simple purposes[16].

The similarities of the retrofitting problems across areas throughout the U.S. supported economies of scale in the supply of standardized parts. Moving the Internet into the mainstream commercial sector did not necessitate building a whole new Internet equipment industry, nor did it require the establishment of many local service, consulting, or maintenance firms. These were already there, supplying goods and services to the universities, business users, and home PC users. With a little modification these same firms could also supply Internet applications. Business users with investments in networking technology, such as LANs or simple client/server architectures, also could adopt basic features of the Internet with little further invention from their suppliers.

[16] For these points see National Telecommunications Information Administrations (1997, 1998) and especially (2000).

More to the point, when the Internet commercialized in the United States in 1992, there was no essential difference between the variety of computing found in New York City, Los Angeles, Omaha or anywhere else in the mainstream. The experience in different cities was similar enough that the availability of an upgrade to the Internet was available to everyone, albeit not in much the same way everywhere. The costs of the upgrade or retrofit differed between locations, but the solution to the engineering issue was sufficiently normalized and understood. In short, the Internet possessed the ability to generate scale economies in the sense that standardization of technical goods enabled ‘thicker markets’ for engineering services to develop in many locations. The national IT consulting houses, such as Accenture or KPMG, also found it worthwhile to invest heavily in specialists with technical knowledge about how to apply TCP/IP across a variety of circumstances[17].

The Internet achieved scale economies in another sense that had unforeseen dynamic consequences. The Internet allowed any user in any location within the U.S. to communicate with any other who made the same upgrade to TCP/IP compatibility. This turned out to be important for building an installed base of applications and sharing them. It also helped get the Internet started in the first place. The commercial Internet did not start from scratch in 1992. Many applications had already started accumulating prior to commercialization. The Internet was attractive to use because it enabled communication with a large set of other computer users at other locations[18].

That potential for a large scale computer-based communications network, in turn, looked inviting for vendors who would eventually start businesses affiliated with building Internet infrastructure. The unanswered question between 1992 and 1995 was whether the benefits of scale economies would still motivate activity in many locations or just in a few homes and businesses of the *cognoscenti*. The earliest demonstration of the power of scale came in the early 1990s. When Tim Berners Lee developed html and the URL in a laboratory in Switzerland in 1989, it spread quickly throughout the computing world, the U.S included. When a team of University of Illinois undergraduates developed a browser called Mosaic in 1993, insiders thought it might be possible to diffuse it to nontechnical users. Mosaic set records for how fast it spread among students and nonstudents alike.

Demand-side scale economies arose because the Internet appealed to more than technical users. It also appealed to a typical nontechnical user. The basis for standardized mass-market applications was e-mail, instant message, and

[17] The economics of standards has long emphasized how unified technologies can support large market activities. See David and Greenstein (1990) for a review of this literature. See Kahin and Abbate (1995) for the state of discussion at the birth of the commercial Internet. See Rohlfs (2001) for a general treatment of the issues and a particularly cogent application to the development of the Internet (Chapter 13).

[18] By 1992, 1 million of hosts existed.

eventually, browsing. Market demand for simple uses arose almost everywhere and induced entry from new entrepreneurs taking advantage of the growing network (See Kenney [2003] for more analysis of these events). This demand was so large that it overcame the forces that might have otherwise confined the Internet to a few locations.

2.5. Summary

At a general level, the location of commercial Internet infrastructure had straightforward determinants. Like all information infrastructure, it was devoted to providing services to users in particular places. It was a local economic good because the service could only be delivered if it were supported by appropriate physical and human capital. These inputs were in a particular place, which shaped the costs and quality of the services received.

Yet, such a general description only goes so far. It overlooks the specific forces that shaped the diffusion of Internet infrastructure in the United States after commercialization. For example, what precise role would economies of scale or economies of density play in the availability of the Internet to urban and rural customers? The general description also does not resolve the uncertainty about competing economic factors. What organizational form would commercialize this technology and, once it became deployed on a national scale, would it resemble the organizations found inside the research universities? These were among the many open questions at the time of commercialization.

3. The spread of commercial Internet access[19]

It begins by discussing the spread of commercial Internet access providers, which subsequently became known as ISPs. It will primarily focus on dial-up providers, deferring discussion about backbone and broadband providers to a later chapter.

The deployment of Internet as a mass-market service depended on the commercial prospects of dial-up firms. Commercial vendors of Internet access first began sprouting in the early 1990s. Their growth accelerated in 1995 and exploded thereafter. In the latter half of the 1990s had three remarkable features:

1. They grew rapidly, inducing tens of thousands of new businesses to form around Internet technology. This infrastructure, in turn, attracted the creation of a whole value chain of support activities—such as hosting services, millions of web pages and business applications—thereby achieving the trappings of mass-market status.

[19] This section substantially borrows from perspectives in Greenstein (2001) and Coffman and Odlyzko (2001).

2. The Internet supplier market spread rapidly: the Internet quickly became nearly pervasive across locations, which attracted considerable attention because it was a unique historical achievement. Rarely do new commercial technologies spread this rapidly.
3. ISPs offered a wide variety of business services. No consensus arose about the optimal combination of service lines to carry, which is indicative of uncertainty about the appropriate business model for commercializing the service.

3.1. The activity of commercial suppliers

Internet Service Providers require a physical connection because the architecture of the Internet necessitates this physical connection. Both under the academic and commercial network, as shown in Figure 1, the structure of the Internet is organized as a hierarchical tree. Each layer of connectivity is dependent on a layer one level above it. The connection from a computer to the Internet reaches back through the ISP to the major backbone providers. The lowest level of the Internet is the customer's computer or network.

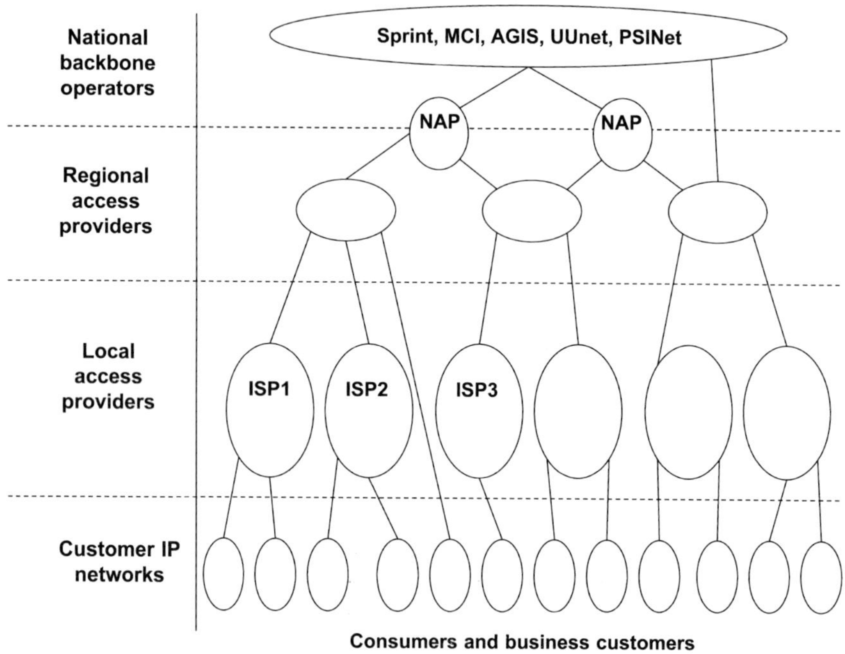

Fig. 1. The Internet as a Hierarchical Tree.

These are connected to the Internet through an ISP. An ISP will maintain its own subnetwork by connecting its POP's and servers with IP networks. These local access providers get their connectivity to the wider Internet from other providers further upstream, either regional or national ISP's. Regional networks connect directly to the national backbone providers. Private backbone providers connect to public (government) backbones at network access points.

The basic commercial transaction for dial-up Internet access and data transport did not raise prohibitive technical issues. Most often it involved repetitious and on-going transactions between vendor and user. A singular transaction arose when the vendor performed one activity, setting up Internet access or attaching Internet access to an existing computing network.

If the ISP also operated the access for the user, then this ongoing operation provided frequent contact between the user and vendor and frequent opportunity for the vendor to change the delivery of services in response to changes in technology and user needs. This worked well for users because in many cases an ISP was better educated about the technological capabilities than the user. In effect, the ISP sold that general knowledge to the user in some form that customized it to the particular needs and requirements of the user. At its simplest level, this provided users with their first exposure to a new technological possibility while educating them about its potential. Within a few years, little or no customization was required for most households[20].

For example, these processes could be seen in the simplest details an ISP would set up for a new user. The most predominant means of communications is e-mail. The e-mail equivalent of bulk mail is called a listserv, where messages are distributed to a wide audience of subscribers. These listservs are a form of conferencing that is based around a topic or theme. Usenet or newsgroups are the Internet equivalent of bulletin board discussion groups or forums. Messages are posted for all to see and readers can respond or continue the conversation with additional postings. Chat rooms and IRC serve as a forum for real-time chat. The ISP could help the user learn to use such facets of the Internet.

Such help became a facet of the Internet access business. For example, building on its previous dial-up business, America On-line developed comparatively simple instructions for its users. It also developed online help that new users valued. Famously, it helped 'Instant-messaging' gain popularity among mass-market users—even though the basic idea is quite old in computing science and equivalent functions even existed on the Internet in other forms.

ISPs also helped users gain familiarity with other basic tools of the Internet. Some tools have been supplanted, but the most common are WWW browsers,

[20] Technology transfers varied from the complex to the simple. The complex could involve issues, such as whether the user wanted to access company servers remotely. The simple could involve tailoring access to the generation of the PC operating system owned by the user.

gopher, telnet, ftp, archie, and wais. Browsers and content have grown in sophistication from the interface developed by Tim Berners Lee. Similar remarks could be made as new functions came online, such as news and entertainment, ecommerce, application hosting, videoconferencing, and so on.

Especially with a business user, access often included more than exposure to the Internet, by providing the installation, maintenance and training, as well as application development. These types of transfers of knowledge and extensions of service typically involved a great deal of nuance, often escaped attention, and yet, were essential to developing infrastructure markets as an ongoing and valuable economic activity. The technology lent itself to this small-scale customization because it was easy to adapt to PC use or available LAN technology. The basic technical know-how did not differ greatly from routine knowledge found in the computing services sector prior to commercialization.

Because these general technical skills were widely available as parts of other existing computer and communications technologies, the knowledge necessary for applying it to the Internet quickly became widely available in all major urban areas (Greenstein, 2001; Mowery and Simcoe, 2002). The only locations lacking in supply were those with quite deficient support structures for a local PC service market. Only the poorest urban areas or smallest and most remote cities were deficient in this way. And even some of them—though not all—did develop adequate commercial support for such services within a few years.

In other words, at the outset of commercialization there was no doubt that a sophisticated and wealthy user could get Internet access in the Washington, DC area or the San Francisco area or, for that matter, in any major city where an early leader in providing service, IBM GlobalNet, choose to locate a point of presence. The open question concerned much smaller cities that are comparatively isolated, cities under a half a million population, but as big as places, such as Peoria, IL or Reno, Nevada.

As will turn out, the maturity and standardization of the technology largely did not become geographically concentrated in a few places. For all intent and purposes, technological maturity did not become a barrier to the spread of basic access after 1995. As it will be discussed below, these themes manifest in myriad ways in the dial-up market in the mid to late 1990s.

3.2. Government policy encouraged a diverse geographic supply

Some NSF decisions and FCC regulatory decisions aided the geographic dispersion of the commercial industry. When the NSF took over stewardship of the Internet backbone, it invested in developing a system of address tables and address systems (for routing TCP/IP based messages) that could be scaled. Data-exchange points remained organized around the cooperative engineering principles used within the NSF days, favoring no particular location, nor any particular participant in the operating network. As a technical matter, interconnection with the public switch network did not pose any significant engineering challenges for dial-up firms[21].

The NSF had encouraged uncoordinated commercial development through the absence of any directional edicts. Accordingly, Internet access evolved in an extremely decentralized market environment. Aside from the loosely coordinated use of a few de facto standards (such as the World Wide Web consortium, see e.g., Kahn [1995], or Kahin and Abbate [1995]), government mandates after commercialization were fairly minimal. The ISPs had little guidance or restrictions and were therefore able to both tailor their offerings to local market conditions and follow entrepreneurial hunches about growing demand.

Furthermore, decades of debate in telephony had already clarified many regulatory rules for interconnection with the public switch network, thereby eliminating some potential local delays in implementing this technology on a small scale. By treating ISPs as an enhanced service and not as competitive telephone companies, the FCC did not pass on access charges to them, which effectively made it cheaper and administratively easier to be an ISP (Oxman, 1999; Canon, 2001). The FCC's decision was made many years earlier for many reasons and extended to ISPs in the early 1990s with little notice at first, since most insiders did not anticipate the extent of the growth that would arise by the end of the 1990s[22]. As ISPs grew in geographic coverage and revenues and threatened to become competitive voice carriers, these interconnection regulations came under more scrutiny (Sidak and Spulber, 1998; Weinberg, 1999). More about this will be discussed below.

In summary, a lesson for all insiders became widely understood by 1996–1997: technical issues associated with building and operating ISPs were not difficult to solve. As a result, commercial factors, not the distribution of technical knowledge among providers, largely determined the variance of development of the basic dial-up access market within a few years after commercialization.

[21] This is a mild simplification. Vendors did have to learn the boundaries of what was possible, but the engineering challenges turned out to be comparatively small (Werbach, 1997; Oxman, 1999).

[22] The Federal Communication Commission (FCC) first classified ISPs as enhanced service, similar to other on line services, such as bulletin boards. This decision built on top of many others, developed over several decades, designed to allow interconnection to the local telephone network by providers of customer premise equipment and enhanced service providers. This is a long story, often told triumphantly, sometimes as if the effects on the Internet were the intended consequence (Kahin, 1997; Werbach, 1997; Oxman, 1999; Cannon, 2001). It is fair to say that when these rules were first developed, nobody was doing it in order to encourage the Internet. Their extension to the new commercial firms called ISPs was a natural extension of the treatment of bulletin board providers, who were a considerably smaller economic force in the 1980s than ISPs became in the 1990s. Other rules could have discouraged ISP entry (e.g., by requiring different ISPs to pay to universal service funds on top of subscriber line or business line charges) or imposed a different cost structure (e.g., per-minute charges).

3.3. The founding of commercial organizations

Shortly after the Internet's commercialization, only a few commercial enterprises offered national dial-up networks with Internet access, and they mostly targeted the major urban areas. Pricing varied widely (*Boardwatch Magazine*[23]). Most of these ISPs were devoted to recreating the type of network found in academic settings or modifying a commercial bulletin board with the addition of backbone connections, so interconnection among these firms did not raise insoluble contracting or governance problems. These ISPs also were devoted primarily to dial-up; some ISPs attempted sophisticated data transport over higher-speed lines, where the regulatory issues could be more complex and where competitive local exchange competitors were developing the nascent market.

Very quickly ISPs learned that low-cost delivery required locating access facilities close to customers because of telephony pricing policies across the United States. The U.S. telephone system has one pervasive feature; distance-sensitive pricing at the local level, but no charge for extremely short distances. In virtually every state of the country, telephone calls over significant distances (i.e., more than 30 miles) engender per-minute expenses, whereas local calls for very short distances are usually free. Hence, Internet access providers had a strong interest in reducing expenses to users by providing local coverage. Indeed, it is probably more accurate to say that the notion of locating in such a way to take advantage of free pricing had already been understood in the bulletin board industry, so it was not hard to understand or reapply to ISPs.

Access over dial-up lent itself to small-scale commercial implementations. Several hundred customers could generate enough revenue (at the basic unlimited access price of $20 per month) to support physical facilities and a high-speed backbone connection in one location, so scale economies were not very binding. The marginal costs of providing dial-up services were low and the marginal costs of expansion also fell quickly, as remote monitoring technology made it inexpensive to open remote facilities. The marginal costs to users of dial-up service were also low in response, involving only incremental changes for organizations that had experience with PC use or LAN technology. It was easy to generate revenue in subscription models, where a commercial firm withheld availability of access unless payment was made. Hence, the economic thresholds for commercial dial-up service turned out to be feasible on a very small scale, which was encouraging to small firms and independent ISPs.

[23] The earliest advertisements for ISPs in *Boardwatch Magazine* appeared in late 1993, growing slowly until mid-1995, at which point *Boardwatch* began to organize their presentation of pricing and basic offerings. There was an explosion of entry in 1995, with thousands being present for the next few years.

Many firms, such as IBM, AT&T, and AOL, wrote up and implemented plans to access businesses and home users on a national scale, but the economic advantage of large branded identities did not preclude the entry of small-scale firms, at least not at first. Contemporaries noted that the largest firms played an important role as 'certifiers' during the early years. This observation was frequently made about IBM's provision to business and AT&T's entry, which came sooner than any other large established firms' entry into provision for homes. These actions helped to ratify and legitimize the market for business and home Internet access (Maloff, 1997). The online service providers—Prodigy, Genie, CompuServe, MSN, and AOL—also began converting to Internet service around 1995 too, with some providing service earlier than others[24].

The failure of known firms to dominate that market (or induce a quick shake-out) was interpreted by many observers as a sign that small-scale firms had a viable opportunity to provide this new service as long as demand continued to grow (Greenstein, 2001). Indeed, thousands of small entrepreneurial ventures also grew throughout the country and gained enough market share to sustain themselves. New entrants, such as Erols, Earthlink, Mindspring, MainOne, Verio, and many others, gained large market positions. Some of these positions were sustained and others were not. Private label ISPs also emerged when associations and affiliation groups offered rebranded Internet access to their members. These groups did not own or operate an ISP, instead their access was being repackaged from the original ISP and rebranded by the association. These firms could survive on relatively low market shares, though, to be sure, they were not very profitable either. By the end of the 1990s AOL's standardized service came to dominate market share, but that still left tens of millions of potential users for others.

The last successful mass-market entrants were the 'free' ISPs. The free-ISP model emerged in late 1998 and grew rapidly in 1999, offering free Internet access in exchange for advertisements placed on the users' screen. These firms eventually signed up several million households. NetZero became the largest of these. It converted to a hybrid free/charge model after its merger with Juno Online in 2001.

A few statistics will illustrate the trends. In one of the earliest Internet 'handbooks,' Krol (1992) lists 45 North American providers (8 have national presence). In the second edition of the same book, Krol (1994) lists 86 North American providers (10 have national presence). Marine et al. (1993) lists 28 North

[24]These conversions received considerable publicity at the time. For example, AOL's struggles with the competition from the Internet received the much media attention. Its full conversion to the Internet came comparatively later than others. It finally adopted unlimited access to AOL content and Internet content in November 1996—i.e., converting from'pricing per service' and 'pricing per minute' to so called 'all you can eat pricing'. The large user response overwhelmed AOL's facilities. It took them several months to increase their modem/server capacity to handle the large volumes of calls and longer session lengths.

American ISP's and 6 foreign ISP's, which is also about the first time *Boardwatch Magazine* begins to carry priced advertisements for ISPs. Meeker and Dupuy (1996) reports that there are over 3000 ISP's, and the Fall 1996 *Boardwatch Magazine's Directory of Internet Service Providers* lists 2934 firms in North America. By January 1998 the same magazine lists 4167 ISPs. Of course, the growth occurred due to the entry of small ISPs, each of whom had a tiny fraction of the market.

Another statistic also represents this growth. When Downes and Greenstein (2002) did their first survey of dial-up Internet access, they compiled a list of 12,000 local phone numbers for calling the Internet in the fall of 1996. By the fall of 1998, when they did their last compilation, the whole list had swollen to over 65,000 local phone numbers. By that latter date every metropolitan area in the U.S. had competitive coverage by then, as did many small towns.

3.4. *Coverage by dial-up providers*

Growth and entry brought about extraordinary results. Downes and Greenstein (1998, 2002) have constructed statistics about the density of location of ISPs at the county level for the fall of 1996 and 1998[25]. See Figures 2 and 3 for illustration of ISP entry in 1996 and 1998.

ISPs tend to locate in all the major population centers, but there are also some providers in rural areas[26]. Their findings also illustrate the importance of changes over time. Many of the areas that had no coverage in the fall of 1996 were covered by the fall of 1998. Many of the areas that had competitive access markets in the early period were extraordinarily competitive in the latter period. This pattern seems to follow a pattern found in other new technologies that increasingly standardize over time, from urban areas into rural ones (Rogers, 1995).

Downes and Greenstein (2002) show that more than 92 percent of the U.S. population had access by a short local phone call to seven or more ISPs by 1998. Less than 5 percent did not have any access. Almost certainly the true percentage of the population without access to a competitive dial-up market is much lower than 5 percent. In other words, with the notable exception of some low-density areas, ISP service was quickly available almost everywhere.

The lowest density and poorest areas of the United States did lag, however, bearing signs of permanent retardation of development. In a study of the Appalachians and some areas with histories of poor communications service, Strover et al. (2002) examine ISP presence in the states of Iowa, Texas, Louisiana, and

[25] For further documentation of these methods, see Downes and Greenstein (2002) or Greenstein (2001).

[26] In a light moment this pattern motivated (the late) Zvi Griliches-known for his research on the diffusion of new strains of corn—to observe that the map from 1996 showed that a good predictor for the absence of Internet Service Providers was the presence of hybrid corn seed.

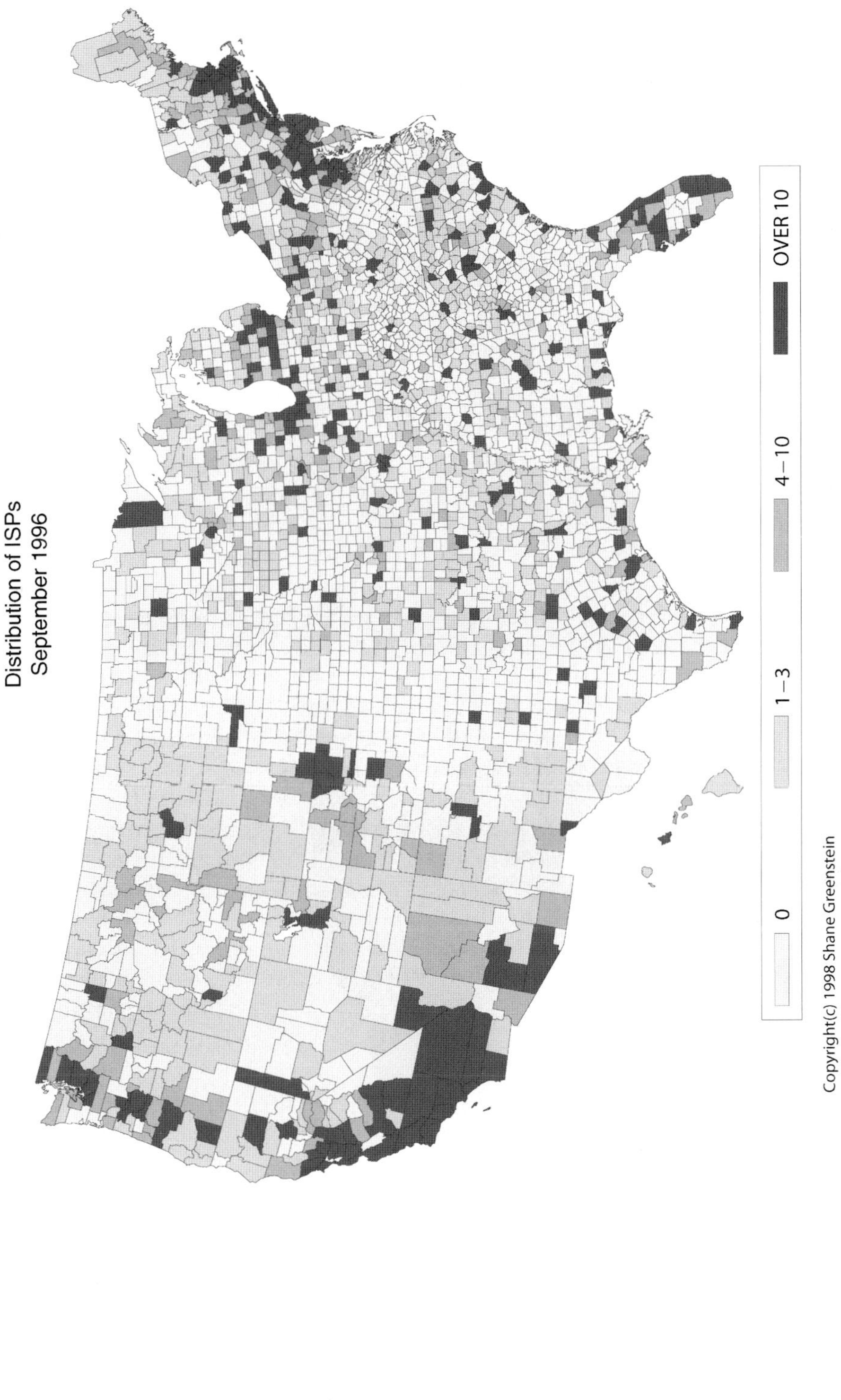

Fig. 2. Distribution of ISPs, September 1996.

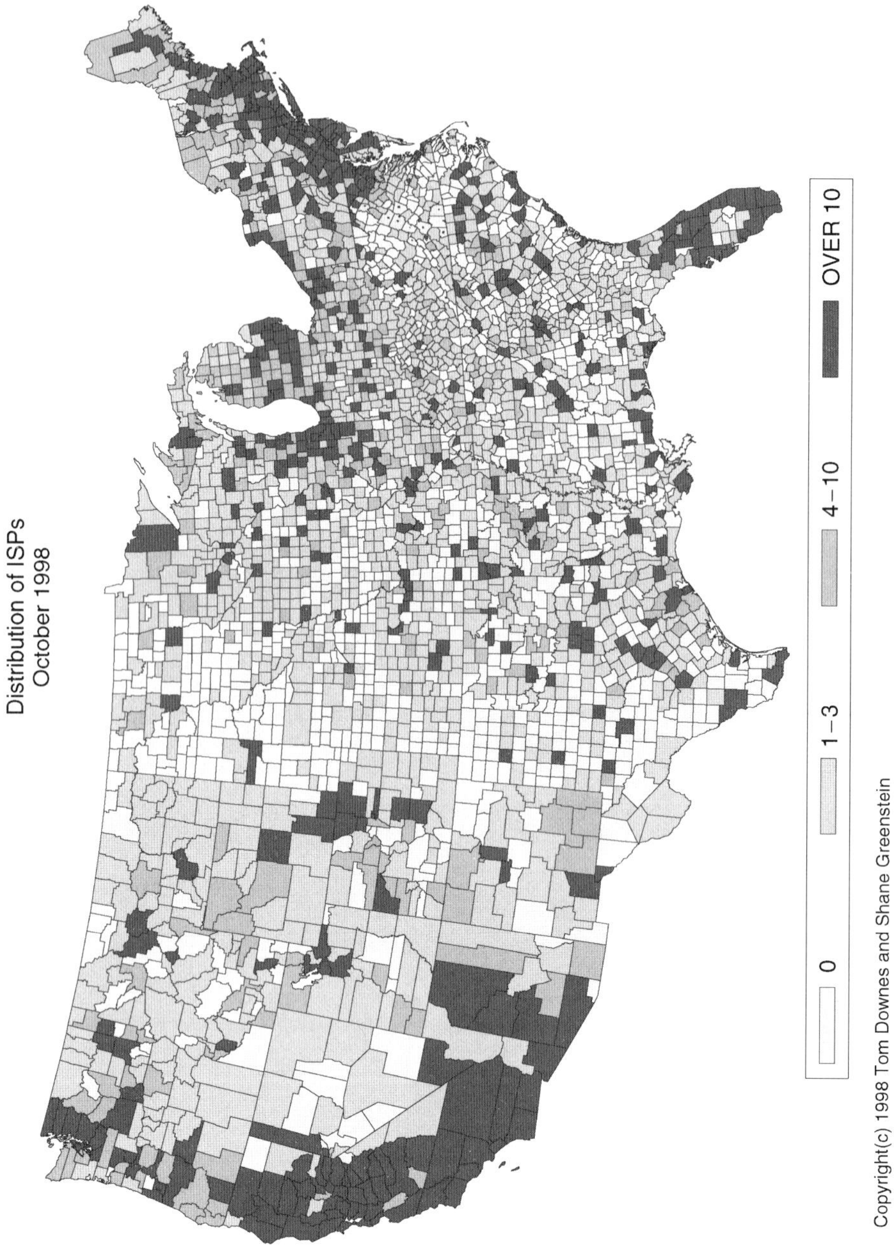

Fig. 3. Distribution of ISPs, October 1998.

West Virginia and determine the availability and nature of Internet services from ISPs for each county. They find that the rural areas suffer significant disadvantages for Internet service. Such counties often lack competitive supply of providers. Even when suppliers exist, they sometimes provide limited services or focus on specific segments, such as business users, not households. It is difficult for any policy—other than subsidy—to overcome these basic economic constraints[27].

Strover (2001) arrives at a comparatively pessimistic assessment, one shared by many observers[28]. She points out that the cost structure for ISPs is unfavorable because of their dependence on commercial telecommunications infrastructure providers, which are reluctant to invest in rural areas due to the high costs necessary to reach what often are relatively few customers. Furthermore, a lack of competition in rural areas among telephone service providers serves to exacerbate the low incentives.

Conversely, there is an optimistic element to ISP rural service, affiliated with the dispersal of E-rate funds. As was previously noted, E-rate funds were dispersed in order to alleviate geographic inequalities in the dispersion of the Internet. Hence some researchers examined the provision of the Internet, not for the home or office, but for libraries and schools. For example, Bertot and McClure (2000) show that library use of the Internet is nearly ubiquitous, irrespective of location[29]. In their survey, the key issues for disadvantageously located libraries revolved around the quality of Internet activity and supporting services. Access was taken for granted nearly everywhere by the late 1990s.

Goolsbee and Guryan (2002) reach a similarly optimistic assessment based on their study of the California schools' access to the Internet. They too show that the discussion had moved to focusing on issues well beyond access. Again, they attribute this to the dispersion of E-rate funds in the late 1990s.

3.5. *Organizational strategy and coverage*

An unexpected pattern accompanied this rapid growth in geographic coverage. First, the number of firms maintaining national and regional networks increased between 1996 and 1998. In 1996, most of the national firms were recognizable—they were such firms as IBM, AT&T, and other established firms that entered the ISP business to provide a complementary part to their existing services (e.g., providing data services to large corporate clients). In 1996 the other recognizable firms were

[27] For example, see Nicholas's (2000) study of the multiple attempts to provide access to rural Texas communities. He shows how the construction of calling area geographic boundaries shapes the entry patterns of ISPs. His close study of regulatory policy in rural Texas shows both the strengths and pitfalls of this policy approach.

[28] See also Garcia (1996), Gillett (2000), Hindman (2000), Parker (2000) or Malecki (2003).

[29] Indeed, earlier reports show that this had been true since the mid-1990s for a high fraction of libraries. See Bertot and McClure (2000) and further studies at http://www.nclis.gov/statsurv/statsurv.cfm for information on the near ubiquity of Internet access in US libraries by 2000.

those who had supported national online consumer services, such as CompuServe, AOL and Prodigy. All were in the process of converting their online service, previously run more like bulletin boards than ISPs, into Internet providers. By 1998, thousands of entrepreneurial firms maintained national networks and few of these new firms were recognizable to anyone other than an industry expert.

There was also a clear dichotomy between the growth patterns of entrepreneurial firms that became national and regional firms. National firms grew geographically by starting with major cities across the country and then progressively moving to cities of smaller populations. Firms with a regional focus grew into geographically contiguous areas, seemingly irrespective of urban or rural features. The growth of standardized contracting practices for renting facilities from other firms in other locations further added to the geographic reach of individual ISPs, even in areas where they did not operate or build facilities.

Most of the coverage in rural areas comes from local firms. In 1996, the providers in rural counties with under 50,000 population were overwhelmingly local or regional. Only in populations of 50,000 or above, do national firms begin to appear. In fall of 1998, the equivalent figures were 30,000 or lower, which indicate that some national firms had moved into slightly smaller areas and less dense geographic locations (Downes and Greenstein, 2002). Figures 2 and 3, taken from Downes and Greenstein provide a visual sense of those patterns in 1996 and 1998.

Local and regional ISPs provided Internet access everywhere, including small rural towns. National firms started with the big cities and slowly expanding into less dense territory. It appears as if it did not pay for many large national providers to provide dial-up service for the rural areas, whereas many small local firms in other lines of business (e.g., local PC retailing) could afford to add Internet access to their existing business. It may also be the case that the local firm may have had an easier time customizing the Internet access business to the unique needs of a set of users in a rural setting.

By 2000, many of these trends came to a halt. Growth in demand slowed, as Internet adoption at households surpassed 50 percent and was approaching 60 percent (National Telecommunications and Information Association, 2002)[30]. Business adoption of basic Internet access for purposes of participating in e-mail and browsing had also reached near saturation (Forman et al., 2003a,b)[31]. Growth of new users was slowing. If suppliers were going to experience growth, it would not be in 'new users,' but in more intensive use by existing users.

[30] Over 65 percent of the US population made use of a computer, while 53.9 percent made use of the Internet from any location.

[31] A survey of business establishments with over 100 employees showed that 88 percent of such establishments participated in the Internet in some form by the end of 2000. This clearly does not hold for small establishments, particularly those with less than 25 employees and especially those with less than 10 (Buckley and Montes, 2002).

As it turned out, a shake-out ensued, in which many small ISPs left the market or sold their assets to other vendors. Several large ISPs, such as PSINet went bankrupt from (what turned out to be) imprudent early expansions. Other ISPs (more commonly) merged with each other, perceiving that bigger firms had better chances for survival. Suhonen (2002) identifies 86 publicly known mergers in 1999, 152 in 2000, and 81 in 2001. There were many more than these[32].

Neither there is evidence that geographic coverage declined, nor would that have been predicted for a technology with such small economies of scale in deployment. To be sure, the set of vendors decreased, in what many observers described as an inevitable and overdue consolidation of suppliers (*Boardwatch*, 2001).

It is hard to get a definitive sense of how many suppliers exited due to bankruptcies. Many sales and mergers were not publicized. Moreover, the industry for tracking the ISP industry underwent a shake-out along the same time as the industry, so there were no consistent reporting norms over time. Here are some indications of the type of change that occurred: in January, 1999, *Boardwatch* lists 4511 U.S. ISPs. By 2001 that had fallen to approximately 3500 (Suhonen, 2002). *Boardwatch* lists just over 2500 for 2002. Most of this decline came from exit of small firms.

This decline in ISPs is plausible for another reason. After 2000 many other facets of the Internet experienced a decline in the rate of growth and a decline in the degree of optimism about future prospects. To be clear, this should not be interpreted as a decline in the level of demand for Internet service, but as a decline in the expected rate of growth of demand.

3.6. Variety and quality over geography

As with many business ISPs pursued activities to develop more than just one source of revenue. Many ISPs provide services that compliment the physical connection. The most important and necessary service is an address for the user's computer. All Internet packet traffic has a 'from' and 'to' address that allows it to be routed to the right destination. An ISP assigns each connecting user with an address from its own pool of available addresses.

ISP's offer other services in addition to the network addresses. These may include e-mail servers, newsgroup servers, portal content, online account management, customer service, technical support, Internet training, file space and storage, web-site hosting, and web development and design. Many larger ISPs also bundle Internet software with their subscriptions. This software is either private-labeled or provided directly by third parties. Some of it is provided as standard part of the

[32] For example, MainOne and Verio both were known for buying other ISPs in attempts to grow into a national ISP. Neither was commercially successful.

ISP contract and some of it is not. Some ISPs also recommend and sell customer equipment they guarantee will be compatible with the ISP's access equipment.

It is obvious from industry publications that ISPs vary in the array of service they offer and the quality with which they provide them. How should this be measured? It is difficult to characterize the variance in either offerings or quality of a new industry. It is even more difficult to get information about the variance of provision in different geographic locations. The measurement challenges were particularly severe because official government reporting agencies had not standardized the classification of activity[33].

A few studies provide insight into the interplay of geography and service quality. For example, Augereau and Greenstein (2001) look at the determinants for upgrade decisions for ISPs. They examine upgrades from dial-up-only service to 56K modem or ISDN service by 1997. In their model, they look for firm-specific and location-specific factors that affect firms' choices to offer more advanced Internet services. Their main finding is that "the ISPs with the highest propensity to upgrade are the firms with more capital equipment and the firms with propitious locations." The most expansive ISPs locate in urban areas. They further argue that this could lead to inequality in the quality of supply between ISPs in high-density and low-density areas.

Greenstein (2000a,b) examines the variety of service offered by ISPs with different location coverage. His work focuses on the offering of additional services, such as hosting, networking, web design, and high-speed access. Why do firms make different choices? Differences across firms arise when decision makers face different demand conditions, quality of local infrastructure, and labor markets for talent, or when they inherit firm assets of differing quality. These create a variety of economic incentives for adapting Internet infrastructure to new uses and applications.

Small ISPs in different parts of the country have very different propensities to offer additional services. Of the 1764 ISPs surveyed, 26.9 percent offer frontier access (i.e., they support provision of speeds higher than typical dial-up access at 56K), 22.9 percent offer networking services, 23.5 percent offer hosting services, and 38.6 percent offer web design services. Of the 325 ISPs primarily found in rural areas, 12.0 percent offer frontier access, 11.0 percent offer networking services, 13.8 percent offer hosting, and 23.3 percent offer web design services. The propensities for rural ISPs are between 40 and 60 percent lower in each category. Surveys of firms with national coverage (primarily in urban areas) find patterns similar to those small ISPs devoted to urban areas.

[33]Some of these difficulties were coincident with the US conversion to North American Industry Classifications in 1997. See E-Stats, as maintained by the US Census, at http://www.census.gov/eos/www/ebusiness614.htm. This activity begins tracking electronic commerce around 1999. The earliest official US data on the Internet is the official price index for Internet access in the US. It begins December 1997.

How should these patterns be interpreted? There appears to be significant ISP-specific factors that cluster together with the choice to offer new services. In particular, the size of the ISP, its geographic reach, some key capital investments, and the ISP's focus on particular types of users and non-ISP lines of businesses all predict experimentation with new services. Greenstein (2000a) develops a statistical model that shows that the location-specific variables have less importance than the firm-specific variables except in the provision of high-speed access, where both factors are important. Interestingly, some of the firm-specific variables—whose distribution does vary over geographic space—also help explain some of the observed variance.

These findings are consistent with the view that small ISPs choose strategies for differentiation based on firm-specific growth strategies, limited (possibly) by indivisible assets arising at a local level. For example, in some cities, there are only a few locations for interconnection with an Internet backbone or a local telephone switch. Other equipment for running an ISP, such as a modem bank, could be cheaper at larger scale.

In other words, a small ISP in a rural area faces the limits of scale economies in facets of its operations. Scale economies arise from capital equipment investment and other fixed costs on the back end. Some times these were not severe when small ISPs rely on existing telecommunications infrastructure. Frontier access also can be scale intensive when it requires significant costs to set up. In that case, it may require higher volume of use to be profitable. Therefore, it may not be profitable without large capital investments and a density of potential users, typically businesses comprised of professionals, a situation prevalent in urban areas but apparent only in a few rural areas.

Ancillary services of the sort described above are also subject to some scale economies. Networking and web design require a core mass of business customers to defray the costs of acquiring capital and maintaining sufficient technical expertise. These costs are also defrayed by some economies of scope among these services and between these and other lines of business.

Overall, urban areas get more new services because of two factors: (1) increased exposure to national ISPs, who expand their services more often, and (2) the local firms in urban areas possess features that lead them to offer services with propensities similar to the national firms. That is, high-density areas almost always get some ISP entry, whereas some low-density areas get none or very little. High-density areas see an especially large amount of entry because they experience entry from nearly all the firms with national ambitions. In contrast, little or no entry in a low-density area virtually precludes availability of any complement to basic access. High-density areas benefit from repeated exposure to many ISPs that offer such services. More entrants will lead to more realized numbers of new services, thereby raising the probability of finding one, two, or three instances of new services in a specific location.

Perhaps this is best illustrated in concrete terms. At the turn of the millennium small or isolated rural towns in the U.S. were the only areas at risk for inadequate Internet service, if that had it at all. If they had some service, whether they got adequate service depended on small details about the local conditions. Did the local PC service guru want to run an ISP on the cheap in his back office or would he invest in a large modem bank? Would the local rural telephone company run Internet service if it barely broke even or operated at a loss? Would they only offer operational connectivity or would they also offer services associated with installation, designing web pages, and designing fire walls for local businesses? Did an Internet backbone connection happen to run along a nearby right-of-way, such as a major highway, pipeline or electrical line path? The denizens of some rural places got to experience the full benefits of the services affiliated with the commercial Internet because the right people were there and took the initiative to bring local services to the highest standards possible. In some places this did not happen the local population had to make do with less.

3.7. Summary

The experience of the dial-access industry offers one lens for filtering through the patterns of deployment of the Internet after commercialization. Over a decade there were many changes in the modal form of business. Changes in the delivery of services and changes in user expectations resulted in numerous qualitative changes in the basic service experienced by all users. The structure of the industry also fluctuated and a few dominant providers emerged in the second half of the decade.

In spite of fluctuation, a basic economic pattern emerged. Dial-up access was cheap to deploy because it was built on top of the existing telecommunications infrastructure. It was a retrofit on the phone system, requiring a firm to organize access and a user to invest in modems. In virtually all locations except those with low density the costs of providing this service was minimal. So prices to households stayed comparatively low (around $20 on average). Closely related services grew on top of basic access, supporting the dispersion of Internet access to most potential users in the country. Within a couple of years Internet service was nearly geographically ubiquitous.

4. The location of network backbone

The commercial Internet is comprised of hubs, routers, high-speed switches, POPs, and high-speed high-capacity pipe that transmit data. These pipes and supporting equipment are sometimes called *backbone* for short. Backbone is comprised mostly of fiber-optic lines of various speeds and capacity. However, no vender can point to a specific piece of fiber and call it 'backbone.' This label is a fiction, but a convenient one. Every major vendor has a network with lines that go from one point to another, but it is too much trouble to refer to it as "transmission capacity devoted primarily to carrying traffic from many sources to many sources."

The Internet's backbone connects servers operated by different ISPs. It connects city nodes. It transports data over long distances. The presence and structure of the backbone (potentially) provide information about which cities are playing the most prominent role in the development and diffusion of the Internet. The geography of these connections (potentially) provides insight into the economic determinants of the network. The connection to the backbone and the size of the backbone has an impact on local economic activity since it has an effect on a firm's ability to distribute large amounts of data via the Internet.

What is the proper economic interpretation of the size of backbones going into cities? Comparing the backbones in different regions is not straightforward. Its presence depends on many things such as population size, type of local industry, and other facets of local demand. Its maximum flow rate only gets used at peak times and not most of the day, so statistics about capacity must be interpreted with care. Moreover, nobody would expect connection and bandwidth to be equally distributed across geographic space, so the appropriate benchmark for assessing this geographic dispersion of backbone was subject to debate[34].

Despite this inherent interpretive ambiguity, there is intense policy interest in understanding the geography of commercially supplied backbone. There are concerns that some areas (e.g., small towns) were underserved while others (e.g., major cities) were served too much. There are also concerns that the industry was too concentrated, as well as nearly the opposite, that the competitive situation motivated firms to make impertinent and redundant investments. What is the relevant evidence and how should it be interpreted?

4.1. The geographic features of the backbone

Prior to commercialization the backbone for U.S. networks grew out of the network built for the NSF. The NSFNet was designed to serve the research needs of universities. It connected over one thousand universities in North America by the early 1990s. It had concentrated much of its communications infrastructure around a dozen supercomputer centers and other major research universities, which were distributed to as many different locations. That resulted in a geographically concentrated transmission network with a wide dispersion of access points. Frazer (1995) analyzes the location of NSFNet through its growth. Just prior to commercialization—i.e., in 1991—the key end-points were university and research centers in Seattle, Palo Alto, San Diego, Salt Lake City, Boulder, Lincoln, Houston, Argonne laboratories (Chicago), Urbana-Champaign (central

Illinois), Ann Arbor, Pittsburgh, Ithaca, Cambridge, Princeton, College Park, and Atlanta[35].

The points of access, transmission and capacity concentration in the NSFNet were destined to be different from the concentration that developed after commercialization. The commercial Internet was not managed by a single entity. Under commercialization, in contrast, multiple suppliers made investments designed to support a myriad of targeted customer needs. What economic principles explain the geographic patterns of backbone capacity? What dimension of geographic location shaped the supply conditions facing users in different locations?

Motivated by the desire to examine the relationship between economic activity and Internet backbone, as well as cautiously aware of the uncertainty inherent in cybergeography, I argue that commercialization resulted in a network with two surprising features.

1. Resources for delivering the Internet concentrate in some areas so that there are overlapping networks serving the location. This redundancy leads observers to call the network 'overbuilt.' On close inspection, however, it is difficult to support this claim unambiguously.
2. The U.S. backbone network was built quickly and ahead of demand in some locations. In retrospect the speed with which it was built appears almost reckless to some observers. On close inspection, however, there was an economic rationale to much of this behavior, so, once again, it is difficult to argue that behavior was unambiguously 'reckless.'

4.2. The industrial organization of the commercial backbone

The geographic features of the U.S. backbone need to be understood in terms of the industrial organization of interconnection in the commercial era. Several factors are salient.

First, the commercial Internet developed distinct cooperative institutions for exchanging data. To be sure, this observation has to be qualified for the time period and the activity. As the Internet made its transition into private hands, cooperative principles persisted due to a combination of good policy and—for lack of a better phrase—corporate ethics. An example of good policy: the commercial network retained several of the public exchange points developed by the NSF, and transferred their operation to associations of private firms[36]. An

[35] See http://archive.ncsa.uiuc.edu/SCMS/DigLib/text/technology/Visualization-Study-NSFNET-Cox.html or Frazer (1995, especially p. 33).

[36] For extensive discussion, see Mackie-Mason and Varian (1995), Srinagesh (1997), Bailey and McKnight (1997), Chinoy and Salo (1997), Cawley (1997), Farnon and Huddle (1997), National Research Council (2001). Also see Milgrom, Mitchell and Srinagesh (2000) as well as Besen, Spigel, and Srinagesh (2001) for a more recent update to this line of thinking.

example of ethics: descendents of the NSFNet were managing the networks at MCI, UUNet, BBN, or IBM and elsewhere. None of them tried to exploit their strong leading positions with strategic denial (or overt degradation) of interconnection to new entrants. Such behavior could have affected private entrants with aggressive ambitions, such as PSI (whose management was also descendent) or at a later time, Level3, who built an entirely new national backbone. That is, incumbent firms did not refuse to interconnect entrants as part of a discriminatory policy designed to foreclose entry[37].

Second, the tenor and tone of cooperative behavior started to change in the mid-1990s as the commercial prospects for the Internet became more apparent and demonstrably larger. For example, a few large firms exchanged traffic but did not charge each other for it, nor discriminate on any basis other than size of data traffic. At the same time they announced policies to charge for interconnection. The practice of not charging became known as private peering.

There were a number of facets to these arrangements. Peering meant that firms exchanged data without the hassles of tracking traffic in order to come up with explicit compensation. Several of the largest data carriers exchanged data this way if the volumes were within (depending on the firm) two to eight orders of magnitude of each other. MCI/UUNet was among the largest backbone providers to implement this policy in the mid-1990s. It justified this policy on the basis of its engineering efficiencies[38]. That is, monitoring traffic levels at high volumes was costly. Peering agreements freed both parties from monitoring each others' traffic[39].

While at first the seemingly genteel nature of peering was regarded as a legacy institution from the precommercial era, more concerns arose when the largest firms started peering with only large firms and charging smaller ISPs for access, using contracts of varying lengths and obligations. This meant that the large firms could de facto price discriminate, charging (effectively) higher prices to small firms and lower prices to the firms with whom they peered. Given the potential costs savings of peering, such price discrimination was not necessarily anticompetitive, but it did start to raise questions among regulators about large firms with market power treating rivals differently.

[37] That is not say there were no conflicts of the public and private. See Kende (2000) for a discussion about the entry of Level3 and thei complaints about the peering policies of UUNet. Kende concludes that much of this complaint arose from Level3's dislike for purchasing transit services from UUNet. UUNet refused to peer at no cost until Level3's actual volumes reached their ambitions.

[38] The engineering justification for peering was seemingly straightforward. As long as volumes of traffic exchange were close to each other in magnitude, there was little business sense for either party to closely track volumes of traffic and charge for transit services.

[39] Coffman and Odlyko (2002), p. 20, assert that most traffic by 2002 travels through private peering, in sharp contrast to 7 years earlier, when most traffic traveled through public exchanges.

Moreover, smaller ISPs had to exchange traffic at the public exchange points. This raised the further question about whether public exchange points were as fast as private peering (if these were equal, why did the large firms prefer to peer away from the public exchange?). It also raised issues about whether the users of smaller ISPs received lower quality service because their traffic exchanged at slower rates.

Because the technology was new and because NSF had commercialized the Internet with only general guidelines about Internet connection practices, there was not a clear regulatory policy for thinking about these issues. Yet, in the background was the vague memory of AT&T self-serving interconnection policies almost a century earlier, feeding a larger public policy concern about the ease with which new firms could enter and interconnect.

While there was no overt evidence of such behavior by the largest backbone providers, MCI or later WorldCom, there were always concerns about abuse of market power. Regulators were sensitive to concerns that market leaders would not invest aggressively to serve new areas or show attention to small firms requesting interconnection in difficult circumstances[40]. Such concerns motivated one of the conditions for the merger between MCI and WorldCom, that the merged firm sells some backbone assets[41]. Similar concerns motivated intervention to prevent the proposed WorldCom merger with Sprint.

Yet, this reasoning necessarily oversimplifies the implications for the geographic reach of the Internet. Some elements of interconnection pricing—between two ISPs, for example—were entirely market oriented and mediated by contracts, while other facets were heavily regulated—such as between and ISP and an incumbent local exchange company (ILEC), especially after the 1996 Telecommunications Act. In effect, no participant in this market ever really escaped the effects of this regulation. Some transactions were simply more distant than others. Would efficiency in one part of the system be in society's interest if many other parts of the network were priced according to noneconomic principles? Or would it simply mean that a few large firms (and their predominantly urban users) were benefiting form the distortions imposed on others[42]?

Once the commercial Internet began to explode in size and number of participants with new entry after 1995, a commercial intermediary became prominent and changed the relationship between geographic location and interconnection. Popular web sites, such as Yahoo or AOL, and retailers who saw competitive value in speed, such as Amazon, made deals with cache/mirror sites operated by firms (e.g., Akamai, Digital Island, among others). Different intermediaries operated in different ways, but all had the same purpose—eliminating performance

[40] For theoretical underpinnings see the discussion in Laffont and Tirole (2000), or Laffont, Marcus, Rey and Tirole (2003).

[41] Eventually MCI's Internet backbone was sold to Cable and Wireless.

[42] For discussion, see, e.g., Noam (2001), Noam (2002), Laffont and Tirole (2000), Kende (2000), and Sidak and Spulber (1998).

degradation from national backbone congestion and data handoffs at switching points. Congestion on the national backbone could degrade quality if it slowed the movement of packets from a host site in one place, say Seattle, to a distant place, say New York. Caching eliminated much of the differences in performance between locations in the U.S.—at least for the most popular sites and ISPs and content providers who could afford it (Carr, 2000)[43]. Users in New York downloaded popular material nearly as fast as users downloaded material in Seattle. Users in all but the most isolated small towns could have as fast a service for mainstream applications as users in central cities.

This intermediary's presence was quite important for the location of backbone. These deals eliminated much of the competitive differences between ISPs in different locations and the backbones supporting them. The difference between a local and national ISP came down to software hosted on the ISPs' server and embodied in its connection with other ISPs. To be sure, it only held for the most popular sites and the largest ISPs, but that was still a high fraction of the experience of most users[44].

Finally, one other institution deserves attention: indefeasible rights of use (IRU). Firms can obtain either space in conduits or dark fiber (i.e., fiber *sans* terminal equipment), typically for 20 years at a time. This permits a firm to buy the option to have distribution capacity in a certain location, deferring investments in equipment, switches and other assets affiliated with operations. IRUs became common in the late 1990s, as firms with rights-of-ways in one location leased IRUs to firms in another, and vice versa.

IRUs were important for three reasons. First, it limited the sunk investments a firm had to make in order to secure the option to provide service in an area when its customer base was small in the short run. This was a cost-minimizing strategy in an era of potential growth and demand volatility (Faulhaber and Hogendorn, 2004). Second, it fostered the impression in policy discussion and popular discussion that the U.S. backbone network was more redundant than truly was the case. That is, the independent statements from firms about their network footprint could lead to errors of double counting if IRUs were not taken into account. For example, Hogendorn (2003) shows that such IRUs accounted for more than half the new *growth* in new fiber between 1997 and 2001, though this was

[43] For example, Akamai's network maps the entire Internet every few minutes, identifying points of congestion. If one server is down, data requests are handled by other servers. Then Akamai's servers send the user all the low-bandwidth elements of a page, such as text. It also instructs the user's browser to get high-bandwidth content, such as photographs, from geographically close servers. Akamai's algorithms determine the optimal servers and routes for each end user.

[44] Web site surfing among non-AOL users during the early period of commercialization, as now, was remarkably concentrated. Several hundred sites accounted for the vast majority of the time online. Use of portals alone accounted for more than a quarter of the time online among non-AOL users (Goldfarb, 2004). Since AOL use was at least 40 percent of household use of access, a large fraction of user experience interacted with a mirror/cache site in some form.

commonly not stated in news accounts in the *Wall Street Journal*, among other publications[45]. Third, these types of contracts eased potential entry by the largest firms into different territories, heightening competitive pressures between them. Hogendorn (2003) speculates that this behavior also contributed to the rapid decline in prices for backbone services.

4.3. Interpreting networking practices

The noncommercial Internet had beget a commercial Internet that retained some features, such as peering and end-to-end, and adopted new features, such as commercial caching. How were firm practices to be understood and interpreted? Such questions went to the differences in behavior between the public/private domains in the commercial Internet (e.g., whether the privately managed Internet required public intervention or not because it did or did not accomplish public goals, such as geographic ubiquity).

To the delight of some and the dismay of others, these factors supported a backbone that some called *tiered* (Kahin, 1997; Frieden, 2001). The tiers were associated with size of footprint and volume of traffic. Tiers became a short-handed designation for which firm carried data over long distances and which collected charges from others for transit service. For example, the largest national firms all became tier-1 providers. This included AT&T, IBM (before being sold to AT&T), MCI (whose backbone was sold to Cable and Wireless), UUNet (eventually sold to WorldCom), Sprint, and Genuity, among others[46]. Most regional and local ISPs became lower tier ISPs, purchasing interconnection from one of several national providers.

Not all analysts argued that tiers had a hierarchical interpretation. Others observers—who were skeptical of the rigidity of tiers—called this system a *mesh* (Besen et al., 2001). In this view, participants faced many options for interconnection. For example, many ISPs arranged to use multiple backbone providers (i.e., known as multihoming), significantly diminishing the market power of any particular backbone provider in any given location. Related, dial-up ISPs with large networks (e.g., such as Genuity) could receive calls in one location but back-haul it to another where it is then connected to the Internet. This type of network design gave ISPs multiple options for connecting to the Internet, limited the discriminatory power of any single backbone firm.

Whatever it was called, this system was not associated with overt discriminatory treatment in quality of service across location. That is, the system had remarkably little effect on the quality of applications in different locations, at least at first. The

[45] He also points out that the FCC discontinued collecting data about the state of fiber networks in the US after 1998, so there was no government oversight to correct this common misperception.

[46] Among the others sometimes counted as tier-1 providers include Qwest, IXC, Williams, and Level3. See Kende (2000) or Hogendorn (2003) for different discussions.

engineers called this the preservation of the 'end-to-end' features of architecture (Blumenthal and Clark, 2001; National Research Council, 2001). Performance was not based on either the identity of the end user, the user's location, or the type of application. Stated simply, transmission capacity built anywhere interacted with transmission anywhere else without altering the application.

The preservation of 'end-to-end' partly reflected economies of scale from standardization. Contracts for different bandwidth and priority handled the costs of interconnection between ISPs and backbone firms, but these transactions did not greatly alter the user's experience of the application[47]. Ownership of facilities did not induce discrimination on the basis of the origin or destination of the data or type of application[48]. It was not in any carrier's interest to use nonstandard equipment or operations, nor to vary from the common protocols, lest they lose access to the possibility of imitating innovative practices arising elsewhere.

To be sure, many Cassandra-like observers foresaw several different types of threats to the preservation of this end-to-end feature from several different types of applications and proprietary solutions (and still do see such threats on the horizon)[49]. And, yet, widespread incompatible balkanization or discriminatory interoperability had not yet emerged in the first decade after commercialization[50].

Rather, the presence of end-to-end accompanied several behaviors that shaped the Internet backbone's geographic layout. First, vendors were not discouraged from building geographically overlapping networks, since such action did not alter whether they ultimately interconnected. Related, it also encouraged the use of IRUs for covering new geographic territories, since using such assets was standardized across geographic regions.

Second, vendors also could build the opposite of a national footprint, namely, specialize with regional footprints but still ultimately interconnect. Or, related, a regional firm could choose to build its own facilities in a region, but contract for backbone capacity in other regions in the event that its clientele valued wider geographic coverage. Again, interconnection was assured, so a national footprint was not necessary for survival.

[47] This is somewhat of a simplification. ISPs who contracted for larger capacity and first priority did offer fewer peak-load issues to downstream users. For many uses, such as e-mail, these contracting practices had little impact on the user experience.

[48] See Kende for a careful description of how interconnection takes place. See Besen et al. (2001, 2002), for an argument that the lack of discrimination in interconnection resulted from a combination of peering, multihoming (ISPs using more than one backbone provider) and other behaviors generating multiple alternatives for users. In their view, these features eroded market power of all backbone players, fostering incentives for nondiscriminatory interconnection.

[49] These arguments became especially heated during the regulatory hearings for the AOL-Time Warner merger. Proponents and opponents of 'open access' foresaw either grave threats or little concern on the near horizon.

[50] See the discussion in Blumenthal and Clark (2001), Lemley and Lessig (2001), as well as in Kruse, Yurcik, and Lessig (2001) on why end-to-end may end in the next generation of applications.

Third, the firms with national footprints did not have big advantages in the deployment of end-to-end applications that required a low amount of signal delay, such as virtual private networks, video conferencing, or applications requiring tight security. In other words, there were few highly valued transactions that were more efficiently done under the ownership of a single firm with wide geographic reach.

In closing this discussion about practices, it is important to note that these patterns touch two open questions at the core of network economics. Many economic models of noninterconnecting networks suggest that firms would build geographically redundant facilities in order to compete for customers. Yet, the Internet networks of the late nineties interconnected as well as any observer could imagine, and redundancy still emerged in many locations. In this case, such redundancy seems to have emerged as a result of competition for customers (more on this below) and deliberate government policies to prevent mergers resulting in concentration of ownership. It would seem that redundancy is inevitable under any structure for competition, whether or not networks interconnect. Is there a grain of economic logic to this observation?

Second, economic arguments about natural monopolies highlight the need to eliminate redundant lines in order to have low-cost operations. Indeed, it is a key argument for establishing a single provider of a natural monopoly. Yet, the Internet of the late 1990s was as far from a natural monopoly as any observer could imagine. Entry was rampant, so was growth, and so was redundancy in geographic breadth. Did the dynamic benefits from this entry and growth outweigh the potential costs of such redundancy?

The following sections illustrate these open questions in a variety of dimensions.

4.4. Interpreting the geographic dispersion of capacity

During the first few years of commercialization of the Internet a handful of cities in the United States contained the majority of backbone capacity (Gorman and Malecki, 2000; O'Kelly and Grubesic, 2002). Specifically, San Francisco/Silicon Valley, Washington, DC, Chicago, New York, Dallas, Los Angeles, and Atlanta contained links to the vast majority of backbone capacity. Depending on how it was measured, sometimes the relative ranking varied, but the list of the top seven does not. As of 1997, these seven cities accounted for 64.6 percent of total capacity. By 1999, even though network capacity quintupled over the previous two years, the top seven still accounted for 58.8 percent of total capacity[51].

[51] Gorman and Malecki (2002a) look at each of the ten major backbone networks. They compare them using measures of median download time and the number of routes available to an ISP. Data was provided by the 1998 *Boardwatch Magazine's Directory of Internet Service Providers*. Malecki (2002) provides a summary of this and related research. They do not correct for the double counting identified by Hagendorn (2003).

From the outset, distribution of backbone capacity did not perfectly mimic population distribution within metropolitan regions. Seattle, Denver, Austin, and Boston have a disproportionately large number of connections (relative to their populations), whereas larger cities such as Philadelphia and Detroit had disproportionately fewer connections (Townsend 2001a, b). In addition, the largest metropolitan areas are well served by the backbone, while areas such as the rural South have comparatively few connections (Warf, 2001)[52].

Cities such as Boston and Seattle also experienced favorable outcomes in terms of growth. Grubesic and O'Kelly (2002) measure the speed of growth at metropolitan areas. Their data indicates that areas such as Milwaukee, Tucson, Nashville, and Portland had large growth in POPs at the end of the 1990s.

Despite the growth of the total capacity, maps of Internet backbone for cross-national data transmission have not changed much from 1997 to 2002. Maps of Internet backbone at both early and later periods display similar geographic features in the biggest arteries. A few key cities have the greatest number of connections. The main difference arises in the minor capillaries and passageways of the network, which are more abundant later[53]. To visualize this idea, a few representative Maps from AT&T and UUNet from 2000 are included in Figures 4 and 5.

At first glance, economic reasoning offers a variety of straightforward explanations for these patterns. For example, there are economies of scale in high-capacity switching and transmission, even during a period of high growth. There are also economies of scale in data-interchange[54]. In other words, as with many communication networks, the present backbone network is a hub and feeder system with a few hubs[55]. These hubs retain classic economic features of all traffic hubs: the variable costs of operating the service are lower when their transit depots and switching functions are close to each other.

Classic hub economics may explain why hubs will arise, but it does not explain where they settle in one place and not another. For that latter question, three factors seem particular salient for the Internet. For one, some of the landline telephony firms were also in the business of carrying data traffic (e.g., AT&T, MCI, and Sprint). Their voice and data network configurations would naturally use some of the same equipment, facilities, etc, so the locations of their high-speed lines would not change much, if at all.

[52] Employing graph theoretical analysis, Wheeler and O'Kelly (1999) rank metropolitan statistical areas (MSAs). Their results are qualitatively similar to the ones mentioned above.

[53] See also the maps of the US network at www.telegeography.com. See Dodge and Kitchin (2001a,b), or www.cybergeography.org./atlas/atlas.html.

[54] There is an additional economy of scale at public interchange points in the sense that all parties benefit if more parties partake in interchange at the same location (Srinagesh, 1997; Kende, 2000).

[55] Moss and Townsend (1997) observe that the earliest commercial backbone for the Internet is organized as "hub and feeder system with distinct nodes, contrary to popular notions of a dispersed, chaotic Internet."

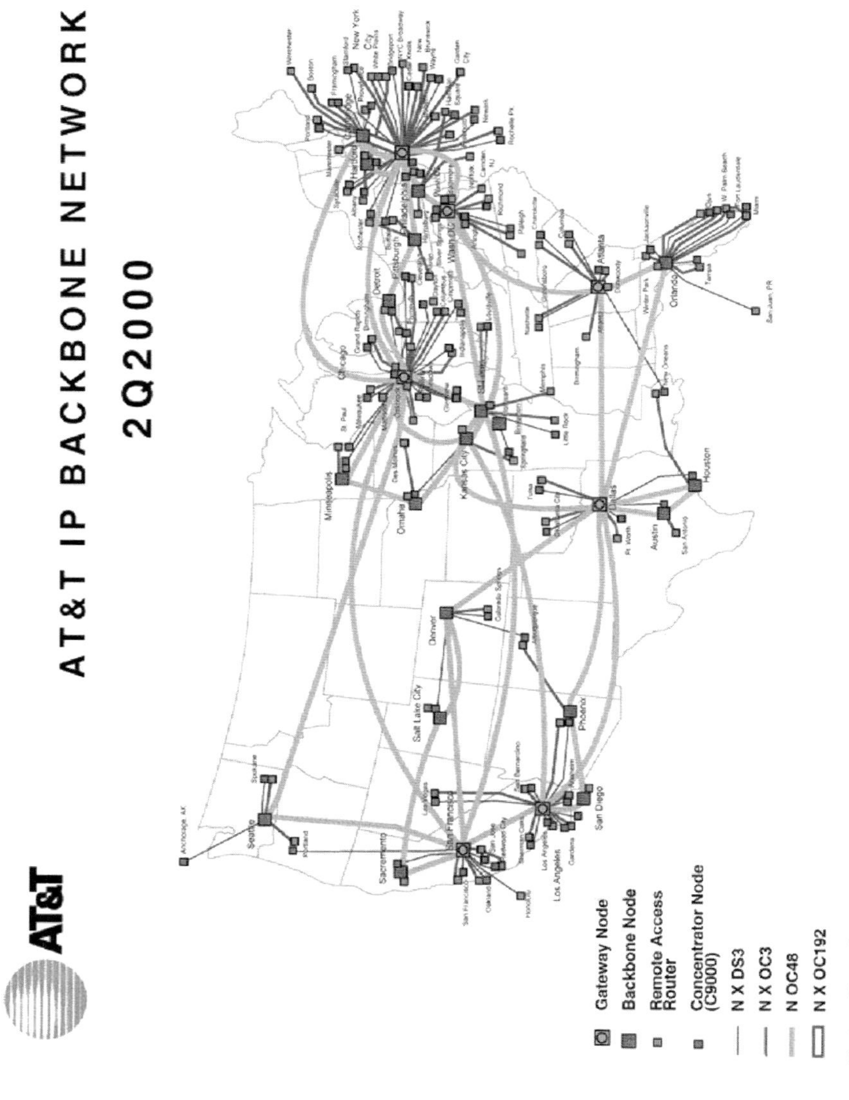

Fig. 4. AT&T IP Backbone Network, 2000.

From the outset, distribution of backbone capacity did not perfectly mimic population distribution within metropolitan regions. Seattle, Denver, Austin, and Boston have a disproportionately large number of connections (relative to their populations), whereas larger cities such as Philadelphia and Detroit had disproportionately fewer connections (Townsend 2001a, b). In addition, the largest metropolitan areas are well served by the backbone, while areas such as the rural South have comparatively few connections (Warf, 2001)[52].

Cities such as Boston and Seattle also experienced favorable outcomes in terms of growth. Grubesic and O'Kelly (2002) measure the speed of growth at metropolitan areas. Their data indicates that areas such as Milwaukee, Tucson, Nashville, and Portland had large growth in POPs at the end of the 1990s.

Despite the growth of the total capacity, maps of Internet backbone for cross-national data transmission have not changed much from 1997 to 2002. Maps of Internet backbone at both early and later periods display similar geographic features in the biggest arteries. A few key cities have the greatest number of connections. The main difference arises in the minor capillaries and passageways of the network, which are more abundant later[53]. To visualize this idea, a few representative Maps from AT&T and UUNet from 2000 are included in Figures 4 and 5.

At first glance, economic reasoning offers a variety of straightforward explanations for these patterns. For example, there are economies of scale in high-capacity switching and transmission, even during a period of high growth. There are also economies of scale in data-interchange[54]. In other words, as with many communication networks, the present backbone network is a hub and feeder system with a few hubs[55]. These hubs retain classic economic features of all traffic hubs: the variable costs of operating the service are lower when their transit depots and switching functions are close to each other.

Classic hub economics may explain why hubs will arise, but it does not explain where they settle in one place and not another. For that latter question, three factors seem particular salient for the Internet. For one, some of the landline telephony firms were also in the business of carrying data traffic (e.g., AT&T, MCI, and Sprint). Their voice and data network configurations would naturally use some of the same equipment, facilities, etc, so the locations of their high-speed lines would not change much, if at all.

[52] Employing graph theoretical analysis, Wheeler and O'Kelly (1999) rank metropolitan statistical areas (MSAs). Their results are qualitatively similar to the ones mentioned above.

[53] See also the maps of the US network at www.telegeography.com. See Dodge and Kitchin (2001a,b), or www.cybergeography.org./atlas/atlas.html.

[54] There is an additional economy of scale at public interchange points in the sense that all parties benefit if more parties partake in interchange at the same location (Srinagesh, 1997; Kende, 2000).

[55] Moss and Townsend (1997) observe that the earliest commercial backbone for the Internet is organized as "hub and feeder system with distinct nodes, contrary to popular notions of a dispersed, chaotic Internet."

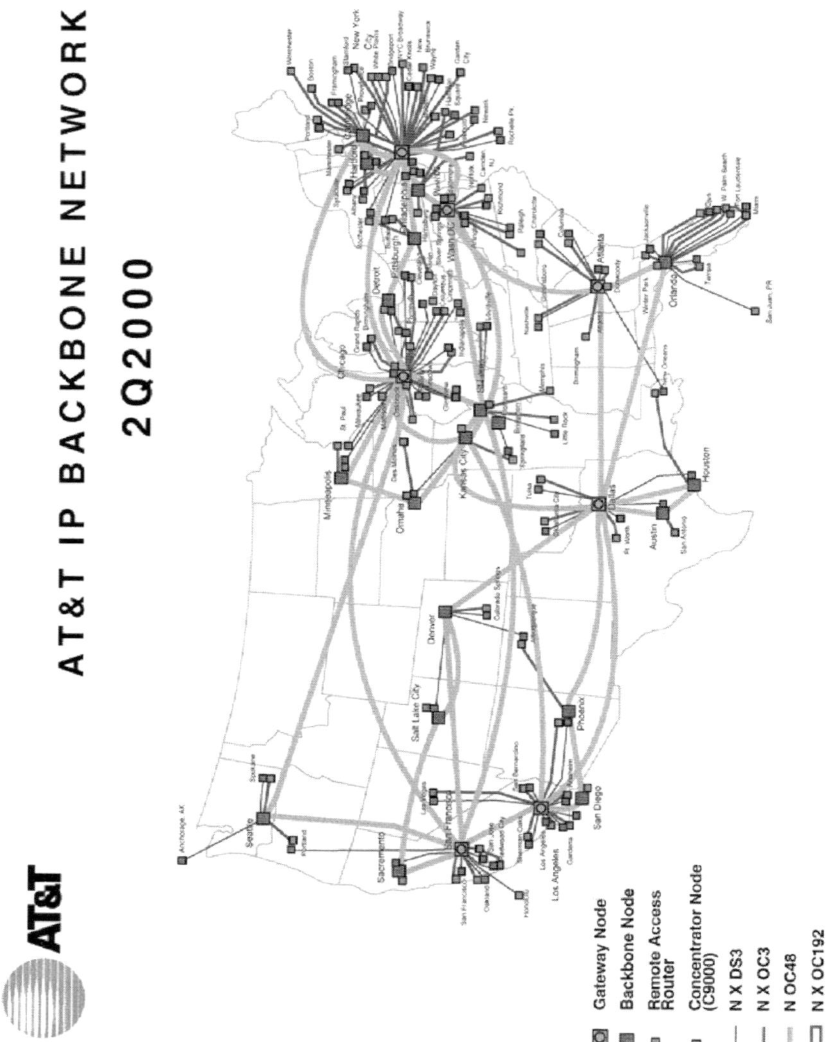

Fig. 4. AT&T IP Backbone Network, 2000.

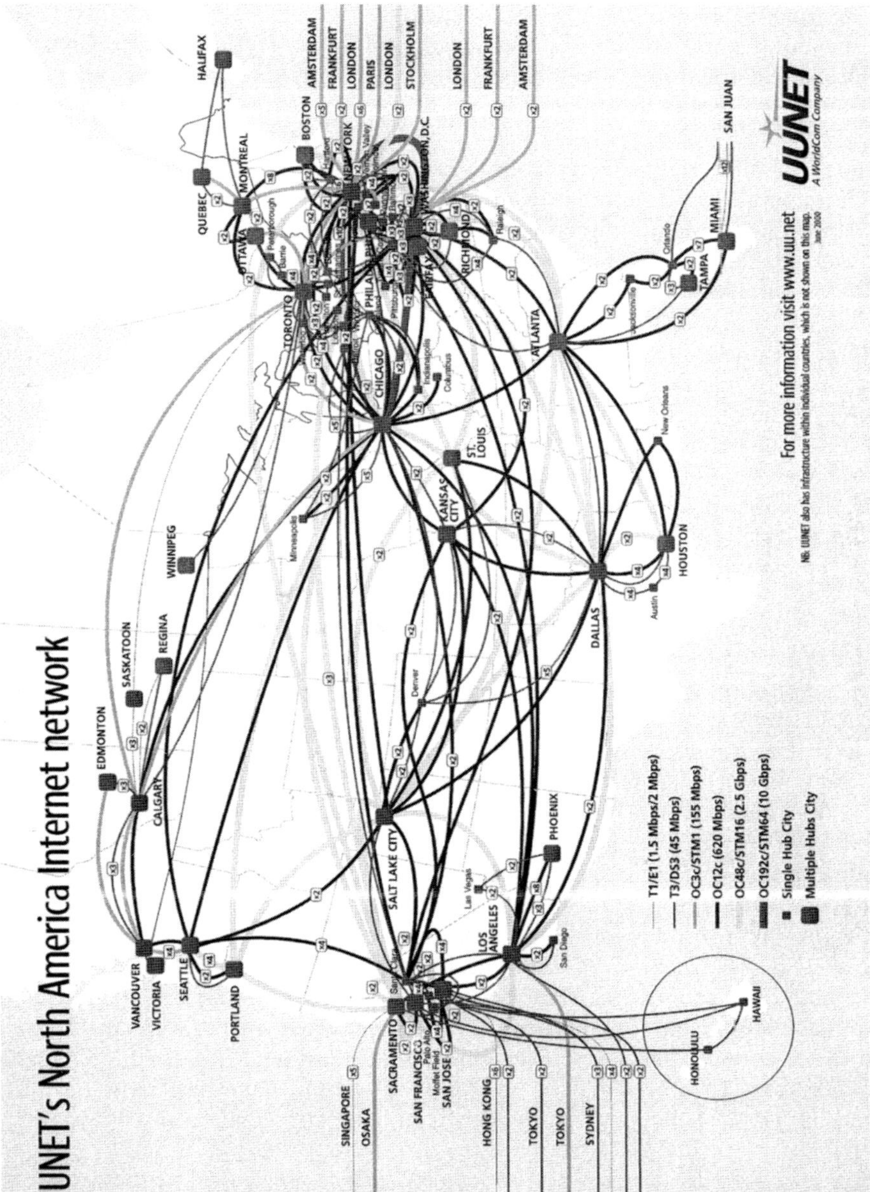

Fig. 5. UUNET's North American Internet Network, 2000.

Next, whether operated by an incumbent, or by a new entrant such as Level3, there are only a few (commonly employed) right-of-ways pathways available for long-distance transmission lines. These are rail lines, highways or preexisting pipelines or electrical lines. Said simply, there are only a few ways to get across the country, so major trunk lines for each commercial network follow similar paths.

Related, similar right-of-way considerations shape the deployment of high-speed lines within any regional area or municipality. Public transit lines, freeways, preexisting underground tunnels, and preexisting telephone/electrical poles and structures act to limit the number of pathways to any given location from any given location. Access to these right-of-ways is available on a nondiscriminatory basis through a licensing process operated in most states[56]. So, again, it should come as no surprise that the same firms employ the same pathway and interconnect to regional networks in a few places.

Many of the choices for the location of NAPs—made by NSF or the earliest entrants—have persisted into the commercial era either as a public exchange point or as a focal location for clusters of private peering. To be sure, some of these points, broadly speaking, would have arisen under any system because their location was determined by the centrality of some places in the midst of traffic flow. For example, data-exchange somewhere near New York or Washington, DC makes sense for traffic along the east coast. One can make similar remarks for Dallas or Atlanta in the south, Chicago in the Midwest or San Francisco and Los Angeles in the West.

One might see related factors shaping the service into some cities. Proximity to major telecommunication centers explains the findings in Tucson and Milwaukee (i.e., near Los Angeles and Chicago, respectively). As another example, some cities, such as Portland, OR, are located between larger cities with high Internet activity (i.e., between San Francisco/San Jose and Seattle). Some cities are simply blessed with centralized locations between many points of traffic, such as Nashville, and, on a larger scale, Chicago.

With a classic transportation hub, the city that acts as the home gets large benefits in the form of jobs, infrastructure, tax revenue, and local support services. Did the location of Internet hubs attract complementary economic activity that led to economic benefits for the local area? In practice, the question is impossible to answer on the basis of what happened in the U.S. in the first decade after commercialization. Some areas that naturally attracted hubs— such as Washington, DC, New York, or San Jose—would have had large network markets anyway. There was high demand and large supply naturally followed.

[56] There is also federal legislation regulating attachment to telephone poles. This legislation has a long history going back to the regulation of Interconnection with AT&T and the diffusion of cable television firms. The 1996 Telecom Act also addressed some of these features.

At the same time it was not at all obvious that a vendor supporting a web page in Akron Ohio got markedly worse service than one in Los Angeles or San Francisco because this depended on matching supply and demand. The key issues concern whether larger network capacity in a backbone (using high capacity OC speeds) makes any difference to a host site or to a user, who ultimately connects at much lower speeds. Since the majority of users connect at much slower speeds than the highest speed backbone, the weakest link may have mitigated any advantage from geographic proximity.

As it turned out, the dot-com crash in the spring of 2000 interrupted the boom in complementary activities, such as the growth of data-warehouses next to fast backbone near Chicago. Moreover, the boom and bust happened so quickly, it also reshuffled the labor market for Internet services and consulting—both up and down. It is also unclear what effect will result from the bankruptcies of some of the firms who provided backbone services. These open questions will only be answered with more time.

Finally, who gets service at all? Since the Internet was experiencing explosive growth it is highly doubtful that any city over a quarter million was not served by several backbone firms. For small cities the relevant issues were similar to that which determined ISP coverage, such as density of demand or proximity to major conduits along highways.

4.5. The economic interpretation of redundancy

Relative to its history and origins, the appearance of redundancy in the backbone was notable. For the Internet, different firms replicated transmission capacity along similar paths to the same cities, thus they potentially serve the same customer. This was not a feature found in the NSFnet[57].

But history probably offers a misleading point of comparison for understanding economic geography with commercial suppliers. The more relevant question is whether this outcome is remarkable in comparison to other industries. In many industries observers *would not* find it surprising that capacity from different firms serves the same customer. Commercial firms in most industries do not coordinate their building plans with each other and, consequently, overlap in their aims. In many industries different firms provide similar products to similar sets of customers. It is important to recognize that some features of the redundancy of the backbone arose from competitive forces introduced into the Internet after 1995, which are otherwise normal for many other markets[58].

[57] The motives for redundancy are quite different in a military network communications network, where the communications lines must survive potential damage to parts of the whole. Only a little research has examined redundancy in this sense (Grubesic et al., 2003).

[58] For an interesting comparison of the differences between factors leading to boom and bust with Internet backbone firms and the classic railroad boom and bust (Hogendorn, 2003).

Several factors were acting simultaneously. They pushed in the same direction and enhanced the incentives to grow quickly: (1) competitive incentives; (2) impatient investment behavior on top of normal competitive behavior; (3) optimistic assessments of future demand growth and (4) beliefs about the strategic advantages of first-mover actions. I address each of these in turn.

For one, the backbone network of the late 1990s embodied features that reflected strongly on the presence of competitive incentives. The U.S. backbone was built in increments by many firms, not by a single firm with monopoly ownership. There was a multiplicity of actors building it and competing with each other. In the late 1990s, notably AT&T, Sprint, WorldCom, GTE, Level3, Qwest, Global Crossing, Cable and Wireless, and Williams, among others, all had well-known plans to build networks with national geographic footprints[59].

Competition enhanced the incentives of each actor to grow quickly, price competitively and experiment broadly. As in many markets competitive incentives induced suppliers to identify and tailor network services to unfilled customer needs. In a few locations this competition increased the competitive options available to potential users and ISPs. Thus, it was quite common for ISPs to seek and sign multiple agreements with backbone partners (*Boardwatch*, various years).

There is no question that the impatient financial environment of the late 1990s provided further strong incentives to grow quickly. Part of this was due to Wall Street's exuberance, as stock prices responded (sometimes without sufficient skepticism) to announcements of building plans, new Internet ventures that generated backbone investments, even when it resulted in significant redundancies between rival suppliers. Certainly this reinforced the previously noted incentives to *announce* plans for expansion, even when it just involved signing an IRU.

And, yet, part of this impatience was real. Businesses did invest in TCP/IP applications. The latter half of the nineties witnessed the single greatest growth in investment in Information Technology in the history of the U.S. Investment grew at 20 percent a year from 1995 to 2000, with almost three quarters of that being business investment[60].

In the face of this growth there was a common perception among suppliers that demand for Internet traffic would grow quickly for many years. Again, it is hard distinguish between the reality, hype and dream that underpinned this demand. The late 1990s gave rise to a pervasive technological deterministic rhetoric in business media outlets. Decision-makers, such as senior executives, did not exactly know what that future was going to look like, but they all believed that technology was going to play a central role in it. So they pursued an array of

[59] For a sense of these plans, see Boardwatch directories, various years.

[60] This includes computer hardware, computer software, communications hardware and instruments. By 2000 computer hardware and software stocks had reached 622.2 billion dollars (1996) (Henry and Dalton, 2002; Price and Mckittrick, 2002).

technology-related initiatives hoping that one or more of these would contribute to company growth or even mere survival. It all translated into extraordinarily loose budgets for information technology projects in business.

Forecasting had been quite sober when the Internet first commercialized (Meerker and Dupuy, 1996), but the restraint was lost in the late 1990s. In some circles, forecasts included seemingly unrealistic growth forecasts—almost without bounds—and in retrospect one has to wonder who could have trusted them enough to invest serious money on such flimsy forecasts. In some circles, consultants backed up projections with statistical samples. For the latter, it is significant that, well prior to its bankruptcy, WorldCom's broadband division had claimed that its Internet traffic was doubling every six months, a canard that became repeated often. While there may have been some truth to this report at one time early in the explosive growth of the commercial Internet, there was little confirmatory evidence of such persistent growth. Yet, this phrase was repeated often well up until the crash of 2000[61]. In other words, traffic grew, but many economic actors such as Wall Street analysts, financial investors, and large firms actually building the networks-believed it would grow even faster than it did.

Finally, there was a common understanding that savvy business practices required laying backbone capacity (i.e., sinking fiber optics into the ground) ahead of buyers' demand for that capacity. Firms perceived that business users of capacity would be hesitant to switch suppliers once contracts were signed. They perceived that small ISPs would not want to renegotiate new contracts with new suppliers. Backbone suppliers saw themselves racing with each other to sign (what at the time was forecast would be) lucrative contracts for services with low variable operating expenses. Hence, all vendors perceived themselves to be in a race to build volume, that is, sign-up users today, which would later foster profitable revenue streams.

This behavior was more consistent with the aggressive practices of venture funded firms, and it was far from the cautious building practices found in the regulated parts of the communication industry. In a regulated industry caution is a virtue, particularly if investment decisions can be second-guessed by regulators. Capacity is built only after it was requested by buyers and approved by regulators. The difference can be stated simply: When a firm is competing for customers, less cautious investment practices are called for. To be sure, a stopping

[61] In retrospect, it appears that this observation had some truth to it in 1997 and 1998, when the Internet first commercialized into a mass market, growing from a small base. While there certainly was growth in data traffic after that, it is unclear when growth rates began to slow down. It is also unclear whether WorldCom or others repeatedly asserted that these observations grew as fast as they asserted due to oversight or, deliberately in order to foster their own business plans. In the heady days of Wall Street boosterism for all things associated with Internet, such assertions were encouraged, and often went unquestioned and untested. For a more skeptical view, see the discussion in Odlyzko (2001), or Sidak (2003).

rule for such aggressive investment is hard to articulate. As in any industry with high sunk costs and (resulting) low variable costs, such aggressive investment is quite risky for any individual firm.

It should be no surprise that this incautious competition between network providers fostered uncoordinated build-outs among rival firms, which, in turn, fostered overlapping footprints and other redundancies in supply at a broad level. This was particularly so when firms signed an IRU, i.e., when they just rented the option to enter without building the entire capacity to operate (Hogendorn, 2003). Not surprisingly, commercial firms did not coordinate their building plans with each other, and they replicated potential capacity along similar paths. Said more concretely, every firm with a national footprint thought it had to be in San Francisco, Chicago, Los Angeles, Washington, DC, New York, and many other major cities. Stated simply, the rapid and redundant build-out of the late 1990s arose from one and the same set of competitive incentives.

4.6. Boom leads to bust in the backbone

A decline in optimism did eventually arrive. This was partly the result of saturation. By 2000 most of the strong ‘first-adopters’ among businesses and households had adopted the Internet[62]. If further growth was on the horizon, it had to occur through capital deepening by existing users, not through addition of new users. Indeed, business investment in IT reached its peak in 2000 and dropped after that[63]. New household use of the Internet also slowed at the same time[64].

The trade press dates the beginning of the decline of optimism at the spring of 2000, when financial support for dotcoms collapsed. Pessimism reached a nadir in the fall of 2001, after the September 11 terrorist attacks shook business confidence in long-term investments. This low plateau continued as the WorldCom financial scandal became publicized in the spring of 2002. Consequently, some backbone providers curtailed the expansion plans they announced in 1999 and previous years. Others left the market altogether.

As of this writing, this down cycle is not over, which leaves an open question about the long-term shape of the U.S. backbone networks. Qwest, Level3, Sprint, Global Crossing, MCI-WorldCom, Williams, PSINet, AT&T and others all invested heavily in redundant transmission capacity during the boom. Most of them

[62] The results in National Telecommunications Information Administration (2000, 2002) indicate that by 2000 most households who had a computer had considered adopting the Internet. Most new growth in adoption was coming from households who needed to buy a computer. See also Goolsbee and Klenow (1999). Forman, Goldfarb and Greenstein (2003a,b) find that most business establishments were participating on the Internet by the end of 2000.

[63] See Price and Mckittrick (2002) or Henry and Dalton (2002).

[64] See National Telecommunications Information Administration (2002). Capital deepening at that point took the form of cannibalization of dial-up with broadband. See the further discussion.

have not fared well. A few, such as PSI and Global Crossing, have exited, selling their assets to others.

Signs of financial distress are evident in many places. Here are some examples: (1) much of that fiber went unlit after its installation (as one would expect if expectations were over optimistic). (2) UUNET, a division of WorldCom, was the largest backbone data carriers in the United States, but the scandals at WorldCom led to its bankruptcy and cessation of investment activity (though the stress did not necessarily signal problems with the Internet division). (3) AT&T bought TCI and MediaOne, invested heavily in upgrading cable networks for purposes of offering household telephone and broadband access, but did not get the type of demand expected. Its CEO had articulated a grand design for the future and he was forced to resign when the strategy clearly failed. The strategy was repudiated and the cable assets were sold to ComCast. (4) PSINet, another early pioneer in building the commercial Internet backbone, overextended itself and had to declare bankruptcy. Its assets were sold to other investors at a severe discount. (5) Severe stress was also felt upstream at switch and equipment makers, such as JDUniphase, Corning, Cisco, Nortel, and Lucent.

While nobody expects consolidation of ownership to completely eliminate redundancies, this downturn will reshape the ownership overflows of data traffic in the future. Will consolidation eventually lead to the emergence of this networks' dominant executive, as Cornell was for the telegraph and Vail was for the telephone? The final configuration of ownership remains in flux, awaiting the results of sales of assets for bankrupt firms, as well as, perhaps, the vision of temperate executives.

4.7. Interpreting a decade of building

Would the U.S. backbone network have experienced different geographic coverage with a different mode for organizing commercialization? The answer depends on whether one looks at the network before or after it is built. In other words, at the outset there was a major concern about whether the network would ever be built at all but little concern for whether it would be oversupplied in one place or another. After the network was built, overlapping footprints and other redundancies appear inefficient, though they were of little concern prior to its building.

A competitive market gives every actor strong incentives to build its networks quickly, price it low, and customize it to user needs. The backbone might not have been built as rapidly in the absence of such competition. Similarly, uncoordinated investment of sunk investments has the potential to lead to price wars in the event of overbuilding. But price wars cannot arise unless firms build their networks in the first place.

It would be surprising if uncoordinated investment resulted in too little infrastructure during a period of sustained demand growth, as had occurred in the late 1990s. Building ahead of demand is a calculated gamble for each provider. Every actor risks winning the same set of future customers. If all actors had an optimistic assessment of the future, then when that optimism declines, all will be caught with excessive assets. Of course, these incentives are further exacerbated if growth forecasts (from rivals or industry analysts) exaggerate true trends, as the forecasts coming from WorldCom did during the late 1990s (Odlyzko, 2001).

There was a sense of inevitability to the price wars that eventually erupted (i.e., the oversupply of capacity compared with demand). High initial demand in a market with uncoordinated investment should lead to a period of intense growth followed by a period of regret. The period of regret is almost inevitable, particularly when the growth is sustained by unbounded optimism and fueled by incautious building of sunken assets used in a race to search for customers.

The concentration of infrastructure in a few cities, while partly the result of classic hub economics, also looks like the natural outcome of the same speculative process during a period of speculation. All firms are rivals, investing in the same corridors for carrying data traffic between major population centers. Each wants to assure that its traffic receives priority from its own operations, not that of its close rivals[65].

4.8. Summary

It is concluded that the concerns about excessive market power on the backbone did not affect the growth and diffusion of Internet infrastructure during the first decade after commercialization. This might have been an issue, and there are many reasons why it did not manifest. The most important was that no firm—notably, not even MCI/WorldCom—amassed enough market share to dictate terms in most places[66]. In most major cities ISPs had many options for sending traffic to the Internet. In sum, it is concluded that interconnection issues have not shaped the geographic dispersion of the Internet in the U.S.—at least, not yet.

The speculative investment behavior of backbone providers underlay an extraordinarily rapid build-out, an outcome that clearly benefited society. This same behavior laid the foundations for a financial bust and restructuring. There is (as yet) little persuasive evidence that it has led firms to massively substitute away from investing in peripheral locations. In addition, there has not been a domino effect of one bankruptcy causing another, even with a bankruptcy from WorldCom, one of the largest backbone providers in the world. In this sense, I also

[65] Similarly, it also suggests that the infrastructure from every major provider serving central locations, such as Washington, DC, should be measured according to benchmarks distinct from those used to measure infrastructure from several providers serving a less central location with much lower aggregate demand.

[66] Clearly this outcome was due, in part, to forced divestiture. As a condition for the MCI and WorldCom merger, MCI's Internet assets were sold to Cable and Wireless. The WorldCom and Sprint merger was (effectively) blocked by the European Commission over these and related issues (before the US Department of Justice registered its stance).

conclude that this boom and bust cycle has not been a central policy issue for the geographic dispersion of the Internet in the U.S.—at least, not yet.

5. The growth of broadband

Internet access is a geographically local and nontradable service. As a dial-up service, it is cheaper if it is part of a local phone call. If it is a broadband service, it is simply not available unless a local supplier has invested in the necessary infrastructure to bring delivery to a home or business[67]. In contrast to dial-up concerns about backbone, the concern for broadband centers on the fast speed, high price, and limited coverage of available service.

By the late 1990s there were three basic delivery modes for broadband. The first was direct supply of high-speed lines, such as T-1 lines. This was prohibitively expensive for all users except businesses, and even then, it was mostly used by businesses in dense urban areas, where the fiber was cheaper to lay. Two other options became available in the latter part of the decade after commercialization: cable Internet or DSL[68].

Three questions shaped analysis on where either provider would make investments in Internet infrastructure:

1. Cable and DSL diffused disproportionately to urban areas initially. What factors encouraged this outcome and was there any indication that this would change?
2. Did broadband have features that would eventually lead it to diffuse with the same speed and geographic ubiquity as found in dial-up Internet access?
3. Did the growth of new regulatory rules under the 1996 Telecom Act alter the geographic diffusion of broadband?

5.1. Why broadband favors urban areas

Broadband access is defined by the FCC as "the capability of supporting at least 200 Kbps in the consumer's connection to the network," both upstream and downstream (Grubesic and Murray, 2002; National Research Council, 2002). Whereas dial-up connection has moved past the frontier stage and is approaching

[67] Advanced telecommunications services for packet switching involve a multiplicity of technologies, such as switching using frame relay or Asynchronous Transfer Mode, as well as Synchronous Optical Network equipment or Optical Carrier services of various numerical levels. Their deployment is not the central focus of this discussion, though it is of importance for supporting broadband networks (Noam, 2001).

[68] As of this writing, there was not a viable wireless high speed access technology, though one appeared poised on the horizon (i.e., 802.11g). For the first decade of commercialization all wireless applications, such as Wi-Fi (802.11b) or the Blackberry (two-way text messaging) used slow speeds.

geographic ubiquity in the United States, broadband access is far from ubiquitously available or adopted. It is much more common to large businesses. Its ubiquity will require quite a build-out because it is a local good in the minute sense-at the level of the block within a city. Sometimes this is called the 'last mile' problem[69].

The appeal to users of broadband is well known. It is faster than dial-up access and 'always on,' and, at the same time, it is typically priced much more cheaply than a T-1 line. In comparison to dial-up use, broadband access enables better applications and more convenient service, enough benefit for some users to justify paying a price higher than for dial-up[70]. As the volume and complexity of traffic on the Internet increases dramatically each year, the value of these features becomes larger. Furthermore, broadband access has appeal to the vendors. It is a distribution mode that enables providers to offer a wider range of bundled communications services (e.g., telephone, e-mail, Internet video, etc.).

Broadband infrastructure favors urban areas because the technology for broadband is more expensive in less dense locations. DSL technology can only extend 18,000 feet (at most) from the central office switch[71]. The radius is noted to be closer to 12,000 feet for high-quality, low-interruption service. Therefore, one should expect that those living outside this radius will more likely suffer from lack of DSL service. Cable systems are also cheaper in high-density areas for the simple reasons that Cable is expensive to lay in low-density areas (Esbin, 1998; Crandall and Alleman, 2002).

A provider of broadband experiences high costs when deploying new services. Lack of preexisting infrastructure is to blame. It is expensive to do as a *de novo* investment, because it requires laying direct lines from trunks to a central switch, which involves the expense of dug-up streets, repeaters, and right-of-way permits if the path has no precedent. This type of investment was so expensive in the mid-1990s that only businesses ever incurred the costs-and usually only in a central city or along an existing trunk line, where the costs were lower. Remote rural establishments simply could not afford the fixed private lines that were needed to deliver high-speed data link-ups.

Broadband to the home is also expensive. It is an expensive retrofit on top of either a telephone network (in the case of DSL) or a cable system (in the case of cable modems). For DSL, the switch and lines must be of high quality and

[69] See, for example, Hurley and Keller (1999), National Research Council (2001, 2002) or Crandall and Alleman (2002) for analysis of last mile issues.

[70] As broadband began to diffuse around 1999/2000, it typically sold for $40 to $60 a month. In comparison, a typical AOL dial up account sold for $23 a month, and plenty of other firms offered cheaper alternatives.

[71] See National Telecommunications and Information Administration (2001) or U.S. Department of Agriculture (2000), for an overview. See Crandall and Alleman (2002), for many studies of broadband deployment and demand.

outfitted with appropriate software. For cable, the entire cable system needs an upgrade. These expenses arise solely from engineering requirements, so they arise even without the regulatory distortions imposed by state and federal regulators.

To be sure, the costs of deploying DSL are not solely a function of engineering costs. Regulatory rules for DSL can alter the incentives to provide it. Local telephone firms faced an additional 'opportunity cost.' Demand for additional telephone lines can be quite profitable, but demand for DSL cannibalizes them. In this setting the resale rules, installation expenses, and other regulatory constraints can raise or lower the net returns to providing DSL at incumbent local exchange carriers. It is difficult to make anything other than this general statement, since the regulatory rules for DSL varied considerably across states.

5.2. The empirical evidence

Was there uneven availability across the country? The FCC estimates that high-speed subscribers were present in 97 percent of the most densely populated zip codes by the end of 2000, whereas they were present in only 45 percent of the zip codes with the lowest population density (NTIA, 2002). It also estimates that one-quarter to one-third of U.S. consumers cannot receive this service. Only 90 percent of the United States has access to cable systems. Most of those who are not reached are located in rural areas. Of those reached, many of those systems require costly upgrades before Internet service is a viable business[72]. This bias can be seen in surveys of use. As of September 2001, 19.1 percent of Internet users had cable modems and DSL. That percentage could be partitioned into central city, urban, and rural rates of 22.0, 21.2, and 12.2 percent, respectively[73].

Research on the diffusion of broadband, almost by definition, must examine diffusion at a very fine level of granularity, such as at the block or zip code or some other neighborhood-related level of granularity. For example, Lehr and Gillett (2000) look at the very early diffusion of broadband to residential areas (primarily in 1998). They compiled a database consisting of communities in the United States where cable modem service is offered and linked it to county-level demographic data. They find that: broadband access is not universal. Only 43 percent of the population lives in counties with available cable modem service[74]. This is quite a low number for their methods, which give a county credit if even a small part of the country is being served. Such access is typically available in counties with large population, high per capita income, and high population density; and there is a notable difference in strategy for cable operators with some being more aggressive than others.

[72] See, for example, U.S. Department of Agriculture (2000).

[73] One should note that the rate of 22.0 percent for central cities is likely biased upward due to the presence of universities in the centers of many cities.

[74] They point out this population is actually closer to 27% (as was stated by Kinetic Strategies), but explain that their data is not fine enough to show this measurement.

In a very data-intensive study, Gabel and Kwan (2001) examine deployment of DSL services at central switches throughout the country, providing a thorough census of upgrade activity at switches. They examine providers' choice to deploy advanced technology—to make broadband services available to different segments of the population. The crucial factors that affect the decision to offer service are listed as: (1) cost of supplying the service, (2) potential size of the market, (3) cost of reaching the Internet backbone, and (4) regulations imposed on regional Bell operating companies[75]. They find that advanced telecommunications service is not being deployed in low-income and rural areas.

Prieger (2001), using very comprehensive data, examines the availability of broadband for both cable modems and DSL. He looks for evidence of red-lining, that is, where broadband carriers avoid areas with high concentration of minority or low-income households. He finds little evidence of such red-lining based on income or against neighborhoods with a concentration of black or Hispanic populations, though the evidence is mixed with respect to Asian or Native American populations. As with others, he finds that broadband is more available in large markets, areas with higher educated populations, and, interestingly, areas with higher Spanish language use. The presence of a Bell company also increases availability, while inner city and rural locations diminish it.

Grubesic and Murray (2004) examine much the same data, but go into deep detail in a few cities, using the closer examination to uncover the drivers of differences from one city to another. They find that density predicts availability if one compares all neighborhoods across the country with each other. But looking inside any given city, the predictors of provision are more varied and not necessarily a function of density. Instead, the wealth and income of the area's residential population is key to understanding demand[76]. Their findings echo Grubesic and Murray's (2002) study of differences in DSL access for different regions in Columbus, OH. They observed that DSL access can be quite inhibited because some of the high income areas are also low-density. This supports one of their counterintuitive findings—DSL is less available in some high-income low-density areas, such as Franklin County, OH, than in some lower income, high-density areas (See also Grubesic [2003]).

In summary, the spread of broadband service has been much slower and much less evenly distributed than that of dial-up. This is not a surprise once their basic economics is analyzed. The broadband ISPs find highly dense areas more profitable due to economies of scale in distribution and lower expenses in build-out. Moreover, the build-out and retrofit activities for broadband are much more involved and expensive than what was required for the build-out of the dial-up

[75] Data was obtained concerning wire centers; also data on DSL and cable modem service availability was collected via web sites and calling service providers. They supplemented it with Census data.

[76] See also Grubesic (2003b) for a detailed study of the demographic make up of such areas.

networks. So within urban areas, there is uneven availability. Thus, even before considering the impact of geographic dispersion in demand, the issues over the cost of supply guarantee that the diffusion process will take longer than dial-up ever did.

5.3. The regulation of broadband suppliers

Every city in the United States has at least one incumbent local telephone provider. The deregulation of local telephony was an attempt to increase the number of potential providers of local voice services beyond this monopoly incumbent, and in so doing, increase the competitiveness of markets for a variety of voice and data services. Deregulation became linked to the growth of broadband in two ways. First, these rules shaped the cost and price of providing broadband. Second, deregulation altered the comparative costs of those who supplied local data services, primarily in urban areas[77]. It was easy to anticipate that these regulations would shape who profited from the diffusion of broadband, though, at its outset, it was an open question what those effects would be. An even bigger open question concerned whether these rules shaped the speed with which broadband diffused.

The new competitor for the deregulated network is called a competitive local exchange company, or CLEC for short. No matter how it is deployed, every CLEC has something in common: each offers phone service and related data carrier services that interconnect with the network resources offered by the incumbent provider (e.g., lines, central switches, and local switches). In spite of such commonalities, there are many claims in the contemporary press and in CLEC marketing reports about CLECs that these differences produce value for end users. In particular, CLECs and incumbent phone companies offer competing versions of (sometimes comparable) DSL services and networking services.

Something akin to CLECs existed prior to the 1996 Telecommunications Act. These firms focused on providing high-bandwidth data services to business. After its passage, the firms grew even more. And CLECs quickly became substantial players in local networks, accounting for over 20 billion dollars a year in revenue in 2000[78]. More to the point, CLECs became the center of focus of the deregulatory movement. Many CLECs grew rapidly and often took the lead in providing solutions to issues about providing the last mile of broadband, particularly to businesses and targeted households. In addition, many CLECs already were providing direct line (e.g., T-1) services to businesses (as was the incumbent local phone company).

[77] See Woroch (2001) for a comprehensive review of the literature. This section is necessarily focused on the issues that shaped the Internet infrastructure market. See, also, Crandall (2001), Crandall and Sidak (2002) or Greenstein and Mazzeo (2003) for exploration of the consequences of deregulation for entrant's behavior.

[78] See New Paradigm Resources Group (2000) or Crandall (2001).

The incumbent delivered services over the switch and so did CLECs. In recognition of the mixed incentives of incumbents, regulators set rules for governing the conduct of the transactions. As directed by the 1996 Telecommunications Act, this included setting the prices for renting elements of the incumbent's network, such as the loops that carried the DSL line[79]. These rental prices were the subject of considerable controversy, as were the precise rules that governed the obligation of the incumbent to the CLEC[80].

For this review, the key question is: Did the change in regulations shape the geographic distribution of Internet access across the United States? The answer is almost certainly, Yes, at least in the short run, though the answer is more ambiguous in the long run. By the end of the millennium the largest cities in the United States had dozens of potential and actual competitive suppliers of local telephone service who interconnected with the local incumbent. By the end of 2000, over 500 cities in the United States had experience with at least a few competitive suppliers of local telephony, many of them focused on providing related Internet and networking services to local businesses, in addition to telephone service (New Paradigm Research Group, 2000). This opportunity extended to virtually all cities with a population of more than one-quarter million, as well as to many cities with a population under 100,000. Very few rural cities, however, had this opportunity except in the few states that encouraged it (Malecki, 2002b; Greenstein and Mazzeo, 2003). So at the outset, the entry of CLECs increased broadband supply somewhat only in urban locations, if it had an effect anywhere.

The entry of CLECs as part of the 1996 Telecom Act had a threefold consequence. First, it encouraged entry by subsidizing it indirectly. For a time, the compensation of CLECs overlapped with the compensation of ISPs. Some ISPs took advantage of rules for compensating telephone companies, a strategy that lost its profitability when the FCC eliminated this distortion in early 1999[81]. Hence, part of the consequence of this entry was temporary, though briefly important.

Second, some CLECs acted as backbone providers for ISPs, thereby spurring the development of local ISP business and allegedly lowering costs for many ISPs. It is unclear how important this was for the geography of the Internet, since many ISPs would have offered service even without the presence of CLECs. The speculation has a basis in fact. While reciprocal compensation may not have influenced the original location of ISPs very much, it had an enormous influence on the

[79] See Gregg (2002) for a review of prices across different states. Also, see Rosston and Wimmer (2001) for review of the determinants of pricing within states.

[80] See, for example, Crandall and Sidak (2002), Sidak and Spulber (1998), and Spulber and Yoo (2003).

[81] See Crandall (2001) for a discussion of the way ISPs took advantage of telephone policies for reciprocal compensation payments. Also see FCC docket No. 99-38, Implementation of the Local Competition Provisions in the Telecommunications Act of 1996, Intercarrier Compensation for ISP-Bound Traffic, released 26 February 1999.

profitability, growth and survival of CLECs and that must have had some influence on subsequent ISP development (even after the FCC order that reduced reciprocal compensation)[82].

Third, it is commonly alleged that CLECs spurred the growth of new Internet services in some areas where other potential providers were slow to deploy it. In some cities, at the outset of the Internet, some local telephone companies were reluctant to diffuse DSL services to potential clients, because DSL would cannibalize existing data-traffic business that used T-1 lines, which was quite lucrative. It is unclear how important this factor was for long run deployment, since many telephone companies entered dial-up and DSL businesses as the demand for the Internet grew, and many would have done so eventually without competitive spurs. Hence, this hypothesis requires more careful statistical testing examining the effect of local regulatory stringencies on incumbent provisioning of new services[83].

It should be added that the 1996 Telecom Act also aided the profitability of cable modem suppliers—in some views this provision of the Act also subsidized the diffusion of broadband to homes. The Act explicitly allowed cable firms to retain their special status outside of the interconnection agreements inherent in common carrier regulation. Many homes in the United States have a cable line running past it. Internet service to homes through cable lines became a second avenue through which the Internet diffused to homes. Since there is a preexisting urban bias to the deployment of cable systems, this mode of delivery necessarily favored urban areas[84].

There was considerable variety of experience around these trends. Zolneirek et al. (2001), followed by Malecki (2002b) and Greenstein and Mazzeo (2003), hypothesize that some basic economics lay underneath the manifest patterns: the need to cover these fixed costs limited the number of entrants in specific locales. That is, because of the presence of fixed costs in each location where they provided services, CLECs required a sufficient number of customers to generate revenue in excess of their recurring expenses, given their operating profit margin. As a result, cities varied in their ability to support entrants[85]. Large cities generated sufficient

[82] If anything, regulatory decisions for reciprocal compensation of CLECs encouraged CLEC entry, which also partly encouraged ISP entry through interconnection with CLECs. Although this is important to incumbent local exchange carriers, one should not exaggerate its effect too much. The scale of this phenomenon grew tremendously in the period between 1996 and 1999, but ISP entry started well before then and continued afterwards until the dot-com bust. Moreover, since CLEC entry was primarily concentrated in dense urban areas, much of this effect was felt in urban areas, which would have had a great deal of ISP entry even without this implicit subsidy to CLECs.

[83] For a beginning on this topic (Mini, 2001; Koski and Majumdar, 2002).

[84] This is a long and complex topic. See, for example, Hausman, Sidak and Singer (2001), Crandall and Alleman (2002) or National Research Council (2002) for an overview.

[85] This arises if there are decreasing returns to scale. It can also arise if there are increasing returns to scale, but Cournot competition or some sort of differentiation between competitors.

revenue to support more entrants better than small cities did. Large cities receive the greatest number of entrants and small cities the least. These patterns were linked to such things as population size, working population, income, and historical patterns of entry (favoring the largest metropolitan areas)[86]. The patterns were also linked to such factors as local regulatory stringency, the identity of the incumbent[87], the difficulty of managing colocation facilities with a hostile or uncooperative ILEC, and the cost of renting local unbundled network elements from the ILEC (subject to regulatory review)[88].

Unfortunately for suppliers, there were finite limits to how much demand was potentially there for their services. The limits became apparent within a few years of the passage of the Act. The residential market never became particularly large for CLECs. Most CLECs became focused on satisfying business demand, where the margins were higher for basic services and the CLECs could tailor their offerings to complex demands. Even then, some CLECs did not realize revenues sufficient to cover the debts incurred building their facilities and marketing their new services. Though the market revenue of CLECs continued to grow, there was a downturn in the specific prospects of many CLECs. This downturn paralleled the downturn among backbone providers and ISPs.

Overall, 2000 was the last year in which there was a consensus of optimistic forecasts about CLEC entry and their expansion. Much of the early enthusiasm was affiliated with anticipated growth in Internet transport or TCP/IP data traffic, development of DSL connections, or affiliated hosting or networking services. It was also affiliated with geographic expansion. By 1999, there was no question that the major cities, such as New York, Chicago, or San Francisco, could support some CLECs. But would CLECs spread to medium-sized cities and smaller locales? As it turned out, the answer was yes, but this was harder to do than was initially thought. With low demand in 2000 and 2001 suppliers reduced their growth plans. Many, especially small CLECs or those dependent on growing demand for broadband and a cooperative ILEC to sell DSL, exited altogether.

[86] Since the facilities-based CLECs must make capital investments in equipment to link their customers, cities with more geographically concentrated residential neighborhoods and business centers may provide CLECs with customers that are less expensive to serve.

[87] Regulators often have different rules for each incumbent carrier within its state. These rules apply to all the areas within that state where the particular incumbent operates (Abel and Clements, 2001). Specific provisions in the 1996 act required incumbents to provide interconnection access to CLEC competitors; however, the incentives to comply with this directive differed across firms. Specifically, Regional Bell Operating Companies (RBOCs) that wanted to enter the market for long-distance services were precluded from doing so until regulators were satisfied that the RBOCs had been sufficiently cooperative with CLECs attempting to interconnect and provide service in their local areas (Shiman and Rosenworcel, 2002).

[88] The rental rate of a local loop typically differs across several density zones within each state (Gregg, 2002).

Any conclusion about this topic comes with caution. The long-term shape of these market actors is currently unclear because the regulatory environment for CLECs is changing. Currently, the FCC is in the midst of rewriting the rules for the rental of unbundled network elements devoted to high-speed data services from incumbent local exchange companies. Courts are in the midst of ruling on the legality of the rewrites being done at the FCC. It is not yet clear how these rule changes will alter the role of CLECs in the geographic development of the broadband markets.

In summary, the era after the passage of the Telecommunications Act of 1996 was a period of uncertainty and flux in the regulatory environment. The rules and regulations certainly shaped the distribution of profits from the supply of broadband. These regulations *might* have speeded up the diffusion of broadband in some urban areas in some states for a short time, especially in the late 1990s, if at all. As yet, there is not yet much evidence these regulations altered the long run geographic properties of broadband—its urban bias. Altering this conclusion requires evidence that one type of firm diffuses broadband faster than another, evidence that no researcher has conclusively presented yet.

5.4. Summary

Economic factors shape the geographic diffusion of broadband. Both DSL and cable Internet are cheaper to deploy in high density areas, so it will continue to be an urban technology. Moreover, the build-out and retrofit activities for broadband are onerous and expensive, so within urban areas there is uneven availability.

The future does not look much different. Broadband will continue to be an urban technology without dramatic changes in regulatory rules or government subsidies. It should also continue until the day that a viable wireless high-bandwidth technology deploys or until the day that someone invents a retrofit to another preexisting set of lines, such as electrical lines.

Regulation can alter the distribution of rents, by making it easier for one party or another to sell and operate high-speed access. I conclude that there is no conclusive evidence yet that the regulations that affect rent distribution have not had large consequences for the geographic dispersion of broadband.

6. The location of business Internet infrastructure services

As with many other general purpose technologies, advances in frontier technology are only the first step in the creation of economic progress[89]. The next step involves the development of complementary services by economic agents. These developments typically need time, invention and resources before outcomes are realized. This principle applies with particular saliency to the Internet, a malleable technology whose form is not fixed across locations. To create value, the Internet must be embedded in investments at firms and households that employ a suite of communication technologies, TCP/IP protocols and standards for networking between computers. Often organizational processes also must change along various points of a distribution chain delivering services to end users.

As there are several ways to look at the features of that chain, it is no surprise that this topic has invited a variety of perspectives. For the sake of brevity, this review will forego comprehensiveness and focus on understanding the geographic features of the layers next to physical infrastructure. Broadly speaking, many business *users* of Internet infrastructure are also *providers* of Internet infrastructure in a different form, usually related to electronic commerce. Even with this narrowing of the topic, many issues arise about its geographic distribution. In these layers are found a wide variety of businesses and applications.

Research on the relationship between location and private investment in Internet infrastructure divides between two kinds of inquiries that operate on two distinct timescales: first, how did existing business establishments react to the diffusion of the Internet? Second, once the Internet became available, how did firms reorganize or relocate in response? Research has begun to make progress on the first question, because there has been sufficient time to observe the first layers of adoption after commercialization. In contrast, research has not made much progress on the second question, because reorganization and relocation of economic activity tends to occur slowly and only over multiple decades.

Befitting a young research topic, the open questions are still quite basic:

1. What are the differences in the predominant patterns of business infrastructure investment between urban and rural locations?
2. Are there any measurable consequences for location of economic activity resulting from differences in investment in Internet infrastructure?
3. What is the evidence, if any, that the diffusion of Internet infrastructure is reshaping the factors supporting urban agglomeration?

6.1. The geography of private investment in IT-overview

The Internet was a malleable technology when it first commercialized in 1992. It needed to be adapted for commercial use. Adaptation was necessarily a local economic activity, resulting from the combination of the local demands of business establishments and the supply constraints of markets for Internet technology infrastructure and services. More to the point, the same technological opportunity (i.e., the commercialization of the Internet across the United States), did not result in the same commercial experience for all establishments in all locations.

As the Internet commercialized, the vast majority of business establishments were faced with a decision about how to react to the availability of new capabilities. As it turned out, American businesses reacted with the largest growth rates in investment in IT in the history of the United States. Stocks of information technology capital grew at a 20 percent annual growth rate from the end of 1995 to the end of 2000[90]. By 2000, computer hardware and software stocks had reached $622.2 billion[91]. The majority of this investment was affiliated with enabling business applications. In 2000, for example, total business investment in IT goods and services was almost triple the level for personal consumption of similar goods[92].

The level of these investment flows is immense and so is the variance across locations[93]. In some locations, the Internet has been adopted across all facets of economic activity, while in other locations adoption is not widespread. What explains this variance? Two distinct views have become prominent.

One view argues that Internet technology requires infrastructure and support services, which are more readily available in urban settings. It forecasts that businesses in urban settings use and deploy Internet applications more frequently than do similar firms in rural settings. It also argues there was little exceptional about the economic geography of the Internet. As with previous frontier IT, most of the productivity benefits from these investments accrued to urban businesses. A contrasting view argues that Internet technology was exceptional, different from all the IT that came before it. According to this view, the Internet decreases coordination costs within firms and between firms, which reduces the importance of distance. Internet technology dramatically reduces the costs of performing isolated economic activity, particularly in rural settings, even when deployment costs are high.

A related debate analyzes how the diffusion of the Internet alters the roles of cities and urban agglomeration in economic life[94]. Again, there are two related viewpoints. One view sees a diffusion process that enhances already existing advantages to urban areas or central cities. The contrasting view observes a diffusion process that diminishes the advantages of urban areas in favor of locations that had previously been considered isolated. The open question concerns the complementarity or substitutability between the Internet and urban agglomeration[95].

[90] This includes computer hardware, computer software, and communications hardware and instruments. See Price and McKittrick (2002) or Henry and Dalton (2002). The growth rates are even higher if communications hardware and instruments are excluded.

[91] These are constant (1996) dollars (Henry and Dalton, 2002).

[92] For 2000, estimated personal consumption of IT goods and services was $165 billion. For business it was $466 billion (Henry and Dalton, 2002).

[93] For an extensive description of this variance (Forman et al., 2003a).

[94] For a general entry into these issues (Cairncross, 1997; Kotkin, 2000; Dodge and Kitchin, 2001; Castells, 2002).

[95] See, for example, Glaeser and Gasper (1997), Kolko (2002), Sinai and Waldfogel (2000).

These two views have deep roots in long-standing debates about the location of economic activity. These led to multiple potential explanations for why advanced IT located in urban areas. These emphasize the costs of adoption, such as (1) availability of complementary information technology infrastructure, such as broadband services, (2) labor market thickness for complementary services or specialized skills, and (3) knowledge spillovers[96]. One other explanation emphasizes that the types of firms found in urban areas are not random. That is, historically IT-friendly establishments may have sorted into areas where costs have previously been low for precursors to Internet technology.

The opposing view emphasizes that establishments in rural or small urban areas derive the most benefit from overcoming diseconomies of small local size. That is, Internet technology substitutes for the disadvantages associated with a remote location. There are several reasons why this may be true. For one, use of Internet technology may act as a substitute for face-to-face communications[97]. Common examples are e-mail or instant messaging. Second, establishments in a rural area lack substitute data communication technologies for lowering communication costs. Third, advanced tools, such as groupware, knowledge management, web meetings, and others also may effectively facilitate collaboration over distances. Some tools enable simultaneous changes to electronic documents by users in multiple locations. Moreover, supply chain management software enables electronic communication of data that would be costly to transmit via phone or through the mail.

Properly answering these questions requires extensive documentation of business use of the Internet. While there has been considerable empirical research on the use of the Internet at households, there has been much less systematic empirical research on the geographic features of Internet technology use by business. Research addressed one of two related goals. First, what is the variance between locations in the average use of Internet technology for some purpose? Second, what is variance in the marginal contribution of the location—instead of some other factor—to the observed outcome?

6.2. *Empirical evidence on domain names*

As the first probe into these issues, several researchers measured the Internet content produced across the United States. For these studies, researchers find the location of each firm with a dot.com Internet address and plot it. Domain

[96] These are closely related to the three major reasons given for industrial agglomeration (Marshall, 1920; Krugman, 1991).

[97] Other authors (Gaspar and Glaeser, 1997) have argued that improvements in information technology may increase the demand for face-to-face communication. In other words, they argue that IT and face-to-face communication may be complements. The implication of this hypothesis is that commercial establishments relocate to urban areas in reaction to technical change in IT.

names are used to help map intuitive names (such as *www.northwestern.edu*) to the numeric addresses computers use to find each other on the network.

Host site counting presents challenges as a measurement of economic activity. It cannot account for the common practice of not physically housing Internet-accessible information at a firm's physical location[98]. So, while it is informative, such evidence is an imperfect way to measure the geographic dispersion of business activity that acts as Internet infrastructure.

If anything, this research has not yet settled on a firm set of hypotheses to test. Authors have tried to use this data to examine classic issues in geography and communications. As Townsend (2001a) argues, for example, with the rise of the Internet, the notion of a global city was questioned, and so too were the two dominant theories on the relationship between telecommunications and urban growth. One theory forecast the centralization of decision-making in 'global' cities, the other forecast wholesale urban dissolution[99]. As yet, there is little evidence for either theory; he argues that we neither see evidence of dominant global cities nor do we see total dissolution. As noted early, it is unsurprising that this debate remains open. It requires evidence that can only come about when establishments relocate, a process that proceeds slowly.

In a pioneering set of studies, Moss and Townsend (2000a,b) analyze the growth rate for domain name registrations between 1994 and 1997, the early years of Internet development. They find that 'global information centers,' such as Manhattan and San Francisco grew at a pace six times the national average, while global cities such as New York, Los Angeles, and Chicago grew only at approximately one to two times the national average.

Examining a mildly later set of registrations, Zook (2000) also finds that San Francisco, New York and Los Angeles are the leading centers for Internet content with regard to absolute size and degree of specialization. Degree of specialization is measured by relating the number of dot.com domains in a region relative to the total number of firms in a region to the number of dot.com domains in the United States relative to the total number of firms in the United States[100].

Kolko (2000) also analyzes such data, but is the first to provide an economic framework in which to understand the outcomes in central and peripheral locations. He considers whether the Internet enhances the economic centrality of major cities in comparison to geographically isolated cities. He argues,

[98] Also, it may be unable to differentiate between various types of equipment. See also Warf (2001) for similar results to those discussed herein, which suggest that there is robustness in some of the observations.

[99] See also Gorman (forthcoming).

[100] Townsend (2001a) argues that the leading cities in total domain names identify the presence of many content providers. He also finds that New York and Los Angeles top the list. He makes much of his findings about Chicago—normally considered along with New York and LA as a global city—which only ranks a distant fifth, far behind the two leaders.

provocatively, that reducing the 'tyranny of distance' between cities does not necessarily lead to proportional economic activity between them. That is, a reduction of communications costs between locations has ambiguous predictions about the location of economic activity in the periphery or the center. Lower costs can reduce the costs of economic activity in isolated locations, but it can also enhance the benefits of locating coordinative activity in the central location. As with other writers, Kolko presumes that coordinative activity is easier in a central city where face-to-face communications take place (Glaeser and Gaspar [1997]).

Kolko's findings do not settle the question, but do raise intriguing issues. As with other researchers, he documents a heavy concentration of domain name registrations in a few major cities. But his framing motivates him to examine the role of proximity vs. remoteness. In particular, he also documents extraordinary per capita registrations in isolated medium-sized cities. He argues that the evidence supports the hypothesis that the Internet is a complement, not a substitute for face-to-face communications in central cities. He also argues that the evidence supports the hypothesis that lowering communication costs helps business in remote cities of sufficient size (i.e., medium-sized, but not too small).

Zook (2001) also examines the locations of the dominant firms in e-commerce. He finds the location of the top Internet companies on the basis of electronically generated sales and other means. His analysis shows San Francisco, New York, and Los Angeles as dominant in e-commerce, with Boston and Seattle not far behind. Many Midwestern cities such as Detroit, Omaha, Cincinnati, and Pittsburgh, as well as many cities in the South are lagging behind those in other parts of the country. Zook (2000) measures the overall size of e-commerce in a region and the E-commerce specialization of a region relative to the number of Fortune 1000 firms headquartered there. He points out that the cities that are major business centers are not necessarily the ones leading in e-commerce adoption. Despite this new approach, once again, he finds a similar ranking of cities and regions.

Another approach followed by Townsend (2001b) analyzes domain name density. This is a ranking of metropolitan areas according to domain names per 1000 persons. Among Los Angeles, New York, and Chicago, only Los Angeles ranks among the top 20 (17th). The full ranking over domain name density indicates that medium-sized metropolitan areas dominate. Many global cities rank highly, as do many medium sized areas, whereas many small metropolitan areas show very low levels of Internet activity. Though statistically simpler than Kolko's analysis, this evidence is consistent with Kolko's view that sufficiently large medium cities benefited from the Internet's diffusion.

A related set of studies analyzes registrations at the level of the neighborhood. This type of study indicates that business needs determine the location of registration activity. A detailed analysis of registrations from the early commercial boom time found that adoption in Manhattan (Moss and Townsend, 1997) and San Francisco (Zook, 1998) was centered on central business districts in those cities. Grubesic's (2003a) detailed examination of Ohio—at the zip code level in

1999—highlights the considerable differences across Ohio's diverse areas. Once again, urban centers of the state have higher activity than the suburbs and the rural counties. At this level of detail it is also apparent that universities can spur further registrations above and beyond that predicted merely by economic factors.

There is a sense in which these findings about domain name registration are not a huge surprise, but it does suggest open questions to explore. Further work needs to be done to understand the links between the location of activity within cities and the concentration of new venture formation, the commuting patterns of highly skilled and educated labor, and the location of other facets of business in dense and nondense locations. Related, now that the dot-com crash has led to many bankruptcies and exit, it is interesting to know whether the geographic distribution of domain names was informative about new ventures only or also about durable business location practices.

6.3. Evidence on urban and rural business use of Internet technology

Another challenge for research arises from confusion about the uses of Internet infrastructure in business. Forman et al. (2003a,b,c) observe there were many purposes for the Internet in business, and show how these different purposes inform studies of the geographic diffusion of infrastructure.

They contrast two purposes, one simple and the other complex. The first purpose, labeled *participation*, relates to activities such as e-mail and web browsing. This represents minimal use of the Internet for basic communications. The second purpose, labeled *enhancement*, relates to investment in frontier Internet technologies linked to computing facilities. These latter applications are often known as *e-commerce,* and involve complementary changes to internal business computing processes.

The economic costs and benefits of these activities are quite distinct, but not highlighted by previous research. Adaptation costs are relevant to the adoption decision for enhancement and largely negligible for participation. The costs of installing enhancement will be more sensitive to increases in density than will participation, while the (gross) benefits from participation will be higher in rural areas than in dense areas. The open question concerns the sensitivity of the 'net benefits' of use to different geographic locations.

Forman (2002) was the first paper to put such distinctions to use. Forman examines the early adoption of Internet technologies at 20,000 commercial establishments from a few select industries. He concentrates on a few industries with a history of adoption of frontier Internet technology and studies the microeconomic processes shaping adoption. He finds that firms with dispersed establishments were more likely to adopt Internet technology for purposes of participation. Rural establishments were as likely as their urban counterparts to participate in the Internet and to employ advanced Internet technologies in their computing facilities for purposes of enhancement of computing facilities. He attributes this to the higher benefits received by remote establishments, which otherwise had no access to private fixed lines for transferring data[101].

Seeking to generalize this type of study to other industries, Forman et al. (2003a) examine all private nonfarm business establishments with 100 or more employees at the end of 2000. This includes a wide array of establishments from manufacturing, service, finance, and for-profit activities. They show that adoption of the Internet for purposes of participation is near saturation in most industries, with some marginal differences for locations. Their findings for enhancement contrast sharply. Establishments in MSAs with over 1 million people are one and a half times as likely to use the Internet for enhancement than are establishments in MSAs with less than 250,000 people.

They conclude that rapid diffusion in participation did not necessarily imply rapid diffusion in enhancement, speculating that diffusion of enhancement will follow a more traditional path than participation. That is, it will take time, innovation, and resources before economic welfare gains are realized. They also concluded that that Internet use in business is widely dispersed across geographic regions. They conclude that research focused on concentration or digital divides—heretofore a central concern of the literature on Internet geography—is a misleading basis for formulating regional economic policy about Internet use in business.

Why do some regions lead and others lag? Forman, Goldfarb and Greenstein (2003c) focus on two factors, the marginal contribution of location to the propensity to use the Internet and the marginal contribution of an establishment's industry. They find that the variance of experience is quite narrow within large MSA—there is little difference between the best area, San Jose, and the worse, Las Vegas. Moreover, the difference within region widens in comparison of large, medium, small and rural MSAs. Hence, the competitive differences between establishments of the same type are greater for smaller metropolitan areas. The difference between the average large MSA area and the worse small MSA area is also great. In extreme situations, location can make a large difference by itself. For example, establishments in small and medium MSAs that are already suffering from slow growth, will typically have low adoption rates of Internet technologies.

Overall, however, an establishment's use of advanced Internet technology is mostly explained by the industry for which it produces. They conclude that the (preexisting) distribution of industries across geographic locations explains much of the differences in average rates in enhancement between locations. Large urban

[101] See, also, Premkumar and Roberts (1999), who identify the state of use of various communications technologies and the factors that influence the adoption of these technologies in rural small businesses in the United States. The authors collected data on 78t organizations in a structured interview process, and their results indicate the importance of many factors. They also express concerns that rural businesses may lack access since Internet infrastructure tends to follow a similar pattern as the interstate road system.

areas lead in the use of advanced Internet applications because the industries that 'lead' in advanced use of the Internet tend to disproportionately locate in urban areas.

This conclusion highlights that some industries are more information intensive than others, and, accordingly, make more intensive use of new developments in information technologies, such as the Internet, in the production of final goods and services. Hence, the geographic dispersion of modern infrastructure will partly depend on whether the information-intensive industries tend to be more geographically concentrated than the less information-intensive industries. Heavy Internet technology users have historically been banking and finance, utilities, electronic equipment, insurance, motor vehicles, petroleum refining, petroleum pipeline transport, printing and publishing, pulp and paper, railroads, steel, telephone communications, and tires (Cortada, 1996).

Forman et al. (2003b) show that establishments from industries with a high propensity to use advanced technology tend to locate in urban areas. They argue that several factors contribute to this persistence of demand for advanced IT at the same establishments in the same industries one decade after another. First, firms are incremental in their approach to investment, compromising between the benefits of frontier and the costs of keeping an existing process, picking, and choosing among those new possibilities that make the most sense for their business. This is consistent with Forman's (2002) finding that installed base of hardware and software applications played a major role in shaping organizations' early decisions to invest in the Internet.

Second, consistent with Cortada (2004), they argue that investment in innovative IT is directed toward automating functional activity or business processes within an organization, such as accounts receivable, inventory replenishment or point of sale tracking. If there is stability of the types of economic activity going on within organizations, then this stability shapes the demand for innovative IT, enhancing the same activities decade after decade. Related, most organizations examine other firms with functions similar to their own and benchmark their own processes against them. Such activity increases the incentives of lead firms to emulate other organizations in the same industry, either close competitors or those with similar supply chains[102].

A related facet of this topic concerns the geographic concentration of labor markets. Because Internet use in business employs highly skilled and educated workers, there are open issues about how the geographic distribution of skilled labor shapes the development of the next layer of Internet infrastructure[103]. Related studies of the labor markets for Internet services are still in their infancy.

[102] For a novel attempt to statistically model these choices, see Fitoussi (2004), who examines the choice over the location of Customer Research Management functions for Fortune 1000 firms.

[103] For a discussion of the role of skilled labor within organizations using advanced IT (Bresnahan, Brynjolfsson and Hitt, 2002; Chapple and Zook, 2002; Feldman, 2002).

This type of research involves analyzing technically adept labor markets, which are characterized by high mobility between firms.

In a particularly striking paper in this vein, Jed Kolko (2002) examines the Internet in light of previous evidence that technology-intensive industries exhibit slower employment convergence than other industries. In this context, *convergence* refers to the tendency of an industry to become more uniformly distributed geographically[104].

Kolko argues that it is not necessarily information technology usage that slows convergence but that the industry hires more educated workers, which causes clusters to persist. Although convergence appears to be slower for Internet technology industries, this conclusion is deceptive. When compared to industries that hire educated workers, Internet technology actually speeds up convergence. Kolko makes a theoretical case as to why certain aspects of the Internet technology industry might converge quickly, namely, knowledge spillovers are less dependent on location and transportation costs are small. Nevertheless, he finds slow convergence[105].

Gorman (forthcoming) takes another approach to this topic by focusing on the top 40 web integration firms in the U.S., such as KPMG or Accenture (Formerly, Andersen Consulting). These firms expanded during the dot-com boom, and Gorman proposes to examine their hiring and employment practices at their home and major branch offices. He argues that this is informative about the distribution of skilled IT labor. He finds a strong bias toward major urban centers, such as New York, San Francisco, Boston, Los Angeles, and Chicago. On one level this is not at all a surprise: it suggests that the skilled labor in these large firms concentrate their home and major branch offices in major cities, where they have access to better IT infrastructure as well as other services, such as large airports. On the other hand, that seemingly obvious pattern is Gorman's main point—that, contrary to utopian visions about the irrelevance of location, the skilled labor for IT did not pick up and largely move to a small or medium sized MSA that many people consider desirable. That place could be Madison, Wisconsin or Durham, North Carolina, or Santa Barbara, California—for Gorman's purpose it does not matter. For a variety of reasons the largest urban centers continue to be the central locations for skilled IT labor markets, even during a boom time.

More research of this type is needed to confirm these observations for more recent experience. For example, as the market for skilled IT labor contracted, how did different areas of the country adjust? As the web enables communications over

[104] For earlier related works about the geography of the computer industry, see Beardsell and Henderson (1999) or Dumais, Ellison and Glaeser (2002). In both of these works, because of longer time periods, researchers have begun to identify the sensitivity of location decisions to labor market features, and vice versa.

[105] His data come from the Longitudinal Enterprise and Establishment Microdata file (1989–1996) and the Current Population Survey.

longer distances, which types of skilled labor becomes more mobile, if any? Which functions move most easily between locations, activities that directly build web infrastructure, such as highly skilled programming, or activities that make use of communications, such as call centers?

6.4. Local Internet and information services

Local information services, such as newspapers, are another important economic activity to examine. These activities not only use Internet infrastructure heavily but also extend it in new ways for local and national readers.

Chyi and Sylvie (1998, 2001) examine the local and long-distance online readership for online newspapers. Data was gathered via a survey of 64 newspapers with online versions. They find that the long-distance market is a substantial submarket constituting about one-third of the online readership, but the local market still outweighs the long-distance market in terms of usage and the online newspapers' targeting intention. They stress the importance of recognizing the difference between their local-market operation and long-distance market. This further argues for the view that local newspapers use the online versions as supplements for enhancing revenue, rather than as substitutes that cannibalize existing revenue streams.

Boczkowski (2002, 2003) provides a close analysis of many of the experiments undertaken by online newspapers during the late 1990s. He documents many of the motives behind local newspapers developing their online divisions, and the consequences that resulted. He shows that most newspapers did not emphasize developing new markets in the sense of finding new customers. Rather, many of these newspapers focused on developing new capabilities that provided new services to their existing customers. This could involve using the interactive capabilities for developing online communities or using new media to enhance coverage in ways that were not viable in a print form. The focus on local customers also followed from the commercial motivation inherent in trying to develop new media tied to ad revenue, much of which came from local sources. The general lesson is that the costs and motives of a local information provider shaped the exploration, extension and direction of development of local Internet infrastructure into information services. The commercial motives of the actors directed this technology toward geographically local services.

In a novel approach to the question, Sinai and Waldfogel (2000) examine what users do on the Internet. They inquire whether the Internet may serve as a substitute for urban agglomeration by leveling the consumption of Internet content between large, variety-laden markets and small, variety-starved markets. The key issues revolve on whether content on the Internet is similarly attractive to persons in large and small markets. They recognize that it can go either way. If the Internet equally aggregates the interests of many users, then those in isolated locations gain. If the Internet offers local, as well as general, information, then it may not necessarily have a role as a substitute for agglomeration. If local on-line content is sufficiently attractive and more prevalent in large markets, then, in some states of the world, it is possible for the Internet to act as a complement to urban agglomeration. Indeed, they find mixed evidence that the Internet acts as both a substitute and complement for urban agglomeration, depending on the type of individual.

Another line of inquiry examines how the Internet alters old definitions for geographic space and redefines urban living spaces. This line of inquiry arises out of an older set of studies examining the impact of the telephone on aspects of urban economic life, such as commuting[106]. There are related questions to ask about the impact of Internet-enabled mobile applications, such as Wi-Fi (i.e., 802.11b or its descendents) or two-way text messaging (i.e., Blackberry or Palm E-mail) or videoconferencing. Similarly, how do these applications affect the use of public spaces, such as airports and cafés, or the travel patterns of particular occupations, such as sales representatives or consulting? This line of inquiry can go in many directions. For a variety of reasons, wireless telephone infrastructure is comparatively ubiquitous across all regions of the U.S. (Gorman and McIntee, forthcoming). An effective mobile application that builds off this infrastructure could quickly become available everywhere[107].

In this vein, Light (1999, 2001) raises questions about how the Internet will alter public life, especially the part of public life taking place in commercial settings. Previously, changes in architecture in cities have been cited as explanations for decline in public life, and scholars are predicting further changes due to the Internet. While people might think the Internet could replace small-town street-corner societies with its virtual communities, Light believes such forecasting is in haste, similar to the way the mall was originally thought to be a replacement for small-town street corners. She points out that the areas of the Internet that act as a location for civic life are generally under strict controls and quite commercialized (online communities, specifically). Light does not dismiss concerns that the Internet may erode public life or alter daily patterns of use of public areas, but she does argue that many of these concerns are a bit overblown—as she believes that perspectives on cyberspace are likely to evolve.

6.5. *Summary*

Has the Internet realized its promise of reducing the importance of location to economic activity? By 2000, participation activities such as e-mail and web browsing had diffused almost everywhere. For complex technologies such as

[106] See, in particular, Kitchin (1998), who discusses the growth of 'back-office' operations, teleworking, and overall employment restructuring.

[107] This was one of the facets that helped the diffusion of the Blackberry, for example. It employs AT&T's wireless infrastructure for mobile text-messaging.

enhancement, in contrast, IT-intensive firms found greater benefits than other firms from pooled resources in large cities.

Does the flow of investment dollars correlate positively with the rankings of location and industries? Were the investment dollars affiliated with the commercialization of the Internet widely dispersed throughout locations and industries in the United States? These questions are important for understanding the impact of the Internet's diffusion of regional growth. They are subject to speculation, and await confirmation with data about investment behavior beyond adoption of infrastructure to support the first generation of e-commerce applications.

Business use of information infrastructure should continue to be a robust research topic. Investment done by private firms rivals or exceeds investments at traditional carriers. Privately owned equipment lives side-by-side with the public switch network and quasi-private Internet carrier network in arrangements without precedent. Moreover, these business networks potentially form the basis of regional comparative advantage. This network arises from the actions of multiple investors, potentially responsive to new concerns. As the commercial Internet moves into its second decade, the geographic properties could be quite different from those more familiar in telephony.

7. Summaries of answers to motivating questions

The NSF began to commercialize the Internet in 1992. Within a few years there was an explosion of commercial investment in information infrastructure in the United States. Research on the geography of the Internet seeks to understand the durable economic processes underneath the Internet's growth and seemingly unique commercial experience.

At the outset of commercialization the economic determinants of the commercial Internet were unknown. Upon examination one can see that a lot of the Internet infrastructure was caused by the same familiar economics that have been seen with other communication networks. This review has touched on familiar economic concepts (i.e., the economies of density and scale in operation, the economies of entry into provisions of services with high sunk costs, the economics of retrofitting technical upgrades on existing infrastructure, and the economics of competitive behavior for growing markets). These have shown up in new places with new sets of economic actors, to be sure, but the analysis required borrowing familiar frameworks from existing communications markets, not a fundamental new set of economic precepts. In that vein, below I summarize answers to the questions that motivated the review.

7.1. Why did near-geographic ubiquity arise after commercialization?

The Internet access business was commercially feasible at a small scale and, thus, at low levels of demand. This meant that the technology was commercially viable at low densities of population, whether or not it was part of a national branded service or a local geographically concentrated service. This partly mimicked the academic experience, where the operations were also feasible on a small scale. That statement alone does not capture all the factors at work, however.

Two important factors prevalent throughout the United States made Internet access feasible in a wide variety of organizational forms, large and small. First, entrepreneurial initiative helped small-scale business opportunities thrive. These businesses included those located in many low-density cities, isolated cities, or rural areas that were not being served by the national firms.

Second, entry was inexpensive. Small-scale implementation depended on the presence of high-quality complementary local infrastructure, such as digital telephony, and interconnection to existing communications infrastructure, which was available due to national and local initiatives to keep the communications infrastructure modern. It also depended on the widespread use of flat rate pricing for telephone service in most locales. Thus, the economies of retrofits took over entry decisions, not the economies of building networks *de novo.*

TCP/IP based technologies can alter the trade-off between frontier applications and costs. Going forward, one should expect similar advantages for new applications that build on preexisting infrastructure. For example, applications that operate within cell phones (e.g., Real Networks) or PDAs (e.g., palm-based or Windows CE based) or laptops (e.g., Wi-Fi) appear so promising because these build on top of preexisting infrastructure.

7.2. Why did market forces encourage extensive growth?

Market forces can customize new technologies to users and implement new ways of delivering technologies. These activities have immense social value when there is uncertainty about technical opportunities and complex issues associated with implementation. As such, this industry offers many examples for illustrating lessons from the economics of market-based experiments.

The ISPs knew about the unique features of the user, the location or the application. This was precisely what was needed to customize Internet access technology to a wide variety of locations, circumstances and users. Removing the Internet from the exclusive domain of NSF administrators and employees at research computing centers brought in a large number of potential users and suppliers, all pursuing their own vision and applying it to unique circumstances. In addition, it allowed private firms to try new business models by employing primitive web technologies in ways that nobody at the NSF could have imagined.

Yet, the location of experiments was necessarily temporary, an artifact of the lack of maturity of the service. As this service matured—as it became more reliable and declined in price so that wider distribution became economically feasible—the geographic areas that were early leaders in technology lost their comparative lead or ceased to be leaders. As such, basic ISP technology diffused widely, to places other than universities, and comparatively rapidly after commercialization.

In contrast, experimentation in the commercialization of the backbone resulted in the build-out ahead of demand. Not only was this a fast build-out, but it also resulted in an oversupply in some locations. As of this writing, it is generally expected that the oversupply of capacity will likely persist for some time, even though demand continues to grow.

The middle of the decade after commercialization also illustrates behavior during times of uncertainty, as well as growth in demand. Firms thought that they were in a race to: (1) meet anticipated demand and (2) build capacity to sell to others. These incentives appeared to be good before the backbone was ever built, since it sped up development. After the network was built, these incentives appear to have been too strong, leaving many firms holding assets with no customers. In retrospect, these were imprudent investments. Such regret is standard for markets after a period of exploratory behavior aimed at resolving commercial uncertainty, so the welfare conclusions are necessarily ambiguous.

7.3. Has the Internet diffused disproportionately to urban areas?

Researchers continue to find two persistent patterns that stand in marked contrast: some see urban areas being disproportionately favored, others find that the technology has spread ubiquitously.

Three factors help reconcile these contrasting assessments. First, transmission technology displays economies of scale. For the foreseeable future, there will be 'big pipes' traveling between 'big switches'. Central locations like Chicago in the middle of the country, New York and Washington, DC in the east, Dallas and Atlanta in the South, and San Francisco and Los Angeles in the West, will continue to have access to big pipes going to it from almost everywhere in the country.

Second, as with many high-tech services, areas with complementary technical and knowledge resources are favored during the early use of technology. This process favored growth in a few locations, such as Silicon Valley, the Boston area or Manhattan—for a time at least, particularly when technologies were young. But did it persist? For a wide variety of uses of the Internet, the answer is simply no. The common perception that the use of the Internet concentrated in a few locations is false, so too is the perception that the revolution dramatically altered the use of IT in only a few locations or a small number of industries.

This technology matured into something standardized that could be operated at low cost in areas with thinner supply of technical talent. In the first generation of

Internet applications, standardization occurred quickly. Just as the Internet moved quickly to locations distance from Universities, so too the first generation of commercial infrastructure did not concentrate in a few large cities. It diffused to virtually every large city and most medium and small cities. In other words, such local factors did not shape the difference between major metropolitan areas as much as it shaped the difference in experience between large MSAs and a few small MSAs or rural areas—not the majority of such small areas, who did just fine. Only a very qualified statement is consistent with the experience of the 1990s, not the alarmist tone found so commonly in discussions about the digital divide.

Third, the next generation of Internet infrastructure, broadband, is a geographically local technology in most of its commercial forms. Unlike the deployment of dial-up, the economics of this deployment strongly favor areas of urban density. Even within that landscape, the build-out has many facets of uncertainty. In particular, regulatory decision-makers at the national and local level continue to hold key levers in determining whether local telephone firms or differentiated competitors (interconnecting with the local network) are the commercially more profitable organizational form for diffusing and commercializing this delivery mode. Accordingly, forecasting the future in this environment is virtually impossible.

7.4. Is the Internet a substitute or a complement for urban agglomeration?

In many respects this often-posed question is not posed very well. Urban agglomeration occurs for many reasons, and only a few of these have direct relationship with the use of information technology. Cities serve multiple purposes to their denizens and the diffusion of a new communications technology will alter some of those reasons, while not changing others at all. The diffusion of the Internet alters both the costs of doing some activities in urban centers and the benefits of doing them elsewhere.

The deeper question is really one about comparative regional advantage, and the competition takes place on multiple margins. Relocation can occur in many different ways. Firms may choose between central cities and suburbs, between urban areas and less dense settings, or, for that matter, between a location within the U.S. and an area outside the boundaries of the U.S.

After a decade of the diffusion of the commercial Internet, it is not obvious that cities have become comparatively less favorable as a location for economic activity. First, cities still retain many of their economic advantages as a location for coordinative activity in comparison to rural areas. Some of this has to do with access to information infrastructure, but most of it arises from access to other city services, such as major airports or transportation hubs, not to mention other shared public goods, such as security services, urban amenities, cultural activities, and so on. Second, cities also retain advantages over rural area due to the thickness of local labor markets for technical talent. For a variety of reasons, cities will continue to be attractive locations for high-technology talent.

It is concluded that, for the foreseeable future, urban areas will serve as the home to the establishments from the industries with long-standing information-intensive demands on the frontier. Perhaps a few key industries will now be able to move some activities outside major urban areas, but it is difficult to see much evidence that the movement on this margin will be dramatic. For example, some activities can move abroad, such as telemarketing, while others can be based out of any U.S. location, such as auditing or aspects of telemedicine, such as reading MRI results. Other financial transactions will also be able to move more easily, conditional on regulatory constraints. That is, the analysis of mobility will take place at this functional level, if at all, and appears not to have broad or sweeping determinants.

The comparison between cities is less obvious. In the comparison of one major city to another (e.g., Los Angeles compared to New York), the diffusion of the Internet probably did not alter the comparative rankings for particular places. Fast communications generates multiple responses. Some activities, such as the financial transactions typical of a corporate headquarters, no longer need to be located at the point of production and can also move to take advantage of complementary factors inside or outside a city, such as land costs, congestion externalities, or the location of complementary services. But where will such headquarters go, to another central city, to a nearby suburb, or abroad? The evidence is sparse, so the answers are still quite speculative.

Further research needs to refine the question. The first effect of a new technology emerges through adoption behavior. As users become accustomed to ubiquitous Internet technology, a second effect emerges. This effect is associated with endogenous location decisions with such infrastructure in place. Whereas adoption behavior can take place on a short timescale (i.e., less than a decade) relocation takes place on a much longer timescale. The full effects on different industries will not be understood until after decades of change.

7.5. *Which policies mattered? Were these effects the intended or unintended consequences?*

The U.S. federal structure is quite unique by International comparison, and gives rise to regulatory tensions that will not arise elsewhere. More broadly, United States regulatory policy imposed a number of constraints on economic actors that shaped outcomes—in the sense that behavior would have been different had policies been different. With that in mind, it is important to catalogue a few policies that contributed to the Internet's comparative ubiquity across the United States.

Some policies had an unambiguous salutary effect, but were unintended consequences of previous decisions. For example, free local phone calls over a short distance had unintended positive effects on the Internet's adoption and ubiquity, particularly when the commercial Internet was quite young. A similar remark could be made about previous national programs to upgrade all switches to digital technology, even those in rural locations, which was done for reasons more related to universal service and because it enhanced the diffusion of complementary digital technologies, such as the fax machine. The Internet came along much later and piggybacked on these upgrades, but would have been difficult without them.

Perhaps the most significant policy—as far as unintended consequences—came from the specter of antitrust law hanging over communications market structure for equipment and carrier services. Prior antitrust decisions—everything from Judge Green's constraining implementation of the divestiture decree to the FCC orders affiliated with deregulating long distance services and customer premise equipment markets—gave the U.S. its eccentric assembly of incumbent suppliers. There were carriers, such as AT&T, WorldCom, Sprint, and the regional Bell operating companies. There were service providers and equipment firms, such as IBM, Cisco, or Lucent, or, for that matter, AOL and other firms with experience in the bulletin board industry. Said succinctly, this was not the IT market of the 1960s or 1970s, when IBM and AT&T dominated so many design decisions and controlled a high fraction of the distribution of equipment, determining how most business users took advantage of the IT frontier.

The 1990s were an era of widely dispersed technical knowledge. The economic actors of the 1990s approached the commercial opportunities afforded by the Internet with quite different experience and capabilities. In the face of market uncertainty over the basic foundations of costs and demand, this variance produced a variety of market based actions. The basic economics of experiments suggests that a variety of views is healthy in an environment where the experts are uncertain because it leads to rapid exploration of a variety of combinations of otherwise unknown demand and cost conditions. U.S. policy fostered this variety by discouraging concentration of decision making in a small number of firms.

At a broader level, it is clear in retrospect that other regulatory decisions also permitted firms and users to experiment. For example, one set of important rules—dating to the FCC's old decisions for Computer I, II, and II—delimited local telephone firm discretion over interconnection between the network and customer premise equipment. These rules eventually resulted in the FCC's mandate for standard physical interconnection between phones and wires, a physical connection that made a PC as easy to hook up to a telephone wire as it is to hook up a facsimile machine to a wire.

A related set of rules delimited incumbent behavior in their relationship with other carriers. These interconnection rules allowed bulletin board providers to interconnect without regulatory oversight or hassle, the type of precedent that allowed the ISP industry to grow rapidly. Related rules provided entrants a measure of protection against exclusionary behavior by incumbents, which permitted backbone providers to deploy sunk assets on a large scale and with confidence that they would go to use. Such rules enabled behavior that developed multiple options for users.

Such an assessment is contingent on understanding a place and time. These rules did not extend to diffusion of broadband over cable and have led to a mixed experience in the diffusion of DSL as well-though, as noted, it is doubtful whether the urban biases of these technologies would differ under any set of rules. In addition, if Internet telephony becomes viable, then the present set of rules are potentially quite disruptive for the U.S. system for assuring universal service in high cost rural areas. Certainly a revision of the revenue base for subsidizing rural telephony would have large consequences for diffusion of related information infrastructure, such as the Internet.

The other set of important rules became embedded in the Telecom Act of 1996. This Act set up the administration of the E-rate program, whose affect was unambiguously expansionary in rural areas and inner-city areas. Beyond this, the contribution of this Act is much harder to assess, at least in terms of its short run geographic impact. To be sure, it is clear that the Telecom Act of 1996, as well as the FCC's and court's interpretation of that Act, had consequences for the experience of CLECs. These decisions shaped the prices paid by CLECs, their ability to offer services such as DSL, and their ability to resolve disputes with incumbent local exchange carriers. But even if the Act had been written and interpreted differently, the particular fortunes of firms might have changed, but not necessarily the geographic dispersion of Internet infrastructure. Hence, this question awaits further analysis.

7.6. Are there lessons for other countries?

Factors unique to the United States shaped the economic geography of the Internet experience in the United States: and, yet, despite these unique features, there are lessons for development of Internet infrastructure in other countries. What was unique about the U.S. experience? As noted, many regulatory issues were first addressed in the U.S. and were addressed in idiosyncratic ways.

In spite of this idiosyncrasy, there are three overriding lessons that have potential to move across international boundaries: first when uncertainty is irreducible, it is better to rely on private incentives to develop mass-market services at a local level. Once the technology was commercialized, private firms tailored it in multiple locations in ways that nobody foresaw. Indeed, the eventual shape, speed, growth, and use of the commercial Internet was not foreseen within government circles (at NSF), despite (comparatively) good intentions and benign motives on the part of government overseers, and despite advise from the best technical experts in the world. If markets are better in this set of circumstances in the U.S., then a similar broad lesson probably applies elsewhere, even with all the differences found elsewhere. This principle holds with particular force for business infrastructure, where opportunities are still being developed.

Second, Internet infrastructure grew because it is malleable, not because it was technically perfect. In many places it is better thought of as a cheap retrofit on top of the existing communications infrastructure. No single solution was right for every situation, but a TCP/IP solution could be found in most places. Hence, the technology appealed to a wide variety of potential users, significantly helping the technology transform from being a niche use into mass-market applications. Stated succinctly, malleability was important because demand-side economies of scale could not be achieved with a perfect technical fix; it could only be achieved with a technology that could be deployed in multiple locations.

Third, there is no optimal combination of unfettered behavior and regulated rules for encouraging Internet growth, but exclusive use of either markets or government will surely fail in most countries. In the U.S. it would not have been possible for any single administrative agency to build and manage such a network under government auspices. Indeed, one of the smartest things the NSF did was decide to give up managing the whole infrastructure, while putting in place a few scalable features, such as an address system and data-exchange points. Similarly, an unfettered system simply could not have worked without government help. For example, the competitiveness of the backbone network within the U.S. almost surely arose due to the novelty of the market and the sheer size of demand. This will not be replicated in many small countries, which will have sufficient demand to support only a small number of backbone providers. In that case, government regulation has a role to play in stopping anticompetitive behavior in interconnection practices.

Acknowledgments

The Elinor and Wendell Hobbs Professor of Management and Strategy, Kellogg School of Management, Northwestern University. The author would like to thank Angelique Augereau, Howard Berkes, Oded Bizan, Pablo Boczkowski,Tim Bresnahan, Paul David, Barbara Dooley, Tom Downes, Chris Forman, Avi Goldfarb, Zvi Griliches, Brian Kahin, Mike Mazzeo, Sumit Majumdar, Eli Noam, Roger Noll, Robert Pindyck, Greg Rosston, Alicia Shems, Pablo Spiller, Dan Spulber, Scott Stern, Greg Stranger, Manuel Trajtenberg, Ingo Vogelsang, Joel Waldfogel, Glenn Woroch, and numerous seminar participants for many useful comments. Jeff Prince provided exceptional research assistance. All errors are the responsibility of the author.

References

Abel, J. and M. Clements, 2001, Entry under asymmetric regulation, Review of Industrial Organization, 19(2), 227–242.

Augereau, A. and S. Greenstein, 2001, The need for speed in emerging communications markets: Upgrades to advanced technology at Internet service providers, International Journal of Industrial Organization, 19, 1085–1102.

Bailey, J. and L. McKnight, 1997, Scalable Internet interconnection agreements and integrated services, in B. Kahin and J. H. Keller, eds., Coordinating the Internet, Cambridge, MA: MIT Press, 309–324.

Beardsell, M. and V. Henderson, 1999, Spatial evolution of the computer industry in the USA, European Economic Review, 43(2), 431–456.

Bertot, J. and C. McClure, 2000, Public libraries and the Internet, 2000, Reports prepared for National Commission on Libraries and Information Science, Washington DC, http://www.nclis.gov/statsurv/2000plo.pdf.

Besen, S. J., P. Milgrom, B. Mitchell, and P. Sringagesh, 2001, Advances in Routing Technologies and Internet Peering Agreements, American Economic Review, May, pp 292–296.

Besen, S. J., J. S. Spigel, and P. Srinagesh, 2002, Evaluating the Competitive Effects of Mergers of Internet Backbone Providers, ACM Transactions on Internet Technology, pp. 187–204.

Besen, S., P. Milgrom, B. Mitchell and P. Sringagesh, May 2001, Advances in routing technologies and Internet peering agreements, American Economic Review, 292–296.

Besen, S., J. S. Spigel and P. Srinagesh, 2002, Evaluating the competitive effects of mergers of internet backbone providers, ACM Transactions on Internet Technology, 187–204.

Blumenthal, M. S. and D. D. Clark, 2001, Rethinking the design of the internet: The end-to-end arguments vs. the brave new world, in B. Compaine and S. Greenstein, eds., Communications Policy in Transition: The Internet and Beyond, Cambridge, MA: MIT Press, 91–139.

Boczkowski, P., 2002, The development and use of online newspapers: What research tells us and what we might want to know, in L. Lievrouw and S. Livingstone, eds., The Handbook of New Media: Social Shaping and Consequences of ICTs, London: Sage, 270–286.

Boczkowski, P., 2003, Emerging Media: Innovation in Online Newspapers, Cambridge, MA: MIT Press.

Boardwatch Magazine, Directory of Internet Service Providers, Littleton, CO.

Bresnahan, T., E. Brynjolfsson and L. Hitt, 2002, Information technology, work organization, and the demand for skilled labor: Firm-level evidence, Quarterly Journal of Economics, 117(February), 339–376.

Bresnahan, T. and S. Greenstein, 2001, The economic contribution of information technology: Towards comparative and user studies, Evolutionary Economics, 11, 95–118.

Bresnahan, T. and M. Trajtenberg, 1995, General purpose technologies: Engines of growth? Journal of Econometrics, 65(1), 83–108.

Buckley, P. and S. Montes, 2002, Main street in the digital age: How small and medium sized businesses are using the tools of the new economy, Washington, DC: U.S. Department of Commerce, Economics and Statistics Administration.

Cairncross, F., 1997, The Death of Distance, Cambridge, MA: Harvard Press.

Cannon, R., 2001, Where Internet service providers and telephone companies compete: A guide to the computer inquiries, enhanced service providers, and information service providers, in B. Compaine and S. Greenstein, eds., Communications Policy in Transition: The Internet and Beyond, Cambridge, MA: MIT Press, 3–34.

Carr, N., 2000, An interview with Akamai's George Comrades, Harvard Business Review (May–June), 118–125.

Castells, M., 2002, The Internet Galaxy, Reflections on the Internet, Business and Society, Oxford University Press.

Cawley, R. A., 1997, Interconnection, pricing, and settlements: Some healthy jostling in the growth of the Internet, in B. Kahin and J. H. Keller, eds., Coordinating the Internet, MIT Press, 346–376.

Chapple, K. and M. A. Zook, 2002, Why some IT jobs stay: The rise of job training in information technology, Journal of Urban Technology, 9(1), 57–83.

Cherry, B., A. H. Hammond and S. Wildman, 1999, aking Universal Service Policy: Enhancing the Process Through Multidisciplinary Evaluation, Mahwah, NJ: Lawrence Erlbaum Associates.
Chinoy, B. and T. J. Salo, 1997, Internet exchanges: Policy-driven evolution, in B. Kahin and J. H. Keller, eds., Coordinating the Internet, MIT Press, 325–345.
Chyi, H. I. and G. Sylvie, 1998, Competing with whom? And how? A structural analysis of the electronic newspaper market, The Journal of Media Economics, 11(2), 1–18.
Chyi, H. I. and G. Sylvie, 2001, The medium is global, the content is not: The role of geography in online newspaper markets, Journal of Media Economics, 14(4), 231–248.
Claffy, K., 2001, CAIDA: Visualizing the Internet, *http://www.caida.org/outreach/papers/2001/caida/.*
Claffy, K., 2000, Measuring the Internet, IEEE Internet Computing, *http://www.caida.org/outreach/papers/2000/ieee0001/.*
Coffman, K. G. and A. Odlyzko, 2002, Growth of the Internet, in I. P. Kaminow and T. Li, eds., Optical Fiber Telecommunications IV B: Systems and Impairments, Academic Press, 17–56.
Compaine, B., 2001,The Digital Divide: Facing a Crisis or Creating a Myth? Cambridge, MA: MIT Press.
Cortada, J. W., 1996, Information Technology as Business History: Issues in the History and Management of Computers, Westport, CT: Greenwood Press.
Cortada, J. W., 2004, The Digital Hand: How Computers Changed the Work of American Manufacturing, Transportation and Retail Industries, New York: Oxford University Press.
Crandall, R. W., 2001, An assessment of the competitive local exchange carriers five years after the passage of the Telecommunications Act, A report for the USTA by Criterion Economics, LLC.
Crandall, R. W. and J. H. Alleman, 2002, Broadband: Should we regulate high-speed Internet access?, Washington DC: AEI-Brookings Joint Center for Regulatory Studies.
Crandall, R. W. and J. G. Sidak, 2002, Is structural separation of incumbent local exchange carriers necessary for competition? Yale Journal on Regulation, 19(2), 1–75.
Crandall, R. W. and L. Waverman, 2000, Who pays for universal service? When telephone subsidies become transparent, Washington DC: Brookings Institution Press.
Cusumano, M. and D. Yoffie, 2000, Competing on Internet Time: Lessons from Netscape and its Battle with Microsoft, New York: Free Press.
David, P. and S. Greenstein, 1990, The economics of compatibility of standards: A survey, Economics of Innovation and New Technology, 1, 3–41.
Dodge, M., 2001, Cybergeography, Environment and Planning B-Planning & Design, 28(1), 1–2.
Dodge, M. and R. Kitchin, 2001a, Atlas of CyberSpace, London: Addison-Wesley.
Dodge, M. and R. Kitchin, 2001b, Mapping CyberSpace, London: Routledge.
Downes, T. and S. Greenstein, 1998, Do commercial ISPs provide universal access? in S. Gillett and I. Vogelsang, eds., Competition, Regulation and Convergence: Current Trends in Telecommunications Policy Research, Lawrence Erlbaum Associates, 195–212.
Downes, T. and S. Greenstein, 2002, Universal access and local internet markets in the U.S., Research Policy, 31, 1035–1052.
Dumais, G., G. Ellison and E. L. Glaeser, 2002, Geographic concentration as a dynamic process, Review of Economics and Statistics, 84(2), 193–204.
Eisner, J. and T. Waldon, 2001, The demand for bandwidth: Second telephone lines and on-line services, Information Economics and Policy, 13(3), 301–309.
Esbin, B., 1998, Internet over cable: Defining the future in terms of the past, Federal Communications Commission, Office of Planning and Policy, Working Paper No. 30.
Farnon, M. and S. Huddle, 1997, Settlement systems for the internet, in B. Kahin and J. H. Keller, eds., Coordinating the Internet, Cambridge, MA: MIT Press, 377–403.
Faulhaber, G. and C. Hogendorn, 2004, The market structure of broadband telecommunications, Journal of Industrial Economics, 48(3), 305–330.

Federal Communication Commission, 2000, Local Telephone Competition: Status as of 30 June 2000, Industry Analysis Division, Common Carrier Bureau, available at *http://www.fcc.gov/Bureaus/Common_Carrier/Reports/FCC-State_Link/recent.html.*

Federal Communication Commission, 2001, Local Telephone Competition: Status as of 30 June 2001, Industry Analysis Division, Common Carrier Bureau, available at *http://www.fcc.gov/Bureaus/Common_Carrier/Reports/FCC-State_Link/recent.html.*

Feldman, M. P., 2002, The Internet Revolution and the Geography of Innovation, International Social Science Journal, 54(1).

Fitoussi, D., 2004, So Close and Yet So Far: Information Technology and the Spatial Distribution of Work, MIT: Mimeo.

Forman, C., 2002, The corporate digital divide: Determinants of Internet adoption, Working Paper, Graduate School of Industrial Administration, Carnegie Mellon University. Available at http://www.andrew.cmu.edu/~cforman/research/corp_divide.pdf.

Forman, C., A. Goldfarb and S. Greenstein, 2003a, The geographic dispersion of commercial Internet Use, in S. Wildman and L. Cranor, eds., Rethinking Rights and Regulations: Institutional Responses to New Communication Technologies, Cambridge: MIT Press, 113–145.

Forman, C., A. Goldfarb and S. Greenstein, 2003b, Which industries use the Internet? in M. Baye, ed., Organizing the New Industrial Economy, Elsevier, 47–72.

Forman, C., A. Goldfarb and S. Greenstein, 2003c, How did location affect adoption of the Internet by commercial establishments? Urban density, Global Village and Industry Composition, NBER Working Paper # 9979, MA: Cambridge.

Frazer, K., 1995, NSFNET: A Partnership for High-Speed Networking, Final Report, 1987–1995, Merit Networks.

Frieden, R., 2001, The potential for scrutiny of internet peering policies in multilateral forums, in B. Compaine and S. Greenstein, eds., Communications Policy in Transition: The Internet and Beyond, Cambridge: MIT Press, 159–193.

Gabel, D. and F. Kwan, 2001, Accessibility of broadband communication services by various segments of the American population, in B. Compaine and S. Greenstein, eds., Communications Policy in Transition: The Internet and Beyond, Cambridge, MA: MIT Press, 295–320.

Garcia, D. L., 1996, Who? What? Where? A look at Internet deployment in rural America, Rural Telecommunications, November–December, 25–29.

Gaspar, J. and Glaeser, E., 1997, Information technology and the future of cities, Journal of Urban Economics, 43, 136–156.

Gillett, S. E., 2000, Universal service: Defining the policy goal in the age of the Internet, Information Society, 16(2), 147–149.

Goldfarb, A., 2004, Concentration in advertising-supported online markets: An empirical approach, Economics of Innovation and New Technology, 13(6), 581–594.

Goolsbee, A. and P. Klenow, 1999, Evidence on learning and network externalities in the diffusion of home computers, NBER working paper # 7329.

Goolsbee, A. and J. Guryan, 2002, The impact of Internet subsidies for public schools, NBER Working Paper #9090.

Gorman, S. P., Forthcoming, where are the web factories: The urban bias of E-business location, Tijdschrift voor Economische Geografie, 93(5), 522–536.

Gorman, S. P. and E. J. Malecki, 2000, The networks of the Internet: An analysis of provider networks in the USA, Telecommunications Policy, 24(2), 113–134.

Gorman, S. and E. J. Malecki, 2002a, Fixed and fluid: Stability and change in the geography of the internet, Telecommunications Policy, 26, 389–413.

Gorman, S. and A. McIntee, Forthcoming, ethered connectivity? The spatial distribution of wireless infrastructure, Environment and Planning A, 35(7), 1157–1171.

Gregg, B. J., 2002, A survey of unbundled network element prices in the United States, Public Service Commission of West Virginia.

Greenstein, S., 2000a, Building and delivering the virtual world: Commercializing services for Internet access, The Journal of Industrial Economics, 48(4), 391–411.
Greenstein, S., 2000b, Empirical evidence on commercial internet access providers' propensity to offer new services, in B. Compaign and I. Vogelsang, eds., The Internet Upheaval, Raising Questions and Seeking Answers in Communications Policy, Cambridge: MIT Press, 253–276.
Greenstein, S. M., 2001, Commercialization of the Internet: The interaction of public policy and private actions, in A. Jaffe, J. Lerner and S. Stern, eds., Innovation, Policy and the Economy, Cambridge: MIT Press, 151–186.
Greenstein, S. and M. Mazzeo, 2003, Differentiation strategy and market deregulation: Local telecommunication entry in the late 1990s, NBER Working Paper 9761, Cambridge.
Greenstein, S. and M. Lizardo, 1999, Determinants of the Regional Distribution of Information Technology Infrastructure in U.S., in D. Orr and T. Wilson, eds., The Electronic Village: Public Policy Issues of the Information Economy, C.D. Toronto, Canada: Howe Institute, 145–180.
Greenstein, S., M. Lizardo and P. Spiller, 2003, The evolution of advanced large scale information infrastructure in the United States, in A. Shampine, ed., Down to the Wire: Studies in the Diffusion and Regulation of Telecommunications Technologies, New York: Nova Press, 3–36.
Grubesic, T., 2003a, Spatial dimensions of Internet activity, Telecommunications Policy, 26, 363–387.
Grubesic, T., 2003b, Inequities in the broadband revolution, The Annals of Regional Science, 37, 263–289.
Grubesic, T. H. and A. T. Murray, 2002, Constructing the divide: Spatial disparities in broadband access, Papers in Regional Science, 81(2), 197–221.
Grubesic, T. H. and M. E. O'Kelly, 2002, Using points of presence to measure accessibility to the commercial Internet, Professional Geographer, 54(2), 259–278.
Grubesic, T., M. O'Kelly and A. Murray, 2003, A geographic perspective on commercial Internet survivability, Telematics and Informatics, 20, 51–69.
Grubesic, T. H. and A. T. Murray, 2004, Waiting for Broadband: Local Competition and the Spatial Distribution of Advanced Telecommunication Services in the United States, Growth and Change, 35(2): 139–165.
Hausman, J. A., J. G. Sidak and H. J. Singer, 2001, Cable modems and DSL: Broadband Internet access for residential customers, American Economic Review, May, 302–307.
Henry, D. and D. Dalton, 2002, Information Technology Industries in the New Economy, Chapter III in Digital Economy 2002, Department of Commerce, at http://www.esa.doc.gov/reports.cfm.
Hindman, D. B., 2000, The rural-urban digital divide, Journalism & Mass Communication Quarterly, 77(3), 549–560.
Hogendorn, C., 2003, Excessive (?) Entry of national telecom networks, 1990–2001, Mimeo, Wesleyan University, http://chogendorn.web.wesleyan.edu.
Huffaker, B., M. Fomenkov, D. Plummer, D. Moore and K. Claffy, 2002, Distance Metrics in the Internet, Presented at the IEEE International Telecommunications Symposium, copy at. *http://www.caida.org/outreach/papers/2002/Distance/*.
Hurley, D. and J. H. Keller, 1999, The First 100 Feet: Options for Internet and Broadband Access, MIT Press.
Jacobs, J., 1969, The Economy of Cities, New York: Vintage.
Juliussen, E. and K. P. Juliussen, 1998, Internet Industry Almanac, San Jose, CA: Peer-to-peer Communications.
Kahin, B., 1992, Building Information Infrastructure, Issues in the Development of the National Research and Education Network, Cambridge, MA: McGraw-Hill Primis.
Kahin, B., 1997, The U.S. National Information Infrastructure Initiative: The Market, the Web, and the Virtual Project, in B. Kahin and E. Wilson, eds., National Information Infrastructure Initiatives, Vision and Policy Design, MIT Press, 150–189.
Kahin, B. and J. Abbate, 1995, Standards for Information Infrastructure, Cambridge, MA: MIT Press.
Kahin, B. and J. H. Keller, 1997, Coordinating the Internet, MIT Press.

Kahin, B. and C. Nesson, 1999, Borders in Cyberspace: Information Policy and the Global Information Infrastructure, The MIT Press.

Kahn, R., 1995, The role of government in the evolution of the Internet, in National Academy of Engineering, Revolution in the U.S. Information Infrastructure, Washington, DC: National Academy Press, 13–24.

Kende, M., 2000, The digital handshake: Connecting Internet backbones, Federal Communications Commission, Office of Planning and Policy, Working Paper No. 32. Washington, DC.

Kenney, M., 2003, The growth and development of the Internet in the United States, in B. Kogut, ed., The Global Internet Economy, Cambridge: MIT Press, 69–108.

Kitchin, R. M., 1998, Towards geographies of cyberspace,, Progress in Human Geography, 22(3), 385–406.

Kitchin, R., 1999, Digital places: Living with geographic information technologies, Progress in Human Geography, 23(4), 647–648.

Kitchin, R. M. and M. Dodge, 2001, Placing' cyberspace: Why Geography Still Matters. Information Technology, Education and Society, 1(2), 25–46.

Kolko, J., 2000, The death of cities? The death of distance? Evidence from the geography of commercial Internet usage, in I. Vogelsang and B. Compaine, eds., The Internet Upheavel: Raising Questions, Seeking Answers in Communications Policy, Cambridge, MA: MIT Press, 73–97.

Kolko, J., 2002, Silicon mountains, silicon molehills, geographic concentration and convergence of internet industries in the U.S., Economics of Information and Policy.

Koski, H. and S. Majumdar, 2002, Paragons of virtue? Competitor entry and the strategies of incumbents in the US local telecommunications industry, Information Economics and Policy, 14, 453–480.

Kotkin, J., 2000, The New Geography: How the Digital Revolution is Reshaping the American Landscape, New York: Random House.

Krol, E., 1992, The Whole Internet, O'Reilly & Associates, Inc.

Krol, E., 1994, The Whole Internet – 2nd Edition, O'Reilly & Associates, Inc.

Krugman, P., 1991, Geography and Trade, Cambridge: MIT Press.

Kruse, H., W. Yurcik and L. Lessig, 2001, The InterNAT: Policy implications of the Internet architecture debate, in B. Compaine and S. Greenstein, eds., Communications Policy in Transition: The Internet and Beyond, Cambridge, MA: MIT Press, 141–157.

Laffont, J.-J. and J. Tirole, 2000, Competition in Telecommunications, Cambridge: MIT Press.

Laffont, J.-J., S. Marcus, P. Rey and J. Tirole, 2003, Internet interconnection and the off-net-cost pricing principle,, RAND Journal of Economics, 34(2).

Lehr, W. and S. Gillett, 2000, Availability of broadband Internet access: Empirical evidence, www.ksg.harvard.edu/iip/access/gillett_lehr.PDF.

Lemley, M. and L. Lessig, 2001, The End of End-to-End Preserving the Architecture of the Internet in the Broadband Era, UCLA Law Review, 48, 925–948.

Light, J., 2001, Rethinking the digital divide, Harvard Educational Review, 71(4), 710–734.

Light, J., 1999, From city space to cyberspace, in M. Crang, P. Crang and J. May, eds., Virtual Geographies, London: Routledge, 109–130.

Mackie-Mason, J. and H. Varian, 1995, Pricing the Internet, in B. Kahin and J. Keller, eds., Public Access to the Internet, Cambridge, MA: MIT Press, 269–314.

Malecki, E. J., 2002a, The economic geography of the internet's infrastructure, Economic Geography, 78(4), 399–424.

Malecki, E. J., 2002b, Location competition in telecommunications in the United States: Supporting conditions, policies, and impacts, The Annals of Regional Science, 36, 437–454.

Malecki, E. J., 2002c, Hard and soft networks for urban competitiveness, Urban Studies, 39, 5–6, 929–945.

Malecki, Edward, J., 2003, Digital development in rural areas: Potential and pitfalls, Journal of Rural Studies, 19, 201–214.

Malecki, E. J. and S. Gorman, 2001, Maybe the death of distance, but not the end of Geography: The Internet as a network, in T. Leinbach and S. Brunn, eds., Worlds of E-Commerce: Economic, Geographical and Social Dimensions, New York: John Wiley and Sons, 87–105.

Mandelbaum, R. and P. A. Mandelbaum, 1992, The strategic future of mid-level networks, in Brian Kahin, ed., McGraq-Hill Primis: Building Information Infrastructure.

Maloff Group International, Inc., 1997, 1996–1997, Internet access providers marketplace analysis, Dexter, MO: Maloff Group International, Inc.

Marine, A., S. Kilpatrick, V. Neou and C. Word, 1993, Internet: Getting started, PTR Prentice Hall.

Marshall, A., 1920, Principles of economics. 8th ed., New York: Porcupine Press.

Meeker, M. and C. DePuy, 1996, The Internet Report, New York: Harper Business.

Milgrom, P., B. Mitchell and P. Srinagesh, 2000, Competitive effects of Internet peering Policies, in I. Vogelsang and B. Compaine, eds., The Internet Upheaval: Raising Questions Seeking Answers in Communications Policy, MIT Press, 175–198.

Milgrom, P., B. Mitchell and P. Srinagesh, 2000, Competitive effects of Internet peering policies, in I. Vogelsang and B. Compaine, eds., The internet upheaval: raising questions seeking answers in communications policy, MIT Press, 175–198.

Mini, F., 2001, The role of incentives for opening monopoly markets: Comparing GTE and BOC cooperation with local entrants, Journal of Industrial Economics, 49(3), 379–413.

Moss, M. L. and A. M. Townsend, 1997, Tracking the Net: Using domain names to measure the growth of the Internet in U.S. Cities, Journal of Urban Technology, 4(3), 47–60.

Moss, M. L. and A. M. Townsend, 1999, How telecommunications systems are transforming urban spaces, in J. O. Wheeler and Y. Aoyama, eds., Fractured Geographies: Cities in the Telecommunications Age, New York: Routledge, 31–41.

Moss, M. L. and A. M. Townsend, 2000a, The Internet backbone and the American metropolis, Information Society, 16(1), 35–47.

Moss, M. L. and A. M. Townsend, 2000b, in D. Janelle and D. Hodge, eds., The role of the real city in cyberspace: Measuring and representing regional variations in Internet accessibility, information, place, and cyberspace, Berlin: Springer-Verlag, 171–186.

Mowery, D. C. and T. S. Simcoe, 2002, The Origins and Evolution of the Internet, in B. Steil, D. Victor and R. Nelson, eds., Technological Innovation and Economic Performance, Princeton University Press, 229–264.

National Research Council, 2001, The Internet's Coming of Age, National Academy Press: Washington DC.

National Research Council, 2002, Broadband, Bringing Home the Bits, Washington, DC: National Academy Press.

New Paradigm Resources Group, 2000, CLEC Report, Chicago, IL.

Nicholas, K. H., 2000, Stronger then barbed wire: How geo-policy barriers construct Rural Internet access, in L. F. Cranor and S. Greenstein, eds., Communications Policy and Information Technology: Promises, Problems, and Prospects, Cambridge, MA: MIT Press, 299–316.

Noam, E., 2001, Interconnecting the Network of Networks, MIT Press.

Noam, E., 2002, Interconnection Practices in M. E. Cave, S. Majumdar and I. Vogelsang, eds., Handbook of Telecommunications Economics, Volume 1, North-Holland: Amsterdam.

Noll, R., D. Older-Aguilar, G. Rosston and R. Ross, 2001, The digital divide: Definitions, measurement, and policy issues, Paper presented at the American Economics Association Meetings, January 2002.

National Telecommunications and Information Administration (NTIA), 1995, Falling through the Net: A survey of the "have nots" in rural and urban America, *http://www.ntia.doc.gov/reports.html.*

National Telecommunications and Information Administration (NTIA), 1997, Falling through the Net: Defining the digital divide, *http://www.ntia.doc.gov/reports.html.*

National Telecommunications and Information Administration (NTIA), 1998, Falling through the Net II: New data on the digital divide, *http://www.ntia.doc.gov/reports.html.*

National Telecommunications and Information Administration (NTIA), 2000, Advanced telecommunications in rural America, the challenge of bringing broadband service to all Americans, See *http://www.ntia.doc.gov/reports.html.*

National Telecommunications and Information Administration (NTIA), 2002, A nation online: How Americans are expanding their use of the Internet, *http://www.ntia.doc.gov/reports.html.*.

Odlyzko, A., 2001, Internet growth: Myth and reality, use and abuse, J. Computer Resource Management, 102, Spring, 23–27.

O'Donnell, S., 2001, Broadband Architectures, ISP Business Plans, and Open Access, in B. Compaine and S. Greenstein, eds., Communications Policy in Transition, The Internet and Beyond, Cambridge: MIT Press.

O'Kelly, M. and T. Grubesic, 2002, Backbone topology, access, and the commercial Internet, 1997–2000, Environment and Planning B: Planning and Design, 29, 533–552.

Oxman, J., 1999, The FCC and the unregulation of the Internet, Federal Communications Commission, Office of Planning and Policy, Working paper 31.

Parker, E. B., 2000, Closing the digital divide in rural America, Telecommunications Policy, 24(4), 281–290.

Premkumar, G. and M. Roberts, 1999, Adoption of new information technologies in rural small businesses, Omega-International Journal of Management Science, 27(4), 467–484.

Price, L. and G. McKittrick, 2002, Setting the Stage: The New Economy Endures Despite Reduced IT Investment, Digital Economy 2002, Chapter I, Department of Commerce, at http://www.esa.doc.gov/reports.cfm.

Prieger, J. E., 2001, The supply side of the digital divide: Is there redlining in the broadband internet access market? AEI-Brookings Joint Center for Regulatory Studies, Working Paper 01–16, December 2001.

Rogers, E. M., 1995, Diffusion of Innovations, New York: Free Press.

Rosston, G. and B. Wimmer, 2001, From C to Shining C' Competition and Cross-Subsidy in Communications, in B. Compaine and S. Greenstein, eds., Communications Policy in Transition: The Internet and Beyond, MIT Press, 241–263.

Shampine, A., 2001, Determinants of the diffusion of US digital telecommunications, Journal of Evolutionary Economics, 11, 249–261.

Shiman, D. and J. Rosenworcel, 2002, Assessing the Effectiveness of Section 271 Five Years After the Telecommunications Act of 1996, in L. Cranor and S. Greenstein, eds., Communications Policy and Information Technology: Promises, Problems and Prospects, MIT Press, 183–215.

Sidak, G., 2003, The failure of good intentions: The WorldCom Fraud and the collapse of American Telecommunications after deregulation, Yale Journal of Regulation, 20, 207–267.

Sidak, G. and D. Spulber, 1998, Cyberjam: The law and economics of internet congestion of the telephone network,, Harvard Journal of Law and Public Policy, 21, 2.

Sinai, T. and J. Waldfogel, 2000, Geography and the Internet: Is the Internet a substitute or a complement for cities? Working Paper, Wharton School of Business, Philadelphia: University of Pennsylvania.

Smarr, L. L. and C. E. Catlett, 1992, Life after Internet: Making room for new applications, in Brian Kahin, ed., Building information infrastructure, McGraq-Hill Primis.

Spulber, D. and C. Yoo, 2003, Access to Networks: Economic and constitutional connections, Cornell Law Review, 88.

Srinagesh, P., 1997, Internet cost structures and interconnections agreements, in L. W. McKnight and J. Bailey, eds., Internet Economics, Cambridge, MA: MIT Press.

Strover, S. and L. Berquist, Developing telecommunications infrastructure: State and local policy collisions, in B. Compaine and S. Greenstein, eds., Communications Policy in Transition: The Internet and Beyond, MIT Press, pp. 221–239.

Strover, S., 2001, Rural Internet Connectivity, Telecommunications Policy, 25(5), 331–347.

Strover, S., M. Oden and N. Inagaki, 2002, Telecommunications and rural economies: Findings from the Appalachian region, in L. F. Cranor and S. Greenstein, eds., Communication Policy and Information Technology: Promises, Problems, Prospects, Cambridge, MA: MIT Press, 317–345.

Suhonen, S., 2002, Industry evolution and shakeout mechanisms: The case of the Internet service provider industry, Dissertation, Helsinki School of Economics.

Townsend, A. M., 2001a, Network cities and the global structure of the Internet, American Behavioral Scientist, 44(10), 1697–1716.

Townsend, A. M., 2001b, The Internet and the rise of the new network cities, 1969–1999, Environment and Planning B-Planning & Design, 28(1), 39–58.

U.S. Department of Agriculture, 2000, Advanced telecommunications in rural America, the challenge of bringing broadband communications to all of America, *http://www.ntia.doc.gov/reports/ruralbb42600.pdf*.

Von Burg, Urs, 2001, The Triumph of Ethernet: Technological Communities and the Battle for the Lan Standard, Stanford, CA: Stanford Univeristy Press.

Waldrop, M. M., 2002, The Dream Machine: J. C. R. Licklider and the Revolution that Made Computing Personal, Penguin Books.

Warf, B., 2001, Segueways into cyberspace: Multiple geographies of the digital divide, Environment and Planning B-Planning & Design, 28(1), 3–19.

Weinberg, J., 1999, The Internet and telecommunications services, access charges, universal service mechanisms, and other flotsam of the regulatory system, Yale Journal of Regulation, 16, Spring, 325–366.

Weinberg, J., 2001, Internet telephony regulation, in L. McKnight, W. Lehr and D. Clark, eds., Internet Telephony, MIT Press.

Werbach, K., 1997, A digital tornado: The Internet and telecommunications policy, FCC, Office of Planning and Policy Working Paper 29, March.

Wheeler, D. C. and M. E. O'Kelly, 1999, Network topology and city accessibility of the commercial Internet, Professional Geographer, 51(3), 327–339.

Woroch, G., 2001, Local network competition, in M. E. Cave, S. Majumdar and I. Vogelsang, eds., Handbook of Telecommunications Economics, Elsevier, 642–715.

Zolnierek, J., J. Eisner and E. Burton, 2001, An empirical examination of entry patterns in local telephone markets, Journal of Regulatory Economics, 19(2), 143–159.

Zook, M. A., 1998, The web of consumption: The spatial organization of the Internet industry in the United States, Paper presented at the Association of Collegiate Schools of Planning 1998 Conference, CA: Pasadena.

Zook, M. A., 2000, The web of production: The economic geography of commercial Internet content production in the United States, Environment and Planning A, 32(3), 411–426.

Zook, M. A., 2001, Old hierarchies or new networks of centrality? The global geography of the internet content market, American Behavioral Scientist, 44(10), 1679–1696.

Zook, M. A., 2002, Being connected is a matter of geography, netWorker, 5(3), 13–16.

Chapter 9

THE ECONOMICS OF THE INTERNET BACKBONE

NICHOLAS ECONOMIDES

New York University, New York

Contents

Handbook of Telecommunications Economics, Volume 2, Edited by S. Majumdar et al.

DOI: 10.1016/S1569-4054(05)02009-9

1. Competition among Internet backbone service providers

1.1. Internet backbone services

The Internet is a global network of interconnected networks that connect computers. The Internet allows data transfers as well as the provision of a variety of interactive real-time and time-delayed telecommunications services. Internet communication is based on common and public protocols. Hundreds of millions of computers are presently connected to the Internet. Figure 1 shows the expansion of the number of computers connected to the Internet.

The vast majority of computers owned by individuals or businesses connect to the Internet through commercial Internet Service Providers (ISPs)[1]. Users connect to the Internet either by dialing their ISP, connecting through cable modems, residential DSL, or through corporate networks. Typically, routers and switches owned by the ISP send the caller's packets to a local Point of Presence (POP) of the Internet[2]. Dial-up, cable modem, and DSL access POPs as well as corporate networks dedicated access circuits connect to high-speed hubs. High-speed circuits, leased from or owned by telephone companies, connect the high-speed hubs forming an 'Internet Backbone Network.' See Figure 2.

Backbone networks provide transport and routing services for information packets among high-speed hubs on the Internet. Backbone networks vary in terms of their geographic coverage. *Boardwatch magazine* has listed the following national backbones[3] in Table 1. Market shares of national backbones are listed in Table 2 based on a 1999 projection. In papers filed in support of the merger of SBC and AT&T as well as the merger of Verizon with MCI, there was mention of two recent traffic studies by RHK. These studies showing traffic for 2004, summarized in Table 3, show a dramatic change in the ranking of the networks, with AT&T now being first and MCI fourth. They also show that now a much bigger share of traffic (over 40 percent) is carried by smaller networks. These latest traffic studies show that the concern of the EU and the USDOJ that the Internet backbone market would tilt to monopoly were proved to be overstated.

1.2. Interconnection

There is wide variance of ISPs in terms of their subscriber size and the network they own. However, irrespective of its size, an ISP needs to interconnect with other

[1] Educational institutions and government departments are also connected to the Internet but do not offer commercial ISP services.

[2] Small ISPs may not own routers and switches, but rather just aggregate traffic at modem banks and buy direct access to a larger ISP.

[3] See http://www.boardwatch.com/isp/summer99/backbones.html. Boardwatch magazine also lists 348 regional backbone networks.

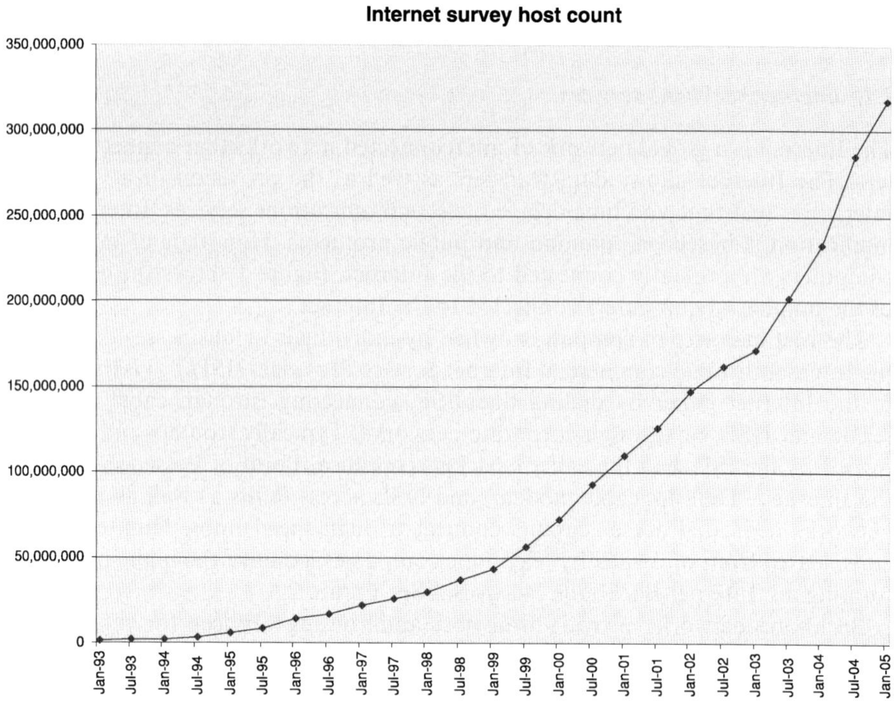
Internet survey host count
350,000,000
300,000,000
250,000,000
200,000,000
150,000,000
100,000,000
50,000,000
0
Jan-93
Jul-93
Jan-94
Jul-94
Jan-95
Jul-95
Jan-96
Jul-96
Jan-97
Jul-97
Jan-98
Jul-98
Jan-99
Jul-99
Jan-00
Jul-00
Jan-01
Jul-01
Jan-02
Jul-02
Jan-03
Jul-03
Jan-04
Jul-04
Jan-05

Fig. 1.

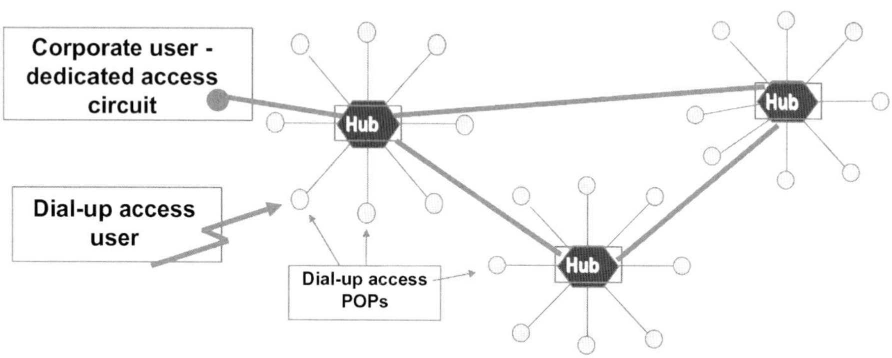
Corporate user - dedicated access circuit
Hub
Hub
Hub
Dial-up access user
Dial-up access POPs

Fig. 2.

Table 1
Partial list of national Internet backbones

@Home Network	Intermedia Business Internet
1 Terabit	Internet Access/GetNet
Abovenet	Internet Services of America
Apex Global Information Services (AGIS)	IXC Communications, Inc
AT&T Networked Commerce Services	Level 3
Cable & Wireless, USA	MCI WorldCom—Advanced Networks
CAIS	MCI WorldCom—UUNET
Concentric	NetRail
CRL Network Services	PSINet, Inc.
Digital Broadcast Network Corp.	Qwest/Icon CMT
Electric Lightwave	Rocky Mountain Internet/DataXchange
EPOCH Networks, Inc.	Savvis Communications Corporation
e.spire	ServInt
Exodus	Splitrock Services
Fiber Network Solutions	Sprint IP Services
Frontier Global Center	Teleglobe
Globix	Verio
GTE Internetworking	Visinet
GST Communications	Vnet
IBM Global Services	Winstar/Broadband
ICG/Netcom Online	ZipLink
IDT Internet Services	

Table 2
Market shares of national Internet backbones

Market Share	1997	1999	2001 (projected in 1999)	2003 (projected in 1999)
MCI WorldCom	43%	38%	35%	32%
GTE-BBN	13%	15%	16%	17%
AT&T	12%	11%	14%	19%
Sprint	12%	9%	8%	7%
Cable & Wireless	9%	6%	6%	6%
All Other	11%	21%	22%	19%
Total	100%	100%	100%	100%

Note: *Hearing on the MCI WorldCom-Sprint Merger Before the Senate Committee on the Judiciary*, Exhibit 3 (Nov 4, 1999) (Testimony of Tod A. Jacobs, Senior Telecommunications Analyst, Sanford C. Bernstein & Co., Inc.), Bernstein Research, *MCI WorldCom* (March 1999) at p. 51.

ISPs so that its customers will reach all computers/nodes on the Internet. That is, interconnection is necessary to provide universal connectivity on the Internet, which is demanded by users. Interconnection services at Network Access Points

Table 3
Carrier traffic in petabytes per month in 2004

Company	Traffic				Market share among all networks
	1Q2004	2Q2004	3Q2004	4Q2004	4Q2004
A (AT&T)	37.19	38.66	44.54	52.33	12.58%
B	36.48	36.50	41.41	51.31	12.33%
C	34.11	35.60	36.75	45.89	11.03%
D (MCI)	24.71	25.81	26.86	30.87	7.42%
E	18.04	18.89	21.08	25.46	6.12%
F	16.33	17.78	17.47	19.33	4.65%
G	16.67	15.04	14.93	15.19	3.65%
Total traffic top 7 networks	183.53	188.28	203.04	240.38	57.78%
Total traffic all networks	313	313	353	416	100%

Note: Data from *RHK Traffic Analysis – Methodology and Results*, May 2005. The identities of all networks are not provided, but it is likely that B, C, E, and F are Level 3, Quest, Sprint, and SBC in unknown order.

(NAP) and Metropolitan Area Exchanges (MAEs)[4] are complementary to Internet transport. In a sense, the Internet backbone networks are like freeways and the NAPs like the freeway interchanges.

Internet networks in two ways:

1. Private bilateral interconnection; and
2. Interconnection at public NPAs.

Private interconnection points and public NAPs are facilities that provide collocation space and a switching platform so that networks are able to interconnect. Network Access Points' services are not substitutes for ISP, or for transport services. Rather, they are a complement to ISP services and to transport services. The NAPs allow networks to interconnect more easily by providing the necessary space and platform.

Interconnection at NAPs is governed by bilateral contracts of the parties. Some NAPs, such as the London Internet Exchange (LINX) facilitate such negotiations by posting a set of common rules and standard contracts, which may be used by its members in their bilateral negotiations. Interconnection of two networks X and Y at a NAP is governed by a contract between networks X and Y. Other NAPs such as the ones owned by MCI do not dictate the terms of contracts between third-party networks[5].

[4] The NAPs run by MCI are called Metropolitan Area Exchanges (MAEs).

[5] In particular, interconnection at a NAP owned or controlled, for example, by MCI, does not imply or require a barter (peering) or transit arrangement between UUNET and networks X and Y.

Table 4
MAEs' capacity growth and utilization

	Capacity (Gbps)			Sales (Gbps)
	1997	1999	January 2000	January 2000
MAE-East	7.6	11.2	19.9	11.4
MAE-West	4.3	11.2	19.9	11.8
MAE-Dallas	N/A	7.5	7.5	2.6

Recently, there has been a significant increase in the number of NAPs as well as expansion and renewal of preexisting NAPs. In 1995, there were only 5 NAPs, MAE East, MAE West, NY (Sprint), Chicago (Ameritech), and Palo Alto (PacBell). In 1999, there were 41 NAPs in the United States (including 5 MAEs), and 40 European NAPs (including 2 MAEs) and 27 Asia-Pacific NAPs[6]. Table 4 shows the capacity expansion of NAPs from 1997 to January 2000. The fifth column of Table 4 shows capacity in January 2000. It is evident that there is very significant spare capacity. A partial list of NAPs in North America and the rest of the world is provided by the Exchange Point Network at http://www.ep.net/ep-main.html[7].

1.3. The transit and peering payment methods for connectivity

Internet networks have contracts that govern the terms under which they pay each other for connectivity. Payment takes two distinct forms: (i) payment in dollars for

[6] Source http://www.ep.net.

[7] The exchange point information net at http://www.ep.net/naps_na.html lists the following NAPs in North America: East Coast: ATL-NAP Atlanta; BNAP – Baltimore NAP; Louisville-nap.net; MAGPI – a Mid Atlantic Gigapop for Internet2; MassachusettsIX; NY6iX – A New York IPv6 exchange; NYIIX – New York International Internet Exchange (Telehouse); Nashville Regional Exchange Point; Nap of the Americas; MetroIX; Philadelphia Internet Exchange; Pittsburgh Internet Exchange; Research Triangle Park; Sprint NAP (Pennsauken NJ); Vermont ISP Exchange; Blacksburg Electronic Village – VA. West Coast: AMAP – Anchorage Metropolitan Access Point; Ames Internet Exchange; COX – Central Oregon Internet Exchange; HIX – Hawaii Internet Exchange; LAIIX – Telehouse Los Angeles; LAAP – A Los Angeles Exchange, includes MAE-LA; Northwest Access Exchange – Portland; OIX – Oregon Internet Exchange; PACIFIC WAVE – Pacific Wave Exchange; SBC-Oakland; SD-NAP – San Diego (Caida); SIX – Seattle Internet Exchange. The South: New Mexico Internet Exchange; IX New Mexico; TTI – The Tucson Interconnect; Yellowstone RIE. The Middle American Exchange Points: CMH-IX – Columbus Internet Exchange; D-MIX – Dayton OH; DIX – Denver Internet Exchange; IndyX – Indianapolis Data Exchange; Nashville CityNet; Ohio Exchange; RMIX Rocky Mountain Internet eXchange; SBC-Chicago STAR TAP (12 GigaPOP); St. Louis, Mo.; Utah REP. Canada: BC Gigapop; CA/NAP Canada/Toronto Exchange; CANIX: Originally CA*net Sponsored; MIX – Montreal Internet Exchange; The Nova Scotia Internet eXchange; Ottawa Internet eXchange; Toronto Internet Exchange.

"transit"; and (ii) payment in kind (i.e., barter, called 'peering'). Connectivity arrangements among ISPs encompass a seamless continuum, including ISPs that rely exclusively on transit to achieve connectivity, ISPs that use only peering to achieve connectivity, and everything in between. Although there are differences between transit and peering in the specifics of the payments method, and transit includes services to the ISP not provided by peering, it should be made clear that these two are essentially alternative payment methods for connectivity[8]. The transport and routing that backbone networks offer do not necessarily differ depending on whether cash (transit) or barter (peering) is used for payment. The same transport and routing between customers of the two networks can be obtained by purchase, or through barter for other transport services.

Under transit, a network X connects to network Y with a pipeline of a certain size, and pays network Y for allowing X to reach all Internet destinations. Under transit, network X pays Y to reach not only Y and its peers, *but also any other network*, such as network Z by passing through Y, as in the diagram below.

X-----------Y-=-=-=-=-=-Z

Under peering, two interconnecting networks agree not to pay each other for carrying the traffic exchanged between them as long as the traffic originates and terminates in the two networks. Referring to the diagram above, if X and Y have a peering agreement, they exchange traffic without paying each other as long as such traffic terminating on X originates in Y, and traffic terminating on Y originates in X. If Y were to pass to X traffic originating from a network Z that was not a customer of Y, Y would have to pay a transit fee to X (or get paid a transit fee by X, i.e., it would not be covered by the peering agreement between X and Y).

Although the networks do not exchange money in a peering arrangement, the price of the traffic exchange is not zero. If two networks X and Y enter into a peering agreement, it means that they agree that the cost of transporting traffic from X to Y and vice versa that is incurred within X is roughly the same as the cost of transporting traffic incurred within Y. These two costs have to be roughly equal if the networks peer, but they are not zero.

The decision as to whether interconnection takes the form of peering or transit payment is a commercial decision. Peering is preferred when the cost incurred by X for traffic from X to Y and Y to X is roughly the same as the cost incurred by Y for the same traffic. If not, the networks will use transit. As is explained below, the decision of whether to peer or not depends crucially on the geographic coverage of the candidate networks.

Generally, peering does not imply that the two networks should have the same size in terms of the numbers of ISPs connected to each network, or in terms of the

[8] Transit customers receive services, such as customer support, DNS services, etc., that peering networks do not receive.

traffic that each of the two networks generate[9]. If two networks, X and Y, are similar in terms of the types of users to whom they sell services, the amount of traffic flowing across their interconnection point(s) will be roughly the same, irrespective of the relative size of the networks. For example, suppose that network X has 10 ISPs and network Y has 1 ISP. If all ISPs have similar features, the traffic flowing from X to Y is generally equal to the traffic flowing from Y to X[10].

What determines whether a peering arrangement is efficient for both networks is the *cost* of carrying the mutual traffic within each network. This cost will depend crucially on a number of factors, including the geographic coverage of the two networks. Even if the types of ISPs of the two networks are the same as in the previous example (and therefore the traffic flowing in each direction is the same), the cost of carrying the traffic can be quite different in network X from network Y. For example, network X (with the 10 ISPs) may cover a larger geographic area and have significantly higher costs per unit of traffic than network Y. Then network X would not agree to peer with Y. These differences in costs ultimately would determine the decision to peer (barter), or receive a cash payment for transport.

Where higher costs are incurred by one of two interconnecting networks because of differences in the geographic coverage of each network, peering would be undesirable from the perspective of the larger network. Similarly, one expects that networks that cover small geographic areas will only peer with each other. Under these assumptions, who peers with whom is a consequence of the extent of a network's geographic coverage, and may not have any particular strategic connotation[11].

In summary, whether two interconnecting networks use peering (barter), or cash payment (transit) does not depend on the degree of competition among backbone services providers. In particular, the presence of peering is not necessarily a sign of intense or weak competition, nor would the replacement of peering by cash pricing necessarily be a sign of diminished or increased competition. Moreover, as the analysis above shows, generally, an ISP's decision not to peer reflects

[9] For example, MCI WorldCom has peering arrangements with a number of smaller networks. See Letter from Sue D. Blumenfeld, Attorney for Sprint Corporation, and A. Richard Metzger, Jr., Attorney for MCI WorldCom, Inc. to Magalie Roman Salas, FCC, CC Docket No. 99-333 (dated January 14, 2000) at p. 20.

[10] Suppose the larger network has 10 ISPs with 10 Websites per ISP and a total of 1000 users, and it interconnects with a smaller network with 1 ISP with 10 Websites and a total of 100 users. For simplicity, suppose that every user visits every Website. Then the smaller network transmits $100 \times 10 \times 10 = 10{,}000$ site-visits to the larger network, and the larger network transmits $1000 \times 1 \times 10 = 10{,}000$ site-visits to the smaller network. Thus, the traffic across networks of different sizes is the same if the types of ISPs and users are the same across networks.

[11] Milgrom et al. (2000) shows how peering (with no money changing hands) can emerge under some circumstances as an equilibrium in a bargaining model between backbones.

its assessment that the average costs of transport within one network are larger than the average costs of transport within the other network. Thus, refusal to peer is not inherently an anticompetitive act; it can be a consequence of some networks being much larger than others in terms of geographic coverage.

1.4. Conduct of Internet backbone service providers

1.4.1. Pricing of transport services in the backbone networks

The author first discusses business conduct of Internet backbone service providers. Structural conditions for Internet backbone services (discussed in the next section) ensure negligible barriers to entry and expansion and easy conversion of other transport capacity to Internet backbone capacity. As discussed in the next section, raw transport capacity as well as Internet transport capacity has grown dramatically in the last four years. Transport capacity is a commodity because of its abundance.

The business environment for Internet backbone services is competitive. Generally, ISPs buying transport services face flexible transit contracts of relatively short duration. Backbones do not impose exclusivity of service on their customers. For example, UUNET (MCI) does not require that it be the exclusive Internet transport provider to its ISP customers.

Often an ISP buys from a backbone bandwidth of a certain capacity that allows it to connect to the whole Internet (through a 'transit' payment). The bandwidth capacity and speed of the connecting pipe vary widely and depend on the demand for transport that an ISP wants to buy from a particular backbone. Price lists for various bandwidth capacities are printed in *Boardwatch magazine*. The strength of competition among the various backbone providers is evidenced in the small, or nonexistent differences in the prices for various bandwidth capacities. For example, Table 5 shows the prices for AT&T and UUNET (MCI) for various bandwidth capacities as reported by the latest edition of Boardwatch magazine (August 1999). Despite the fact that AT&T's backbone business was significantly smaller than UUNET's, their prices are identical for most bandwidths, and when they differ, the differences are very small. Many other providers of various sizes have very similar prices as reported in *Boardwatch magazine*[12].

As the expected growth of the Internet in the mid to late 1990s of 400 percent a year in terms of bits transferred was not realized in the post 1999 period, and instead a growth of only about 100 percent a year was realized, transit prices fell. As an example, Table 6 compares the AT&T prices for the same connectivity in 1999 and 2001.

[12] As *Boardwatch Magazine* reports in the 1999 and subsequent editions, prices for the same connectivity were very comparable for a large array of services among large IBPs.

Table 5
Comparison of early 1999 monthly prices of AT&T and UUNET (MCI) for U.S. DS3s (T3s)

Service	AT&T	UUNET	Price difference = UUNET-AT&T
Burstable 0–6 Mbps	$12,500	$12,000	$500
Burstable 6.01–7.5 Mbps	$14,000	$14,000	$0
Burstable 7.51–9 Mbps	$17,000	$17,000	$0
Burstable 9.01–10.5 Mbps	$19,000	$19,000	$0
Burstable 10.51–12 Mbps	$22,000	$22,000	$0
Burstable 12.01–13.5 Mbps	$26,000	$26,000	$0
Burstable 13.51–15 Mbps	$29,000	$29,000	$0
Burstable 15.01–16.5 Mbps	$32,000	$32,000	$0
Burstable 16.51–18 Mbps	$37,000	$37,000	$0
Burstable 18.01–19.5 Mbps	$43,000	$43,000	$0
Burstable 19.51–21 Mbps	$48,000	$48,000	$0
Burstable 21.01–45 Mbps	$55,000	$55,500	$500

Note: *Boardwatch Magazine's* Directory of Internet Service Providers, 11th Edition, 1999.

Table 6
Comparison of 1999 and 2001 monthly prices of AT&T for U.S. DS3s (T3s)

Service	Year 1999	Year 2001	Percentage price difference $(p_{2001}-p_{1999})/p_{2001}$
Burstable 0–6 Mbps	$12,500	$6550	–47.60%
Burstable 6.01–7.5 Mbps	$14,000	$8150	–41.79%
Burstable 7.51–9 Mbps	$17,000	$9250	–45.59%
Burstable 9.01–10.5 Mbps	$19,000	$10,150	–46.58%
Burstable 10.51–12 Mbps	$22,000	$11,050	–49.77%
Burstable 12.01–13.5 Mbps	$26,000	$11,950	–54.04%
Burstable 13.51–15 Mbps	$29,000	$12,850	–55.69%
Burstable 15.01–16.5 Mbps	$32,000	$13,600	–57.50%
Burstable 16.51–18 Mbps	$37,000	$14,350	–61.22%
Burstable 18.01–19.5 Mbps	$43,000	$15,100	–64.88%
Burstable 19.51–21 Mbps	$48,000	$15,850	–66.98%
Burstable 21.01–45 Mbps	$55,000	$31,050	–43.55%

Notes: *Boardwatch Magazine*'s Directory of Internet Service Providers, 11th and 13th Edition, 1999 and 2001.

1.4.2. ISP multihoming; Additional demand responsiveness to price changes

Internet Service Poroviders are not locked-in by switching costs of any significant magnitude. Thus, ISPs are in good position to change providers in response to any increase in price, and it would be very difficult for a backbone profitably to increase price. Moreover, a large percentage of ISPs has formal agreements that allow them to route packets through several backbone networks and are able to control the way the traffic will be routed (multihoming). Table 7 shows that, in

Table 7
Additional backbone connections held by multihoming ISPs

Year	# ISPs	Number of backbone connections sold to ISPs	Share of additional connections sold to multihoming ISPs
1997	4354	5739	24%
1998	4470	5913	24%
1999	5078	8950	43%

Note: *Boardwatch Magazine*'s Directory of Internet Service Providers, Fall 1997, p. 6. *Boardwatch Magazine*'s Directory of Internet Service Providers, Winter 1998, p. 5. *Boardwatch Magazine*'s Directory of Internet Service Providers, 11th Edition, 1999, p. 4. The last column is calculated as the difference between the third and the second columns divided by the third column, for example, for 1999, (8950–5078)/8950 = 43.26% rounded to 43%.

1999, additional (i.e., second or subsequent) connections sold to multihoming ISPs amounted to 43 percent of all ISP connections to backbones. One of the reasons for the increase in multihoming is likely the decrease in the cost. The cost of customer routers that are required for ISP multihoming has decreased from $10,000 to $2000–$3000[13]. An additional reason for an ISP to multihome is that it increases the ability of the ISP to route its traffic to the lowest-priced backbone, as discussed in the next section.

When an ISP reaches the Internet through multiple backbones, it has additional flexibility in routing its traffic through any particular backbone. A multihoming ISP can easily reduce or increase the capacity with which it connects to any particular backbone in response to changes in prices of transit. Thus, multihoming increases the firm-specific elasticity of demand of a backbone provider. Therefore, multihoming severely limits the ability of any backbone services provider to profitably increase the price of transport. Any backbone increasing the price of transport will face a significant decrease in the capacity bought by multihoming ISPs.

Large Internet customers also use multiple ISPs, which is called 'customer multihoming.' They have chosen to avoid any limitation on their ability to switch traffic among suppliers even in the very shortest of runs. Customer multihoming has similar effects as ISP multihoming in increasing the firm-specific elasticity of demand of a backbone provider and limiting the ability of any backbone services provider to profitably increase the price of transport.

New technologies of content delivery that utilize distributed storage of Web-based content on various locations on the Internet reduce the need for backbone network transport. 'Caching' stores locally frequently requested content. 'Mirroring' creates a replica of a Website. Intelligent content distribution,

[13] Source: *Boardwatch Magazine's* Directory of Internet Service Providers, 11th Edition, 1999.

implemented among others, by Akamai Technologies[14], places its servers closest to the end users inside an ISP's network. Intelligent content distribution technology assesses the fastest route on the Internet for content access, and delivers content faster to end users. Placing content delivery close to end users and optimizing content delivery through intelligent content distribution, caching, and mirroring reduces in effect the demand for Internet transport services and the ability of backbone providers to affect the transit price.

2. Structural conditions for Internet backbone services; Negligible barrier to entry and expansion

2.1. The markets for raw transport capacity and other inputs to Internet transport services

Almost all Internet transport uses fiber-optic transmission capacity, which is based on a well-known and easily available technology[15]. There are no significant barriers to entry in the supply of additional raw transmission capacity. Fiber transmission capacity is essentially fungible, and the same physical networks can be used for the transmission of voice, Internet traffic, and data by using different protocols.

Fiber that will not be needed by an Internet transport supplier can be leased, or sold for nonInternet uses. The same fiber and electronics are used for both circuit switched and packet switched networks, which can each transport both voice and data. Before construction, the operator has a completely open choice between creating either a circuit switched or a packet switched network. Only the interface differs between voice and data applications. Once capacity is in place, there are small costs of converting from one use to the other. Moreover, capacity can be upgraded in small steps so that fiber networks can respond flexibly to increasing capacity requirements.

Fiber capacity has grown rapidly and is expected to grow for the indefinite future. Because there is always new capacity in the planning stage, no operator needs to consider switching the use of existing capacity. As a result, fiber capacity is not in any way a barrier to entry in Internet transport[16].

[14] Akamai was founded in 1998 and made a $234M initial public offering in October 1999. Akamai has industry relationships with AT&T, BT plc, DIGEX, Global Center, GTEI, Lycos, Microsoft, PSINet, Qwest, Real Networks, Telecom Italia, Teleglobe, Universo Online, UUNET, and Yahoo!, among others.

[15] The transport and switching technologies are available from firms that do not sell backbone transport or ISP services.

[16] In the early stages of Internet expansion and given the explosive growth that was anticipated then, the possibility of a future backbone capacity shortage may have bid up the value of firms with installed Internet backbone capacity and may explain the price that WorldCom paid for MFS and implicitly UUNET. This should be seen in the context of a real options analysis. See Economides (1999a,b) and Hubbard and Lehr (2000).

In order to build or expand Internet backbone capacity, besides fiber-optic cable, networks need routers and switches. Routers and switches are readily available from a variety of third-party suppliers. Fiber capacity can be leased, and there is no shortage of capacity that would constrain the ability of smaller networks, or new entrants to expand capacity or enter the market. Fiber networks can add leased capacity, or increase their capacity by deploying new technologies such as Dense Wave Division Multiplexing (DWDM). The construction of fourth-generation fiber-optic networks, deploying the latest technology, promises an abundance of capacity that appears to be able to accommodate the very rapid growth in capacity demand that has been the hallmark of the Internet market to date.

2.2. Ease of expansion and entry

National, international, and regional long-haul fiber-optic transmission capacity has increased very rapidly, both as a result of expansion of networks of incumbents, such as AT&T, MCI, Sprint, and GTE but also as a result of entry of a number of carriers that created new networks, including Quest, Level 3, Williams, and others. The FCC's *Fiber Deployment Update* reports that total fiber system route miles of interexchange carriers increased by two-thirds between 1994 and 1998[17]. After 1998, the FCC discontinued the publication of this report. However, data reported by Besen and Brenner (2000)[18] and Hogendorn (2004) supports the conclusion that the capacity of long-haul fiber is increasing in an accelerated rate.

As evidence of ease of entry, the number of North American ISPs more than tripled in the years 1996–1999, and has continued thereafter. The number of North American backbone providers has grown almost fivefold in the same period. These statistics are shown in Tables 8 and 9.

Bandwidth and equipment costs have decreased and continue to decrease. Hence, access to fiber capacity is unlikely to be an impediment to sellers wishing to upgrade their networks, or to new competitors wishing to enter the market.

2.3. Public standards and protocols on the Internet

In markets, where the incumbent has a proprietary standard and an entering rival must promote an incompatible alternative standard—as in operating systems for personal computers—standards can be used to create a barrier to entry. However, in markets where all rivals use the same public standard, no such barrier exists, or

[17] See Jonathan M. Kraushaar, *Fiber Deployment Update: End of Year 1998*, FCC, Industry Analysis Division, Common Carrier Bureau, Table 1.

[18] *See* Declaration of Stanley Besen and Steven Brenner, March 20, 2000.

Table 8
Growth of ISP industry

Number of North American ISPs	Date
1447	February 1996
2266	May 1996
3747	April 1997
4354	October 1997,
5078	1999

Note: *Boardwatch magazine*, Fall 1997, and 11th Edition, Fall 1999.

Table 9
Growth of U.S. national backbone operators

Number of U.S. national backbone operators	Date
9	Summer 1996
22	May 1997
37	Fall 1997,
43	1999

Note: *Boardwatch Magazine*, Fall 1999. *Boardwatch* acknowledges excluding backbone providers from its directory, which otherwise would have brought the total to 47.

can be created. Rather, the use of a single standard can support unlimited numbers of rivals, as in the market for household fax and telephone appliances today.

The Internet is based on open and public standards and protocols, which are outside the control of any one of the incumbent network operators. These are vital for keeping traffic running smoothly among the extraordinary number of networks comprising the Internet and the diverse mixture of hardware employed by different providers. There is no danger that proprietary standards will emerge in the future since there are well-established mechanisms for extending Internet standards. A proposed new Internet standard 'undergoes a period of development and several iterations of review by the Internet community and revision based upon experience[19] before it is adopted as a standard and published. This whole process takes place under the auspices of the Internet Society, a nonprofit body, managed by the Internet Architecture Board and the Internet Engineering

[19] Scott Bradner, *The Internet Standards Process*, revision 3, Network Working Group (ftp://ftp.isi.edu/in-notes/rfc2026.txt), Section 1.2.

Steering Group, and conducted by the Internet Engineering Task Force. In considering changes in standards, these groups require mandatory disclosure of any proposed change before it gets considered, so no proprietary standard can be introduced[20].

3. Potential for anticompetitive behavior on the Internet backbone

Some have proposed[21] that the existence of network effects creates a grave danger that the Internet backbone will quickly become monopolized once the largest Internet backbone provider becomes 'large enough.' Various theories have been proposed of how this could be done. The author first discusses the general context in which network effects affect competition on the Internet, and subsequently discusses the specific theories.

4. Network externalities and the Internet

Like any network, the Internet exhibits network externalities. Network externalities are present when the value of a good or service to each consumer rises as more consumers use it, everything else being equal[22]. In traditional telecommunications networks, the addition of a customer to the network increases the value of a network connection to all other customers, since each of them can now make an extra call. On the Internet, the addition of a user potentially

1. adds to the information that all others can reach;
2. adds to the goods available for sale on the Internet;
3. adds one more customer for e-commerce sellers; and
4. adds to the collection of people who can send and receive e-mail, or otherwise interact through the Internet.

Thus, the addition of an extra computer node increases the value of an Internet connection to each connection.

In general, network externalities arise because high sales of one good make complementary goods more valuable. Network externalities are present not only in traditional network markets, such as telecommunications, but also in many other markets. For example, an IBM-compatible PC is more valuable if there are

[20] "No contribution that is subject to any requirement of confidentiality or any restriction on its dissemination may be considered in any part of the Internet Standards Process, and there must be no assumption of any confidentiality obligation with respect to any such contribution." *Id*, Section 10.2.

[21] See Cremer, Rey, and Tirole (1998, 2000).

[22] See Economides (1996a, b), Farrell and Saloner (1985), Katz and Shapiro (1985), and Liebowitz and Margolis (1994, 2002).

more compatible PCs sold because, then, there will be more software written and sold for such computers.

In networks of interconnected networks, there are large social benefits from the interconnection of the networks and the use of common standards. A number of networks of various ownership structures have harnessed the power of network externalities by using common standards. Examples of interconnected networks of diverse ownership that use common standards include the telecommunications network, the network of fax machines, and the Internet. Despite the different ownership structures in these three networks, the adoption of common standards has allowed each one of them to reap huge network-wide externalities.

For example, users of the global telecommunications network reap the network externalities benefits, despite its fragmented industry structure. If telecommunications networks were not interconnected, consumers in each network would only be able to communicate with others on the same network. Thus, there are strong incentives for every network to interconnect with all other networks so that consumers enjoy the full extent of the network externalities of the wider network.

The Internet has very significant network externalities. As the variety and extent of the Internet's offerings expand, and as more customers and more sites join the Internet, the value of a connection to the Internet rises. Because of the high network externalities of the Internet, consumers on the Internet demand universal connectivity (i.e., to be able to connect with every Website) on the Internet and to be able to send electronic mail to anyone. This implies that every network must connect with the rest of the Internet in order to be a part of it.

The demand for universal connectivity on the Internet is stronger than the demand of a voice telecommunications customer to reach all customers everywhere in the world. In the case of voice, it may be possible but very unlikely that a customer may buy service from a long-distance company that does not include some remote country because the customer believes that it is very unlikely that he/she would be making calls to that country. On the Internet however, one does not know where content is located. If company A did not allow its customers to reach region B or customers of a different company C, customers of A would never be able to know or anticipate what content they would be missing. Thus, consumers' desire for Internet universal connectivity is stronger than in voice telecommunications. Additionally, because connectivity on the Internet is two-way, a customer of company A would be losing exposure of his/her content (and the ability to send and receive e-mails) to region B and customers of company C. It will be difficult for customer A to calculate the extent of the losses accrued to him/her from such actions of company A. Thus, again, customers on the Internet require universal connectivity[23].

The existence of common interconnection standards and protocols in the telecommunications and the network of fax machines have guaranteed that no service provider or user can utilize the existence of network externalities to create and use monopoly power. Similarly, the existence of common and public interconnection standards on the Internet guarantees that no service provider or user can utilize the existence of network externalities to create and use monopoly power based on proprietary standards. With competitive organization of the Internet's networks, the rising value is shared between content providers and telecommunications services providers (in the form of profits) and end users (in the form of consumer surplus).

4.1. Procompetitive consequences of network externalities

The presence of network externalities does not generally imply the existence of monopoly power. Where there are network externalities, adding connections to other networks and users adds value to a network, so firms have strong incentives to interconnect fully and to maintain interoperability with other networks. Thus, network externalities can act as a strong force to promote competition for services based on interconnected networks[24]. For example, various manufacturers compete in producing and selling fax machines that conform to the same technical standards, and are connected to the ever-expanding fax network. It would be unthinkable that a manufacturer, however large its market share, would decide to produce fax machines for a different fax network that would be incompatible with the present one. In contrast, firms would like to conform to existing standards and fully interconnect to a network so that they reap the very large network externalities of the network.

The incentive to interconnect and to conform to the same standard applies similarly to competitive firms as it applies to firms with market power. Although, as in other markets, firms involved in network businesses may sometimes have

[23] If universal connectivity were not offered by a backbone network, a customer or its ISP would have to connect with more than one backbone. This would be similar to the period 1895–1930 when a number of telephone companies run disconnected networks. Eventually most of the independent networks were bought by AT&T, which had a dominant long-distance network. The refusal of AT&T to deal and interconnect with independents was effective because of three key reasons: (i) AT&T controlled the standards and protocols under which its network ran, (ii) long-distance service was provided exclusively by AT&T in most of the United States, and (iii) the cost to a customer of connecting to both AT&T and an independent was high. None of these reasons apply to the Internet. The Internet is based on public protocols. No Internet backbone has exclusive network coverage of a large portion of the United States. Finally, connecting to more than one backbone (multihoming) is a common practice by many ISPs and does not require big costs. And ISPs can interconnect with each other through secondary peering as explained later. Thus, the economic factors that allowed AT&T to blackmail independents into submission in the first three decades of the 20th century are reversed in today's Internet, and therefore would not support a profitable refusal to interconnect by any backbone.

[24] See also Faulhaber (2004).

market power, that power does not arise automatically from the network, even in the presence of externalities.

4.2. Conditions under which network externalities may inhibit competition

In markets with network externalities, firms may create bottleneck power by using proprietary standards. A firm controlling a standard needed by new entrants to interconnect their networks with the network of the incumbent may be in a position to exercise market power[25]. Often a new technology will enter the market with competing incompatible standards. Competition among standards may have the snowball characteristic attributed to network externalities.

For example, VHS and Beta, two incompatible proprietary standards for video cassette recorders (VCRs), battled for market share in the early 1980s. Because Sony, the sponsor of the Beta standard, chose a pricing and licensing strategy that did not trigger the snowball effect; VHS was the winner. In particular, Sony refused to license its Beta standard, while VHS was widely licensed. Even though VHS was the winning standard, the market for VCRs did not become a monopoly since there are a number of suppliers of VHS-type video equipment. Thus, a standard may be licensed freely or at a low cost, and therefore the existence of a proprietary standard does not preclude competition. Moreover, in many cases a sufficiently open licensing policy will help to win the standards battle, and may therefore be in the interest of the owner of the standard to freely license even its proprietary standards[26].

Economics literature has established that using network externalities to affect market structure by creating a bottleneck requires three conditions[27]:

1. Networks use proprietary standards;
2. No customer needs to reach nodes of or to buy services from more than one proprietary network;
3. Customers are captives of the network to which they subscribe and cannot change providers easily and cheaply.

First, without proprietary standards, a firm does not have the opportunity to create the bottleneck. Second, if proprietary standards are possible, the development of proprietary standards by one network isolates its competitors from network benefits, which then accrue only to one network. The value of each proprietary network is diminished when customers need to buy services from more than one network. Third, the more consumers are captive and cannot easily and economically change providers, the more valuable is the installed base to any

[25] See Economides (2003).
[26] See Economides (1996b).
[27] See Economides (1996a, 1989), Farrell and Saloner (1985), and Katz and Shapiro (1985).

proprietary network. The example of snowballing network effects just mentioned—VHS against Beta—fulfills these three conditions. The next section shows that these conditions fail in the context of the Internet backbone.

5. Network externalities and competition on the Internet

5.1. *Conditions necessary for the creation of bottlenecks fail on the Internet*

The Internet fails to fulfill any of the three necessary conditions under which a network may be able to leverage network externalities and create a bottleneck. First, there are no proprietary standards on the Internet, so the first condition fails. The scenario of standards wars is not at all applicable to Internet transport, where full compatibility, interconnection, and interoperability prevail. For Internet transport, there are no proprietary standards. There is no control of any technical standard by service providers and none is in prospect. Internet transport standards are firmly public property[28]. As a result, any seller can create a network complying with the Internet standards—thereby expanding the network of interconnected networks—and compete in the market.

In fact, the existence and expansion of the Internet and the relative decline of proprietary networks and services, such as CompuServe, can be attributed to the conditions of interoperability and the tremendous network externalities of the Internet. America On Line (AOL), CompuServe, Prodigy, MCI, and AT&T folded their proprietary electronic mail and other services into the Internet. Microsoft, thought to be the master of exploiting network externalities, made the error of developing and marketing the proprietary Microsoft Network (MSN). After that product failed to sell, Microsoft relaunched MSN as an ISP, adhering fully to the public Internet standard. This is telling evidence of the power of the Internet standard and demonstrates the low likelihood that any firm can take control of the Internet by imposing its own proprietary standard.

Second, customers on the Internet demand *universal connectivity*, so the second condition fails. Users of the Internet do not know in advance what Internet site they may want to contact, or to whom they might want to send e-mail. Thus, Internet users demand from their ISPs and expect to receive universal connectivity. This is the same expectation that users of telephones, mail, and fax machines have: that they can connect to any other user of the network without concern about compatibility, location, or, in the case of telephone or fax, any concern about the manufacturer of the appliance, the type of connection (wireline or wireless), or the owners of the networks over which the connection is made. Because of the users' demand for universal connectivity, ISPs providing services

[28] See Kahn and Cerf (1999) and Bradner, *The Internet Standards Process*, revision 3, Network Working Group (ftp://ftp.isi.edu/in-notes/rfc2026.txt), Section 1.2.

to end users or to Web sites must make arrangements with other networks so that they can exchange traffic with *any* Internet customer.

Third, there are no captive customers on the Internet, so the third condition fails, for a number of reasons:

1. ISPs can easily and with low cost migrate all, or part of their transport traffic to other network providers;
2. Many ISPs already purchase transport from more than one backbone to guard against network failures, and for competitive reasons (ISP 'multihoming');
3. Many large Web sites/providers use more than one ISP for their sites ('customer multihoming');
4. Competitive pressure from their customers makes ISPs agile and likely to respond quickly to changes in conditions in the backbone market.

5.2. Bottlenecks such as the ones of the local exchange telecommunications network do not exist on the Internet

There are significant differences between local telephone networks and the Internet, which result in the existence of bottlenecks in local telephone markets and lack of bottlenecks on the Internet. Until the passage of the Telecommunications Act of 1996, the local telephone company had a legal franchise monopoly over local telephony in its territory in most States. Most importantly, the local telephone company monopolizes the fixed wireline connection to customers, especially the residential ones, thereby controlling the bottleneck for access to customers. Such a bottleneck does not exist on the Internet backbone. A number of reasons contribute to this:

1. the cost of connecting an ISP to the rest of the Internet is very low compared to the cost of connecting every house to local telephone service;
2. the location of an ISP is not predetermined, but can be placed most conveniently within a geographic area;
3. the elasticity of supply for Internet transport services is high (i.e., there are no barriers to expansion);
4. there are negligible barriers to entry on the Internet; and
5. Internet demand growth and expansion are exponential, driven by expanding market and geographic penetration, and by the introduction of new applications[29].

The only bottleneck in the Internet arises out of the control of the first/last mile of the local telecommunications network, by incumbent local exchange

[29] Demand grows yearly at about 100 percent. The number of North American ISPs more than tripled in 3 years. Up to 2003, demand growth has been overestimated as 400 percent per year both by the U.S. government and most providers of backbone connectivity, including MCI-WorldCom.

carriers, since this first/last mile is used by the majority of users to connect to the Internet.

In summary, an analysis of network externalities shows that network effects cannot create barriers to entry for new networks on the Internet or barriers to expansion of existing ones. It is also showed that network effects on the Internet do not create a tendency to dominate the market or tip it toward monopoly. On the contrary, network effects are a *procompetitive force* on the Internet, providing strong incentives to incumbents to interconnect with new entrants. In the next sections is discussed in detail the competition on the Internet.

6. Strategies that a large IBP might pursue

There are two main ways in which a large Internet backbone connectivity provider could attempt to exercise market power and harm consumers:

1. **Price increases**. It could raise the price of network services across-the-board to all customers, including replacing peering with transit sold at a high price; alternatively, it could selectively increase price to one or few networks;
2. **Raising rivals' costs or degrading interconnection without changing price(s)**. It could selectively degrade the quality of interconnections with competing networks, in an effort to make their networks less attractive and divert traffic to itself.

As is explained in the next section, neither of these courses of action is likely to be profitable on the Internet backbone.

6.1. Raising the price of transport

The simplest exercise of market power by a large firm would be to raise the price of its transport services. In addition, the company might refuse to continue peering with some networks and to charge them transit fees instead. The ability of a company to profitably depeer other networks is equivalent to the ability of a company to increase the price of transport. Depeering does not mean cutting off a customer from the network or charging an infinite price to the customer; it does not mean refusal to deal. A price increase would create profit opportunities for the large internet backbone provider's (IBP's) rivals in the transport market, and is also likely to induce entry.

Internet backbone providers sell transport as a bandwidth of a certain capacity that allows an ISP to connect to the whole Internet. If a large Internet backbone connectivity provider were to increase the prices it charges to ISPs for such capacity, ISPs would promptly switch to other backbone providers. Thus, an increase in transit price by a large IBP would decrease its sales sufficiently to make such a price increase unprofitable.

ISP connections to multiple backbones are very common. Forty-three percent of all ISP connections to backbones were sold as *additional* connections to ISPs who connected to more than one backbone. A multihoming ISP can easily and at a low cost limit the size of its purchases from an IBP that increases the price of transport. Thus, the presence of multihoming increases the firm-specific elasticity of demand of IBP transport services and creates a bigger demand response to IBP price increases. This makes it even more likely that the firm-specific demand response to a price increase will be sufficiently negative to render a contemplated price increase unprofitable.

If the large Internet backbone connectivity provider's strategy were to impose equal increases in transport costs on all customers, the response of other backbone providers and ISPs will be to reduce the traffic, for which they buy transit from the large IBP and to instead reroute traffic and purchase more transit from each other. Thus, in response to a price increase by the large Internet backbone connectivity provider, other IBPs and ISPs reduce the traffic for which they buy transit from the large IBP down to the minimum level necessary to reach ISPs that are *exclusively* connected to the large IBP. All other IBPs and ISPs exchange all other traffic with each other bypassing the large IBP network.

Figures 3 and 4 show the typical reaction of an increase in the price of a large IBP, and illustrate why the strategy of increasing price is unprofitable. Consider, for example, a situation where, prior to the price increase, 4 ISPs (1 to 4) purchase transit from IBP 0, which considers increasing its price. Two of these ISPs (ISP 2 and ISP 3) peer with each other. This is illustrated in Figure 3. Internet Service Provider 1 and ISP 4 buy transit capacity for all their traffic to IBP 0, and the other 3 ISPs, whereas, ISP 2 and ISP 3 buy transit capacity for all their traffic to ISP 0, ISP 1, and ISP 4.

Now suppose that, IBP 0 increases its transit price. In response, ISP 1 and ISP 4 decide to reduce the traffic for which they buy transit from IBP 0, and instead to reroute some of their traffic and purchase more transit from ISP 2 and ISP 3, respectively. See Figure 3. Because of the peering relationship between ISP 2 and ISP 3, all traffic from ISP 1 handed to ISP 2 will reach ISP 3 as well as ISP 4, who is a customer of ISP 3. Similarly, by purchasing transit from ISP 3, ISP 4 can reach all the customers of ISP 1, ISP 2, and ISP 3. Thus, in response to the price increase of IBP 0, each of the ISPs 1, 2, 3, and 4 will reduce the amount of transit purchased from the IBP 0. Specifically, each of the ISPs buys from IBP 0 only capacity sufficient to handle traffic to the customers of network 0. This may lead to a considerable loss in revenues for IBP 0, rendering the price increase unprofitable. The big beneficiaries of the price increase of IBP 0 are peering ISPs 2 and 3, who now start selling transit to ISPs 1 and 4, respectively, and become larger networks.

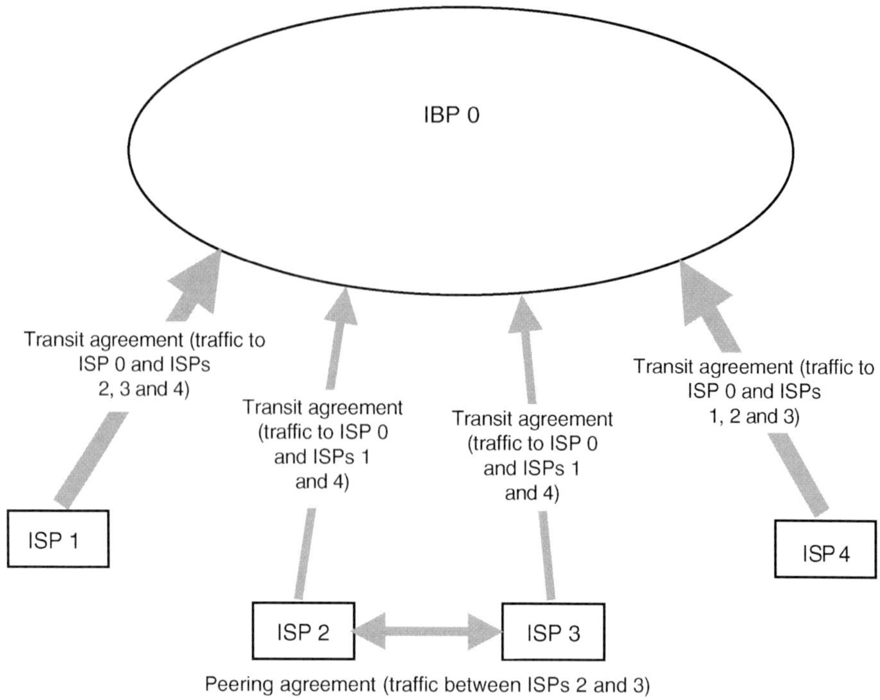

Fig. 3.

In response to a price increase by the large IBP, rivals would be able to offer their customers universal connectivity at profitable prices below the large IBP's prices. In the scenario described in the example above, market forces, responding to a price increase by a large network, reroute network traffic so that it is served by rival networks, except for the traffic to and from the ISPs connected exclusively with the large network. The rivals purchase the remaining share from the large IBP in order to provide universal connectivity. Thus, the rivals' blended cost would permit them to profitably offer all transport at prices lower than the large IBP's prices, but above cost.

A direct effect of the increase in price by the large network is that: (i) ISPs who were originally exclusive customers of the large IBP would shift a substantial portion of their transit business to competitors and (ii) ISPs that were not exclusive customers of the large IBP would also shift a significant share of their transit business to competitors' networks, keeping the connection with the large

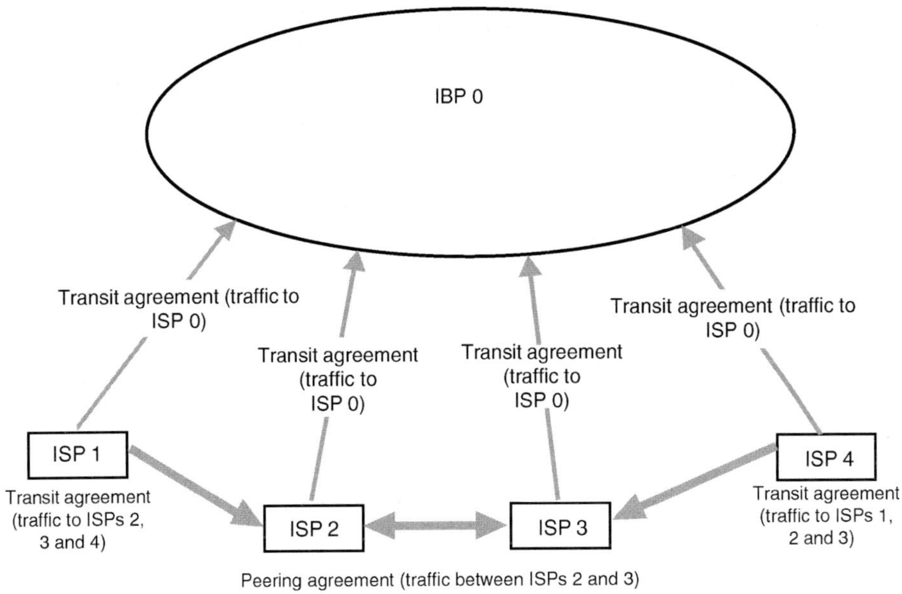

Fig. 4.

IBP only for traffic for which alternate routes do not exist, or for cases of temporary failure of the rivals' networks.

6.2. Discriminatory price increases directed simultaneously against all backbone rivals

Here I consider the possibility that a large IBP might try to displace its rivals by charging them more than it charges ISPs who are not rivals in the transport business. It is believed that this form of price discrimination is particularly unlikely. Of all customers, rivals in the transport business—major backbones and smaller regional networks—are the best positioned to avoid the use of the large IBP's network if it is more expensive than the alternatives. Even the smaller rivals are large enough that the transactions costs of establishing alternative connections are unimportant in relation to the cost increases for transport that could be avoided by making new deals.

6.3. *Raising rivals' costs and degrading connectivity*

Alternatively one may consider the possibility that the large IBP would find it profitable to raise the nonprice costs of rivals by reducing the connectivity it provides with other IBPs[30]. 'The first observation regarding the 'raising rivals' cost' or 'degradation' strategy applied to clients is that as a matter of economics, it is *always* preferable to a firm to increase price rather than increase the nonprice costs of rivals. A firm can choose a price increase that will have the same effect as increasing the costs (or reducing the benefits) of its clients, and it is able to collect extra revenue through the price increase while, if it just degrades the product it receives no extra revenue. Another difference from the traditional raising rivals cost theory is that on the Internet backbone often imposing a quality decrease on a rival simultaneously results in a quality decrease on the perpetrator. This is because the quality degradation affects both the services demanded by the target network (its clients connecting to the perpetrators network) as well as the services demanded by perpetrators' network (its clients connecting to the target networks). As discussed in detail in the next section, because of the network feature of connectivity, such a degradation cannot be confined to the target, but it also simultaneously affects negatively the perpetrator.

6.3.1. *Terminating interconnection simultaneously with all rivals (refusal to deal)*

The author first considers the extreme case in which a large IBP terminates interconnection with all rivals. This setup is equivalent to the large IBP increasing the rivals' price to infinity if they were to interconnect with it.

Termination by the large IBP of interconnection with a network customer has a bilateral effect. It prevents the other network's customers from reaching any customer of the large IBP, and it prevents the large IBP's customers from reaching any customers of the other network. Whatever the relative sizes of the two networks, customers of both networks are harmed. If the large IBP's network has more customers than the interconnecting network, then the termination strategy will affect the large IBP's network as much, or more than the interconnecting network.

Termination of interconnection would deny the large IBP's customers the universal connectivity sought by every customer, and would have devastating effects for the large IBP. Its customers—larger Web sites and the ISPs specializing in end-user services and Web hosting—would seek new transport providers to make up for the large IBP's inability to deliver universal connectivity. The loss of business is likely to make termination of interconnection highly unprofitable.

[30] Hausman (2000) at paragraph 53.

This is a good demonstration of the procompetitive effects of network externalities in the Internet. Each network, including a large network, has a more valuable product if it interconnects with other networks. Termination of interconnection would severely lower the value of the large network's service because it would shrink the connectivity the company offered.

6.3.2. Degrading interconnection simultaneously with all rivals

Alternatively, it has been suggested that a large IBP would degrade interconnection with all rivals without terminating service[31]. However, a large IBP could always make more profit by charging more for interconnection than by offering poor service. There is always a price level that has the equivalent harmful effect on customers as a program of degradation. The higher charge puts money in the seller's pocket immediately; degradation does not. Because, as the author has concluded, a large IBP would not find it profitable to raise transport charges; it follows immediately that it would suffer even more from degrading service.

In a similar vein to the earlier discussion, even if a large IBP decided to degrade interconnections rather than raise price, degrading interconnections would impose a cost on it that is comparable to the cost imposed on the rivals. In total, the large IBP's customers would experience the same level of degradation in terms of the traffic sent to, or received from, the other networks as would the other networks' customers.

Some have argued that the effects of degraded interconnections would be less severe for a large IBP than for the other networks because of the large IBP's size. In this line of argument, if the traffic is isotropic[32], a large number of Internet interactions will be within the network of the large IBP, and these interactions will be unaffected by degradation of interconnection. According to this theory, the rest of the Internet networks (with the smaller total number of customers if the large IBP has more than 50 percent of Internet customers) will suffer more than the larger network; it follows that the large IBP can then attract the customers of other networks[33].

This argument is based on the assumption that Internet users do *not* require universal connectivity. This, however, is factually incorrect. Internet users demand to be able to reach every node of the Internet, in a similar way that telecommunications customers demand that they be able to reach anyone connected to the telecommunications network, no matter where the receiving party is located, which local exchange carrier he/she subscribes to, and who carries the long-distance call.

[31] Hausman (2000) at paragraph 53.
[32] Isotropic traffic is generated when every user initiates the same number and type of Internet interactions with every other user.
[33] Hausman (2000) at paragraph 53.

Since users demand universal connectivity on the Internet, no network, however large, can afford not to offer universal connectivity. Therefore, no network would decide to degrade connections with the rest of the Internet networks unless the degrading network was certain that *all* ISPs *not* connected to it would immediately react to the degradation by instantaneously switching to the degrading network. This instantaneous switching is extremely unlikely to happen. Instead, many ISPs would reduce rather than increase use of a network that is degrading the quality of interconnections for a significant amount of Internet traffic. And, as long as there are ISPs who have not switched to the degrading network, all customers of the degrading network suffer. Each one of these customers of the degrading network is receiving connectivity significantly below his expectations of universal connectivity, and is now willing to pay less for it. Thus, the loss in value from degradation is comparable on both sides of the degraded interconnections, and can in fact be higher for the larger network. This means that a large network can only harm its rivals by harming itself by just as much or more.

Degradation of interconnections, like termination of interconnections, sacrifices the benefits of network externalities. It would result in a loss of value in the large IBP's Internet businesses because it would limit its customers' ability to interact with the rest of the Internet. A rational business would not take this step. Because there are limited switching costs and negligible barriers to expansion and entry, transport customers would switch to other networks or new entrants rather than tolerate a degraded interconnection and alienate their customers. Networks monitor the quality of service aggressively on behalf of their end users and Website customers, and they are able to identify and react to problems that would result from deliberate degradation of interconnection.

6.3.3. Sequential attacks on rivals

Some authors have claimed that although a raising-rivals'-costs strategy is unprofitable against all rivals; it would be profitable if applied sequentially to one rival at a time[34]. In this line of thought, a large IBP would degrade interconnections by targeting rivals and ISP customers one after the other. Cremer et al. (2000) raise a number of anticompetitive concerns for networks that obey the following assumptions:

1. consumers do *not* demand universal connectivity;
2. there is an installed base of clients (ISPs) of Internet backbone networks who cannot migrate to other providers.

Under these assumptions, Cremer et al. (2000) argue: (a) a large IBP network has an incentive to introduce incompatibilities and to degrade interconnection with one rival, but not with all rivals, (b) even small differences in network size will

[34] *See*, Hausman (2000) at p. 54 and Cremer et al. (2000).

Table 10
Contrast between the assumptions of Cremer et al. (2000) and Internet facts

Issue	Assumptions of Cremer et al. (2000)	Internet facts
Consumers' preferences for connectivity	Consumers do not demand universal connectivity	Consumers demand universal connectivity[a]
Consumers' willingness to switch Internet provider	No switching by existing customers	Easy customer migration
Effects of congestion on network performance	Interconnection is totally degraded when capacity is slightly exceeded	Networks have spare capacity; in situations of congestion, quality falls proportionally with congestion

[a]For example, the European Union Commission, in its Statement of Objections to the MCI-WorldCom merger recognized the lack of validity of Cremer et al.'s (2000) first assumption in stating that 'the demand for Internet connectivity continues to be universal in scope' (at p. 81).

lead to a spiral of ever-increasing dominance by a larger IBP network, since dominance is defined by size, (c) large IBP networks will refuse to cooperate with small networks, and (d) in the case, where switching costs are low, large IBP networks will still be able to dominate small networks[35].

The Internet violates the assumptions of Cremer et al. (2000), as is described later. And, since the fundamental assumptions of Cremer et al. (2000) diverge in fundamental ways from the reality of the Internet, the conclusions of Cremer et al. (2000) do not apply to competition on the Internet. These differences are summarized in Table 10.

The claim of Cremer et al. (2000) that a large IBP network will degrade interconnection with a targeted rival, is empirically invalid as explained in next section. The fact that such behavior has not occurred on the Internet backbone despite significant differences in market shares among the various backbone providers should be sufficient proof that Cremer et al. (2000) are discussing a different network from the Internet. Moreover, on the Internet we have observed a trend in the opposite direction (i.e., toward interconnection and full compatibility). Proprietary networks that preceded the commercial Internet, some dominant in their realm, such as AOL, CompuServe, Prodigy, MCI Mail, AT&T Mail, and MSN chose not to remain incompatible networks, but instead accepted full compatibility as parts of the Internet.

The results of Cremer et al. (2000) are indeed extremely sensitive to variations in the assumptions made. The assumption that ISPs are not allowed to migrate to

[35] Cremer et al. do not identify the structural conditions under which they expect the anticompetitive behavior described in the earlier paragraph would occur.

other backbones, which is presented by Cremer et al. (2000) as 'conservative,' is not only unsubstantiated but also critical to support its claim of dominance of a large IBP network and of degradation of interconnection. When consumer migration is allowed within the framework of the Cremer et al. (2000), there is no dominance or 'snowballing.' It is shown in the Appendix that, with exactly the same assumptions of Cremer et al. (2000), except now allowing customer migration, the market equilibrium shows no network dominance by any firm, and no network has an incentive to degrade interconnection.

Cremer et al. also state that multihoming will not diminish the incentives or ability of a dominant firm to engage in serial degradation. They base this on an unrealistic network model setup. In Cremer et al.'s (2000) targeted degradation model, a large IBP network (network 1) cuts interconnection with network 3, while the only other remaining IBP network (network 2) interconnects with both networks 1 and 3, but is prevented (by assumption) from offering transit to the targeted network 3[36]. Cremer et al. (2000) allow for multihoming only between networks 1 and 3; they do not allow multihoming across other networks. Thus, multihoming *a-la*-Cremer et al. (2000) shields some customers of network 3 from the effects of targeted degradation but has no other effect. In the reality of the Internet, multihoming is available to customers of all networks, and large percentages of customers of all networks utilize it. If the interconnection between networks 1 and 3 were severed: (i) customers of network 1 that multihome with network 2 would shift their traffic to network 2 to gain access to network 3, thus reducing the capacity they would buy from network 1, causing the targeted degradation to be even less profitable for network 1, (ii) customers that multihome with all three networks would also shift their traffic to network 2, since network 2 is the only one that provides universal connectivity, and (iii) customers of network 3 that multihome with network 2 would increase the capacity of transit they buy from network 2, so that they are able to gain access to network 1. Thus, in the real Internet, the existence of multihoming: (i) makes targeted degradation even less profitable for the targeting network since it results in a steeper demand response and (ii) makes the nondegraded network(s) stronger competitors of the targeting network. In conclusion, the presence of multihoming makes it even less likely that targeted degradation will ever occur. Table 11 summarizes the differences in the results when customer migration is allowed.

A key conclusion of Cremer et al. (2000) is that the largest network will use targeted degradation of rival networks. But, targeted degradation is unprofitable for a large network that would initiate it because:

1. ISP clients of the targeted network are likely to switch to third IBP networks that are unaffected by the degradation; it is very unlikely that any will switch to

[36] The assumption of Cremer et al. that network 2 (or other third networks in a more general setting), will not sell transit to the targeted network is totally unreasonable.

the degrading IBP network because it is itself degraded, and cannot offer universal connectivity; there is no demand reward to the large IBP network;
2. Degradation of interconnection hurts all the ISP customers of the targeting IBP network as well, since they lose universal connectivity; these customers of the large network would now be willing to pay less to the large network; this leads to significant revenue and profit loss;
3. After losing universal connectivity, customers of the large IBP network are likely to switch to other networks that are unaffected by degradation and can provide universal connectivity; this leads to even further revenue and profit loss for the degrading network;
4. Multihoming ISPs would purchase less capacity from the large IBP network, or even terminate their relationship with the large network, which, through its own actions sabotages their demand for universal connectivity; this further reduces demand and profits for the degrading network; the same argument applies to multihoming customers of ISPs;
5. As the large IBP network pursues target after target, its customers face continuous quality degradation while the target's customers face only temporary degradation; this would result in further customer and profit losses for the large IBP network;
6. Prospective victims would seek alternative suppliers in advance of being targeted by the large IBP network; the scheme cannot play out the way it is proposed;
7. The degradation scheme is implausible in its implementation. How large do networks need to be to become serial killers? Why have we not observed this behavior at all?
8. There is no enduring change to the number of competitors in a market caused by serial degradation in a market with negligible entry barriers; the eliminated rival is likely to be replaced by another.

Table 11
Contrast between the results of Cremer et al. (2000) and results when customer migration is allowed

Issue	Claims by Cremer et al. (2000)	Results when customer migration is allowed
Strategic power	Dominance by 'large' network	Equal bargaining power among networks
Dynamic effects	'Snowballing' or 'tipping' leading to monopoly	Equilibrium at equal market shares; no 'snowballing' or 'tipping'
Willingness of providers to interconnect	Even a slightly larger network will refuse to interconnect with other networks	Network externalities and demand for universal connectivity force networks to interconnect

The reasons why the strategy of targeted degradation would be self-defeating: first, degrading interconnections with networks that have an alternative way to send and receive traffic through a second network connection with another network would lead to a quick response by the rivals of routing almost all of their traffic through the second network, and would therefore be undesirable to a large network. Figures 3 and 4 above illustrated the rerouting of traffic in response to a price increase by a large IBP. The response of competitors and clients of an IBP that degraded interconnection would be very similar to the responses of rivals and clients to a price increase by the large IBP as shown in Figures 3 and 4. Moreover, a target network is likely to enter into new peering and transit arrangements with other networks that would further divert traffic from the degrading IBP. The target network could buy transit from other networks, whose connectivity with the large IBP's network is intact, and avoid all degradation problems. Thus, in response to degradation, traffic is routed away from the degrading IBP so that the culprit loses customers, traffic, and profits.

Second, as explained earlier, inequality in size does not imply inequality in the value of the damage sustained by two interconnecting networks as a result of a degraded interconnection. Suppose that the large IBP degraded its interconnection with a much smaller network. If traffic were spread evenly across all customers (end users and Websites), the reduction in service quality experienced by each of the large IBP's customers may be smaller than the reduction in service quality experienced by each of the smaller rival's customers. Some argue that this implies that ISPs connected to the targeted rival would then switch to the large IBP, and therefore the degradation strategy is 'successful' in attracting customers to the large IBP. This argument is based on the assumption that Internet users do *not* require universal connectivity, an assumption that is factually incorrect. Since Internet users demand universal connectivity, no network would decide to degrade a target network unless the degrading network was certain that *all* ISPs of the target network would immediately react to the degradation by instantaneously switching away from the target network. This instantaneous switching is extremely unlikely to happen. The target network is likely to establish new peering and transit relationships with other networks and utilize its multihoming arrangements to divert traffic away from the degraded interconnection and minimize the effect on its customers. After all, since the target network is the only one with degraded connectivity to the large IBP's network, the target network can easily buy transit service from other networks, which have full connectivity to the large IBP's network and avoid all degradation problems. And, as long as there are ISPs of the target network who have not switched to the degrading network, the users of the ISPs connected to the large IBP will suffer significantly as a result of the degradation. If the large IBP were to degrade its interconnection to a target network, the customers of the large IBP will be willing to pay less for the degraded service, and the large IBP would lose profits, even if the degradation strategy were 'successful' in attracting customers to it. After all, a larger number of customers of the large IBP would experience a reduced service quality than the potential number of customers that the large IBP could attract from the small target ISP[37]. Thus, the commercial impact of the serial degradation on the large IBP in terms of profit loss would be significant.

Third, the large IBP's customers are anything but captives. Business and individual end users and Website operators are sensitive to the quality of the service they receive. The large IBP could not use its customer base as a tool for harming rivals because it would lose the customer base in the process. Customers would switch to another network in response to a reduction in service quality. A degraded interconnection reduces the quality of the service that the large IBP's customers receive, and if they could not get reliable and quick access to popular Websites served by the network rival whose connection was degraded, these customers would move to other networks whose connection with the victimized network was unimpaired. Therefore, picking rivals one by one would not reduce the damage of this strategy to the large IBP.

Fourth, as already discussed, a significant number of end-user service providers have connections with more than one transport provider and most large content providers have connections with a number of networks. Even if the serial killer argument were correct for traffic that went to ISPs that were exclusively connected with the large IBP, and somehow the large IBP benefited from degradation of quality to these ISPs, the degradation of quality of the large IBP network would lead multiple connection ISPs to move traffic away from the large IBP and terminate their relationship with the large IBP.

Fifth, by targeting rivals sequentially (rather than all at once), the large IBP might limit the size of the damage to itself at any point in time, but it would be just as large in total. Moreover, over a period of time, the serial degradation strategy hurts more a customer of the large IBP than a customer of any targeted network.

If the serial degradation strategy is pursued, the large IBP's customers would experience constant problems in connecting to Websites not served by the company, while each victim would face only temporary quality degradation. For example, suppose that, over a period of a year, a large network sequentially degrades interconnections for 4 months for each of three smaller competitors. Then customers of the larger network will experience degradation over the course of all 12 months, but customers of each of the smaller networks will not experience degradation for 8 months of the year. The continuous quality degradation experienced by customers of the larger network is at least as great as that occasionally experienced by customers of smaller (target) networks.

[37] As explained earlier, even if the merged company is 'successful' in making customers leave the target network, it is likely that most of the customers leaving the target will not switch to the merged company because of the merged company's network also faces a quality degradation.

Sixth, the serial killer scenario assumes that the purchasers of Internet transport services have a passive response to the plan as it unfolds. After each victim falls, they switch their transport business to the predator, knowing perfectly well that the ultimate result will be higher prices for transport services. In fact, the rational response would be the opposite. As the plan developed, the prospective victims would take action to avoid becoming victims at all. They would seek alternative suppliers for the majority of their Internet connectivity, cutting back purchases from the large IBP to the bare minimum.

Seventh, the 'serial killer' scenario is totally implausible in its implementation. Its proponents have left a number of key questions unanswered. For example, for how long will the large IBP target a network before switching to its next victim? How does the large IBP hide from its customers the increasing degradation in its service to them? How large do networks need to be to find it desirable to be serial killers? Why have we not observed this behavior at all? How do the proponents of the serial killer theory explain why the degradation of connectivity would happen in the future but has never happened up to now?

Eighth, the serial degradation strategy would be impossible to execute in practice, because new networks are coming into existence all the time. By the time that the large IBP had degraded interconnection with one network, the number of alternatives will have multiplied. In a market with negligible barriers to entry, there is no gain to eliminating one set of rivals because they will be replaced by another.

Ninth, the role of customer mobility that helps in maintaining competition in Internet transport. Larger customers already have multiple connections to the Internet and all customers can switch suppliers easily. Many ISPs have multiple connections to IBPs. Advocates of the serial killer scenario have suggested that customer mobility may contribute to the potential success of the serial killer strategy, because the customers of the targeted IBP will abandon that IBP quickly and fully[38]. This theory is incorrect because it disregards the incentives of multihoming customers and of other customers of the large IBP to switch their traffic away from the large IBP in response to the degradation.

A multihoming ISP who is a customer of the large IBP (which initiates the connectivity degradation of the small IBP in the serial killer scenario) will also observe the degradation. Such an ISP will have an incentive to switch most of its traffic away from the two affected IBPs (large and small) to a third network. The ISP that switches traffic to a third network will now buy less transit from the large IBP. This provides incentives for the large IBP not to engage in degradation. The existence of multihoming implies that ISPs can easily reduce the amount of transit they buy from the large IBP in response to even small degradation of

[38] *See*, id at p. 57.

quality. Thus, multihoming decreases the incentive for a large IBP to degrade connectivity.

In conclusion, serial degradation is no more likely than simultaneous degradation. It would lower, not raise, the large IBP's profits.

7. Conclusions

The commercial Internet is one of the most important innovations in telecommunications and computing of the last 50 years. This ubiquitous data network based on low-level public technical standards has displaced well-established sophisticated high-level networks and has grown to reach a very large percentage of computers worldwide. At the core of the ability of the Internet to provide transport services lie the Internet backbones. The Internet backbone market has quickly grown to extremely high capacity of transmission and has surpassed the transmission capacity of the traditional long-distance network. Despite ups and downs, including the dot com boom and bust, and the WorldCom accounting scandal and bankruptcy, the Internet backbone market has shown robust competition. The dire predictions of the European Union Competition Authority in 1998 and 2000, that the Internet would be dominated by a single firm that would impose its own standards and refuse to interconnect rival backbones, have failed to materialize.

Appendix

A. 1. Duopoly

Reexamine the duopoly model of Cremer et al. (2000), Section 4, keeping all the assumptions of the model, except one: Allow customers in the installed base of each network to migrate to the other network if price and quality considerations so warrant. Thus, in the modification, the size of the network (sales) and the installed base coincide. All the symbols are the same as in Cremer et al. (2000) except that now output of firm i is q_i rather than $q_i + \beta_i$

In particular, as in Cremer et al. (2000), assume two interconnected Internet backbone networks, $i = 1, 2$ with θ in [0, 1] being the quality of interconnection between the networks and v signifying the importance of connectivity. Cremer et al. (2000) assume that backbone i has an installed base of captured customers β_i, who do not respond to prices and would not sign up with a backbone other than i at any price. Backbone i also has q_i customers, who respond to prices. Assuming that the quality of interconnection within a backbone is $\theta = 1$, Cremer et al. (2000) define the 'quality' of service of backbone i corresponding to its ability to reach customers as

$$s_i = v\left[(\beta_i + q_i) + \theta(\beta_j + q_j)\right] \tag{1}$$

As mentioned earlier, in the analysis in Cremer et al. (2000), customers β_i and β_j are not allowed to change providers. Cremer et al. (2000) show (Proposition 1) that under these conditions, for some parameter values, the larger of the two backbones chooses a lower interconnection quality (θ) than its rival, and that the quality of interconnection that the larger backbone chooses decreases in the difference between the captured customers of the larger and the smaller networks who are not allowed to change providers.

If, alternatively, all customers are allowed to buy service from a competing backbone, that is the number of captive customers is zero, $\beta_i = \beta_j = 0$, then Cremer et al.'s (2000) Equation (1) that defines the quality of good i becomes

$$s_i = v\left(q_i + \theta q_j\right).$$

Equation (2) remains as in Cremer et al. (2000). Equations (3) and (4) defining willingness to pay for each backbone become

$$q_i + q_j = 1 - (p_i - s_i) = 1 - \left(p_j - s_j\right),$$

$$p_i = 1 - \left(q_i + q_j\right) + s_i = 1 + v\left(q_i + \theta q_j\right) - \left(q_i + q_j\right),\ i = 1, 2.$$

Profits of firm (backbone) i are

$$\Pi_i = (p_i - c)q_i = \left[1 + v\left(q_i + \theta q_j\right) - \left(q_i + q_j\right) - c\right]q_{i.}$$

Maximization with respect to q_i results in the best response of firm i to the sales of the opponent:

$$q_i = R_i(q_j) = \frac{1 - c - q_j(1 - v\theta)}{2(1 - v)}.$$

Cournot equilibrium sales (network sizes) are then

$$q_i^* = q_j^* = \frac{(1 - c)}{3 - v(2 + \theta)}.$$

Notice that, for all θ, that is whatever the degree of network interconnection quality, both networks have exactly the same size, $q_i^* = q_j^*$. Thus, when customers in the installed base are allowed to migrate across networks, contrary to the results of Cremer et al. (2000), *there is no network dominance at the market equilibrium*.

Equilibrium profits of the two networks are equal:

$$\Pi_i^* = (1 - v)\left(q_i^*\right)^2 = (1 - v)\left(q_j^*\right)^2 = \Pi_j^*.$$

It follows that both networks have the same incentive to increase quality:

$$\frac{d\Pi_i^*}{d\theta} = \frac{d\Pi_j^*}{d\theta} > 0.$$

Thus, contrary to Proposition 1 of Cremer et al. (2000), both networks have equal and positive incentives to maintain a high quality of interconnection between them.

A. 2. *Merger analysis*

Examine the merger analysis model of Cremer et al. Section 6, keeping all the assumptions of the model, except one: Allow customers in the installed base of each network to migrate to the other network if price and quality considerations so warrant. As in the duopoly model above, the size of the network (sales) and the installed base coincide so that all the symbols are the same as in Cremer et al. (2000) except that now output of firm i is q_i rather than $q_i + \beta_i$.

Cremer et al. (2000) start with four networks of equal sizes. In the original equilibrium, all networks have equal sizes and profits. After a merger between two of them, there are three networks in the market. In the 'targeted degradation' scenario of Cremer et al. (2000), network 1 severs its interconnection to network 3, while maintaining full interconnection to network 2. Networks 2 and 3 are fully interconnected, but network 3 is not allowed to use network 2 for transit to network 1. Cremer et al. (2000) show that, for some parameters, the merged firm will prefer to follow the 'targeted degradation' strategy (Proposition 6).

However, if one alternatively assumes that the installed base of each network is allowed to migrate to the other network if price and quality considerations so warrant, the 'targeted degradation' result of Cremer et al. (2000) is reversed. Specifically, calling q_i the sales of firm I, after a merger between two of the four networks, there are now three networks in the market. Assuming no degradation, their prices and profits are

$$p_i = 1 - (q_1 + q_2 + q_3) + v(q_1 + q_2 + q_3), \ \Pi_i = (p_i - c)q_i, i = 1, 2, 3,$$

Equilibrium quantities, prices, and profits without degradation are

$$q_i^* = \frac{(1-c)}{4(1-v)}, \quad p_i^* = \frac{(1+3c)}{4}, \quad \Pi i^* = \frac{(1-c)^2}{16(1-v)}.$$

Now consider the 'targeted degradation' scenario of Cremer et al. as described above. In this scenario, network 1 severs its interconnection to network 3, while maintaining full interconnection to network 2. As in Cremer et al. (2000), although networks 2 and 3 are fully interconnected, it is assumed that network 3 is not allowed to use network 2 for transit to network 1. Then prices and profits are:

$$p_1 = 1 - (q_1 + q_2 + q_3) + v(q_1 + q_2), \ p_2 = 1 - (q_1 + q_2 + q_3) + v(q_1 + q_2 + q_3),$$

$$p_3 = 1 - (q_1 + q_2 + q_3) + v(q_1 + q_2), \ \Pi_i = (p_i - c)q_i.$$

Using superscript 'd' to denote degraded interconnection, equilibrium quantities, prices, and profits are

$$q_1^d = q_3^d = \frac{(1-c)}{2(2-v)}, \quad q_1^d = \frac{(1-c)}{2(2-v)(1-v)},$$

$$p_1^d = p_3^d = \frac{1+3c-v(1+c)}{2(2-v)}, \quad p_2^d = \frac{(1+3c-2vc)}{2(2-v)},$$

$$\Pi_1^d = \Pi_3^d = \frac{(1-c)^2(1-v)}{4(2-v)^2}, \quad \Pi_2^d = \frac{(1-c)^2}{4(1-v)(2-v)^2}.$$

Now compare profits of network 1 with and without targeted degradation of interconnection. It is easy to show that profits of network 1 without degradation are higher than profits of the same network with targeted degradation:

$$\Pi_1^* = \Pi_1^d = \frac{(1-c)^2(4-3v)v}{16(2-v)^2(1-v)} > 0,$$

since $0 < v < 1$. Therefore, contrary to Cremer et al. (2000), if all consumers are allowed to change providers if prices and qualities so warrant, network 1 (the largest one) finds it profitable *not* to use 'targeted degradation.'

References

Boardwatch Magazine, 1999, Directory of Internet Service Providers, 11th Edition.

Boardwatch Magazine, 1997, Fall.

Cremer, J., P. Rey and J. Tirole, 2000, Connectivity in the commercial internet, Journal of Industrial Economics, 48, 433–472.

Cremer, J., P. Rey and J. Tirole, 1998, The Degradation of Quality and the Domination of the Internet, Appendix 5 in the Submission of GTE to the European Union for the merger of MCI with WorldCom.

Besen, S. M. and S. R. Brenner, 2000, Declaration to the FCC, March 20.

Economides, N., 1989, Desirability of compatibility in the absence of network externalities, American Economic Review, 79(5), 1165–1181.

Economides, N., 1996a, The economics of networks, International Journal of Industrial Organization, 14(2), 675–699, lead article. Prepublication electronic copy available at http://www.stern.nyu.edu/networks/94–24.pdf.

Economides, N., 1996b, Network externalities, complementarities, and invitations to enter, European Journal of Political Economy, 12, 211–232. Prepublication electronic copy available at http://www.stern.nyu.edu/networks/Networkexternalities_EJPE_1996.pdf.

Economides, N., 1999a, The Telecommunications Act of 1996 and its impact, Japan and the World Economy, 11, 455–483, lead article. Prepublication electronic copy available at http://www.stern.nyu.edu/networks/98–08.pdf.

Economides, N., 1999b, Real Options and the Costs of the Local Telecommunications Network, in J. Alleman and E. Noam, eds., Real Options: The New Investment Theory and its Implications for Telecommunications Economics, Regulatory Economics Series, Boston: Kluwer Academic Publishers. Prepublication electronic copy available at http://www.stern.nyu.edu/networks/real.pdf.

European Union Commission, 1998, Statement of Objections to the MCI WorldCom Merger.

Farrell, J. and G. Saloner, 1985, Standardization, compatibility, and innovation, Rand Journal of Economics, 16, 70–83.

Faulhaber, G., 2004, Bottlenecks and Bandwagons: Access Policy in the New Telecommunications, Handbook of Telecommunications, Amsterdam: Elsevier Publishers.

Hausman, J., 2000, Declaration to the FCC.

Hogendorn, 2004, Excessive Entry of National Telecom Networks, 1990–2001, NET Institute Working Paper, available at http://www.netinst.org/Hogendorn.pdf.

Hubbard, G. and W. Lehr, 2000, in I. Vogelsang and B. M. Compaine, eds., Telecommunications, the Internet, and the Cost of Capital in The Internet Upheaval—Raising Questions, in Seeking Answers in Communications Policy, Cambridge, MA and London: MIT Press.

Kahn, R. E. and V. G. Cerf, 1999, What Is The Internet (And What Makes It Work), December 1999, at http://www.wcom.com/about_the_company/cerfs_up/internet_history/whatIs.phtml.

Katz, M. and C. Shapiro, 1985, Network externalities, competition and compatibility, American Economic Review, 75(3), 424–440.

Liebowitz, S. J. and S. E. Margolis, 1994, Network externality: An uncommon tragedy, The Journal of Economic Perspectives, 133–150.

Milgrom, P., B. M. Mitchell and P. Srinagesh, 2000, Competitive Effects of Internet Peering Policies, in I. Vogelsang and B. M. Compaine, eds., The Internet Upheaval—Raising Questions, Seeking Answers in Communications Policy, Cambridge, MA, and London: MIT Press.

RHK, 2005, Internet Traffic Analysis—Methodology and Results, May 2005.

Further reading

Armstrong, M., 1998, Network interconnection in telecommunications, Economic Journal, 108, 545–564.

Bradner, S., The Internet Standards Process, revision 3, Network Working Group at ftp://ftp.isi.edu/in-notes/rfc2026.txt.

Carter, M. and J. Wright, 1999, Interconnection in network industries, Review of Industrial Organization, 14, 1–25.

Competition Policy in Network Industries: An Introduction, in Dennis, Jansen, ed., *The New Economy and Beyond: Past Present and Future*, Edward Elgar (2006), pre-publication copy at http://www.stern.nyu.edu/networks/Competition_Policy.pdf.

Economides, N., 1998, The incentive for nonprice discrimination by an input monopolist, International Journal of Industrial Organization, 16, 271–284. Prepublication electronic copy available at http://www.stern.nyu.edu/networks/1136.pdf.

Economides, N., 2000, Declaration to the FCC, May 21.

Economides, N., G. Lopomo and G. Woroch, 1996, Regulatory pricing policies to neutralize network dominance, Industrial and Corporate Change, 5(4), 1013–1028, Prepublication electronic copy available at http://www.stern.nyu.edu/networks/96–14.pdf.

Kende, M., 2005a, Declaration of Michael Kende to the FCC in the Verizon-SBC merger.

Kende, M., 2005b, Reply Declaration of Michael Kende to the FCC in the Verizon-SBC merger.

Laffont, J.-J., P. Rey and J. Tirole, 1998a, Network competition: I. Overview and nondiscriminatory pricing, RAND Journal of Economics, 29, 1–37.

Laffont, J.-J., P. Rey and J. Tirole, 1998b, Network competition: II. Price discrimination, RAND Journal of Economics, 29, 38–56.

Laffont, J.-J., S. Marcus, P. Rey and J. Tirole, 2001, Internet interconnection, and the off-net cost pricing principle, American Economic Review, May 2001.
Laffont, J.-J. and J. Tirole, 2000, Competition in Telecommunications.
Liebowitz, S. J. and S. E. Margolis, Network Effects, in M. Cave, S. Majumdar, and I. Vogelsang, eds., Handbook of Telecommunications Economics, pp. 75–96, Amsterdam: Elsevier Publishers.
Noam, E. M., 2001, Interconnecting the Network of Networks, Cambridge, MA: MIT Press.
Odlyzko, A. M., 2003, Internet traffic growth: Sources and implications, at http://www.dtc.umn.edu/~odlyzko/doc/itcom.internet.growth.pdf.
Schwartz, M., 2005a, Declaration of Marius Schwartz to the FCC in the SBC-AT&T merger.
Schwartz, M., 2005b, Reply Declaration of Marius Schwartz to the FCC in the SBC-AT&T merger.

Chapter 10

PRICING TRAFFIC ON INTERCONNECTED NETWORKS: ISSUES, APPROACHES, AND SOLUTIONS

ALOK GUPTA

University of Minnesota

DALE O. STAHL

University of Texas at Austin

ANDREW B. WHINSTON

University of Texas at Austin

Contents

Handbook of Telecommunications Economics, Volume 2, Edited by S. Majumdar et al.

DOI: 10.1016/S1569-4054(05)02010-5

1. Introduction

Congestion based usage pricing for management of IT resources has been discussed for several decades. Congestion-based usage pricing relies on pricing the resources based on negative externalities created by overuse of shared resources. However, a whole array of fundamental issues plagued productive discussions about its feasibility in computing environments. The research presented in this chapter summarizes developments that have helped move the discussion on Internet pricing from analytical economic modeling to real-time computability and implementation. We also present comparative analysis of various approaches from the perspective of total benefits achievable via the use of networks to issues of investment incentives in network capacity.

The need for pricing arises from the congestion that data networks experience when excess demand causes packets to be dropped packets, which results in a loss of service quality for the users of these networks. Ironically, one of the most exciting phenomena of the last decade (i.e., the dramatic transformation of the Internet from being an academic and research network into a part of American folklore), has exasperated the congestion problem. In Section 2, we review the problems associated with the congested resources, especially from an economic perspective. We also present brief summaries of historical approaches used by computer scientists to deal with the congestion problems in computer networks.

Arbitrary pricing strategies of infrastructure owners and access providers complicate the issue of managing traffic and providing incentives for appropriate network resource usage. Infrastructure owners charge flat fees for access; access providers use either a price based on time of usage, or, as recent trends indicate, a flat monthly fee. Neither takes the volume of traffic into account. This leads to a total lack of any incentives for providers and users to choose appropriate levels of network use. For instance, the provider of a slow network pays a flat fee for access to a fast network such as Sprint's regardless of usage; so this provider dumps all the traffic onto the fast network as quickly as possible, the so-called 'hot potato' approach (*The Economist*, 19 October, 1996). The major network providers suffer from providing higher bandwidth, and have no incentive to invest in further infrastructure. At the user level, a curious teenager and a researcher pay the same flat rate for Internet access; the former downloads terabytes of video clips, blocking the computations the latter may be trying to execute on a remote supercomputer. Mission-critical applications have neither guarantee of precedence over others or any expected level of performance. Thus, users lack incentives to control misuse of resources, providers lack incentives to manage congestion, and to invest in new capacity. In Section 3, we present a brief comparison of several implemented and suggested pricing strategies and discuss the desired characteristics of a pricing mechanism that can effectively deal with the congestion to provide better quality of service.

New technological developments, such as next generation Internet Protocol IPv6 (Deerin and Hinden, 1995), are trying to address the performance issue by moving away from best-effort, first-come first-serve approach by differentiating traffic based on priority classes and associating priorities with different application classes. However, such solutions still do not consider the criticality of the usage context and are still prone to misuse. For example, just because a video stream requires better response time, it does not mean a recreational video should be preferred over a simple text stream being used for a stock purchase. Other technological solutions such as raising the bandwidth, or building new networks to segregate traffic may work in the short term; but eventually congestion will catch up and the burstiness of the Internet traffic will still cause congestion at times even with a larger bandwidth, as new lanes and new highways do not eliminate traffic jams during rush hours. There is increasing recognition of the fact that the issue is one of resource allocation, and needs to be addressed by a convergence of the economic and technological viewpoints so as to ensure the streamlined functioning of a virtual economy.

Economic theory can be gainfully applied to develop mechanisms for managing network resources in a manner that is incentive compatible for both users and providers. The focus needs to be on managing resources, rather than providing better resources while existing misallocation persists. The application of economic principles is not restricted to Internet; rather, they are applicable to allocating resources efficiently over computer networks in general, such as distributed computing systems. Indeed, recent research in Computer Science has been increasingly drawing from Microeconomic theory to design resource allocation schemes (e.g., see Ganesh et al., 2001; Marbach, 2001; Wang and Schulzrinne, 2001). Any such scheme trying to allocate network resources in a globally optimal fashion will incur prohibitive overhead costs. It will not be feasible to collect and process information over all the nodes in a huge network to allocate capacity for each request, nor will it be possible to model such a scheme. As such, we advocate decentralized mechanisms that yield significant increase in benefits and performance, and undertake to demonstrate their feasibility. Section 4 presents the central contribution of this chapter by describing a general decentralized priority pricing mechanism and discusses its benefits as well as its computational feasibility. This section also discusses the investment incentives and the role of users' value in determining the appropriateness of a given pricing scheme for optimal investment incentives.

The present day Internet is divided into an array of different logical constructs via the use of overlays. A collection of services can define their own logical routing network by abstracting the physical topology of the network. Such approaches provide logically manageable search and routing from application perspective. In addition, the dearth of IP space and affordability of multiple ISPs has made multihoming a logical choice for most organizations. Multihoming, allows for tantalizing prospects in terms of smart routing where, for example, based on responsiveness of its ISPs, a customer can intelligently choose to route its traffic. Section 5 describes the new challenges and opportunities in network traffic management and the generalizability of the model presented in Section 4. Finally, concluding remarks are presented in Section 6.

2. Congestion, overuse, and economic implications

The spread of the World Wide Web (WWW), Networked Computing (NC), and Electronic Commerce (EC) have been crucial factors in creating problems of congestion on the Internet. These developments have been too fast for the evolution of the required infrastructure and protocols, always leaving the useable bandwidth short of the desired levels. The challenge is how to facilitate all this traffic, of a nature and volume not anticipated by the original design of the Internet. Indeed, the inefficiencies in infrastructure investments and in handling traffic are showing up in widespread congestion, frustrating delays, and a lack of applications that require guaranteed Quality of Service (QoS). Congestion results in losses to the tune of billions of dollars to the economy every year, besides affecting network performance and discouraging infrastructure investments. Further, unpredictable performance inhibits EC investments.

These delays are due to several reasons:

- Traffic has been growing almost at double the rate of the number of users.
- The advances in high-bandwidth technologies are not able to match this pace.
- New technologies are allowing richer multimedia information to be transferred over the Net, and to execute applications over the network.
- Exponential growth of routing tables. For example, the growth rate for active BGP entries in Internet core routers is estimated to be over 40 percent annually since 1998. Larger routing tables adversely affect the routing performance, increasing the congestion problem. (http://www.potaroo.net/papers/ietf/draft-iab-bgparch-02.html)
- The providers face only fixed one-time costs of setting up the network, and are essentially unregulated.
- The providers also lack any incentives for managing congestion better.
- Internet users lack incentives to economically use bandwidth, which currently serves as a public good.

It is not that the issue of congestion has been overlooked. There have been several approaches that have been proposed in the literature to deal with congestion in computing systems. A brief review of these approaches is provided in the next subsection.

2.1. Approaches to deal with congestion in computing systems

Given a fixed capacity in the short run, the approaches to minimize congestion rely on resource management techniques such as load balancing. This subsection discusses resource management approaches, both in centralized and distributed computing systems.

Many resource management algorithms in centralized systems devise a benefit/cost equation to represent their service goals and resource requirements and then attempt to maximize the benefit/cost of their decisions. Examples include LRU cache replacement, which approximates the optimal policy of replacing the page that will not be used for the longest period of time, and decay usage CPU scheduling (Hellerstein, 1993), which approximates the optimal shortest job first policy. Although this approach can work well for small subsystems, it requires the system designer to understand the entire subsystem well enough to devise an optimization criterion. Unfortunately, criteria for different subsystems are not generally easy to formulate. If a system designer wants to manage resources shared by several different subsystems, she must devise the more complex optimization equations for the larger combined system.

Recently, several centralized resource management techniques have been borrowing concepts from microeconomic theory. These algorithms continue to take the approach of optimizing the benefit/cost of their resource usage, but the economic theory provides additional leverage. For example, phrasing the benefit/cost calculation in terms of a common currency that can be shared across subsystems makes it easier to compose the optimization criteria of different subsystems. Patterson et al. (1995) devised optimization criteria that balance the use of memory buffers by the prefetching and caching subsystems of a machine. In a similar vein, lottery scheduling (Waldspurger and Weihl, 1994) and stride scheduling (Waldspurger and Weihl, 1995) allow CPU cycles to be assigned according to tickets that could correspond to a revenue stream in an economic model.

Most load balancing approaches in distributed systems rely on three basic techniques: global knowledge, randomization, and feedback. Most global knowledge based process migration systems, including Amoeba (Mullender, Rossum, Van Renesse, and Van Staveren, 1990), Condor (Litzkow and Solomon, 1992), Charlotte (Artsy and Finkel, 1989), Sprite (Douglis and Ousterhout, 1991), and process lifetime algorithm (Harchol-Balter and Downey, 1996) send the load level of their machines to a central server that directs jobs to unloaded machines. This central server limits the scalability of this approach.

Randomization can reduce the amount of global knowledge needed to make good decisions. For example, it is well known that after placing n balls uniformly randomly into n bins, the fullest bin holds $O(\log n/\log \log n)$ balls with high probability; a small amount of additional load information can improve this result to $O(\log \log n)$ (see Azar et al., 1994; Adler et al., 1995). Systems have exploited either oblivious randomization, or randomization plus localized load information both to balance load across processors (see Barak and Shiloh, 1985; Eager et al., 1986), and to distribute cache pages across memories (Dahlin et al., 1994; Feely et al., 1995; Sarkar and Hartman, 1996). Random early detection (RED) routers illustrate an additional advantage of randomization; it can help avoid performance oscillations sometimes caused by global synchronization from deterministic algorithms (Floyd and Jacobson, 1993).

Another common technique for distributed load balancing is feedback. This approach is exemplified by the Ethernet (Metcalfe and Boggs, 1976) and TCP (Jacobson, 1988) exponential backoff schemes to deal with network congestion. The load information used by some of the randomized algorithms listed above could also be characterized as feedback information. A key question in distributed networking is how to partition resources between real-time and nonreal-time traffic. The most common current approach gives absolute priority to multimedia traffic at the expense of 'best effort' traffic by transmitting multimedia traffic using IP and best effort traffic using TCP (Eriksson, 1994). A better approach is to partition bandwidth so that both real-time and best-effort traffic are guaranteed some fraction of the system's bandwidth (Goyal et al., 1996), but for these algorithms to work well the system must determine a good partitioning of resources between the traffic classes.

While the resource allocation approaches described above consider economic metrics, they do not address the resource allocation problem using economic rationale. Instead, isolated devices or techniques are borrowed from economic theory and used in rationalizing trade offs. To leverage the benefits of economic approaches, it is essential that there be theoretical support for the approach and that the participants have sufficient incentives to use the mechanisms, and use of such mechanisms should lead to sustainable equilibria. The research in computer science, for the large part, treats all service requests equally in allocating resources, and does not account for the value or precedence of individual jobs. Economics, on the other hand, lets the forces of supply and demand operate over markets, in achieving equilibria with efficient resource allocation and maximized benefits. Such approaches are more relevant in the context of the privatization of the network infrastructure and the advent of EC. Load balancing in distributed systems will also need to maximize user benefits accounting for their preferences. In resource allocation issues in networked computing, the locality of computation is the crucial choice variable, and computing research has not really looked at how decisions on this would be made, and why. In the next subsection, we discuss the economic issues related to resource allocation in computer networks as well as the problems and challenges in applying economic theory to manage computing resources.

2.2. Economic implications: Tragedy of the commons problem

There are serious implications of congestion from economic perspective. Congestion can be looked as the result of what is termed in economics as a 'tragedy of the commons.' A tragedy of the commons arises when a common resource (referred to as public good in economic terms) is degraded by overuse. Well-known examples are ocean fisheries, urban roads, air, and water. Whenever a public good does not belong to some legal entity empowered to manage the resource through usage restrictions and/or fees, there are inadequate incentives for individual users to restrict usage to the socially optimal level. Consequently, the public good deteriorates (or the public bad swells), and the public suffers a loss relative to the potential social benefits of the commons.

However, applying economic solutions to the Internet is complicated by various factors. First, since infrastructure investments are sunk (one-time fixed costs), and the marginal cost of data transport is essentially zero, the equilibrium price suggested by economics (based on marginal costs) is zero. Indeed, infrastructure providers with flat-fee prices admit that they do not recover the capacity investments from users; rather, it is voice-line users that typically bear these costs. The access fees merely act as steady revenue, and are often dictated by competitive considerations in trying to survive with giants such as AOL, MSN, AT&T, etc. (Lewis, 1996). Second, network resources are in effect public goods, and congestion is a potential negative externality. As no one has any incentive to reduce congestion, and as there is no limit on the utilization of resources, the capacity is misutilized, resulting in loss of social welfare, similar to say, air pollution. Thus, as discussed earlier, a tragedy of the commons results. In such a case, prices should include the marginal social cost of congestion, and the consumer will utilize the resource only if his value exceeds the social cost of usage plus his private cost of time. Multiple ownerships, instead of improving the state of affairs by changing the public good to a private property, can still cause a tragedy of the commons (Gupta et al., 1997a).

Congestion on the Internet is a present and potentially paralyzing public bad. Today, Internet services include e-mail, WWW, and limited real-time audio/video services, and a variety of Peer-to-Peer (P2P) services. The evolution of these services indicates a desire for a perfectly interoperable communication network, where any kind of information can be digitized and then shared, transmitted, or stored. The result of increased availability of user-friendly interfaces and high speed of access coupled with higher awareness about these services has resulted in serious congestion problems on the Internet.

The instinctive economic solution to a tragedy of the commons is to assign property rights. Hence, it might appear to some observers that the transfer of the NSF backbone to the private sector, being an assignment of property rights over network hardware, was the solution to the problem of congestion. However, this assignment of property rights falls short of solving the problem. Each owner of network infrastructure will seek a pricing structure that maximizes its profits subject to the pricing structure of competing network owners and the price-sensitive demands of users. The resulting noncooperative pricing structures will not necessarily be (and generally will not be) socially optimal (Scotchmer, 1985a,b). This gives rise to an interoperability problem of a different kind (i.e., the problem of providing incentives to different entities) to provide hardware/

protocol interoperability. Note that this is not simply the question of standards but of economic incentives to design and implement standards. To alleviate the problem, several researchers have suggested using prices to regulate the traffic on the Internet. In the next section, we discuss some of these pricing strategies.

3. Pricing approaches for the Internet traffic

Most suggested pricing approaches for Internet traffic are based on the idea of reducing negative externalities, or allocating resources from the perspective of allocative efficiency (i.e., allocating the resources) to applications/users that value it the most. In addition, many pricing strategies account for the different data transfer rates that different applications require, resulting in what is termed as a 'multiservice class' network. One possible way to provide different classes of services is to provide different priority levels on demand. Bohn et al. (1994) propose a voluntary choice of priority level for different applications, and penalize users if they don't make appropriate choices. However, such a system lacks incentives for providers to implement such a system, as it will only mean increased overhead costs for them. Further, priority levels are based on the type of application, such as e-mail or FTP; the value of the job may vary regardless of the type of application. In any such system, individual providers should have incentives to match the service priority levels offered by other networks, as most requests have to be serviced by multiple networks.

The ideas of managing traffic from a resource allocation viewpoint, and that traffic should be differentiated according to the level of performance required by the users reflect current trends in Internet technology. Internet protocol IPv6 allows users to specify their priorities by using a header field in each packet. Router technology is moving in the direction where traffic can be differentiated at router level.

Pricing the different priority levels would provide the users with the incentive to make choices depending on the value of individual jobs: the economic solution would be to set the price equal to the incremental social cost of congestion; the optimal congestion toll (Vickrey, 1969). The users will compare the value of a job with the total private cost (congestion toll + private cost of wait), and voluntarily make appropriate service level choices. Further, socially optimal pricing provides increased revenues for the provider and better capacity utilization. This viewpoint of pricing for services over the network is consistent with the emerging trends in EC, and would thus enable a more meaningful discussion of network management. Furthermore, pricing could be used as a mechanism to assign priority class in next generation Internet protocol IPv6, instead of using application based priority assignment. In pricing based priority assignment, the priority for a data stream is based on the value a user associates with it, rather than what application is associated with the data stream.

A pricing scheme for managing resources should have theoretical support, should be operational, should be able to adjust prices in real time to account for the transient demand, should have manageable overhead costs such as administration, accounting, monitoring, implementation costs, and should be adequately tested with models and experiments. Cocchi et al. (1991) propose a pricing scheme that has fixed or static prices. This scheme assumes that user demand is not affected by cost and requires information on the entire network to compute prices. This information requirement makes the implementation of such a pricing scheme difficult due to high-overhead costs in large networks. MacKie-Mason and Varian (1995) address the congestion issue directly by a pricing mechanism based on spot auctions at each node, where each packet bids for the value of the job, and the bids are resolved so that the best strategy for users is to bid the correct value. Bidding is a static, one-shot process, and as such each auction will only account for packets present at that node, while packets arriving a nanosecond later can also cause congestion. More importantly, in TCP-IP networks, constituent packets of same job may pass through different routes, and the value of any individual packet clearing a node is contingent on the remaining packets of that transaction. The price paid at a node may go wasted in many cases. Any attempts to address this may render the system infeasible (Gupta et al., 1996b,c). In addition, from a theoretical perspective, as Stahl (2002) shows, the first- and second-price auctions do not produce economically efficient allocations. While it does not imply that there is no auction-like mechanism that can support the socially optimal allocation, the required mechanism would have to be considerably complicated, involving more sophisticated strategies, such as time-dependent and state-dependent bids. Furthermore, the design of the mechanism is likely to depend on the details of the queuing protocol (first-in-first-out vs. round-robin, and noninterruptible vs. interruptible). Given the complication of real-world setting such as the Internet, it is unlikely that we will be able to solve for optimal allocations, let alone implement optimal solutions.

Another recent article by Afeche and Mendleson (2004) presents an auction-based approach to perform priority-based allocation for a single server. Their priority queuing models assume a continuum of priority levels and an atomless distribution of users with respect to the user valuation of services. In other words, the optimal approach will be to have a priority class for each value level. Given the computational burden of implementing a large number of priority classes, most systems will have only a finite number of levels. In such a case, users may inflect negative externalities on other users with the same priority level, and these externalities will not be reflected in the equilibrium bids in the model of Afeche and Mendleson (2004). Hence, the auction model will not result in a socially optimal allocation, unless a special reserve price is set for each priority level. Moreover, when users have diverse attitudes regarding delays, even an auction with a continuum of priority levels cannot support the socially optimal allocation.

An alternative to auction mechanisms would be to set optimal prices that can support a near-optimal allocation. In the next section, we present an overview of the dynamic pricing approach along with the characteristics of the computational approach to implement the theoretically derived prices in real time.

4. A generic model of priority pricing for data networks

The traditional view of the noncommercial Internet has already been replaced by the realization that the Information Superhighway will entail complete reliance on a global network infrastructure owned and operated by private companies interested in making a profit. In this new profit-oriented world, telecommunication infrastructure companies will explore different ways of providing a variety of user services of the appropriate quality, taking care not to overprovision the network, and thereby create excessive costs. To date, communication services have typically been divided into voice- and video vs. Internet services, and companies were for the most part equipped to handle one or the other. There are clear-cut differences between the two. The former requires an allocation of bandwidth from origin to destination, which we can call connection-oriented communications, while the latter is based on connectionless or 'best effort' communication.

Network traffic management is complicated by the unique characteristics of this market. The data communications networks operate in a completely distributed environment with little, or no precedence or inheritance relationships. Furthermore, the demand for services is highly irregular, and even if there were a central governing body, it would be impossible to centrally manage and allocate computing resources in this setting. Another important dimension is the need to manage the network traffic in real time, since by definition the data communication services are time constrained, and the demand structure is quite dynamic. One mechanism for facilitating efficient resource usage and allocation as well as the management of a multiservice, multiprotocol data communication network is the development of pricing schemes for information products and services such as business applications, entertainment, and educational material. Gupta et al. (1997b) identified the following characteristics for a viable pricing approach:

1. It should reduce the excess load from the users' end without wasting network resources (for example by congesting routers), when users can reschedule without any loss in their value;
2. It should take the impact of current load on the future arrivals into account;
3. It should preferably be coarser than packet level pricing so that it is easier and less-costly to implement;
4. The load status of the network nodes (routers, gateways) should be taken into account in users' decision-making process;
5. It should be implemented in a completely decentralized manner, thus not requiring any central governing body and allowing for interaction with other possible resource management schemes;

6. It should result in effective load management;
7. It should have multiple priorities to take into account the different qualities of service required by different applications and users; and
8. It should be implemented in such a way that users have incentives to make the right decisions, and service providers have incentives to provide the right level of service.

As in virtually all areas, the development of appropriate pricing schemes for the electronic marketplace is in its fledging stage. Consumers, on the one hand, are only interested in obtaining the services/products, and could care less whether they are delivered through the airwaves or using ISDN or POT lines. As long as the products are delivered when required and in the correct form, they will be satisfied. Companies, on the other hand, must take into consideration the cost of network services (i.e., data transmission), the cost of the informational products/services themselves, and fixed costs for both the network service providers and the information providers. Appropriate pricing is also related to management of cross-traffic among different parts of the network that may be owned by different entities. Without being motivated by appropriate prices, infrastructure service providers will have little incentive to fulfill the requests of customers who do not subscribe to their network.

4.1. Model description

Gupta et al. (1996a) developed a model of the Internet as an economic system that provides a theoretical foundation for deriving priority prices that maximize net user benefits taken as a collective. Quite simply, the essential idea is to charge a user an amount that is proportional to his/her usage of the network. Specifically, the idea is to charge as much as others on the network would have valued a particular amount of usage at a given point in time. Over and above this 'base rate,' in the case of real-time or urgent applications, the need for faster access—and, thus higher priority—is taken care by the use of priority pricing.

In the model a rental price, a priority premium, and an expected waiting time are associated with each processor. These prices are dependent upon the size of services desired, while user requests for these services will depend on the benefits to the user and the costs. Given the prices and the anticipated waiting time, a user evaluates and selects cost-minimizing service schemes. Figure 1 provides a schematic representation of the model. Optimal service demands are then translated into demands on individual processors. Note that this model subsumes the current thought of thinking about overlay networks, where high-level routing is developed based on service description and the underlying topology is abstracted to simplify search purposes.

Research shows (Gupta et al., 1997c) that there is a generically unique welfare-maximizing allocation, and the rental prices are set at a level that supports this optimum as a 'stochastic equilibrium.' That is: (i) user flow rates are optimal for

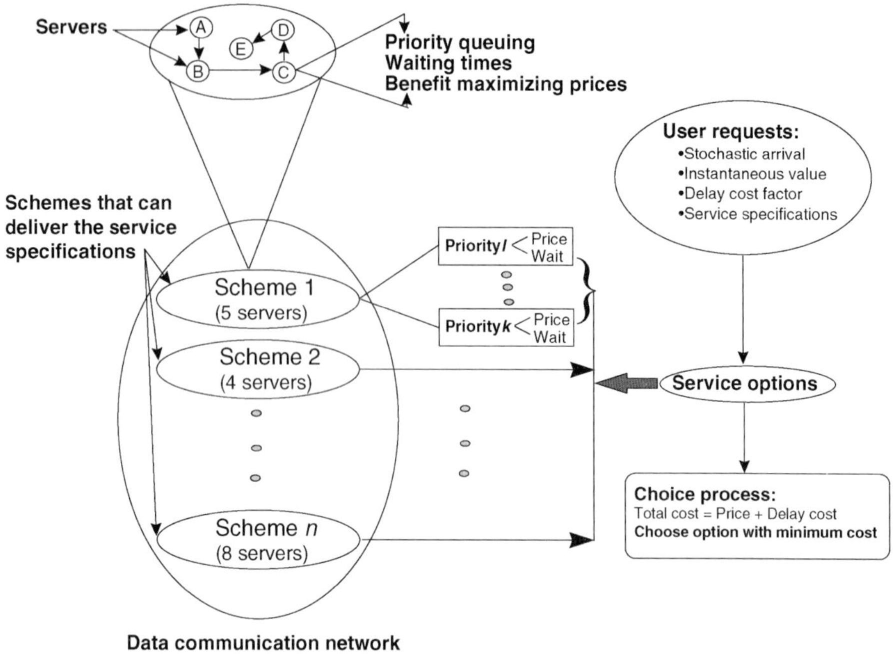

Fig. 1. Schematic representation of the network priority pricing model.

each user given the rental prices and anticipated waiting times; (ii) the anticipated waiting times are the correct *ex ante* expected waiting times given the flow rates; and (iii) the aggregate average flow rates are equal to welfare-maximizing rates.

A key feature of this model is that the pricing mechanism can be implemented in a decentralized manner with relatively little information overhead, since price calculations can be decentralized to each server and user decisions can be decentralized to their client machines.

With the addition to the model of a priority pricing capability, a rationale and mechanism exist to take into account the urgent needs of users. In fact, a priority pricing scheme is the only way to provide multiple levels of performance for different types of services in an interoperable network that will be able to deliver all types of service without significant deterioration in service quality (see Bohn et al., 1994; Shenker, 1995; Gupta et al., 1997b for more detailed discussion on multiservice class networks).

And, finally, in order to be able to adjust prices according to fluctuating demands and to provide automatic management of resources, a real-time pricing mechanism is an essential feature of the model. Although in a stochastic system prices may never be exactly equal to the optimum, if they converge quickly toward the optimum, significant benefits can still be realized. Since it will be practically impossible to predict or judge the service demands on global networks, a real-time pricing mechanism is the key.

Although the area of real-time problem solving is still in its infancy, we developed pioneering simulation models that show how this can be accomplished (see Gupta et al., 1997b, c for details of the simulation model). Note that many researchers, such as Marbach (2001), incorrectly infer that we are suggesting that prices should be changed every second. While its true that the maximal benefits would be reaped if optimal schedule of prices is available for each request, our approach provides complete flexibility in making the decision to change prices. For example, bounds can be established and prices changed only when estimated prices go above or below these bounds. The key is that the robustness of the price computation mechanism provides flexibility to pick a system at any state and steer it towards optimal direction. While the focus of the following description is infrastructure pricing, it is also a valuable tool to explore the impact of other relevant issues such as regulatory policies, alternative product pricing strategies, capacity and design, and competition. Simulations can be used to evaluate the alternatives and may be extremely valuable in anticipating problems and devising solutions. In the next subsection, we briefly provide some key results from simulation study.

4.2. Simulation and results

The most important questions answered by our simulation of the Internet (Gupta et al., 1994) are: What are the benefits of a properly designed pricing approach, and who will benefit? The results presented here reflect two different information conditions. First, the results for the free-access policy are based on perfect information regarding the waiting times. This is the 'best case' scenario for implementation of the free-access policy because users first check where they can get the fastest service, and the information they receive is exact. However, providing perfect information is not practical in reality given the excessive cost involved in computing new information for each new request. Also, since several users can submit requests at virtually the same time, the waiting time information would still be invalid even if it was feasible from a financial perspective. However, it is virtually impossible to make good future predictions with *free access,* if information is periodically provided, negating the realized benefits. Even though this in itself justifies the use of the model's pricing scheme, we are also interested in other performance measures, such as the benefits delivered by the network, all other things being equal. Because of this, our free-access results are based on a perfect information scenario. We also looked at priority pricing under a periodic updating scenario.

4.2.1. Benefits of priority pricing

Figure 2 below illustrates the results of comparison of a nonpriority pricing scheme under a free-access system and a fixed-price scheme. Both the net and customer benefits are shown under varying load conditions, using the perfect

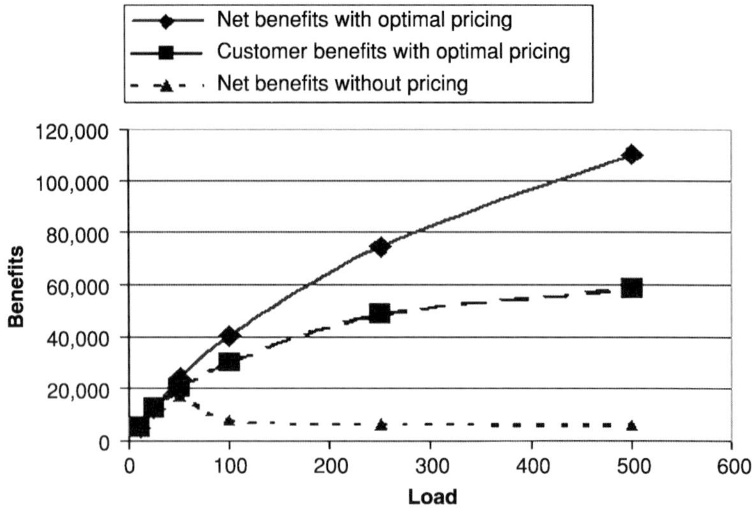

Fig. 2. Net and customer benefits with and without pricing.

information scenario. By 'net benefits,' we refer to the benefit to the system as a whole, which is equal to the aggregated users' value of the services rendered minus the loss of value due to the delay. By 'customer benefits,' we refer to the net benefit less the price charged by the service providers. (Note that in the case of free-access net benefits = customer benefits.)

Somewhat surprisingly our results suggest that this may be a win-win situation on the aggregate level. In other words, both consumers and service providers may benefit from using appropriate prices. Users benefit because their requests are delayed less, resulting in the availability of timely and hence useful information. Providers benefit because of the revenues generated and the optimal usage of the network.

When we take the simulation one step further and compare the benefits between a relatively simple two-priority system and a nonpriority system, this time using the *periodic update* scenario (i.e., the information regarding the waiting time is not exact), the results clearly favor the two-priority system. As Figure 3 shows, this holds true as regards both net benefits and customer benefits and over varying load conditions.

4.2.2. Responsiveness of real-time pricing

To explore the responsiveness of price adjustment process, we varied the arrival process during the simulation and collected the data on how responsive price adjustment process was. In addition, we also explicitly modeled *fractal* demand (see Gupta et al., 1999) by drawing each new arrival from a different distribution while keeping the long run average arrival rate fixed. Figure 4 presents results from such a simulation. Note that system responds very quickly to the changed

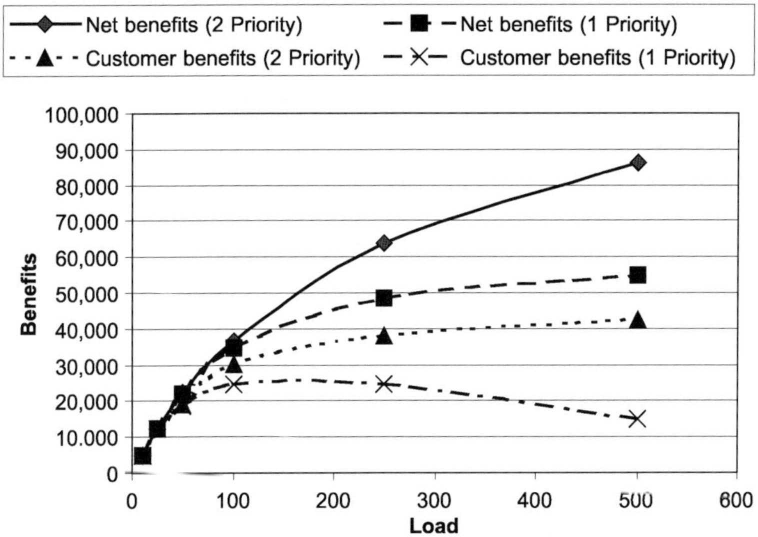

Fig. 3. Net and customer benefits with two-priority pricing under periodic update.

demand patterns, and the benefits from fractal demand are virtually indistinguishable from the case when demand is not explicitly modeled as a fractal demand.

4.2.3. *Effect of competition*

Another related issue is the question of what effect competition will have on the pricing of products and services. Different parts of the public networks are or will be owned by different entities, and service providers already compete fiercely to win customers over. It is unlikely that all of these will use the same pricing scheme. Evaluating all the possibilities is a monumental task—and may not result in an analytic equilibrium anyway. In reality, there may be many big service providers—we can think of them as electronic department stores—and the pricing strategies they employ may be complex, continuously evolving, and inherently dynamic. If left unregulated, pricing strategies may emerge that can create a monopolistic market and a single multiservice class network, and may turn out to be infeasible. On the other hand, hastily drawn legislation with little understanding of its impact may end up providing the same results.

We explored the effect of competition among networks. In Gupta et al. (2001a), we modeled two identical networks that employ different pricing strategies. We explored three different cases: (i) where both competitors use optimal pricing policies; (ii) where both competitors use a fixed-fee system; and (iii) where one competitor uses optimal prices and the other fixed fee. Among several other interesting results, we found that people who use network for smaller jobs benefit more with the priced network, the people requesting larger jobs get less individual benefits at least at moderate level of congestion as shown in Figures 5 and 6. At

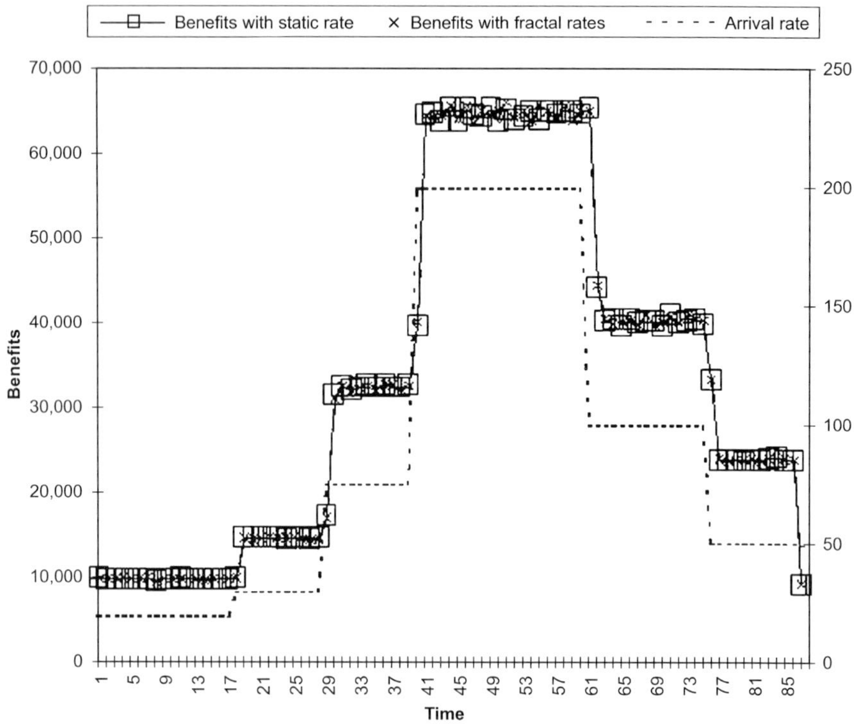

Fig. 4. Responsiveness of pricing mechanism.

higher level of congestion, some individuals with low valuations prefer free systems but individuals with high valuations prefer to pay; therefore, the benefits from free system and mixed system are similar as shown in Figure 6. However, in such cases a network environment where less time-sensitive customers can go to fixed-fee access and more time-sensitive customers to priced part (i.e., the case (iii) described above), is more beneficial since the customers can self select the level of service they want.

4.2.4. Investment issues

Believers in unlimited bandwidth (see, e.g., Gilder, 1997) at zero prices contend that pricing mechanisms that manage network congestion are irrelevant since they deal with a problem that will soon not exist, and that such pricing policies would only provide a disincentive for capital investment since they discourage usage and are contrary to customer's desires for price simplicity (Odlyzko, 2001). One of the main arguments from opponents of the implementation of any congestion regulating mechanisms for public computer networks is their belief that congestion on such networks is not a long-term issue. This belief is founded on the notion that a so-called 'bandwidth bonanza' is bound to happen within the next 2–3

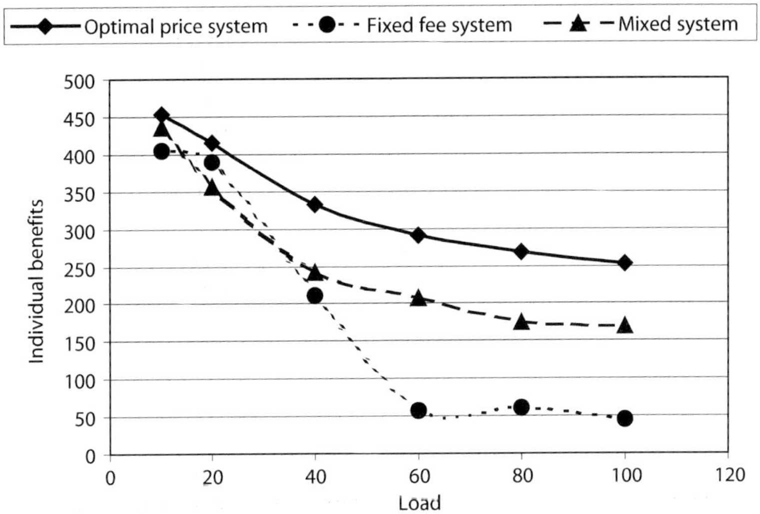

Fig. 5. Benefits for customer requesting smaller jobs.

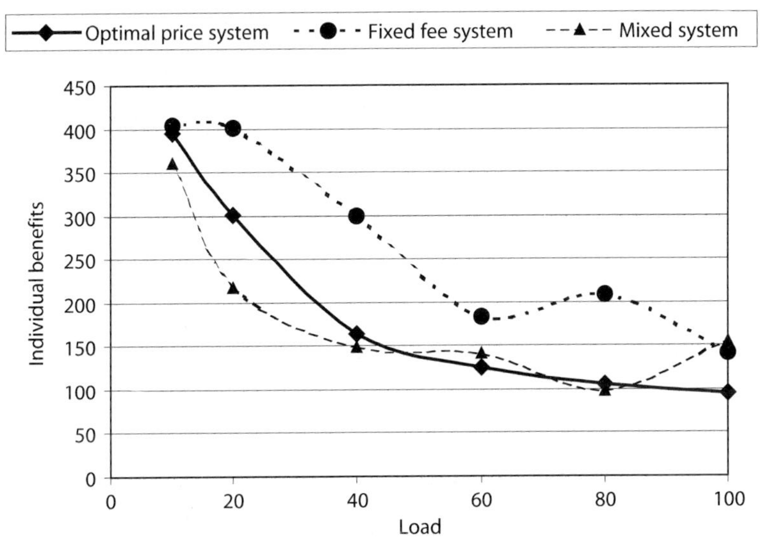

Fig. 6. Benefits for customer requesting larger jobs.

years. In Gupta et al. (2001b), we examined the incentives to invest in capacity with optimal pricing and with fixed-fee pricing. The research finds evidence that:

> The optimal level of system-wide capacity is higher under congestion-based pricing when the average value of user's request is less than the per-unit cost of the capacity. When that value/cost threshold is reached, building of a large system will be the optimal strategy under flat-rate pricing.

> Congestion-based pricing will result in an earlier deployment and a more gradual expansion of a system. In other words, until the users start putting high value for a service relative to the cost of its delivery, congestion-based pricing is necessary for providing investment incentives. However, once a high valuation is attached to the network usage (as compared to capacity costs), fixed-price access will provide sufficient incentives to expand the capacities as required.

It is interesting to note that in voice communications we are witnessing precisely this phenomenon. However, for data communication networks, we are traversing in the reverse direction, and we believe, is the root cause of delay in widespread, sustainable availability of broadband access and QoS guarantees. In this research, we also found that the profit maximizing capacity will be lower than the social value maximizing levels of capacity, where profits are zero.

4.2.5. Estimating user delay costs

One of the main contributions of this research stream is the demonstration of computationally feasible decentralized pricing approach and its robustness. Naor (1969), Mendelson (1985), Mendelson and Whang (1990), Westland (1992), Li and Lee (1994), Lederer and Lode (1997), Lin et al. (2002) have proposed that levying congestion-based tolls can result in optimal allocation of resources. However, none of these researchers discuss computational aspects of these pricing mechanisms. All these studies assume that consumers' demand characteristics (in terms of their delay costs) are known. However, Shenker et al. (1996) criticize these approaches and claim that the consumers' delay costs are fundamentally unknowable, and that even though, theoretically, these pricing mechanism are incentive compatible, the consumers will not provide their delay costs. In Gupta et al. (2000), we addressed the issue of estimating these (fundamentally unknowable) delay costs based on observed users' choice. Figure 7 presents results that compare

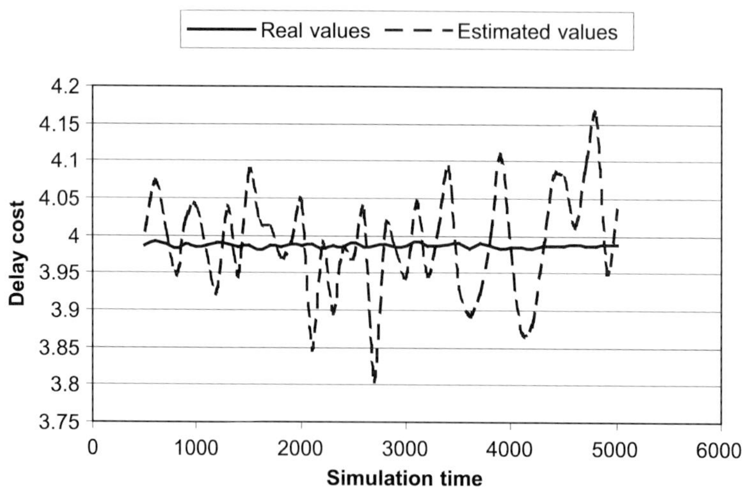

Fig. 7. Estimating delay cost based on observed consumer choices.

the average *real delay costs* and the *estimated delay costs*. Note that the estimated values are within the range of 3.8–4.17, which is within 10 percent of the actual mean. In addition, we showed that using the estimated delay costs results in minimal efficiency loss in terms of system-wide benefits.

This section presented characteristics of a real-time priority pricing approach for decentralized resource allocation and results from extensive simulations that explore a variety of issues from responsiveness of dynamic pricing approach to the issue of infrastructure investment. In the next section, we discuss how the evolving challenges in the Internet infrastructure and service provisioning can be addressed by the model presented in this section.

5. Future issues and challenges

The Internet is experiencing a significant transition from primarily a network of desktop computers to a network of variety of connected information devices such as cell phones, GPS in automobiles, and other networked devices. These advances have created significant opportunities as well as challenges in managing large-scale networks. On one hand, the explosion of devices and multihoming practices—where a physical network may use IP fragments from several distinct ISPs—have resulted in exponential growth in the number of entries in routing tables of core Internet routers. On the other hand, new paradigms such as overlay networks are defining service-based logical architecture for the network services that make locating content and routing more efficient. In this section, we discuss how the dynamic pricing model presented in the previous section can address some of the challenges presented by the future evolution of the Internet and its services.

5.1. Overlay networks

An overlay network is a logical network, which consists of a collection of nodes placed at strategic locations in an existing network fabric (Jannotti et al., 2000). Two nodes that are next to each other in an overlay network may be geographically far apart. Overlay networks are perhaps best understood in the context of P2P Internet applications, which have recently been popularized through file sharing applications such as Napster (Napster protocol specification. http://opennap.sourceforge.net/napster.txt), Gnutella (The gnutella protocol specification. http://www.clip2.com), and Freenet (Clarke et al., 1999).

For example, in Napster, a central server stores the index of all the available files. An user has to query the central server using a file name or other search criteria. The central server then obtains a list of IP addresses, which can be used to locate other user machines that have the requested file Thus, while the first generation of P2P systems used decentralized storage mechanisms, the process of locating content was still very centralized, and most of the current P2P designs are not scalable. Several research groups have independently proposed a new generation of scalable P2P systems such as Tapestry (Zhao et al., 2004), Pastry (Rowstron and Druschel, 2001), Chord (Stoica et al., 2001), and CAN (Ratnasamy et al., 2001). These systems use a distributed hash table (DHT) functionality where files are associated with a key (produced, for instance, by hashing the file name), and each node in the system is responsible for storing a certain range of keys. By using such a system, essentially, a new logical service network is generated on top of the actual network topology, which is usually referred to as an overlay network. The DHT defines a virtual routing table for the purpose of a specific service, while the actual routing still involves the IP routing at the infrastructure level. In other words, while overlay networks provide a mechanism to enable users to control their routes by relaying through overlay nodes, the route between two overlay nodes is still governed by the underlying routing protocol (e.g., see Savage et al., 1999).

While most discussions regarding P2P applications are around intellectual property rights issues and file sharing, overlay networks have a potential for significant business applications. Some potential applications of overlay networks are discussed below:

- *Collaborative Work*: While the WWW has allowed easy access to a variety of computing artifacts using a browser, locating and maintaining different versions of documents, software, etc., presents a significant challenge. Tightly integrated document repositories, versioning tools, and other collaborative applications can be implemented using overlay networks for a geographically dispersed group for seamless network operations.
- *Distributed Resource Management and Access*: Overlay networks can allow large and geographically dispersed companies to locate their content close to users. In other words, data can be made available at the edge of individual networks of the geographically distributed organization. For example, synchronized data can be made available in a distributed fashion instead of always loading it from central server, or by using explicit updating schemes that may make the data stale. In addition, computational resources of the organization as a whole can be tapped to provide businesses with large-scale computer processing capabilities.
- *Task Automation Using Software Agents*: Overlay network can allow business networks to dynamically work together using intelligent agents. Agents reside on various computers within the organization and can automate tasks and manage computing tasks such as prioritization of tasks, searches, coordination of activities, synchronization of information, and detection of anomalies such as viruses.

The priority pricing model presented in Gupta et al. (1996a) uses a generic model of network services, where each service can be provided by using a variety of different schemes and several services may share a given resource. Therefore, it can be interpreted as a model of an overlay network. The distance between nodes in this overlay network is proportional to the throughput time required to obtain the desired service, plus the externality imposed by service delivery. Optimal allocations are minimum distance routes in the overlay network, and these can be implemented by putting a surcharge on the service equal to its externality cost. In contrast, current implementations of overlay networks define distance as only the throughput time, ignoring the externalities. This is a serious oversight for two reasons. First, obviously ignoring externalities leads to an inefficient allocation of network resources. Second, high levels of congestion and rapid user response to that congestion can destabilize the network, causing a major breakdown. In contrast, with smooth (but dynamic) pricing, congestion is reduced and network traffic converges to stable efficient levels.

The generality of our model can also be used to study alternative overlay designs in terms of traffic, user benefits, and infrastructure costs.

5.2. Multihoming and smart routing

Multihoming (Smith, 2001) is becoming increasingly popular method for connecting to the Internet for businesses and ISPs. Multihoming refers to the Internet access arrangement, where a user has multiple access links to the network. This is typically achieved by obtaining multiple external connections to other networks by using a number of upstream ISPs as service providers.

Since multihoming allows a user to use IP address blocks from several ISPs, or in other words an ISP provides its IP blocks to several users, small address fragments are inserted into the global routing tables at the Internet core routers with a distinct connection policy that differs from that of any of the upstream neighbors of the IP block. This has resulted in a sharp rise in the size of the routing tables of the Internet core routers. For example, since 1998, the core routing tables have been growing at a rate of 40–45 percent every year (see BGP analysis at http://bgp.potaroo.net). At the same time, multihoming has the potential for providing significant benefits of reliability and performance of communications to its users. One of the ways in which users can benefit by multihoming is by actively distributing their traffic among its multiple access links to different ISPs. This is often referred to as *smart routing* (see e.g., Roughgarden and Tardos, 2002; Qiu et al., 2003).

Clearly, the distribution of traffic among different ISPs requires optimization of benefits to the users. Tangmunarunkit et al. (2001) argued that routing hierarchy and policy-based routing results in path inflation, which results in suboptimal performance. Smart routing allows end users to choose the route of their traffic (at least partially). If the choice of the route is based on appropriate metrics, both users and the ISPs can benefit from smart routing (Dai et al., 2003). Akella et al. (2003) show the potential benefits of smart routing and argue that selecting the right set of providers can yield significant performance improvement.

In contrast to multihoming in which the smart router is owned and controlled by the user, bundled services are provided by innovative entrepreneurs, such as Sockeye Networks GlobalRoute service, who connect many ISPs to a smart router and sell the managed service to users. In such a scheme, differentiated service levels can be offered and priced accordingly. As Dai et al. (2003) indicate, smart routing can improve network user surplus compared to having the option of just one ISP. Furthermore, network users tend to subscribe more capacity, since the traffic condition information provided by smart routers can increase the marginal benefit of network service. This has the further impact of improving network utilization level, which might have positive impacts on network service market where bandwidth is usually left unused. Further, with the diffusion of smart routing technology, ISPs can potentially charge higher prices since their services are no longer pure substitutes but provide differentiated service based on the performance of their networks.

5.3. *Channelling*

The Internet Engineering Task Force (IETF) has proposed several new service models to offer better network service than best-effort, including integrated service (Intserv) (see e.g., Braden et al., 1994) and differentiated service (Diffserv) (see e.g., Blake et al., 1998). Guaranteeing a higher QoS requires load control (or traffic shaping) by the users' ISP. A common method of load control is to reserve a portion of the network capacity for 'premium' services, and to set the supply of tokens so the reserved capacity is sufficient to provide the higher QoS for all tagged packets (see e.g., Nichols et al., 1998, 1999). More generally, the network capacity would be divided into a number of channels, each providing a designated QoS and requiring an associated token to be admitted to that channel.

The idea of providing QoS on the Internet by partitioning the network into several logically independent channels has been proposed and studied extensively in computer networking (see e.g., Odlyzko, 1999). This approach is called Paris Metro Pricing (PMP), which was initially introduced in the Paris Metro System. With this method, service providers will post different prices for the access rights to different channels and expect that the more expensive channels will be less congested. Therefore, users who care more about service quality can pay more for access to the more expensive channels and enjoy better service. Although it is widely known that channeling could introduce inefficiencies in bandwidth usage, experiments have shown that many real-time applications, that cannot run properly in today's Internet, work well in a PMP system.

Stahl et al. (2003) show that creating multiple QoS levels via channelization is an inefficient use of resources. An important implication of this research is that: although PMP can improve the usability for some real-time applications, the overall costs of abandoning statistical multiplexing will actually offset the benefits.

6. Conclusions

The types of Internet applications are undergoing radical changes. Many existing applications have introduced multimedia and interactive components, such as Digital Video over IP (DVIP) in Internet streaming services. The convergence of delivering television-quality images over the Internet extends interactivity in many forms, such as customized advertising and interactive voting. End users not only receive high-quality video, but also get personalized information and the ability to direct the program through interaction. Compared to traditional plain services, such as Web surfing and e-mails, multimedia and interactive entertainment have increased the demand of bandwidth and generated more traffic bursts.

Likewise, audio/video and interactive services, such as Voice-over-IP, Video-over-IP, and teleconferencing, expose end users directly to unpredictable network traffic. Applications such as streaming media are more sensitive to delays and jitters, which have negligible effects on the QoS of the traditional applications. When critical information is needed for user interaction to synchronize their states, network delays may change the outcome of distributed applications. For example, in massive multiplayer games, if the event of triggering a player's rifle is delayed, that player may get killed and lose the battle.

When many traffic-intense applications are deployed on the Internet, the traffic management will become more complicated. Since the number of potential users is very large and the usages are not coordinated, the quality of services will be the top issue for all applications. This problem will not be solved by Next Generation Internet (NGI) technologies, such as Internet2. These engineering architectures are created in a closed environment with dedicated and subsidized bandwidth. Although there will be faster infrastructures and new technologies to control network traffic, network resources will not be sufficient to support all applications. When a service needs higher priorities and premium qualities in a real application, computation protocols cannot assess the value of traffic. Without proper pricing mechanisms to manage quality, it is impossible to allocate network resources efficiently.

Although dynamic pricing may not be attractive at the packet level, the next generation applications on the Internet will require more enforcement of traffic management. Since even sporadic problems of network delay could harm the growth of deployment of the new applications on the Internet, the issues of quality of service are undoubtedly important and need to be solved before these traffic-intense applications are widely deployed.

We provided an overview of the benefits of dynamic priority pricing for network traffic management. Some specific benefits of the approach include:

- Economic theory-based approach maximizes system wide benefits using decentralized priority pricing.
- A real-time price computation mechanism converges quickly to near-optimal prices.

- Predictive ability based on transient (nonsteady state) information.
- Ability to infer user demand characteristics from observed choices.
- Ability to analyze consumer incentives and choice under competitive scenarios for public policy analysis.
- The generality of the approach allows applications in emerging paradigm of overlay networks.
- The pricing approach can be used for smart routing purposes.

However, a lot more could and should be done before actual implementation. In future, we plan to develop a test bed for pricing implementations in various applications and use real-world processes to price and manage the test-bed network.

Acknowledgment

This research was funded in part by the National Science Foundation, # IRI-9509914 and IIS-0219825, but does not necessarily reflect the views of the NSF.

References

Adler, M., M. Chakrabarti, M. Mitzenmacher and L. Rasmussen, 1995, Parallel Randomized Load Balancing, Proceedings of the Twenty-seventh ACM Symposium on Theory of Computing.

Afeche, P. and H. Mendleson, 2004, Priority auctions vs. uniform pricing in a queueing system with a generalized delay cost structure, Management Science, 50(7), 869–882.

Akella, A., B. Maggs, S. Seshan, A. Shaikh and R. Sitaraman, 2003, A Measurement-based Analysis of Multihoming, Proceedings of ACM SIGCOMM '03, Karlsruhe, Germany, August 2003.

Artsy, Y. and R. Finkel, 1989, Designing a process migration facility: The Charlotte experience, IEEE Computer, 22(9), 47–56.

Azar, Y., A. Broder, A. Karlin and E. E. Upfal, 1994, Balanced Allocations, Proceedings of the Twenty-sixth ACM Symposium on Theory of Computing, 593–602.

Barak, A. and A. Shiloh, 1985, A distributed load balancing policy for a multicomputer, Software Practice and Experience, 15(9), 901–913.

Blake, S., D. Black, M. Carlson, E. Davies, Z. Wang and W. Weiss, 1998, An Architecture for Differentiated Services, Internet RFC 2475. Available on Internet:http://smakd.potaroo.net/ietf/idres/draft.ietf-diffserv-arch/

Bohn, R., H. Braun and S. Wolff, 1994, Mitigating the Coming Internet Crunch: Multiple Service Levels Via Precedence, Technical report, San Diego Supercomputer Center, University of California at San Diego.

Braden, R., D. Clark and S. Shenker, 1994, Integrated services in the Internet architecture: an overview, Internet RFC 1633, Available on Internet:http://www.rfc-editor.org/rfc/rfc1633.txt.

Clarke, I., O. Sandberg, B. Wiley and T. Hong, 1999, Freenet: A Distributed Anonymous Information Storage and Retrieval System, Freenet White Paper. http://freenetproject.org/freenet.pdf.

Cocchi, R., S. Shenker, D. Estrin and L. Zhang, 1991, Pricing in computer networks: Motivation, formulation, and example, IEEE/ACM Transactions on Networking, 1(6), 614–627.

Dahlin, M., R. Wang, T. Anderson and D. Patterson, 1994, Cooperative Caching: Using Remote Client Memory To Improve File System Performance, Proceedings of the First Symposium on Operating Systems Design and Implementation, 267–280.

Dai, R., D. Stahl and A. Whinston, 2003, The Economics of Smart Routing and QoS, Proceedings of the Fifth International Workshop on Networked Group Communications (NGC'03).

Deerin, S. and R. Hinden, 1995, Internet Protocol, Version 6 (Ipv6) Specification, Technical Report [On-Line], IETF, Available on Internet, http://www.globecom.net/ietf/rfc/rfc1883.shtml.

Douglis, F. and J. Ousterhout, 1991, Transparent process migration: Design alternatives and the sprite implementation, Software Practice and Experience, 21(7), 757–785.

Eager, D., E. Lazowska and J. Zahorjan, 1986, Adaptive load sharing in homogeneous distributed systems, IEEE Transactions on Software Engineering, 12(5), 662–675.

Eriksson, H., 1994, MBone: The multicast backbone, Communications of the ACM, 37(8), 54–60.

Floyd, S. and V. Jacobson, 1993, Random early detection gateways for congestion avoidance, IEEE Transactions on Networking, 1(4), 397–413.

Ganesh, A., L. Koenraad and R. Steinberg, 2001, Congestion Pricing and User Adaptation, Proceedings of IEEE Infocom [Online], Available on Internet: http://www.ieee-infocom.org/2001/.

Gilder, G., 1997, On the bandwidth of plenty, IEEE Internet Computing, 1(1), 9–18.

Gupta, A., D. O. Stahl and A. B. Whinston, 1994, Managing The Internet as an Economic System, Technical Report [On-Line] University of Texas, Austin. Available on Internet: http://cism.bus.utexas.edu/ravi/pricing.ps.Z.

Gupta, A., D. O. Stahl and A. B. Whinston, 1996a, An economic approach to network computing with priority classes, Journal of Organizational Computing and Electronic Commerce, 6(1), 71–95.

Gupta, A., D. O. Stahl and A. B. Whinston, 1996b, A priority pricing approach to manage multiservice class networks in real time, The Journal of Electronic Publishing [On-Line], 2(1), Available on Internet: http://www.press.umich.edu/jep/econTOC.html.

Gupta, A., D. O. Stahl and A. B. Whinston, 1996c, Economic issues in electronic commerce, in R. Kalakota and A. B. Whinston, eds., Readings in Electronic Commerce, Readings: Addison-Wesley, 197–227.

Gupta, A., D. O. Stahl and A. B. Whinston, 1997a, A stochastic equilibrium model of Internet pricing, Journal of Economic Dynamics and Control, 21, 697–722.

Gupta, A., D. O. Stahl and A. B. Whinston, 1997b, Priority pricing of integrated services networks, in L. McKnight and J. Bailey, eds., Internet Economics, MIT Press, Cambridge, MA, 323–352.

Gupta, A., D. O. Stahl and A. B. Whinston, 1997c, The Internet: A future tragedy of the commons? in H. Amman, B. Rustem and A. B. Whinston, eds., Computational Approaches to Economic Problems, Kluwer Academic Publishers, Dordrecht, The Netherlands, 347–361.

Gupta, A., D. O. Stahl and A. B. Whinston, 1999, The economics of network management, Communications of the ACM, 42(9), 57–63.

Gupta, A., B. Jukic, D. O. Stahl and A. B. Whinston, 2000, Extracting consumers' private information for implementing incentive compatible Internet traffic pricing, Journal of Management Information Systems, 17(1), 9–29.

Gupta, A., L. L. Linden, D. O. Stahl and A. B. Whinston, 2001a, Benefits and costs of adopting usage based pricing in a subnetwork, Information Technology and Management, 2(2), 175–191.

Gupta, A., B. Jukic, M. Li, D. O. Stahl and A. B. Whinston, 2001b, Estimating Internet users' demand characteristics, Computational Economics, 17, 203–218.

Goyal, P., H. Vin and H. Cheng, 1996, Start-time Fair Queuing: A Scheduling Algorithm for Integrated Services Packet Switching Networks, Proceedings of the ACM IGCOMM '96 Conference on Applications, Technologies, Architectures, and Protocols for computer Communication, 157–168.

Harchol-Balter, M. and A. Downey, 1996, Exploiting Process Lifetime Distributions for Dynamic Load Balancing, Proceedings of the SIGMETRICS Conference on Measurement and Modeling of Computer Systems.

Hellerstein, J., 1993, Achieving service rate objectives with decay usage scheduling, IEEE Transactions on Software Engineering, 19(8), 813–825.

Jacobson, V., 1988, Congestion Avoidance and Control, Proceedings of the ACM SIGCOMM '88 Conference on Applications, Technologies, Architectures, and Protocols for Computer Communication.

Jannotti, J., D. Gifford, K. Johnson, M. Kaashoek and J. O'Toole, Jr., 2000, Overcast: Reliable Multicasting with an Overlay Network, Proceedings of the Fourth Symposium on Operating Systems Design and Implementation, 97–212.

Lederer, P. and L. Lode, 1997, Pricing, production, scheduling, and delivery-time competition, Operations Research, 45, 407–420.

Lewis, P., 1996, All you can eat price is clogging Internet access, New York Times.

Li, L. and Y. Lee, 1994, Pricing, and delivery-time performance in a competitive environment, Management Science, 40, 633–646.

Lin, Z., P. Ow, D. O. Stahl and A. B. Whinston, 2002, Exploring traffic pricing for the virtual private network, Information Technology and Management, 3, 301–327.

Litzkow, M. and M. Solomon, 1992, Supporting checkpointing and process migration outside the UNIX kernel, Proceedings of the Winter 1992 USENIX Conference, 283–290.

MacKie-Mason, J. and H. Varian, 1995, Pricing the Internet, in B. Kahin and J. Keller, eds., Public Access to the Internet, Prentice-Hall; Englewood Cliffs, NJ.

Marbach, P., 2001, Pricing differentiated services networks: Bursty traffic, Proceedings of IEEE Infocom [On-Line], Available on Internet: http://www.ieee-infocom.org/2001/.

Mendelson, H., 1985, Pricing computer services: Queuing effects, Communications of the ACM, 28(3), 312–321.

Mendelson, H. and S. Whang, 1990, Priority pricing for the M/M/I queue, Operations Research, 38(5), 870–883.

Metcalfe, R. and D. Boggs, 1976, Ethernet: Distributed packet switching for local computer networks, Communications of the ACM, 19(7), 395–404.

Mullender, S., G. van Rossum, R. van Renesse and H. van Staveren, 1990, Amoeba—A distributed operating system for the 1990s, IEEE Computer, 23(5), 44–53.

Naor, P., 1969, On the regulation of queue size by levying tolls, Econometrica, 37, 15–24.

Nichols, K., S. Blake, F. Baker and D. Black, 1998, Definition of the Differentiated Services Fiel(DS Field) in the IPv4 and IPv6 Headers, Internet RFC2474, Available on Internet: http://www.ietf.org/rfc/rfc2474.txt.

Nichols, K., V. Jacobson and L. Zhang, 1999, A Two-bit Differential Services Architecture for the Internet, Internet RFC2638, Available on Internet: http://www.faqs.org/rfcs/rfc2638.html.

Odlyzko, A., 2001,Paris Metro Pricing for the Internet, Proceedings of ACM Conference on Electric Commerce, ACM Press, NY, 140–147.

Odlyzko, A., 2001, Internet pricing and the history of communications, Computer Networks, 36, 493–517.

Patterson, R., G. Gibson, E. Ginting, D. Stodolsky and J. Zelenka, 1995, Informed Prefetching and Caching, Proceedings of the ACM Fifteenth Symposium on Operating Systems Principles, 79–95.

Qiu, L., Y. Yang, Y. Zhang and S. Shenker, 2003, On selfish routing in Internet-like environments, Proc. of ACM SIGCOMM, Germany, August 2003.

Ratnasamy, S., P. Francis, P. Handley and R. Karp, 2001, A scalable content addressable network, ACM SIGCOMM.

Roughgarden, T. and E. Tardos, 2002, How bad is selfish routing? JACM, 49(2), 236–259.

Rowstron, A. and P. Druschel, 2001, Pastry: Scalable, Distributed Object Location and Routing for Large-Scale Peer-to-Peer Systems, IFIP/ACM International Conference on Distributed Systems Platforms (Middleware), Heidelberg, Germany, 329–350.

Sarkar, P. and J. Hartman, 1996, Efficient Cooperative Caching Using Hints, Proceedings of the Second Symposium on Operating Systems Design and Implementation.

Savage, S., T. Anderson, A. Aggarwal, D. Becker, N. Cardwell, A. Collins, E. Hoffman, J. Snell, A. Vahdat, G. Voelker and J. Zahorjan, 1999, Detour: A case for informed Internet routing and transport, IEEE Micro, 19(1), 50–59.

Scotchmer, S., 1985, Two-Tier Pricing of shared facilities in a Free-Entry equilibrium, RAND Journal of Economics, The RAND Corporation, 16(4), 456–472.

Scotchmer, S., 1986, Non-anonymous crowding: the core with a continuum of agents, Harvard Institute of Economic Research, Discussion Paper No. 1236.

Shenker, S., 1995, Service models and pricing policies for an integrated services Internet, in B. Kahin and J. Keller, eds., Public Access to the Internet, Prentice-Hall, Englewood Cliffs, NJ, 315–337.

Smith, P., 2001, BGP Multihoming Technique, NANOG 23, http://www.nanog.org/mtg–0110/smith.html, October.

Stahl, D., R. Dai and A. Whinston, 2002, An economic analysis of multiple internet qos channels, The Economics of the Internet and E-commerce, in M. Beye, ed., Chapter 10, Elsevier. University of Texas, Austin.

Stoica, I., R. Morris, D. Karger, M. Kaashoek and H. Balakrishnan, 2001, Chord: A scalable peer-to-peer lookup service for Internet applications, ACM SIGCOMM, San Deigo, CA, 149–160.

Tangmunarunkit, H., R. Govindan and S. Shenker, 2001, Internet path inflation due to policy routing, Proceedings of SPIE ITCom, Denver, CO, August 2001.

Vickrey, W., 1969, Congestion theory and transport investment, American Economic Review, 59, 251–261.

Waldspurger, C. and W. Weihl, 1994, Lottery Scheduling: Flexible Proportional-share Resource Management, Proceedings of the First Symposium on Operating Systems Design and Implementation, 1–12.

Waldspurger, C. and W. Weihl, 1995, Stride Scheduling: Deterministic Proportional-share Resource Management, Technical Report TM–528, Massachusetts Institute of Technology Laboratory for Computer Science.

Wang, X. and H. Schulzrinne, 2001, Pricing network resources for adaptive applications in a differentiated services network, Proceedings of IEEE Infocom [On-Line], Available on Internet: http://www.ieee-infocom.org/2001/.

Westland, J., 1992, Congestion and network externalities in the short run pricing of information systems services, Management Science, 38, 992–1009.

Zhao, B., L. Huang, J. Stribling, S. Rhea, A. Joseph and J. Kubiatowicz, 2004, Tapestry: A resilient global-scale overlay for service deployment, IEEE Journal on Selected Areas in Communications, 22(1), 41–53.

Further reading

Mendelson, H., D. Clark, D. Estrin and S. Herzog, 1996, Pricing in computer networks: Reshaping the research agenda, Journal of Telecommunications Policy, 20(3), 183–201.

Chapter 11

TOWARD AN ECONOMICS OF THE DOMAIN NAME SYSTEM

MILTON MUELLER

Syracuse University School of Information Studies

Contents

Handbook of Telecommunications Economics, Volume 2, Edited by S. Majumdar et al.

DOI: 10.1016/S1569-4054(05)02011-7

1. Introduction

In August 2000, an arbitration panelist of the World Intellectual Property Organization (WIPO) ordered the transfer of the domain name barcelona.com to the Barcelona City Council. The domain name had been registered by a New York-based start-up company, Barcelona.com, Inc., back in 1996. The company planned to develop the domain into an Internet portal for Barcelona residents, tourists, and enthusiasts. Barcelona.com was in the process of negotiating with potential investors when the City filed a complaint under the Internet Corporation for Assignment Numbers' (ICANN) Uniform Domain Name Dispute Resolution Policy.

The Barcelona City Council based its complaint on the city's many registered trademarks that included the word 'Barcelona,' many of which had just been registered in 1996 or 1995. Although Spanish trademark law prohibits the registration of place names, the City claimed that its trademarks, such as '*Barcelona Excelentísimo Ayuntamiento de Barcelona*,' could be confused with the domain name barcelona.com. The City Council also claimed that the entrepreneurs were cybersquatters who had registered the name in bad faith only to sell it to the city. World Intellectual Property Organization panelist Marino Porzio agreed, and ordered the Internet entrepreneurs to hand over their names. Its prospective portal business was destroyed.

Years of court challenges and appeals over the right to the name followed. Finally, on June 2, 2003, a U.S. Court of Appeals denied all of the City Council's arguments regarding trademark rights, bad faith use of the domain and cybersquatting. A Spanish municipality was told by a U.S. court that it had no rights to the name under Spanish or U.S. law. It was declared a reverse domain name hijacker, and ordered to restore the name to the original registrant. On August 16, 2003, the domain was transferred back to Barcelona.com, Inc.

The Barcelona.com conflict was only one of thousands caused by the rise of the domain name industry after the rise of the World Wide Web (WWW) in 1993. The rise of a domain name market has produced conflicts over trademark rights, personal names, and misspellings of famous names, legal jurisdiction, and the creation of new generic top-level domains (TLDs), such as *.web*. In essence, the Internet domain name space created a new kind of value, and law and institutions had to make rapid and major adjustments to human efforts to capture that value. This chapter focuses on the economics of the domain name system. The Internet has commoditized identifiers on an unprecedented scale, creating global identifier-related services markets with their own unique economic characteristics. In the market for names, technical functions interact with semantic meaning in complex and sometimes explosive ways.

The term 'identifier' refers to a string of characters or symbols that can be used to give a name to users or objects on a network, such as servers, routers, printers, documents, or multimedia content. Unique identifiers serve a critical role in most networks or online environments (Berners-Lee, 1996, 1998; Paskin, 1999). By uniquely identifying an object on a network, names make them retrievable or accessible (Saltzer, 1993). A telephone number is an identifier, and so is an ISBN number (the international code for books used by publishers). An Internet Protocol (IP) address is an identifier used by routers to determine how to get packets from their origins to their destination. This chapter, however, concentrates exclusively on Internet domain names. What makes domain names of particular interest is that they (unlike IP addresses or ISBN codes) have evolved into things that can be bought and sold in a market. The supply of domain names to Internet users generates an estimated $2.5 billion annually in revenues. Despite the economic significance of this market and its centrality in provoking policy debates about Internet governance, there is surprisingly little literature about the economics of domain names (FTC, 1998; Rood, 2000; Mueller, 2002a).

Domain names are composed of alphanumeric characters in a hierarchical structure. It was the rise of the WWW after 1994 that created most of the demand for domain name registrations (Mueller, 2002a, Chapter 6). Three-level domain names became synonymous with Website locators (URLs). Domain names began to appear in television ads, on the sides of buses, on business cards. For a short time, guessing the domain name associated with an organization in order to find its Website became a common way of navigating the Internet. In late 1995, responding to the rush of Web-inspired domain name registrations, the National Science Foundation authorized the U.S.-based domain name registry Network Solutions, Inc. (NSI) to charge annual fees for registering and supporting domain names. The market for registrations doubled in size annually during the Internet boom of the second half of the 1990s[1]. During its frenzied peak in the first two quarters of 2000, revenues grew at a 150 percent annual growth rate. Responding to artificial scarcity in the supply of names, robust secondary markets for domain names began to evolve. Speculators registered hundreds of names and advertised them for sale to the highest bidder.

In line with the general depression in Internet-related markets, the number of domain name registrations worldwide declined by about 8 percent from October 2001 to June 2002, as large numbers of speculative and promotional registrations in the generic TLDs .com, .net, and .org were not renewed. Domain

[1] See NSI Annual Report (2000).

Table 1
Growth of .com, .net, and .org domain name registrations

Date	# Registered domains
July 1996	488,000
Jan. 1997	828,000
July 1997	1,301,000
Jan. 1998	2,292,000
July 1998	3,282,117
Jan. 1999	5,504,151
July 1999	9,098,066
Jan. 2000	13,402,448
July 2000	23,864,611
Jan. 2001	33,045,397
July 2001	37,539,541
Jan. 2002	30,225,000
July 2002	29,255,166

name registrations in most country-code TLDs, however, still grew at double-digit rates, and so did the new TLDs .info and .biz. The domain name market as a whole returned to positive with modest growth in the last quarter of 2002. About 46.5 million domain names are registered worldwide as of the third quarter of 2002.

The significance of domain names goes beyond their relatively modest economic niche in the market for Internet-related services. The Domain Name System (DNS) sits at the heart of the Internet infrastructure. Its technical functions are critical to the ability of users to establish connections to each other, just as consistent use of telephone numbers is critical to interoperability of the telephone system. Because control of the top level of the domain name hierarchy gives its administrator considerable power over the domain name industry and its customers, commercialization of domain name services stimulated major institutional changes in international governance arrangements (Mueller, 2002a). The U.S. government and key corporate actors in the Internet business desired to place domain name regulation into a global jurisdictional framework and to avoid control of the DNS by traditional intergovernmental organizations (NTIA, 1998). This led to the 'privatization' of the regulatory responsibilities, policy-making, and coordinating functions associated with domain names. Much of this happened because the semantic properties of domain names led to legal and political conflicts over trademark rights (and other types of claimed rights to names), resulting in new international rules and rulemaking mechanisms for arbitrating rights to names.

Aside from that, how things are named on the Internet—a problem that involves many actual or proposed standards and services in addition to DNS—profoundly affects the Net's usability and efficiency. Many of the economic and policy issues initially raised by DNS are being or might be repeated by the growth of other naming systems (Bechtold, 2002). Thus, experience with DNS may be taken as an example (for better or worse) of what economic issues to anticipate with different naming designs[2]. The DNS protocol also may play a role in new information and telecommunication services. The ENUM protocol, for example, uses the DNS to map telephone numbers to IP addresses, facilitating efficient interconnection of IP telephony systems and traditional telephone networks using E.164 numbering as identifiers (Huston, 2002; Hwang and Mueller, 2002). Already, controversies regarding DNS economics and policy have spilled over into the implementation of ENUM (Cannon, 2001; McTaggart, 2001; Rutkowski, 2001).

The market for domain names and related services will grow as the Internet diffuses. The institutional changes provoked by the growth and commercialization of DNS will have a lasting impact on communication industries. It is important, therefore, to understand how the DNS functions, the economic properties of domain names as resources, the supply of domain name services, and the industrial organization characteristics of the domain name supply industry and its regulator, the ICANN.

The exposition is divided into the following parts. First, a basic technical description of the assignment and resolution of domain names is provided. The next section describes the demand characteristics of domain names. Next, the characteristics of the supply industry are examined. Finally, the chapter looks at the policy issues associated with the industry. Most of the focus is on the economics of demand and supply, which are the least-analyzed aspects of the domain name problem. A great deal of policy has been made and debated with little application of economic theory and no empirical research into the underlying economics. One purpose of this chapter is to identify and call attention to those areas where research is needed.

[2] For example, the 'Handle' system of digital object identifiers also has a centralized root structure like DNS, which may lead to governance issues if it becomes widely adopted. On the other hand, the handle system is designed to permit unlimited expansion of the top level (see Kahn and Wilensky, 1995).

2. Technical description of DNS

Domain names most likely have gone unnoticed by economists because before one can understand the economics one must make sense of a complicated Internet protocol known as DNS and its implementation by Internet service providers (Mockapetris, 1987; Albitz and Liu, 1992). What follows is a brief overview of the technical structure of DNS. The DNS protocol specifies a name space, permitted structures and characters for names, formats for storing data associated with names, and a mechanism for retrieving that data in client–server interactions. The two basic aspects of domain names are assignment (the process of delegating exclusive control of a name within the DNS name space to a responsible party) and resolution (the process of discovering what IP address is associated with a given domain name).

2.1. The name space and name assignment

Domain names are alphanumeric strings used to name computers on the Internet. Domain names are most often seen as parts of e-mail addresses and Web URLs. In the e-mail address vogelsan@bu.edu, bu.edu is a domain name registered by Boston University and vogelsan is a user ID assigned by BU. The name www.nokiagirls.com is a three-level domain name that functions as a Web URL. Domain names are organized hierarchically, with each level representing the delegation of responsibility from one organization or entity to another. The starting point of the hierarchy is known as 'the root,' which is unnamed. For example, the operator of the .com domain was delegated the right to register names under .com from the authority over the DNS root; the owner of the nokiagirls domain was delegated that authority by the .com domain, and the Web server www was so named by the owner of the nokiagirls domain. In the most commonly used notation, dots separate the different levels of the hierarchy. The label farthest to the right is termed the 'TLD,' the next level to the left is known as the second-level domain, and so on down the hierarchy. The DNS was designed to permit deep hierarchies but in actual practice it is rare to see more than four levels used, and most of the real work is done by name servers at the top and second levels.

In purely technical terms, the supply of unique domain names is inexhaustible. Within any single TLD, second-level domain names can use 63 character spaces and 37 available characters, meaning that there are 37^{63} possible combinations of characters available. This is an unimaginably large number, and that array is available in one level of the hierarchy alone. The DNS protocol permits the hierarchy to go down more than 100 levels. Moreover, there could be thousands more TLDs. Of course, huge parts of the available pool of names are undesirable from a human point of view, being either meaningless or too long to be usable, or both.

To function properly as a global identifier, each domain name must be unique and exclusively assigned. Early attempts to have one central naming authority coordinate all name assignments broke down under the weight of constant increases in the number of host computers on the network (Vixie, 1994). The DNS solved this problem by creating a *hierarchical* name space and delegating the authority to assign domain names down the levels of the hierarchy. In this arrangement, a central authority is responsible for assigning unique and exclusive names at the top of the hierarchy. Each assignee of a top-level name then takes responsibility for coordinating the assignment of unique and exclusive second-level names. The owner of a second-level name can in turn assign unique third-level names, and so on.

2.2. *Resolution, name servers, and BIND software*

As unique identifiers, domain names are sometimes referred to as Internet 'addresses.' But the *real* addresses that guide the routing of data packets are IP addresses, 32-bit binary numbers uniquely assigned to a host. The domain name serves as a higher-level, meaningful name. In order to route information across the Internet a domain name must be mapped to its corresponding IP address, a process known as 'resolving a domain name,' or as 'resolution.' That layer of indirection serves two purposes. First, semantically meaningful names are easier to key in and remember than strings of numbers. Second, higher-level names can provide a more stable, persistent identifier than an IP address. The IP addresses assigned to groups of computers are determined by the topology and routing practices of the Internet, which tend to change relatively frequently as networks grow, shrink, or are reorganized. By separating a computer's public identifier from its IP address, and relying on some mapping function to translate between the two, a network manager can give computers stable names, which can be used consistently even when their IP addresses change.

The DNS can be characterized as a highly distributed global database, wherein the responsibility for entering, updating, and serving up data on request is handled by hundreds of thousands of independently operated computers known as *name servers*. At the top of the DNS hierarchy, the root-level authority registers TLD names, and the root servers handle requests to resolve TLDs. The 250 or so TLD name registries operate name servers that contain lists of all second-level domain names registered under that TLD, and pointers to two or more name servers capable of operating that domain. Thus an economist might view the DNS as a distributed network of compatible name servers, the function of which is to enable assignment and resolution of globally unique names.

Whenever end users use a domain name, their e-mail or browser client (or some other software) must send a query out to resolve the name into an IP address. Typically, one resolves a name by sending a query containing the domain name to the nearest name server operated by an organization or its' Internet Service Provider (ISP). That name server first checks to see it if has already stored the answer to the query in its cache after responding to a previous query for the same domain name. Thus, the relevant records for the most popular domain names, such as google.com and yahoo.com, will already be stored and there will be no need to waste time and cycles searching for it. Caching is critical to the scalability of DNS. Most DNS queries never need to go out into the Internet; if they did, the reliability and scalability of the Internet would be impaired. Only if the local name server does not know the answer, or if an earlier record has expired, must it initiate the process of resolving the name.

The resolution of a domain name follows the same hierarchical structure as the assignment of names. A query first goes to the root, which tells the resolver which TLD name server to query; the resolver then queries the appropriate top-level name server, which tells it which second-level domain name server stores the records for that domain, and so on.

In principle, anyone can write software that implements the DNS protocol. In reality, an implementation software known as Berkeley Internet Name Domain (BIND) became established in the 1990s and dominates the industry. Berkeley Internet Name Domain is open-source software but its revision and release are controlled by a group of Internet Engineering Task Force (IETF) veterans organized as the Internet Software Consortium. The BIND's dominance has important economic consequences, particularly regarding the potential for competing DNS roots. The software release contains a list of the IP addresses of the official root servers, which are used as default values by the name server unless those values are manually altered by the name server administrator.

2.3. The root servers

To make sure that all domain names on the Internet are unique and can be globally resolved, there must be a coordinated source of information about what TLDs exist and at what IP addresses their name servers can be found. That function is performed by the root administrator(s). The root administrator(s) determines what names exist at the top of the DNS hierarchy. The list of recognized TLDs and associated records of their IP addresses is known as the *root zone file*. As we shall see in Section 4.1, this technical coordination function provides a point of leverage that can also provide control over market entry and the conduct of suppliers and consumers. The root is the one point of centralization on what is otherwise a highly distributed system.

There are 13 official root servers. The existence of multiple root servers provides redundancy in case one root server crashes; they are also geographically distributed in order to reduce query time and expense for users in different parts of the world[3]. The number of root servers is technically limited, however, to a maximum of 13[4]. The limit on the number of root servers has important economic consequences. Until some new way of coordinating a larger number of root servers is invented, Internet growth increases the load on existing root servers rather than providing an incentive to supply additional root servers.

3. The demand for domain names

Domain names combine technical and human functions. On the technical side, they provide a unique symbol, similar to a telephone number, which facilitates the identification of objects on the network and helps guide the movement of information from its source to its destination. As words, symbols or phrases that are *meaningful* as well as unique, they also perform psychological and communicative functions, such as making an address easier to remember or more likely to attract attention. The semantic aspects of demand introduce radically different economic properties, and therefore must be distinguished from the technical aspects and analyzed separately.

3.1. Demand for technical functions

In order to have an Internet identity end users need an assignment of a unique alphanumeric string devoted to their exclusive use, as well as continuous maintenance of public database records reserving that string and storing the data needed to resolve the name. This aspect of the service may be called the *registration service*. Users of Internet domain names also need a service that, when queried

[3] In a recent coordinated denial of service attack on the root servers, 7 of the 13 root servers were disabled, yet most Internet users noticed nothing, or a slight increase in waiting time. David McGuire and Brian Krebs, 'Attack on Internet Called Largest Ever,' Washingtonpost.com, October 22, 2002.

[4] The limit comes from the ability of a single UDP packet to carry the IP addresses of all the servers.

at any time by other users of the public Internet, responds with the information needed to resolve the name. This is called *name service* or *name resolution* service. Registration and name resolution services are needed regardless of whether the domain name is meaningful or not. As noted earlier in the technical section, the layer of indirection provided by a domain name is valuable quite apart from its semantic functions. In this regard, it is worth noting that for hundreds of millions of non-English-reading Internet users around the world, the ASCII characters used by the standard DNS protocol are basically meaningless, being based on an alphabet that is mostly illegible to them.

For small-scale consumers, the supply of registration and name resolution services can be efficiently bundled with the provision of basic Internet access and Web hosting services. (However, this may increase switching costs, as the ISP has operational control of the name and any attempt to change ISPs might affect the name's functionality. There is a longer discussion of switching costs in Section 3.1.1.) Larger consumers often self-provide their own DNS servers or contract with separate specialized vendors for the various components of the service (registration, name service, Internet access, and hosting).

3.1.1. Portability and sunk costs

The costs of the technical services associated with domain names are quite low relative to the total costs of having an Internet presence. Yet the value of an established name can be enormous, when it has been used for a long time and numerous directories, links or publicity materials direct traffic to the name. Consumers of domain name registration services incur sunk costs when they invest in publicizing their unique identifier (e.g., Website or e-mail address). This restricts their ability to easily change names, which may affect their ability to change service suppliers. As a name (e.g., firm.com) becomes well known by external parties or embedded in their stationery, directory listings, and so on, changing to a new domain name (e.g., firm.biz) can involve substantial switching costs. Additional investments must be made to inform consumers of the new name. The change may result in foregone sales to consumers who fail to become aware of the new name in time, another switching cost. While it is technologically possible, even easy, to automatically redirect traffic from the old domain name to a new one, this requires retaining a subscription to the old domain for some time. It is theoretically possible, therefore, that a supplier could raise the price to locked-in customers.

In a (non-empirical) study of the significance of sunk costs on competition in the registry market, the U.S. Federal Trade Commission's Competition and Economics Bureaus wrote:

> The economic analysis of markets with switching costs has identified a number of factors that, in appropriate circumstances, can diminish the ability and the incentive of a supplier to act opportunistically with respect to its locked-in customers: ... (1) the extent to which prospective customers are aware of the possibility of supplier opportunism; (2) the extent to which customers have effective means (e.g., enforceable long-term contracts) to protect themselves against opportunism; (3) the intensity of competition among suppliers; and (4) the importance of reputation and repeat business to suppliers' current and future profits (FTC, 1998).

All four of these factors apply in some degree to the domain name market, although their effectiveness has been limited by artificial constraints on competition imposed by ICANN and the U.S. Department of Commerce. Most customers, particularly those most dependent on their domain name identities, are aware of the prospect of supplier opportunism (some small-scale casual consumers may not be). Long-term contracts are an option, and have become more available as the market has become more competitive. The contractual terms would become more favorable to consumers with additional competition among registries. Competition among suppliers is potentially almost unlimited, although the regulatory restrictions on entry discussed below limit it. And of course, in a subscription-based service, such as domain name registration, repeat business and reputation are critical.

Number portability has been a major policy issue in telephony and there is a significant amount of economic literature on switching costs in telephone numbers (Andeen and King, 1997; NERA, 1998; Reiko and Small, 1999; Farrell and Klemperer, 2001; Gans et al., 2001; Viard, 2003). In telephony numbers were assigned to connectivity providers (i.e., carriers) in blocks. Changing one's carrier meant changing one's number, creating a substantial switching cost for local subscribers. Implementation of number portability (in 800 numbers, Local Number Portability and Mobile Number Portability) required investment in and operation of a database-driven number recognition system common to all carriers. The key policy issue is how the costs of this added infrastructure are distributed. The empirical and theoretical literatures indicate that number portability facilitates consumer choice and carrier competition (Gans et al., 2001; Viard, 2003). Several authors, however, have raised doubts about the ability of mandated number portability to make the market as a whole more efficient, and have emphasized the need to vest users with property rights in their identifiers (Reiko and Small, 1999).

By way of contrast, under the DNS protocol names have always been virtual resources grounded in a shared database. Thus, domain names have always been portable across connectivity providers unless consumers were ill-informed and allowed their ISP to register the domain in the ISP's name. Moreover, customers have always had stronger, more transferable rights in domain names than in telephone numbers. Even when telephone number portability is implemented, customers cannot reserve them in unlimited quantities without use, cannot engage in free reassignment or sale of them, and have a very limited ability to select a specific number. Domain names in the most popular open TLDs, in contrast, can be reserved without use and resold, and the selection of the desired string is entirely in the hands of the customer. Thus, an important part of the demand for domain names stems from the ability they give consumers to obtain unique identifiers that are portable across connectivity providers, and whose value can be captured and exploited more easily by the user.

Nevertheless, regulation of the domain name market has provided consumers with an additional layer of portability. In the generic TLDs and some country codes, regulations separate the wholesale 'registry' part of the service from the retail 'registrar' functions. This makes domain names portable across registrars as well as across ISPs. (It is not possible to make domain names portable across registries without radical changes to the DNS protocol.) This aspect of the domain name market is a supply side phenomenon and will be discussed in more detail in Section 4.2.2.

3.1.2. Network externalities (demand-side economies of scope)

Anyone with control over the configuration of a DNS client or name server has the ability to choose which computer they treat as the root of the hierarchy for resolution of domain names[5]. To some, this suggests that the DNS root need not be a centralized monopoly and that competition could prevail in the provision both of the definition of the root zone file and the operation of root servers. There are, however, strong demand-side network effects in the choice of a DNS root that undermine competition. When Internet user A registers a domain name, she almost always wants that address to be compatible with the DNS implementations of all existing and future communication partners (which we will refer to as N). 'Compatibility' in this case means that any N who encounters A's domain name will, after sending the appropriate queries to the name servers to which N is connected, be returned the *correct* answer about the IP address associated with A's domain name. Incompatibility means that no answer, or an incorrect answer, is returned, which makes resolution of the name impossible.

Competing DNS roots can result in varying levels of incompatibility (Table 3). In general, incompatibility can arise in two ways. First, if A's and N's name

[5] In actual practice, default values are set in the implementation software, which for 90 percent of the world's computers is a software known as BIND (see below).

assignments are derived from different, uncoordinated roots, A's domain name may not be globally unique. If it is not unique, there is a chance that the name servers that handle user *N*'s query will confuse it with some user other than A who has adopted the same name, and return the wrong information. Second, even if the name is unique, the name servers that handle user *N*'s query may not know where to find the information needed to resolve A's name if it does not begin its query process at the same root as A.

Obviously, the value of a domain name to A increases as the scope of its compatibility widens. An identifier that cannot be resolved by other users is as useless as a telephone set without any connectivity. If only a fragment of A's universe of communicating parties can resolve the name, A will have to carefully manage and restrict his use of the identifier and utilize another identifier for use with communication partners who employ an incompatible root. Because so much of the communication on the Internet takes place among parties who do not know each other and have no prior relationships that would allow for coordination, the compatibility barriers created by fragmentation of the name space are high. There is very little reason to ever want to utilize a domain name that is not globally compatible. A single root increases the chance that all domain names will be compatible. Thus, network service providers who want to offer universal connectivity to their customers or clients have a strong incentive to converge on a single DNS root. Consumers of domain name services have a strong incentive to select their names from a single name space provider with a dominant market share or a monopoly. Of course, competing roots could be coordinated in certain ways, and the effect would be identical to that of an interconnection agreement among competing telecommunication networks (Mueller, 2002c).

3.1.3. Pure technical demand vs. semantic or human demand

To conclude the analysis of technical demand factors, in a market unsullied by human factors, the URL xyz.ispname.at/users/cs/hjjfg23s.htm is a perfect substitute for www.mywebsite.org. In other words, there is no significant difference between a long and meaningless URL and a URL based on a domain name registered specifically to contain a meaningful reference to a person or a Website's content. Similarly, in a purely technical market for domain names, the e-mail address me@xx3c.gobbledygook.ispname.com is a perfect substitute for me@myname.com. Of course, when human factors are taken into account, the economic value of those pairs of names is quite different (see below). It follows that if demand for domain name services was based entirely on their technical functions, the market for domain names would be much smaller than it is now. Most individuals and organizations would need only one domain name, and even the largest organizations would need only a few. Domain name users would be willing to get all the additional names they wanted by extending the hierarchy under their existing domain name(s), or by extending the directories of Web URLs to the right. There would be little incentive to register names specific to products, people, or a variety of other potential identities (if they were hosted on the same machine). There would be no incentive to speculate in domain names, no incentive to register multiple names for the same purpose, and no incentive to use TLD names to differentiate users or applications. Indeed, there would be little need for end users to participate in the selection of their domain names at all. Assignments could be done by machines, as users would be indifferent as to the specific string they received as long as it was unique and exclusive.

3.2. Demand for semantic functions

Most of the market value—and the political and social complications—of domain names stem from the semantic aspects of demand and related human factors considerations. As explained below, the demand for a memorable identifier, the desire to economize on the length of the domain name, concerns about authenticity, and the demand for names that will attract Web traffic or aid in search and navigation functions have been powerful influences on the market for domain names.

As the market for domain names has matured it has become increasingly clear that individuals and organizations demand multiple online identities rather than merely one. The multiplication of identities is not a function of imperfect compatibility among systems, as is often assumed. Rather, it reflects end users' different roles and their need to organize their increasingly numerous and complex online interactions. A person on the Internet may have an Internet identity at work, a different Internet identity at home, a hotmail address for when they are on the road, and so on. They probably have multiple identities within their home ISP account, corresponding to different family members or different online persona. They may want to use one identifier on public Websites and e-mail lists and another one for private, priority e-mail in order to foil spammers. They may want to disguise themselves to access illicit Websites, or take on new identities for fun. A company wants distinct identifiers for its corporate name, its product and brand names, its departments, task forces, etc. Online identities are created and destroyed by the tens of thousands on a weekly basis. The demand for distinctive identity and the way in which computer technology makes it inexpensive to create, disseminate and erase identifiers combine to fuel semantic demand for names.

The discussion below isolates four aspects of demand related to semantic and human factors:

- Names as memorable identifiers;
- Economizing on the length of the name;
- Names as authenticators;
- Names as search tools or navigation aids.

3.2.1. The demand for memorable identifiers

The primary driver of domain name demand is the need of Internet users for *memorable identifiers*. The registrant of a domain name uses the name(s) to advertise and communicate their Internet location, and the registrant's communication partners store them or remember them for use when they want to find the registrant's Internet location. Users select names to convey something about themselves (e.g., a cat-lover may adopt paw as a domain if given the choice, whereas a critic of corporations might value registering a business' name under a .sucks TLD to post critical information). The demand for meaning, memorability, or catchiness in domain names all assume that the names are visible to and used by members of the public, rather than by machines alone. The further domain names recede from public view and the more the recognition and storage of Internet identifiers is performed by machines, the less important these functions become as a determinant of demand. However, there are no known models for incorporating semantics into the demand function for identifiers. Here is an interesting problem for a theorist.

The demand for meaning or memorability profoundly affects the consumer's relationship to the available supply of names. It means that the vast majority of available unique strings are of no interest to an individual consumer. Consumers are only interested in combinations of characters that are semantically related to the use they plan to make of the name. We can refer to those clusters of meaningful combinations of strings as 'semantic clusters' (e.g., a small-time restauranter named Joe would be interested in joe.com, joes.com, joes-eats.com, joes.food, and many others). Demand-side competition for the right to occupy domain name assignments occurs when there is some degree of overlap between the semantic clusters of two or more users. The alternatives available to prospective registrants are not homogeneous, perfect substitutes. The logic of substitution is based on subjective factors of interpretation; it is fuzzy, not discrete.

The existence of meaning also leads to selection of domain names by customers rather than assignment by suppliers. Of course, in exceptional cases users of driver's licenses and telephone numbers are allowed to select vanity license plates or particular telephone numbers, too. Because the domain name space is so vast, however, it can accommodate consumer selection of their preferred identifier much more readily than other identifier spaces. Consumer selection is the rule rather than the exception.

One of the key problems facing this aspect of demand is the existence of different language scripts (Hotta, 2001; Katoh, 2001). For a very large and growing portion of the Internet's user base[6], ASCII-based domain names are hardly meaningful and memorable. This has led to demand for domain names based on the more complex Unicode standard, which can accommodate most other scripts. Experimental implementations of multilingual domain names, involving proprietary browser plug-ins, are operational now but they are not globally compatible. Prodded by market and political pressures, the IETF has reluctantly made an effort to figure out how to define a standard for internationalized domain names while retaining global compatibility[7]. Adding scripts involves such basic changes to the DNS protocol, however, that severe compatibility issues arise in any transition. Internationalizing domain names involves a long-term standards migration. On the other hand, making available domain names in non-ASCII characters would greatly expand the demand for registrations, by expanding the population to whom memorable identifiers were available, and by increasing the usability of the names. The localization of domain names would also, however, create many new opportunities for conflicts over name rights (see Section 5.2).

Because the human capacity for memory is limited, and because time is scarce, shorter domain names tend to be more valuable than longer ones, all else being equal. Despite the attempt of some trademark lobby groups and trademark holders involved in domain name litigation to portray domain names as both powerful keywords and the legal equivalent of source identifiers, most companies utilize truncated, easier-to-remember and type-in domain names as their primary online identity. Merrill Lynch may have registered merrilllynch.com for defensive purposes, but the domain name it really uses in its publicity materials is ml.com. Likewise, the Ford Foundation relies on fordfound.org, and has not even registered fordfoundation.org.

In this calculus of memorable identifiers, the supply of clusters of names meaningful to and desired by any individual user expands linearly with the addition of TLDs. However, the meaning and hence the value of second-level

[6] It is estimated that by 2003, two thirds of all Internet users will be non-English speakers. Over 90 percent of the world's population speaks a primary language other than English.

[7] For the charter and schedule of the 'IDN' (internationalized domain name) working group of the IETF, see http://www.ietf.org/html.charters/idn-charter.html

domains is dependent upon the meaning or qualities associated with the TLD (e.g., the Disney Corporation is unlikely to be interested in any names under a .sex TLD whereas a porn site operator would have both legal and business reasons to avoid registering under a .kids TLD).

3.2.2. Names as authenticators

As a complement to and extension of memorability, names can aid in the performance of authentication functions. That is, in addition to establishing a mnemonic linkage between the character string and its owner, it can also suggest or guarantee that the owner is an entity of a certain type or status. The .edu TLD, for example, is (mostly) restricted to organizations that are accredited 4-year degree-granting institutions of higher education in the U.S[8]. Thus, if one sees a domain name with .edu at the end, and *if* one is familiar with the policies of the .edu registry, one knows more about the registrant than one would otherwise. An example of a more modest form of authentication would be a country-code TLD that requires that the registrant be domiciled in the country referenced by the TLD. Authentication is always a matter of degree, and the level of trust or credibility that can be provided by domain name registration authorities is necessarily limited.

Linking domain name registrations to verification or authentication functions makes domain name registration much more expensive and the number of registrants smaller, which is why restricted domains are not as common as one might expect. Registries that perform the additional functions are required not only to reserve a unique string and provide name service, but must also engage in verification and enforcement functions that require human judgment, manual processing of registrations, and enforcement procedures or dispute resolution mechanisms. The problem is complicated even more by political and regulatory issues. One could, for example, imagine a .xxx domain for pornographic content or a .kids TLD for children's content. But who will define and apply the criteria for deciding what is 'pornographic' and what is 'suitable for kids'? Of course, private sector firms have a long history of providing ratings of video games and movies on a self-regulatory basis. Such signals lower the search costs of consumers by narrowing down the choices of consumers into appropriate ranges and converging the expectations of suppliers and consumers. However, in mass media industries these self-rating systems achieve uniformity of application by emerging from collective action among dominant industry players, often incented by the threat of government intervention. Internet domain names are global, not

[8] The standards for .edu registration were just expanded to include Community Colleges.

national, in scope, and the range of activities embraced by them are much more heterogeneous than movie or video game ratings. In a free marketplace, domains can acquire a reputation just as a brand name or a movie rating category can. But they can also lead to political friction over divergent expectations or attempts by politicians to impose the classifications on unwilling parties.

Domain names as authenticators overlap with efforts to make top-level or second-level domains some kind of taxonomy or categorization of registrants. Country codes are the most obvious example of a categorization scheme imposed on the domain name space. As an indication of the limitations of this approach, however, it should be noted that many country code TLDs do not limit registrations to businesses or individuals located in that country. Many country codes restrict second-level domains to categories, such as ac or edu for academic institutions, co or com for commercial organizations, etc., and do not allow individual organizations or people to register in the second level. Most of the sorting of registrants into these categories is performed by the registrants themselves when they register, although some registries issue policies and may enforce them if deviations are brought to their attention.

The problem with taxonomic approaches to the domain name space is that the categories of interest are manifold and change constantly over time. No classification scheme will ever be exhaustive. New categories will always arise that will seem more meaningful or important than the old ones. For example, there is great interest now in a .blog TLD because of the proliferation of a new genre of communication known as 'Web logs.' But the 'blog' category simply did not exist 3 years ago. Due to the switching costs discussed earlier, domain names cannot be quickly or easily resorted into new categories as things change. Thus, it may be unwise to attempt to link identifiers, for which persistence and memorability are key virtues, with content-based classifications, which might shift over time or be the site of competing approaches. Unless a central authority has complete control over who registers in which domain—something that seems undesirable to most Internet users—the mapping of domain name registrants to specific categories will be imperfect, particularly given the large number of legacy registrations in existing domains. Moreover, it is unlikely that any categorization scheme will be truly global in effect (i.e., any scheme that works at the psychological and user behavior levels in sub-Saharan Africa or Indonesia will probably not work well in Scandinavia or the U.S.).

3.2.3. Demand for search and directory functions

In addition to the memorable identifier function, part of the demand for domain names is derived from their ability to perform search functions. Domain names can perform search functions when they are 'guessable' and can be used to locate

Websites (or other online services) for the first time. In the paradigmatic case of using a domain name as a search aid, an Internet user who wants to find AT&T's Website may guess that the address will be att.com and type that into the URL bar of her Web browser. (Note that even this seemingly simple guess requires knowledge on the part of the user that the '&' character is not allowed in DNS.) In effect, the second-level domain name is used as a keyword. Early browsers reinforced this practice by automatically making .com the default TLD whenever a word was typed into the URL window. Obviously, search engines and portals are substitutes for these aspects of domain naming and improvements in the quality of Internet searching services has begun to greatly diminish the demand for domain names as search keywords.

Reliance on guessing was a response to unique and temporary conditions that emerged from 1994 to 1997 when the WWW was new[9]. At that time the WWW was overwhelmingly American and Internet content was overwhelmingly English. The number of registered domain names was 1 percent of the number registered now. Moreover, during the initial rush of WWW growth, domain name registrations were concentrated in the .com TLD to a degree that was excessive and historically unique. This concentration happened because .com was the only commercially oriented TLD that accepted registrations from anyone at a low price. Most of the other TLDs were restricted or much more expensive at that time. At the end of 1996, more than 88 percent of all domain names in the U.S., and just over 75 percent of all domain name registrations in the world, were under .com. The state of search engine technology and service was much less developed than it is now. Under these historically unique conditions, the search strategy of 'guessing' a domain name was reasonably effective (although there are no generalizable empirical studies on how often users did this).

Virtually every commentator who has remarked on the use of domain names for search and directory functions has agreed that domain names make lousy directories or search tools. Domain names require a precise match between the search term and the domain name to succeed; most searches involve guesses that are highly unlikely to be exact matches. Domain names must be unique, but there can be multiple legitimate users of the same names or search terms (e.g., 'Apple' can refer to a music recording company, a fruit, a computer company). The sheer number of second-level domains makes guessing increasingly less viable. Instead of the 1 million or so registered domain names in 1997, there are now nearly 50 million registered domain names. As the number grows, the viability of

[9] The WWW protocol was developed at a Swiss physics research institute (CERN) in 1990–1992. The first free, easy to use Web browsers only appeared in 1993; Netscape appeared in 1994.

guessing declines. Instead of 75 percent of all registered domains being concentrated in the .com TLD, now less than half of them are. There are over 2 million registrations in the new generic TLDs .info, .biz, and .us. These domains are growing more rapidly than .com. Expansion of the number of generic top-level names and distribution of more registrations into the country code TLDs undermines the ability of a second-level name to function as a keyword. As this happens, the importance of the search function of domain names declines in relative importance to the memorable identifier function. As the domain name universe and the Internet grow, guessing domain names may seem as benighted as attempting to guess phone numbers. Higher-level tools, such as search engines, act as direct substitutes for domain name-based guessing or trial and error. Here again, however, there is a paucity of empirical research. The cross-elasticity of demand for domain names and search engines has never been measured or even studied systematically.

Nevertheless, the function of domain names as search tools has had and continues to have a profound effect on the political economy of domain names. The prospect of capturing Web traffic stimulated speculative registration of names, including some bad faith or deceptive registrations of well-known trademarked names. It also encouraged companies and brand owners to engage in systematic defensive registrations. On the assumption that large numbers of users might expect to find them at those domains, a company might try to preemptively register all of the names in the semantic cluster associated with a particular concept or source identifier. It might also attempt defensive registration of the same names across multiple TLDs. Obviously this served as a major stimulus to the domain name registration market within the established TLDs. But it also led to efforts to regulate the market and restrict the supply of new TLDs.

4. Domain name supply

The supply side of the domain name industry consists of three basic elements: the root server operators, registries, and registrars.

4.1. Root servers

Root servers are the name servers for TLDs. ICANN, with the U.S. Department of Commerce looking over its shoulder, is the registry for TLDs. The root servers' technical, political, and economic distinctiveness is derived from the importance of the starting point, the root, in the name space tree. No TLD can operate unless the information needed to resolve it is entered into the root zone file and advertised to the public Internet via the root servers. Thus, the root zone file is an important point of policy control in the DNS, because it controls supply and market entry in the DNS name space. Adding TLDs (and authorizing registries to operate them) expands the supply of names and (possibly) the number of competitors in the domain name registration market[10].

At the time of this writing, there is no centralized and uniform method of operating the root servers. The information contained in the root zone file is controlled by ICANN's decision making process and ultimately by the U.S. Department of Commerce. But the actual operation of the root servers is performed by a heterogeneous group of entities from the Internet technical community who inherited operational control of specific root servers prior to the Internet's commercialization. One of the root servers is operated by ICANN itself. Two are operated by VeriSign. ICANN and VeriSign are both directly or indirectly beholden to the U.S. government. Other root servers are in the hands of universities, federal military or scientific agencies, Internet technical veterans or ISPs. Most do not have formal contracts with ICANN or the U.S. Government. All but three of the root servers are in the U.S. (Table 2).

So far ICANN has failed to execute contracts with any external root server operators. The operators' compliance with policy changes ordered by ICANN is essentially voluntary, although compliance is strongly influenced by the need for a coordinated, unified name space. Any attempt by one root server operator (or a sub-coalition of them) to enter their own TLD names unilaterally or to refuse to accept changes proposed by the others would risk fragmenting Internet connectivity. There are strong professional norms among the involved communities against such a course of action, as well as the likelihood of more direct pressures from industry and government.

4.1.1. DNS root as commons

The DNS root is a commons. Anyone who sets up standard-implementation DNS can query the root without being charged. With the exception of ICANN, which taxes the registries and the registrars it accredits, none of the root server operators receive any compensation from the users of the DNS service. Financial support for the root servers comes from the institutions that operate them. The prospect of

[10] Note that new names can be assigned to existing registry operators. In that case, a new TLD would not expand the number of registry competitors, but it would expand the choice of names available to consumers.

Table 2
Root servers

Name	Geographic location	Operating institution
A	USA – East	VeriSign Global Registry Services
B	USA – West	Information Sciences Institute
C	USA– East	Cogent Communications (an ISP)
D	USA – East	University of Maryland
E	USA – West	NASA Ames Research Center
F	USA – West	Internet Software Consortium
G	USA – East	U.S. DOD Network Information Center
H	USA – East	U.S. Army Research Laboratory
I	Stockholm, Sweden	Autonomica
J	USA – East	VeriSign Global Registry Services
K	London, Great Britain	Réseaux IP Européens (RIPE) Network Coordination Centre
L	USA – West	ICANN
M	Tokyo, Japan	WIDE Project, Keio University

charging for DNS queries does not seem to be feasible, although this is a good topic for new economic research[11].

DNS root servers share all the familiar advantages and disadvantages of a commons. The absence of charging minimizes the transaction costs of interacting with the DNS root. The programs and networking structures associated with the Internet can (at least for now) take its availability for granted. But there is no incentive for Internet users to ration or control their usage of the root. A minor glitch in widely distributed Microsoft software generated millions of queries to the root that could not be resolved, wasting root server capacity (Brownlee et al., 2001). As the size of the Internet grows and new DNS applications such as ENUM

[11] Technical experts insist that end users lack basic forms of control over when, where, and how their resolvers issue DNS queries, making any charge-per-query scheme unfeasible. Users might be subjected to massive manipulation and abuse, as operators who charged for queries could disseminate programs or links that generated large numbers of queries, or that provoked queries by the machines of uninvolved third parties. Aside from those problems, the value of any individual transaction might exceed the transaction costs of defining a pricing, exchange, and collection mechanism. However, these problems have been solved at lower levels of the hierarchy. Rather than saying it is 'impossible' to charge for root service, it would be more accurate to say that the DNS industry and many forms of Internet operation have been organized around the assumption that the root is a commons, and altering that would involve sweeping adjustments.

Table 3
Compatibility relations among competing DNS roots

Origin of domain name assignment	*Origin of domain name query*	
	Users of root-I	**Users of root-C**
(a) Competing root (Root-C) supports established TLDs of incumbent root (Root-I)		
Root-I	Compatible	Compatible
Root-C	Incompatible	Compatible
(b) Competing root (Root-C) is uncoordinated with incumbent root (Root-I)		
Root-I	Compatible	Incompatible
Root-C	Incompatible	Compatible

are created, the query load on the root servers will increase. Because the number of coordinated root servers is fixed at 13 it is conceivable that growth in query load can outrun the technical capabilities of the root servers at some point. But this too is an area where new modeling and empirical research is needed.

4.1.2. Root administration as service to TLD registries

While for end users the DNS root is a commons, the central coordinator of the root zone file must also provide critical services to registries. The root zone file must contain correct, up-to-date information about the IP addresses of the TLD name servers. When that information changes, the TLD registries must be able to request changes from the root authority and have them executed promptly. These exchanges of information must be highly secure to make sure that malicious redirection of DNS queries, capable of disabling entire TLDs, does not take place.

At the lower levels of the hierarchy (e.g., in the domain name registrant—registry relationship), these record updates are contract-based services provided for a fee. Currently, however, ICANN does not treat this relationship as a service to paying customers. Instead, it seeks a governmental relationship between the TLD registries and itself, wherein the registries sign contracts recognizing and accepting ICANN as policymaking authority with the power to regulate and tax the registries. This conflict of visions over the role of the root administrator has led to considerable friction between ICANN and the country code registries.

4.1.3. Competing DNS roots

We have already noted that on the demand side, consumers of DNS services have strong incentives to converge on a single root. On the supply side, however, there are very few economic barriers to the creation of a competing root authority or root server operators. The root servers are just name servers that happen to be high-capacity because of the reliance of the global Internet on them. Any other network of 13 high-capacity name servers could perform the same function. The level of investment required to replicate all 13 would be in the tens of millions of U.S. dollars.

There is one salient barrier to entry aside from demand-side network effects. BIND software contains a 'hints' file that contains a list of the IP addresses of the 13 root servers. When a name server is started, it needs to establish contact with a root server so that it knows it can send queries there. The hints file tells the ordinary name server operator where to look for root servers. It is possible to alter this file manually. Any attempt to mount a coordinated shift to a new DNS root, therefore, would have to convince those hundreds of thousands of name servers to manually reconfigure their software, or convince the stewards of BIND software to include the new root server addresses, or both. The hints files embeds the IP addresses of the dominant root in all BIND implementations, making the dominant root the default value and thus giving it the advantage of inertia as well as network externalities.

4.2. Registries and registrars

Below the root level of the domain name hierarchy, registration, and name resolution services are supplied by firms that operate on a subscription fee basis. Typically, domain names are sold in 1- or 2-year subscriptions although as the market has become more competitive longer term contracts have appeared. Most of the market is supplied by firms that are commercial for profits. Some registries are operated by nonprofit consortia of Internet service providers or other members of the local Internet community. Still others (usually in developing countries) are run by government ministries or state-owned entities.

4.2.1. Registries

In essence, registries operate public databases that assign names under TLDs and provide, in real time, the name resolution data needed to use the names for communication over the Internet. Domain name registries maintain records, known as *zone files*, of second-level domain name registrations under their TLD. They record which second-level domains have been occupied and who has occupied them, and provide name service for those registrations. In addition, they maintain and may offer publicly a 'WHOIS' service, which allows Internet users or private parties to type in a domain name and see the name and contact information of the person who registered it. Registries may also perform the functions of accepting customer orders for specific names, maintaining customer accounts, billing customers, accepting changes from customers, notifying them of expiration, and so on. In the major generic TLDs regulated by ICANN, these 'retail' functions must be separated from the 'wholesale' functions of maintaining the zone files (see Section 4.2.2).

Each TLD is by definition a unique string. Under the current technical implementation of the DNS protocol, there can only be one .com and only one registry operator for .com. This raises the question whether registries compete with each other or are 'natural monopolies.' There is a need for greater empirical research on the cross-elasticity of demand across different TLDs. However, experience with both new TLDs and country code registrations indicate that Internet users contemplating a *new* registration do view registries as competitive alternatives to each other. There is also anecdotal evidence of gradual shifts of Websites from one TLD to another[12]. Registries who wish to compete with .com can offer TLD names that either convey the same basic semantic meaning as .com, such as .biz, or that target specific subsets of the commercial world, such as .shop or .law. Registry competition for new registrations thus involves a form of product differentiation; competition is imperfect but there is sufficient cross-elasticity of demand to make a registration in one TLD a competitive alternative to registration in another. Indeed, competition across registries may exist even when the semantic similarity between TLDs is minimal. Although no formal study has been done, observation of country code registration data suggests that countries that impose numerous restrictions on registering in their ccTLD have a much higher portion of organizations who register .com names. In Japan and France, country code registries known for their highly restrictive policies toward registration, more firms are registered in .com and other gTLDs than in the country code TLDs. Countries with more accommodating policies, such as .de and .uk, have greatly outstripped the French and Japanese TLDs in number of registrations and have fewer companies per country registered in .com. Thus, it is clear that many Japanese and French registrants view gTLDs and specifically .com as an alternative to their country code, despite the absence of any semantic relationship.

[12] For example, the popular 'State of the Domain' Website operated by Snapnames, which supplies domain name industry data, has shifted from sotd.com to sotd.info, and the former domain no longer resolves.

Generic TLD registries, such as .com, .net, and .info are regulated via 'contracts' with ICANN. As there is no alternative to the ICANN-managed DNS root, it would be more accurate to describe this as a *licensing* regime similar to the licensing of broadcast stations by the U.S. Federal Communications Commission. When a registry is assigned a TLD, it signs a lengthy contract with ICANN that contains, among other things:

- A cap on the wholesale price that can be charged per name-year;
- Commitments to abide by ICANN-developed regulatory policies, especially those regarding dispute resolution, WHOIS availability, and other regulations favored by intellectual property interests;
- Detailed regulations regarding its technical and commercial relationship to registrars (see Section 4.2.2);
- Commitments to reserve or remove specific names from the market;
- Prohibitions on certain kinds of conduct (e.g., interconnecting with alternate roots, introducing new services without ICANN approval);
- Rates of taxation to support ICANN.

On the other hand, most country-code TLDs have no formal contracts with ICANN, having received their TLD assignments prior to its existence. ICANN and the U.S. lack the authority to compel their participation in the 'contractual' regime, given that any threat to eliminate an entire country code from the root would not be credible. While ICANN and national governments have begun to develop a method for redelegating country code TLDs, this is still an open, and highly sensitive, area of policy.

Although there has been no nonproprietary empirical research on this topic, an intuitive understanding of the DNS protocol and database operation suggests that there may be economies of scale up to a certain point in the operation of a registry. Economies of scale would come as increases in the number of registrations under a TLD consumed unused capacity in the existing server and operations infrastructure. As the peak load of the server infrastructure was exhausted, however, new investments would have to be made. As a TLD becomes widely used worldwide, geographic distribution of the server infrastructure must take place to maintain quality of service; this may lead to diseconomies of scale. TLD name servers face the same 13-server constraint as the root servers. The dominant registry operator VeriSign has developed a 13-server, geographically distributed infrastructure to support .com, .net, and .org. It has leased capacity on that infrastructure to some country code operators and ENUM service providers, indicating that it is more efficient to use existing capacity more intensively than to build a new one.

There are also likely to be significant economies of scope when new TLD names are added to existing registry infrastructure. Any given TLD name is likely to appeal to only a portion of the marketplace. A diversified bundle of TLD names, appealing to different types of registrants and different usage patterns, is likely to make more intensive usage of the fixed investment. Adding an additional TLD name to the list of names served is a trivial expense; it involves a one-time entry in a list. Just as magazine and music publishers find it efficient to develop a repertoire of titles and artists to distribute their risk, so it is likely that registry operators would benefit from spreading their load and risk over multiple TLD offerings. But this, too, is just speculation as no scholarly empirical research or modeling has been done in this area.

4.2.2. Vertical separation of 'registry' and 'registrar'

In some cases the wholesale and retail functions of the registry are separated. This may happen voluntarily (e.g., when the registry is organized as a consortium of retail domain name registration businesses or ISPs who want to keep the database operator in a subordinate role and removed from the retail market[13] or when registry operators do not want to enter the registrar market[14]). In other cases it is required by regulation (e.g., both the U.S. Department of Commerce's contracts with VeriSign and ICANN's contracts with most new generic TLD registries required the implementation of a 'shared registration system' (SRS) that allowed multiple competing registrars equal access to the registry database). Moreover, the ICANN regime makes it impossible for customers to deal directly with registries; they must go through a class of intermediaries known as 'accredited registrars.' In a shared registry, registrars are retail firms that do not operate the database but can enter registrations into the registry database. Registrars handle the customer relationship (billing, receiving, and executing changes in the customer account, notifying the customer of pending expirations, etc.).

Vertical separation is supposed to improve the competitiveness of the market by giving customers a choice of retail service providers and introducing price competition for the right to service accounts. Customers can transfer names from one registrar to another, so that registrars compete actively for customer accounts. Under the ICANN regime, the price of the registry (wholesale) is regulated but the price of the registrars (retail) is not. This form of regulatory mandated

[13] Examples are Nominet U.K., the registry for the country code .uk, and Afilias, the registry operator for .info and .org.

[14] Neustar, the registry operating .biz and .us, does not operate in the registrar market.

competition is quite similar in form to the mandated separation of local and long distance telephone service in the U.S. following the AT&T divestiture. Both required the creation of a new technical interface to provide 'equal access' to what is perceived as a monopolized facility (the registry database). In both cases, prices for a required input are regulated while retail prices are not. In both cases, the vertical separation greatly reduced barriers to entry, allowing hundreds of firms to enter the market and bid retail prices down much closer to the floor set by regulation. But in both cases, the ease of entry and the artificiality of the interface created problems of 'slamming,' or unauthorized transfers of the customer's account (see Section 5.3). Just as the local and long distance separation was promoted as a remedy for the historical dominance of a single supplier (AT&T), so the registry–registrar split was a response to the historical dominance of the domain name market by NSI. In the late 1990s, NSI (now VeriSign) controlled nearly 90 percent of the global domain name market, and its ability to adequately service millions of .com registrants was questioned, with many reports of poor service. The major difference, however, is that while there are major economic and technical barriers to new entry in the local exchange market, there are minimal economic or technical barriers to new entry in the registry market. ICANN's reluctance to permit new TLDs and new operators to enter the market is the only real barrier. Once the registry market becomes more competitive, the need for compulsory separation of the registry and registrar functions becomes questionable. Customers who prefer the additional choice and control would be able to select registry services that include a variety of competing registrars; others may prefer integration of the functions for simplicity or efficiency (Tables 4 and 5).

4.3. The secondary market

Semantic demand for domain names led to the development of a increasingly well-organized secondary market (i.e., a market, in which domains are registered by one party and warehoused rather than used, and then auctioned or sold for a negotiated price to a third party). The practice of name speculation is clearly identifiable as a product of semantic demand because the additional value of the name over the subscription price is based entirely on the name's meaning. Also, the market has also moved to capitalize on the residual or inertial traffic-generating capabilities of existing domains, leading to battles over the control of expiring domain names. Data from Matthew Zook on the rate of name expiration indicates that in the second half of 2001, about 10 percent of all registered domain names changed ownership annually, and about 18 percent expired annually (The

Table 4
Registry market share

Name	Share (%)	Registrations (millions)
Verisign (.com, .net)	53.78	25.04
DENIC (Germany)	11.73	5.46
Afilias (.info, .org)	7.10	3.32
Nominet (.uk)	7.10	3.30
Neustar (.biz, .us)	2.40	1.11
Total for top five	82.11	38.23
World total		46.55

Table 5
Registrar market share

	Rank		Market share		Number of registrations	
Name	Q4 2000	Q3 2002	Q4 2000 (%)	Q3 2002 (%)	Q4 2000	Q3 2002
VeriSign registrar	1	1	53.0	30.1	14,474,754	8,795,276
Tucows (OpenSRS)	3	2	7.4	10.4	2,011,880	3,044,787
Register.com	2	3	12.3	9.8	3,379,237	2,863,004
MelbourneIT	5	4	3.5	5.6	969,423	1,619,257
Bulkregister	4	5	6.6	4.7	1,812,582	1,372,454
CoreNIC	6	12	3.5	2.0	967,185	1,290,309
All others	–	–	13.5	37.5	3,748,857	10,206,005

expiration rate would be highly sensitive to industry conditions and cannot be extrapolated.) (Table 6).

The emergence of a secondary market for domain names was a predictable product of basic economic characteristics of the supply of registry services. Due to semantics, the value of any individual unit within the vast pool of names under a TLD such as .com will vary greatly. And yet, registries with their automated procedures of registration were for the most part unwilling and probably unable to engage in accurate price discrimination reflecting variations in value. Thus, domain name registrations were sold in the primary market at a low, uniform price. Any user willing to invest a few thousand dollars could register multiple names and attempt to capitalize on the widespread gap between cost and potential

Table 6
Data relevant to the secondary market

Changes in domain status	Three months (%)	Monthly (%)	Yearly (%)
Changes in ownership			
No change in ownership	93.0	–	–
New owner	2.5	0.8	9.8
Expired domains	4.6	1.5	18.3
	100.0	2.3	28.1
Changes in registrar			
No change in registrar	92.7	–	–
New registrar	2.7	0.9	10.7
Expired domains	4.6	1.5	18.3
	100.0	2.4	29.0
Ownership & registrar changes			
Same owner, same registrar	91.6	–	–
Same owner, new registrar	1.4	0.5	5.5
New owner, same registrar	1.1	0.4	4.6
New owner, new registrar	1.3	0.4	5.2
Expired domains	4.6	1.5	18.3

Source: Matthew Zook, Zooknic Internet Intelligence. Based on a sample of 50,000 randomly selected names with observations in June and September 2001.

value[15]. Add to this the artificial restriction on the supply of new TLDs that was in place from 1996 until 2001, which both restricted alternatives and reinforced the search value of domains within .com, and the recipe for name speculation was complete. Most of the speculative activity was in the .com domain, although copies of important or popular .com domain names began to be registered in .org and .net and even some country codes as speculative activity increased.

It is theoretically possible to capture the gap between cost and value by holding auctions for initial assignments in the primary market (for a proposal along these lines, see Kahin, 1996). One registry, the .tv TLD, experimented with auctions in the primary market. It allowed prospective registrants to enter their name of choice and then returned a price that purported to reflect interest in the name. This pricing mechanism did not work because the auctions were not thick,

[15] When the 'hunch' about the name's potential value was more than a guess name speculation could become an abusive practice. For example, a certain class of name speculators would comb business journals for news of mergers and acquisitions, and as soon as the deal was publicly announced they would register the name of the combined company (e.g., ExxonMobil.com).

simultaneous, nor organized by a neutral party. Rather than organized auctions, the .tv initial assignments resembled more an algorithm for negotiation with the registry, with asymmetric information due to the registry's collection of past bids or expressions of interest. Indeed, the registry operator could (and for all we know, did) 'bluff' about the auction value of names, as there was no transparency.

The value of expiring names is another area of the secondary market. Many Internet users would like to have a secure, predictable method of queuing up for expiring domain names. As an example, if a company has just trademarked a new product it wishes to name 'Agile' and agile.com has been held for 2 years by a home user who thought it was a cool domain name to have, the company wants to stand first in line when or if the name expires. The company may not want to contact the home user to negotiate a transfer due to the uncertainty and potential for opportunism surrounding the market. The company does not know whether the registrant plans to renew the name or not. If it contacts the registrant and offers to buy it before it expires, the registrant has been tipped off that the name has value, and the registrant might exploit that to renew the domain and set a high sale price. The waiting registrant thus cannot know whether it stands to win or lose by initiating a transaction directly with the registrant.

Prior to mid-2002, the market for expiring domain names was a common pool wherein competing registrars sold domain name recovery services to end users and then stormed the registry with 'add' commands when the name expired. The same recovery service could be sold by any competing registrar, making attempts to capture the name a matter of brute technical force. The 'add storms' that afflicted the registry infrastructure were a pure example of a tragedy of the commons. The expiration system gave secondary market makers an incentive to inundate the registry with requests for the expiring name regardless of its effect on the registry infrastructure.

Economic analysis would suggest holding an auction for the name upon its expiry. An auction would assure that the name went to whoever valued it the most. The problem is: who should serve as the auctioneer? The prior registrant is not the best choice due to the information and opportunism problems mentioned earlier. Aside from that, efficient auctions require organization and publicity, both of which cost money. Unless the incumbent registrant is an organized domain name brokerage, she will not be prepared to operate an auction. If the registrant uses the WHOIS record to advertise that the name is for sale, he runs afoul of ICANN's trademark-protecting dispute resolution rules, which make it possible to take away names from people who registered them purely for resale. Probably the most efficient and direct entity to serve as the auctioneer is the registry itself. The registry knows when the name expires and has direct control over when it is released. A centralized auction that incorporates all prospective bidders is more efficient at equilibrating supply and demand than a bunch of smaller, fragmented auctions. However, such an auction would step over the boundaries of the artificial separation of registry and registrar functions, which is a linchpin of the ICANN/U.S. Department of Commerce regime. In other words, registries would be selling names at retail, and making higher margins to boot. A registry auction would create a distribution of wealth favorable to registries rather than registrars, yet registrars are more numerous and politically in favor than registries in ICANN's regime.

In Fall 2002 ICANN permitted VeriSign to enact a mild version of a registry auction known as the Wait Listing Service (WLS)[16]. In this service, companies interested in an expiring domain can pay their registrar a market price to gain 'first dibs' on a domain name when it expires. The registrar in turn will pay VeriSign registry a \$35 per name wholesale rate (with rebates of \$7 or \$11 depending on volume) to put their customer first in line upon expiration. Almost a year after it was first proposed, ICANN approved the service, but not without adding a regulation that may have completely undercut its business value. ICANN required the registry to inform existing domain name subscribers when their names had been waitlisted (thus encouraging them to renew the name and bargain directly with the interested party).

A great deal of the interest in expiring names comes from Internet services that make their money selling 'hits' to advertisers. These companies have recognized that for months and possibly even years after the expiration of a domain name, there may still be users or outdated links that direct traffic to that domain. By gaining control of a once-popular domain name that has expired, these businesses can capture a traffic stream and expose them to advertisements that may result in billable hits.

5. Economic policy issues

Building upon the analysis in the previous sections, the framework can now be applied to the analysis of current policy issues. It should be noted that some of the most important policy problems are *institutional* in nature. The DNS industry is relatively young, and the attempt to create a global, private sector 'self-regulatory' system around it is even younger. Unlike many policy analyses, therefore, this

[16] For a description of the WLS services, see 'Domain Name Wait Listing Service,' VeriSign Global Registry Services, March 2002. http://www.icann.org/bucharest/vgrs-wls-proposal-20mar02.pdf

discussion cannot take for granted an established, stable governance framework that uses known methods to establish policies or known techniques for defining and enforcing regulations. There are also problems and issues regarding the formation of the institutions themselves: how they make decisions, the proper scope of their authority, the incentives to participate in the regime, and so on. Such a discussion falls outside the scope of this paper, however. The next sections survey the economic policy issues in the domain name industry in a way that basically takes the ICANN-based institutional framework for granted. Some indications of the institutional problems that pervade the treatment of many of the policy issues are provided.

5.1. New TLDs: Expanding supply

ICANN was created to manage the root of the domain name system. That function implies making policy decisions about how many TLDs are added, what pace they can be added at, what criteria will be used to determine who gets the available assignments, and how to resolve competing applications for the same new TLD, and so on.

Policy conflict over adding new TLDs is one of the issues that led to the creation of ICANN in the first place (NTIA, 1998; Mueller, 2002a, Chapter 6). Calls for name space expansion are not, however, based on a 'shortage' of domain names *per se*. As noted earlier, the DNS name space can handle all conceivable technical demand without any change. If users were willing to register and use extremely long names, meaningless names, and names that go five or more levels down into the name space hierarchy, there would be no need for expansion of the supply of TLDs. Likewise, if the ownership and operation of current registry services were perfectly satisfactory to everyone, there would be no need to permit the entry of new registry operators. The debate over new TLDs is really a debate about the degree to which DNS administration should respond to human factors, user demand, and competition policy concerns.

There are many potential sources of demand for new TLD names. Users may want more choices regarding the identity they project online. For example, a .blog TLD might attract numerous registrants from the growing Web loggers community. It is also possible that new entrants might have valid ideas about how to provide targeted services better than incumbent operators. For example, the .name TLD is targeted at personalized domain names, but their business model and policy restrictions are unattractive to many registrants. An alternative TLD for personal names would add competition and choice to the market. Or users may want to move their name up in the DNS hierarchy for ease of use and/or greater control over the management of their name and the name space below it. A corporation to whom online identity is essential, such as AOL or Amazon.com, may decide that it wants to 'in-source' its DNS management functions completely by running its own TLD, just as many companies privatized their network management functions in the 1970s. Some groups of organizations may want to establish a controlled name space, analogous to .edu for U.S. universities, to promote authenticity of online identity. Thus, adding TLDs will have a major impact on: (a) the variety and usability of identifiers; (b) competition among registries; and (c) the ability of firms or industries to control their digital identity.

Surprisingly, after 4 years of existence ICANN has not defined a timetable or a rule-based method for responding to requests for new TLDs. Indeed, it has not even set in motion a proceeding to define a method of routinizing this function. Instead, it has approached additions in an ad hoc manner. In year 2000, it held a long and controversial process that added seven new TLDs. That round received 44 applications from bidders willing to pay a U.S.$50,000 nonrefundable fee to be considered. Before making its seven awards, mostly to firms with close political and organizational connections to ICANN's management and Board, ICANN announced that the whole new TLD process was an experiment or 'proof of concept,' the results of which would be monitored. No timetable for the beginning or end of this monitoring process was fixed. Late in 2002, ICANN's CEO suddenly announced that he thought it would be a good idea to add three more restricted TLDs. No one knows where this number came from.

ICANN's unwillingness to tackle the problem of new TLDs in a systematic manner is caused by three factors: (a) it is a self-regulatory system, in which incumbent operators have a great deal of influence over policymaking, and incumbents have little incentive to define an open entry process that would subject themselves permanently to the prospect of additional competition[17]; (b) brand-name holders, trademark interests and holders of .com domains that are important and valuable fear that additional TLDs will only result in the need to defensively register their existing second-level domain names in order to preempt their occupation by others, including trademark infringers; and (c) it emerged from a technical culture in which policy decisions affecting the demand and supply conditions of an industry are often confused with design decisions, which can be

[17] In an article in *Computerwire* in October 2002, ICANN CEO, Stuart Lynn was reported as saying "the last round of new TLDs, which included .info, .pro, and .name, were introduced largely due to market demand, but some people think this is no longer an appropriate reason to introduce new domains." The 'some people' referred to are incumbent registries, who believe that the market is saturated and do not want additional competitors.

centrally planned to create a local optimum. All three pressures act to maintain artificial scarcity in the supply of TLDs.

Earlier in the debate there were attempts to suggest that there are technical obstacles to the addition of new TLDs. The IETF has never created a working group to systematically analyze or define any technical constraints on TLD additions or their impact on the operation of the root zone. It is an irrefutable fact, however, that the original root administrator and one of the designers of the DNS, the late Dr. Jon Postel, proposed adding 50 new TLDs a year for 3 years in a row back in 1996. While this plan was rejected, its failure was not attributable to technical concerns about expansion of the root zone. During the early and mid-1990s, as country code TLDs were being delegated, the root zone was expanding by at least 10 TLDs per year, and for several years in a row more than 20 were added each year. Key figures within the IETF, such as Paul Vixie and Karl Auerbach, have pointed out that the root servers use the same technology as the name servers for the TLDs. In that respect, the root zone file is no different from any other DNS zone file. As there are millions of functioning registrations in the *.com*, *.net*, *.org*, *.de*, and *.uk* zone files and those zones work reliably, there could be millions of TLDs. However, the scalability of the system depends in large part upon the existence of hierarchy, so almost no one supports having that many TLDs.

There is one important difference about the root zone, however, errors or corrupted files at the root level could have more harmful consequences for Internet users than mistakes that occur lower in the hierarchy. An erroneous root zone file could result in the inaccessibility of entire TLDs until the problem was fixed, whereas the effects of a corrupted TLD zone file would be more localized. Thus, there is a valid technical concern about limiting the *rate at which the root zone changes* in order to minimize the risk of errors in the root zone. While some Internet technologists believe that root zone changes could and should be automated, more conservative traditionalists believe that the root zone should continue to be altered by hand and subject to human inspection before being released and propagated to the root servers. Even the adherents of this most conservative view, however, believe that 30–90 additions and changes in the root zone file made in batch mode at a specific periodic rate, such as annually or every 6 months, are safe. Thus, it is conservative to note that 50 or so new TLDs could be added annually to the root zone (see e.g., Hoffman, 2002).

The only other technical concern raised by the addition of new TLDs is the degree to which TLD additions increase the query load on the root. Because the number of root servers is fixed at 13 and there are technical limits on the capacity of each root server, there is some cause for concern about increasing root server load. However, because the DNS relies so heavily on caching at lower levels of the name server hierarchy, there is no simple, linear relationship between adding TLDs and increasing root server load. The consensus position within IETF seems to be merely that the number of TLDs should be finite rather than infinite, such that the hierarchical character of name resolution and assignment is maintained. As IETF Chair Fred Baker put it,

> If we can add one TLD (and we obviously can), we can add 1000 TLDs to the [root zone] table. How that relates to [root-server] lookups for those TLDs is indeterminate from the fact that they are there. How many lookups, for example, do .tv or .info get? The fact that we added seven TLDs does not mean that we have even *changed* the root server load, much less multiplied it by something. How much additional load we would get is a *business* question: how many new computers, with people using them (and what would they be doing, and what sites would they therefore be accessing), will be added to the Internet because this TLD is in operation[18]?

It is widely agreed that ICANN's initial addition of seven new TLDs and the revitalization of the already existing .us TLD have led to no discernable change in root load or root server behavior. In sum, TLD additions have little direct bearing on what is a more fundamental question, which is how the DNS is able to scale with the growth of the Internet. The growth of the Internet, and new DNS applications, such as ENUM, have more relevance to root server load than TLD additions *per se*.

Expansion of the number of TLDs is really a policy issue rather than a technical one. ICANN's unwillingness to define a stable process for adding TLDs, and its reliance on sporadic 'beauty contests' (i.e., merit assignment procedures) to assign new domains are produced in part by amateurism on the part of ICANN management and in part by political pressures.

5.2. Domain name—Trademark conflicts

When the WWW accidentally made domain names into global public identifiers, it fomented property rights conflicts over who had the right to specific name assignments. The most important form of conflict occurred over trademark rights. National laws and international treaties recognize various forms of exclusivity in names, based on registration and (sometimes) use of the name in commerce as a source identifier. It is possible to register and use domain names in a way that constitutes trademark infringement or passing off. A domain name corresponding to a trademark, registered and used by an unauthorized person, may lead a user to a Website that deceives or confuses her into thinking that she is dealing with the

[18] E-mail to author, October 11, 2002.

trademark owner. For example, if someone other than America Online registers http://www.aim5.com and displays information and logos related to AOL Instant Messenger (AIM) Version 5 in order to attract advertising revenue or sell goods, classic concepts of trademark infringement apply.

There were two problems with applying trademark law to domain name registration, however. First, the costs of initiating a problem were absurdly low, while the costs of prosecution were high. Domain names in most registries were assigned on a first come, first-served basis at low annual fees. The economics of domain name supply did not support prechecking of applications; indeed, most registration processes were automated after 1995. Thus, a potentially abusive registrant could acquire a domain name corresponding to a trademark in a matter of hours for less than U.S.$100. The cost of initiating a legal dispute, however, started in the tens of thousands of dollars. Second, the scope of a domain name was global, but trademark jurisdiction is mostly national. Thus, the transaction costs of enforcing property rights in the domain name space were high. High transaction costs often harmed innocent registrants as well as trademark owners. Many legitimate registrants of generic terms or names that happened to be trademarked by someone (e.g., prince.com or clue.com) found themselves assaulted by lawsuits that required hundreds of thousands of dollars to defend. This threat, which came to be known as 'reverse domain name hijacking' was cynically employed by some trademark counsel in order to acquire valuable DNS real estate at below market cost.

Domain name—trademark conflicts led to an institutional innovation: the creation of a new global system of dispute resolution over name assignments (Froomkin, 2002). ICANN's Uniform Domain Name Dispute Resolution Policy (UDRP) created a mandatory system of arbitration wherein multiple dispute resolution service providers compete for the business of complainants (i.e., trademark holders initiating a dispute). The policy emerged from a WIPO report and ICANN's domain name supporting organization (WIPO, 1999). The UDRP dramatically lowers the cost of initiating a dispute and is often (though not always) faster than traditional courts. As basic economics would suggest, lowering the cost of dispute resolution led to huge increases in the quantity supplied. Whereas domain name-related court cases number in the hundreds at best, as of November 2002 approximately 7500 proceedings involving 12,000 domain names have been initiated under the UDRP.

The UDRP now functions as a kind of global law in name-rights, at least within the domain name space. While the stated goal of the new system was to translate *existing rights* recognized by law into a new arena, the move from territorial jurisdiction based on national court systems to global jurisdiction based on contracts of adhesion with registrars has necessarily led to substantive changes in the nature of name rights.

The new regime has strengthened the exclusivity of trademark owners and sometimes coined entirely new rights to names. Research has shown that the right of complainants to select the dispute resolution service provider leads to forum-shopping, which rewards trademark-friendly providers and weeds out any providers who interpret the policy in a pro-defendant way (Geist, 2001; Mueller, 2001). Research also confirms that the UDRP has created new rights in names, notably in the area of rights of personality (Mueller, 2002b) WIPO and various national governments have been eager to use the new regime to create expanded rights to names in various areas, such as geographical indicators, country names, non-proprietary pharmaceutical names, and names of international governmental organizations (WIPO, 2001).

5.3. Competition policy

Some of the most difficult and technical policy issues in the domain name industry involve competition policy. Concerns about fostering competition and ameliorating market dominance have shaped the U.S. government's approach to the problem since 1997. At that time, the primary concern was the monopoly on commercial generic TLDs by the U.S. government contractor NSI. (NSI was later acquired by VeriSign and henceforth will be referred to as VeriSign.) VeriSign's dominance was an unanticipated result of a decision by an educational/scientific agency, the National Science Foundation, which authorized it to begin charging for domain name registrations in 1995 with no thought for competitive alternatives.

The U.S. government could have responded to VeriSign's dominance of the market in two distinct ways. One would have been to authorize the entry into the market of new registry operators with exclusive control of new TLDs. The other would have been to regulate it as a dominant provider and impose upon it the vertical separation of the registry and registrar functions, leaving VeriSign's dominance of the registry market in place but creating managed competition in the retail market for registrar services. For better or worse, the U.S. chose the latter path. It failed to authorize new TLDs (and hence stifled nascent registry-level competition) for more than 5 years of the Internet's most rapid commercialization and growth. Instead, the USG opened up the dominant .com domain to registrar competition, thereby making .com names even more popular and thus prolonging and supporting VeriSign's dominance of the registry market. Due to the stickiness of domain name registrations, the effects of that delay of competitive entry cannot be reversed.

Registry–registrar relationships are now a sensitive area for regulation that affects both consumer protection and competition policy. The separation of registry and registrar was supposed to make domain names easily portable across registrars and thereby make the market more competitive. However, ease of movement of the account across registrars also makes it relatively easy for a domain name version of 'slamming' to take place. Unethical registrars can transfer domain name registrations without the authorization, knowledge or consent of the registrant. The threat of slamming in turn makes it possible for dominant registrars that are losing market share to impose more restrictive practices on requests for transfers of accounts. As a result, ICANN (or some other regulator) must play a significant role in defining procedures that consumers can rely on when they want to transfer domain names from one registrar to another[19].

Interesting competition policy questions also arise in relation to the secondary market.

Several major registrars have through acquisitions gained control over major domain name resale brokerages. By forward integration into the secondary market, a major registrar can undermine the registry–registrar split, because expiring names might be retained by the registrar and sold on the secondary market for an auction price instead of being released into the common pool where any registrar has an equal chance to sell it. VeriSign as noted earlier promoted a 'wait list' service as an alternative to the 'first come, first served/brute force' method of grabbing expiring names. This too would tend to favor a supplier with a larger market share because it would allow a supplier with a large but declining market share to generate additional revenue from registrations they already controlled. However, these alleged distortions in the market are nowhere near as significant as the long-term restriction on competitive entry into the registry market caused by the de facto moratorium on new mass market-oriented TLDs. Indeed, one could argue that with a more competitive registry market regulation of the registry–registrar interface would be entirely unnecessary.

As part of its agreement to allow VeriSign to continue in control of the .com domain in 1999, the U.S. Commerce Department required VeriSign to divest itself of its registrar subsidiary after a certain period of time. VeriSign's market share in

[19] See the report of the Names Council Task Force on Transfers, 'Policies and Processes for Gaining and Losing Registrars,' which states in its executive summary that "actions by 2–3 of the largest Registrars means that choice is simply 'on hold.' Despite several attempts at forging voluntary agreement on the necessary changes, the record shows that there are still problems with the 'portability' of domain registrations—customer choice remains limited." http://www.byte.org/nc-transfers/final/.

the registrar segment had declined so rapidly, however, that when the period expired VeriSign and ICANN entered into a new bargain, under which VeriSign was permitted to keep its registry business but was required to divest itself of the .org TLD and possibly also the .net registry. In divesting VeriSign of .org ICANN's management decided at the outset that it would be safest not to award it to any new entrant, but only to applicants that already ran a registry with 500,000 or more registrants. That policy decision effectively restricted the field to two incumbents, Afilias, the registry consortium that it had already awarded the .info TLD in the first round of expansion, and Neustar, the registry operator that it awarded the .biz TLD in the first round of expansion. In 2002 ICANN awarded the .org registry to a partnership of the Internet Society and Afilias. The registry market remains highly concentrated (Table 4).

5.4. WHOIS and privacy policy

Associated with the DNS protocol is the WHOIS service, a protocol that allows one to look up the name and contact information of the registrant of a domain name. The WHOIS service was developed in the early days of the Internet when contact information associated with a domain might be needed to resolve technical problems.

As domain names became economically valuable after 1995, WHOIS also became a popular way to find out which domain names were taken, who had registered them, and the creation and expiration date of the registration. Economic value also brought with it domain name—trademark conflicts. Trademark holders discovered that they could perform searches for character strings that matched trademarks, and pull up many of the domain name registrations in the generic TLDs that matched or contained a trademark. This automated searching function proved to be so valuable in instigating litigation and/or UDRP complaints that the trademark interests began to demand that the WHOIS functions be institutionalized, expanded, and subsidized. The first WIPO Domain Name process recommended that the contact details in a WHOIS record be contractually required to be complete, accurate and up to date, on penalty of forfeiture of the domain name (WIPO, 1999, para. 73). The intellectual property interests also demanded 'bulk access' to the WHOIS data of domain name registrars (i.e., the right to purchase the complete list and contact data for all of a registrar's customers in one fell swoop).

When ICANN was created its power to accredit registrars gave it the ability to implement the transformation of WHOIS demanded by the intellectual property interests. The registrar contract was drafted in accordance with WIPO's policy suggestions. By signing the accreditation contract, registrars commit themselves to the provision of a public WHOIS service for free, and to the supply of bulk access to their WHOIS records for a regulated price[20].

Intellectual property holders now want WHOIS functionality to be expanded so that data can be searchable by domain name, the registrants' name or postal address, technical or administrative contact name, NIC handles[21], and IP addresses. They also want searches to be based on Boolean operators or incomplete matches, as well as exact string matches. Further, they are requesting that the results of searches not be limited to a certain number (VeriSign only returned 50 records at a time). Moreover, they want this expanded capability to be subsidized (i.e., they want it to be considered a part of the public Internet infrastructure and not a value-added service that they would have to pay for). Not content with the already massive reduction in transaction costs brought about by the existence of a single, integrated name space that can be searched using automated tools, they want to shift the costs of policing and monitoring the trademark-domain name interface onto users, registries, and registrars.

The issue is no longer exclusively one of trademark surveillance and protection. Copyright interests now view expanded WHOIS functionality as a way to identify and serve process upon the owners of allegedly infringing Websites. Moreover, public law enforcement agencies, notably the U.S. Federal Bureau of Investigation (FBI), have become deeply interested in the use of WHOIS to supplement their law enforcement activities. The intent is to make a domain name the cyberspace equivalent of a driver's license. Unlike the drivers' license database, however, this one would be publicly accessible to anyone and everyone.

The growth and commercialization of the domain name system, and political pressure from intellectual property interests, continue to have a major impact on the evolution of WHOIS. The technical challenges involved in providing an integrated lookup system across the gTLD name space(s) have been magnified by the introduction of competing registrars who share access to the same registry, and by the addition of new TLDs. For the first year or so after the introduction of registry sharing, WHOIS service was fragmented across registrars (i.e., one had to know which registrar a customer used to register a name in order to be able to look up the name). Some limits have been placed on technically abusive

[20] See Section 3.3.

[21] The NIC handle is a short, unique alphanumeric code that a registry assigns to a domain name holder when the registrant registers a name. People who use different names might use the same NIC handle in the WHOIS record.

data-mining techniques that have been directed at the WHOIS databases of registries and registrars[22].

In response to pressure from major corporate trademark and intellectual property holders, the U.S. Commerce Department is pushing for universalizing and globalizing the level of WHOIS service formerly associated with the VeriSign monopoly. Efforts are underway to create a 'universal WHOIS' that would provide access to registrant information within country code TLDs as well as the generic TLDs. The revised Verisign-Commerce Department contract of May 2001 requires the company to continue 'development and deployment of a universal WHOIS service that allows public access and effective use of WHOIS across all registries and all TLDs at the direction of the Department.'

6. Conclusions

A rich series of economic and policy problems have been raised by the emergence of a domain name market. Economists could profitably devote more attention to issues of naming and identity and the associated industries of registration and name service. There is a need both for modeling and for empirical research. The ICANN's loosely organized and amateur policy development apparatus is badly in need of professional research into policy options and the consequences of prior policies.

References

Albitz, P. and C. Liu, 1992, DNS and BIND in a Nutshell, Sebastopol, CA: O'Reilly & Associates.

Andeen, A. and J. L. King, 1997, Addressing and the future of communications competition: Lessons from telephony and the Internet, in B. Kahin and J.H Keller, eds., Coordinating the Internet, Cambridge, MA: MIT Press, 208–257.

Bechtold, S., 2002, Governance in namespaces, TPRC 2002. The 30th Research Conference on Information, Communication, and Internet Policy, Alexandria, VA.

Berners-Lee, T., 1996, The myth of names and addresses, World Wide Web Consortium, 2002.

Berners-Lee, T., 1998, Cool URIs don't change, World Wide Web Consortium, 2002.

Brownlee, N., K. Claffy et al., 2001, DNS measurements at a root server, San Diego, California: Cooperative Association for Internet Data Analysis (CAIDA).

Cannon, R., 2001, ENUM: The collision of telephony and DNS policy, TPRC 29th Annual Research Conference on Information, Communication, and Internet Policy, Alexandria, VA, USA.

[22] See 'Investigation Report by Verisign Global Registry Services,' In the matter of *Register.com, Inc. v. Verio Inc.*, 00-Civ-5747 (BSJ).

Farrell, J. and P. Klemperer, 2001, Coordination and lock-in: Competition with switching costs and network effects, 2002.

Froomkin, M., 2002, ICANN's "uniform dispute resolution policy"—Causes and (partial) cures, Brooklyn Law Review, 67, 605–717.

FTC, (1998). Comment of the staffs of the Bureaus of Economics and Competition of the Federal Trade Commission on improvement of technical management of Internet names and addresses, Washington, DC.

Gans, J. S., S. P. King and G. Woodbridge, 2001, Numbers to the people: Regulation, ownership and local number portability, Information Economics and Policy, 13(2), 167–180.

Geist, M., 2001, Fair.com?: An examination of the allegations of systemic unfairness in the ICANN UDRP, Ottawa, Canada: University of Ottawa Faculty of Law.

Hoffman, P., 2002, Reforming the administration of the DNS root, ICANN-notes, April 25, Council for National Research Initiative, available at http://www.proper.com/ICANN-notes/dns-root-admin-reform.html.

Hotta, H., 2001, Technology and policy aspects of multilingual domain names, Geneva: International Telecommunication Union 56.

Huston, G., 2002, ENUM—Mapping the E.164 number space into the DNS," The Internet Protocol Journal, 5(2), 13–23.

Hwang, J. and M. Mueller, 2002, The economics of ENUM over cable broadband access networks, Syracuse University School of Information Studies, The Convergence Center.

Kahin, B., 1996, Auctioning global namespace: A new form of intellectual property, A new vision of universal service, CIX/Internet Society workshop on Internet Administrative Infrastructure, Washington, DC.

Kahn, R. and R. Wilensky, 1995, A framework for distributed digital object services, May 13, 1995, Council for National Research Initiative available at http://www.cnri.reston.va.us/home/cstr/arch/k-w.html.

Katoh, M., 2001. Report of the Internationalized Domain Names Internal Working Group of the ICANN Board of Directors, Internet Corporation for Assigned Names and Numbers.

McTaggart, C., 2001, E Pluribus ENUM: Unifying international telecommunications networks and governance, 29th Research Conference on Information, Communication, and Internet Policy, Alexandria, VA.

Mockapetris, P., 1987, Domain names—Concepts and facilities, Internet Society, RFC 1034.

Mueller, M., 2001, Rough justice: A statistical assessment of ICANN's uniform dispute resolution policy, The Information Society, 17(3), 153–163.

Mueller, M., 2002a, Ruling the Root: Internet Governance and the Taming of Cyberspace, Cambridge, MA: MIT Press.

Mueller, M., 2002b, Success by default: A new profile of domain name trademark disputes under ICANN's UDRP, Syracuse, NY: The Convergence Center.

Mueller, M., 2002c, Competing DNS roots: Creative destruction or just plain destruction? Journal of Network Industries, 3(3), 313–334.

NERA, 1998, Feasibility study & cost benefit analysis of number portability for mobile services in Hong Kong, HK, National Economic Research Associates Economic Consultants (NERA) and Smith System Engineering for the Office of the Telecommunications Authority (OFTA).

Network Solutions, Inc. Annual Report for 2000.

NTIA, 1998, Management of Internet names and addresses, Washington, DC: National Telecommunications and Information Administration, U.S. Department of Commerce.

Paskin, N., 1999, Towards unique identifiers, Available form the author at n.paskin@doi.org.

Reiko, A. and J. Small, 1999, The economics of number portability: Switching costs and two-part tariffs, 1–30.

Rood, H., 2000, What's in a name, What's in a number: Some characteristics of identifiers on electronic networks, Telecommunications Policy, 24(6–7), 533–552.

Rutkowski, A., 2001, The ENUM golden tree: The quest for a universal communication identifier, Info, 3(2), 97–100.

Saltzer, J., 1993, On the naming and binding of network destinations, Internet Society, RFC 1498.

Viard, V.B., 2003, Do switching costs make markets more or less competitive? The case of 800-number portability, Stanford Graduate School of Business Research Paper 1773 R1.

Vixie, P., 1994, External issues in DNS scalability, The global information infrastructure, Annenberg Washington Program Washington, DC, Science, Technology and Public Policy Program John F. Kennedy School of Government, Harvard University.

WIPO, 1999, The management of Internet names and addresses: Intellectual property issues, Geneva: World Intellectual Property Organization.

WIPO, 2001, The recognition of rights and the use of names in the Internet domain name system: Report of the Second WIPO Domain Name Process, Geneva: World Intellectual Property Association.

Chapter 12

BOTTLENECKS AND BANDWAGONS: ACCESS POLICY IN THE NEW TELECOMMUNICATIONS

GERALD R. FAULHABER

University of Pennsylvania

Contents

Handbook of Telecommunications Economics, Volume 2, Edited by S. Majumdar et al.

DOI: 10.1016/S1569-4054(05)02012-9

1. Introduction[1]

The essential facilities doctrine (bottleneck access) has been at the heart of antitrust since *United States v. Terminal Railroad Association* (1912)[2], in which the Court held that a 'bottleneck' (i.e., essential) facility that could not feasibly be duplicated must be shared among rivals. This doctrine has been upheld over the intervening years (e.g., *Otter Tail Power Co. v. United States* (1973) and *MCI Communications Corp. v. AT&T Corp.* (1983)), although the doctrine is not beyond dispute (e.g., H. Hovenkamp, 1994). Most famously, *US v. AT&T* (1982) was based entirely on the concept that the telephone local-access line (the local loop) was a bottleneck facility; divesting the Operating Companies that owned the local loop from the AT&T long-distance company would thus remove the incentive to use the local loop to disadvantage AT&T's long-distance competitors. Most recently, the Telecommunications Act (1996) requires the Regional Bell Operating Companies to offer to all potential competitors the constituent parts of its local loops (unbundled network elements) at cost-based prices, based on the premise that the local loop is an essential facility. More recently, 'open access' for high-speed Internet access has become an active policy debate; the FTC's remedy in the recent AOL-Time Warner merger and current FCC proceedings are based on the idea that cable modems may indeed be a bottleneck facility.

Interconnection (bandwagon access) has been at the heart of the regulation of network industries since the early decades of the 20th century. The end of the 19th century saw somewhat disruptive practices in railroad and telephone, in which dominant players refused to transit traffic to or from their competitors' customers. For example, customers of Keystone Telephone Company in Philadelphia, a local-exchange provider, could neither receive nor complete calls to customers of Bell of Pennsylvania, the dominant local-exchange provider, which in turn was connected to AT&T's unique long-distance service. The telephone industry developed into local monopolies dominated by the Bell System and subject to state regulation. Regulators required Bell to interconnect with the remaining Independent telephone companies, and regulated the interconnection prices as well. Railroad regulators encouraged railroads to enter into 'interlining' agreements, whereby each party agreed to accept traffic from other railroads destined for their

[1] Portions of this chapter appeared in Faulhaber (2002).

[2] "A consortium of 14 of the 24 railroads that shipped freight across the Mississippi River at St. Louis got control of the terminal facilities at each side of the river. The Supreme Court, while assuming that the operation of these facilities as a single entity was the most efficient way to operate them (i.e., they comprised a natural monopoly), held that the Sherman Act required the consortium to provide access to the terminal facilities to the 10 other railroads on nondiscriminatory terms." Richard Posner (1995).

customers. Central to the interconnection requirement is that such networks exhibit 'network effects' (or 'bandwagon effects;' alternatively 'network externalities'), in which the value of the network to each customer increases as the number of customers increases[3]. In the telephone network, this is very easy to see: the more customers you can call (or be called by), the more valuable is the network. Similarly with rail transport: the more customers you can ship to (or receive from), the more valuable is the network to you. Should a firm decide not to interconnect with its competitors, then all customers lose value as they now have fewer customers they can reach. Of course, smaller competitors are disadvantaged more than large competitors, suggesting that refusal to interconnect may be a strategic decision by larger firms. More recently, the FCC's remedy in the recent AOL-Time Warner case regarding Instant Messaging reflected a concern that interconnection/interoperability will be at least as important in the new telecommunications as in the old, perhaps even more so.

Network effects were extensively studied in the mid-1980s[4]; initially, network effects were thought to be limited to two-way communications systems ('direct' network effects), but this later work identified 'indirect' or 'virtual' network effects based on complementary goods. This latter category includes videotape players, in which the more customers own a particular format (say, VHS), the more tapes are produced in that format, and therefore, the more benefit each customer derives from the service. This chapter is devoted to direct network effects and most relevant to interconnection and access issues.

Some industries exhibit both essential facilities and network effects; the wireline telephone industry is a case in point. Other industries may have network effects without an essential facility (such as Instant Messaging), and some have essential facilities without network effects (such as railroads). In each case, possible anticompetitive actions can lead these and related industries to (near-) monopoly, and may require some form of government intervention (either regulatory or antitrust) to prevent these anticompetitive actions.

The focus of this chapter is simple: the two forms of access (bottleneck access and bandwagon access) result from very different market failures and call for very different remedies. Even in industries, in which both forms of access problems exist, solving one will not solve the other. For example, solving the interconnection problem (by regulatory fiat) in telephone has not solved the bottleneck local-loop problem. Similarly, even if the local loop had competitive alternatives (i.e., not a bottleneck), a solution to the interconnection problem would still be required, both price and entry. Not only might a dominant provider deny interconnection to smaller players, but even a small player may charge very high

[3] The network externality was first introduced into the economics literature by Rohlfs (1974). For a recent survey with real-world examples, see Rohlfs (2001).

[4] For an excellent and accessible summary of this research, see Katz and Shapiro (1994).

call-termination rates to other carriers for access to its customers, for whom it is the only connection (see Armstrong, 2002).

2. Essential facilities (Bottleneck access)

Traditionally, an essential facility is in fact a physical facility[5]. The telephone local-loop, terminal facilities in St. Louis, a rail line to the mouth of a coal mine, all constitute essential facilities. Also, essential facilities cases tend to be vertical cases; a vertically integrated firm may own an essential facility through which upstream firms with whom it competes must reach their customers. The question at the heart of essential facilities cases is what makes the facility so essential, and how does it differ from simple monopoly?

A simple example is that of a harbor suitable for shipping for which one individual owns all land surrounding the harbor. Any firm that wished to use the harbor must use this individual's land and any improvements, such as roads and docks. Assume that this harbor owner also owns firms in the neighborhood that ship from this harbor. The owner of the harbor could be expected to charge a monopoly price, more likely discriminatory prices, to those who wished to use the harbor and may refuse to deal with firms that are its direct upstream competitors. The pricing of the monopolist is limited only by: (i) the value of the service he is providing to his customers; and (ii) the cost of alternative methods to his customers of obtaining equivalent services. If the nearest harbor is hundreds of miles away over unimproved roads, then that alternative cost could be quite high, allowing the monopoly harbor owner to extract substantial rents from his customers. Should the value of harbor use to his customers be very high, then the rents will be correspondingly high, particularly if the monopolist is adept at price discrimination.

In this case, the essential facility is truly nonduplicable; there is no feasible alternative to ocean shipping beyond this unique resource. However, it raises a fundamental issue: how is this different from simple monopoly? Since antitrust law does not forbid monopolies that are earned (as opposed to merger to monopoly or monopolization via anticompetitive practices), why should the ownership and exploitation of this essential facility not be treated as a simple monopoly? Of course, it may be judged that the exploitation of a particular monopoly has pernicious spillovers, such as drastically reduced economic development within the harbor's natural economic zone due to high pricing of access to shipping. If so,

[5] This has not stopped antitrust cases based on nonphysical 'essential' facilities. In Blue Cross et al. v. Marshfield et al. (1995), Blue Cross claimed inter alia, that the Mansfield Clinic and HMO had monopolized medical services in northern Wisconsin, and was therefore an essential facility to which they and their subsidiary CompCare should be provided access. The trial judge held in their favor, but this portion of the verdict was overturned on appeal in the 7th Circuit.

this suggests a criterion for determining if an 'essential facility' should be subject to scrutiny: is the monopoly an earned monopoly, or is there a compelling external, social, or economic reason (beyond mere monopoly exploitation) for forcing access?

But this is not sufficient; granting access to the harbor for the harbor owner's upstream competitors need not help with these spillovers, if the harbor owner simply charges a monopoly price to all. In order to solve this broader problem, both access and price would have to be controlled[6], a job more suited to regulators and less suited to antitrust authorities. We note that there is a long and extensive literature that calls into question the efficacy of such regulation in promoting broader economic and social aims[7]. Unless antitrust authorities are willing to pursue such regulatory solutions, the imposition of access requirements on owners of essential facilities would appear to have only the effect of redividing the monopoly rents that accrue from these facilities among competitors, with little positive impact on other agents such as customers. This suggests a second criterion for scrutiny: is there a mechanism in place for monitoring and regulating the price and terms of trade of access?

Even in the case of a truly essential facility such as the harbor example, the question of how essential is the facility still has teeth. If the surrounding economic zone produces low-value bulk goods for export such as grain or coal, then sea transport is indeed essential. However, is it essential that grain or coal have to be produced here? If production shifted to high-value small goods such as microprocessors or drugs, then the preferred mode of transport is airfreight, not ocean shipping, and the essential nature of the harbor facility fades. If it is essential that grain and coal be produced here (because of natural endowments, say), is it essential that they be sold overseas? If trains or coal slurry pipelines could reach domestic markets easily, then the essential nature of the harbor facility fades. This suggests a third criterion for scrutiny: do final customer have no feasible way of

[6] More than just price is at issue. A number of difficult questions must be addressed if access to an essential facility is granted: (i) do all competitors have such access? How about customers? If only a limited number of competitors are granted access, how is it decided which ones get access? (ii) The quality of service to the competitors must be monitored; are competitors getting service as good as the bottleneck owners' upstream operator? (iii) Do competitors receive installation and maintenance services as promptly and thoroughly as the bottleneck owners' upstream operator? (iv) During congestion periods, are competitors treated the same as the integrated operator? (v) Do all competitors receive the same choices of price and service quality? The authority that regulates access must monitor these issues and enforce its regulations in a contentious, litigious environment, see Areeda (1990).

[7] Another solution is public ownership of the essential facilities, an option that may have even more practical difficulties than public regulation of private essential facilities. However, many port facilities in the U.S. are run by publicly managed Port Authorities (such as New York and Philadelphia), apparently with some success.

circumventing the use of the facility, such as changing production plans, accessing new markets, etc.?

In examples more realistic than the harbor owner, other factors come into play in an essential facilities analysis. In the case of the telephone local loop, it is not the case that this facility is nonduplicable. It is actually quite easy to build a parallel local loop; the technology is well known and indeed quite old. The problem is that the local loop is subject to extreme economies of scale, and is thus thought to be a 'natural monopoly.' Having two or more local loops serving the same homes and businesses was thought to be unnecessarily duplicative and wasteful, because of the extreme scale economies[8]. The premise is that the market could not support more than one local loop. This suggests a fourth criterion for scrutiny: are there no new technologies for the foreseeable future that could substitute for the essential facility?

The market evidence for this is dubious. In the 1890s, most towns and cities in the U.S. were served by two or more local-exchange companies, each with their own local loops. Until the Bell System introduced long-distance service and refused to interconnect with its independent competitors, the market easily supported multiple local-exchange carriers, using a technology (the twisted pair) that differs little from that in place today. Further, until the 1996 Telecommunications Act, almost all states granted monopoly franchises to their incumbent telephone companies, suggesting that legal protection against potential entrants may well have been necessary. And lastly, telephone prices to residential users have traditionally been subsidized by higher rates to business users and long-distance access charges, a situation that kept residential local-access prices below cost, and therefore unattractive to entry.

In addition, the role of technology in assessing how essential is the facility is important. For example, at the time of the AT&T cases, wireless services were just beginning to emerge with their potential to substitute for the local loop. More recently, wireless has become far more ubiquitous; should wireline services (which use the local loop) become substantially more expensive, it would be reasonable to expect customers to abandon wireline service for wireless only. This phenomenon is already occurring in Europe, in which wireline marginal prices are considerably higher than in the U.S.

In both *MCI Communications v. American Tel. & Tel. Co.* (1983) and *United States v. American Tel. & Tel. Co.* (1982), the Courts found the local loop to be indeed a bottleneck facility. In the latter case, the remedy was the most extensive divestiture in U.S. history, up to the date of this writing. Many industry analysts and observers (e.g., Kaserman and Mayo, 2002) view this divestiture as a success in bringing competition to the long-distance market; though some do not agree

[8] In the sense that having two gas stations at an intersection is not considered duplicative and wasteful, but simply the result of healthy competition (and no scale economies).

(see MacAvoy (1996)). This success supports the hypothesis that the local loop is (or at least was) an essential facility; and that separating ownership of the bottleneck from ownership of the competing upstream long-distance firm achieved the policy objectives of the case.

In the AT&T case, the monopoly over the bottleneck local loop was not earned but rather granted by regulators as a franchise to the Bell System[9]. This was not a situation, in which the monopolist won its favored position fair and square. Further, the failure of regulation to control the vertical power of the Bell System to leverage its local-access monopoly as been documented elsewhere (see Faulhaber (2003)). The remedy of divestiture, radical though it was, seems in this case to justify antitrust intervention based on an essential facilities premise.

Nevertheless, the burden of proof for essential facilities cases should remain high. In order for the Court to find an essential facility and grant access to competitors, the analysis above suggests that each of the following criteria be answered affirmatively:

1. Was the monopoly not earned, or is there a compelling external, social, or economic reason (beyond mere monopoly exploitation) for forcing access?
2. Is a mechanism in place for monitoring and regulating the price and terms of trade of access (see footnote 6)?
3. Do final customers have no feasible way of circumventing the use of the facility, such as changing production plans, accessing new markets, etc.?
4. Are there no new technologies for the foreseeable future that could substitute for the essential facility?

In a world in which capital is costly, economic options are few, and technology unlikely to provide new alternatives, the essential facilities doctrine may have a place. But this best describes the world of *United States v. Terminal Railway Association*; the AT&T case represents a unique combination of facts that for this case appears to justify the use of the essential facilities doctrine. Looking forward to a world of inexpensive and readily available capital, temporary technology-based monopolies that could be overturned by next generation systems, customers with lots of options, it is difficult to see a justifiable essential facilities case being successfully prosecuted[10].

A recent case in point is the Federal Trade Commission's (2000a) conditional approval of the AOL-Time Warner merger in December 14, 2000. The FTC

[9] The original Bell patents gave strong protection to Bell's business to about 1892; their expiration was the occasion of substantial entry into the local-loop business. Later, Bell's patents on long-distance telephone technology gave them the ability to again control the local loop. However, the advent of regulation and its grant of franchise monopoly to the local telephone companies made them impervious to competitive entry.

[10] Spulber and Yoo (2003) raise similar issues regarding bottleneck access, as well as asserting that mandating access to an incumbent's facilities raises constitutional issues under the takings clause.

imposed conditions on the parties that required 'open access' to the IP channel[11]. on Time-Warner's cable systems to Internet Service Providers (ISPs) other than AOL. The FTC's analysis of the merger focused on increased market power in the broadband ISP market and in the broadband transport market. It states, "...the merger will increase the ability of the combined firm to unilaterally exercise market power in Time Warner cable areas..." (Federal Trade Commission (2000b)). It also notes that the only viable competitor to cable modems was DSL, a similar transport service offered by telephone companies, and that the merger would inhibit DSL's deployment as a competitive alternative to cable modem in the transport market. This could occur if a merged AOL-Time Warner distributed AOL only over its own cable broadband facilities, and denied competing DSL providers access to AOL's popular service[12]. While not explicitly mentioning the essential facilities doctrine, much of the order reflects a view of the cable IP channel as an essential facility. Certainly the decision to force open access is an essential facility-type remedy.

In fact, the remedy is not quite open access as envisioned by open access's most fervent proponents. The AOL-Time Warner is not required to open its IP channel to all comers; it is required to open to at least three other ISPs, one of which is Earthlink (that signed a contract with Time Warner prior to FTC approval of the merger) and the other two must be approved by the FTC. Prices and terms of trade are to be 'similar' to those embodied in the Earthlink agreement, which was thoroughly vetted by the FTC. To ensure compliance with these regulations, the FTC appointed a 'monitor trustee,' whose ongoing job is to monitor AOL-Time Warner's performance and report to the FTC. In short, the FTC has taken on the job of regulating AOL-Time Warner's open access going forward.

The author takes the liberty of considering this condition an essential facilities condition, and asks if it would meet the criteria set forth above. Can we answer all of the questions in the affirmative?

1. It is certainly the case that the monopoly[13] of Time Warner cable was not 'earned' by superior performance, but rather by being granted local franchises in each of their service territories. So, the first question can be answered in the affirmative.

[11] Cable systems that offer broadband Internet access using cable modems do so by committing some portion of their cable capacity (typically, 6 MHz out of a total of 750 MHz) to a two-way IP (Internet Protocol) channel for transmitting and receiving Internet traffic to the cable modem. The cable modem is connected to the customer's PC that interprets the information and commands and displays the results on the customer's screen (either a browser, e-mail program, or other client software). This cable capacity is referred here as the IP channel.

[12] Hausman, Sidak, and Singer (2001a) call this 'conduit discrimination.'

[13] Ignoring for the moment the potential competition from DSL for data, and the real competition from Direct Broadcast Satellite for video.

2. There does exist no mechanism for monitoring and regulating the price and terms of trade of open access. The most likely candidate for this job was the Federal Communications Commission, which had no interest in (in fact, recoiled at the mere suggestion of) this job. As a result, the FTC took on this job itself. This question cannot be answered in the affirmative.
3. Final customers have other options; they can continue to use narrowband services and they can use broadband services at their place of employment[14]. Individual municipalities can deploy broadband systems, as has been done in a number of U.S. cities. This question cannot be answered in the affirmative.
4. New technologies for providing broadband access to customers are legion. Not only is DSL a real competitive threat but also wireless could well evolve into a threat as could satellite providers such as Starband. Providing open access may thus reduce the incentives to deploy other technologies to bypass this essential facility. This question cannot be answered in the affirmative.

This suggests the FTC's case, considered as an essential facilities case, was not a good one. However, the FTC did not explicitly consider this an essential facilities case, though they did impose an access condition, redolent of the essential facilities doctrine.

3. Network effects and interconnection (Bandwagon access)

Direct network effects industries such as two-way communications present unique public policy problems for both regulators and antitrust authorities. The earliest research on network effects was performed in the context of the old Bell System monopoly, in which a single firm controlled all telephone customers. The issue addressed was pricing telephone service to reflect the presence of the network externality: the total value of a customer joining the network was not only the private benefits (upon which the customer based his/her decision to join), but on external benefit of being able to call this customer that is conferred on others by his action. The result was a justification for charging a below-cost price for access to encourage subscribership[15]. To achieve 'universal service,' the regulatory sobriquet sometimes used to describe this phenomenon. Initially, the link between network effects and interconnection was not an issue of economic study.

[14] There is some controversy over how substitutable narrowband and broadband services are. Clearly, for services such as e-mail, they are almost perfect substitutes, and for others, such as online video, they are not. See Hausman, Sidak, and Singer (2001b) for the argument that narrowband and broadband are not good substitutes.

[15] Of course, the issue is more complicated; below-cost pricing of access must be compensated for by above-cost pricing elsewhere in the system, such as long-distance prices, which could itself discourage subscribership.

Interconnection was a staple of telephone regulation post-World War I. The newly dominant Bell System had agreed to stop acquiring independent telephone companies and to interconnect with them, in return for which states moved toward a regulated monopoly model for the industry[16]. Local-exchange companies no longer competed for local customers, and franchise monopoly, controlled by a state (and after 1934) and Federal regulator, became the industry standard. The requirement to interconnect grew out of the brutal competitive period during which Bell refused interconnection with independent companies, driving down their market value and eventually acquiring them on the cheap. At that time, refusal to interconnect was perceived as an aggressive even anticompetitive strategy. However, early antitrust authorities did not prosecute the early Bell System for monopolization; rather, the problem was viewed as a regulatory problem. Similar issues arose in the context of railroad regulation as well.

As wireless voice networks developed, the regulatory mandate that telephone companies had to interconnect was extended to include the new wireless carriers. Local-wireline carriers, such as Verizon, are required to interconnect with wireless firms that compete with their own wireless carriers. Customers of AT&T, Sprint, and Nextel wireless can complete calls to Verizon customers, both wireless and wireline. More recently, competitive local-exchange carriers (CLECs) have entered local-exchange markets, and the incumbent carriers (such as Verizon) are required to interconnect with the CLECs as well. This is, of course, a mutual obligation; the CLECs and wireless firms must complete the calls of Verizon wireline customers to their customers as well. Clearly, the regulators were aware of the anticompetitive potential of a large wireline monopoly refusing to interconnect to a small CLEC or fledgling wireless firm.

The price of interconnection is clearly as important as the requirement to interconnect, and interconnection pricing has occupied the FCC for more than two decades. Initially, interconnection pricing focused on the price charged by incumbent local-exchange carriers (ILECs) to long-distance carriers such as AT&T, Sprint and MCI to originate/complete their customers' calls. More recently, interconnection prices to wireless firms, paging firms and CLECs have occupied center stage at the FCC, which has launched a proceeding aimed at achieving a unified approach to 'intercarrier compensation' (Federal Communications Commission (2001b); see also DeGraba (2000) and Atkinson and Barnekov (2000)).

The rise of the Internet as an alternative communications medium in the early 1990s brought new networks into being that were not subject to telecommunications regulation, and therefore not subject to the injunction to interconnect. After the National Science Foundation withdrew from the Internet backbone network business, a number of private firms took up the function, collectively providing the core of the Internet's nonlocal-transport function of shuttling messages from

[16] This history has been recounted often; see e.g., Faulhaber (1987) and Brock (2002).

clients to servers and back again. Each of these Internet backbone providers had agreements to interconnect (or 'interoperate') with each other. The largest five did so on the basis of 'peering,' in which traffic was exchanged and no money changed hands. Since traffic was relatively balanced, it apparently was not necessary to incur the expense of actually counting minutes and charging; the large backbone firms simply exchanged traffic[17], as discussed in Kende (2000) and Milgrom et al. (2000).

The advent of wireless telephony, Commercial Mobile Radio Service (CMRS) presented two interesting interconnection regulation problems to the FCC: (i) should the traditional wireline telephone companies be required to interconnect with CMRS carriers? and (ii) should the CMRS carriers be required to directly interconnect with each other? The FCC answered the first question in the affirmative, recognizing the powerful network effects in the voice telephone market held by the incumbent telephone companies. But it answered the second question in the negative, noting that no CMRS possessed market power, so no interconnection mandate was needed (Federal Communications Commission (2000)). The CMRS-to-CMRS market more nearly parallels that of the Internet backbone market, in which no player is dominant.

The growth of the Internet and the growth of wireless outside the traditional strictures of regulation and their market-based solution to interconnection could well lead one to believe such restrictions aren't necessary to ensure interconnection, despite the experience of the early 20th century in telephone and regulations requiring interconnection in that industry. However, the explosive growth of Instant Messaging, first introduced by AOL, suggests some caution is in order.

Instant messaging is a text-based means of near-real-time communication between customers who have signed up for the service. Customer *A* can send an 'instant message' to customer *B* and receive an immediate reply, thus carrying on a text conversation. Instant Messaging fits the classic definition of a network effects business, in that the value to each customer depends upon how many other customers can be reached. This service has roots in the original Unix system, in which users on the same server could exchange messages in conversation mode. Instant messaging differs from e-mail in that it is a true dialog, operating in synchronous rather than asynchronous mode. The AOL introduced the service as a feature for its customers in 1989. However, it became wildly popular when AOL added the 'buddy list' feature in 1996. This feature displays a small window that lists all the customer's (self-designated) 'buddies' with an indication of whether or not each buddy is online. This feature ensured that if the customer sends an IM to a particular buddy, the customer knows the buddy would receive it immediately and could reply. With the advent of the buddy list, IM became one of the most

[17] However, the larger backbone firms have 'transit' fees for smaller networks that are charged a per-message fee for use of the larger backbone network.

popular features of AOL. In 1997, AOL took the unusual step of offering AOL Instant Messenger (AIM) as a stand-alone free download on the Web for non-AOL subscribers[18]. The AIM customers could IM both other AIM customers and AOL subscribers using IM, and vice versa; this is referred to as interoperation, equivalent to interconnection.

In 1999, several firms established competitive IM services, including Microsoft, Yahoo!, Otigo, and Tribal Voice. However, these competitors quickly learned that few customers wanted to sign up for a service that couldn't IM with AOL's huge customer base, reported at the time to be over 30 million subscribers. These competitors designed their IM clients to be compatible with AOL's IM protocols (which they earlier had published on the Web), to ensure their IM clients interoperated with AOL's AIM/IM[19]. The AOL chose to interpret these attempts to interoperate as 'hacking,' and took immediate steps to block such attempts. During the summer of 1999, various attempts by competitors were temporarily successful but quickly blocked by AOL. By year's end, AOL had demonstrated that it had both the will and the capability of blocking all such competitor attempts to interoperate with its AIM/IM services.

AOL was criticized in the press for refusing to interoperate with other IM services. Indeed, many contrasted AOL's call for open access in the pending AT&T-MediaOne merger with their refusal to interoperate their AIM/IM services with competitors. The AOL's response was that they had safety and security concerns on behalf of their customers, and were unwilling to expose them to nuisance or even pornographic IMs from non-AOL sources. They did indicate that they would interoperate with competitors, but only when technical protocols were agreed to ensure the safety and security of their IM customers.

Is the continuing refusal to interoperate/interconnect with competitors spring from a heartfelt desire to protect AOL's customers from evil messages? Or is perhaps a return to the early days of the telephone industry when refusal to interconnect was used aggressively by the Bell System against its independent rivals?

[18] All other features of AOL, such as chat rooms and proprietary content, are provided only to AOL subscribers. The AIM represents the first and only (to the author's knowledge) departure from this policy.

[19] The AOL IM model was replicated closely by the new competitors, who introduced services almost identical to that of AOL's. The feature sets of AOL, Microsoft, Yahoo, and other IM services differ only minimally. Incremental feature improvements are built into each upgrade, but competitors immediately match these new features. Absent the future development of a proprietary 'killer feature' by one of the competitors, IM appears to be a market, in which product differentiation does not play a significant role.

3.1. Network effects and corporate strategy

In general, there may be many producers of a network effects good, competing amongst themselves for customers and perhaps cooperating to help grow the overall market. However, strategic competition is somewhat different in a network effects market than in normal markets, in several important ways:

1. Firms will often choose to adhere to a common standard and interconnect with each other. In this way, customers achieve maximum benefit, as they are in the largest possible network. An example is the current arrangement among the largest Internet backbone providers to interconnect (without cost) to each other to carry Internet traffic. However, firms may have difficulty differentiating their product or service in such a market. Further, innovation may be difficult, in that it is constrained by the existing standard and existing interconnection agreements.
2. Firms may choose a different standard or not interconnect with their competitors, perhaps believing that the superiority of their product more than offsets the network effects loss for customers. Firms with a technologically advanced product may opt for this strategy, in order to supplant an existing standard or system for their own advantage.
3. In some cases, other firms can provide 'converters,' which permit customers to access an otherwise incompatible system. For example, in the mid-1980s, owners of Macintosh computers were unable to read floppy disks created on DOS systems. As the popularity of the PC grew, Apple built DOS-compatible disk drives into their systems. Another example is the software template built into Microsoft Word that permits it to exhibit the precise keyboard behavior of the rival WordPerfect word processor. WordPerfect customers were used to a complex series of keyboard combinations for performing actions; switching to another word processor such as Word could render this human capital valueless. To overcome this asset barrier, Microsoft built in a 'converter' to make Word behave just like WordPerfect. Note that converters are one-sided, in that Microsoft could unilaterally implement its converter for WordPerfect without first obtaining permission from the producers of WordPerfect.
4. In other cases, interconnection requires consent of both parties, and often a formal business arrangement. It is technically possible (though currently illegal) for one telephone company to refuse to interconnect with another telephone company, refusing to accept and complete calls originated by customers of the other firm or hand off calls originating with its own customers.
5. For many markets involving network effects, interconnection and standards are mandated by regulation. For example, in the telephone system, interconnection is mandated by law and interconnection prices are regulated by the Federal Communications Commission and state regulators.

The *scope* of the network effect matters greatly. If you join a communications system, in which you want to communicate with (say) only people in your work group, then adding other customers to your communications network has no value to the work group, and may actually decrease value. In this case, the network effect spans only a fairly small group; in such cases, the communications system may be designed and/or chosen by a representative of that group (such as the telecommunications manager or the IT manager), so everyone who ought to be on the system is on the system. In systems that have a local scope, interconnection among the relevant customer group can often be assured by the customer group itself (such as a firm, a club, or a school). Economists refer to this as *internalizing* the network externality. If the full scope of the network effect is a small, otherwise coherent group, then the group can internalize the externality.

In fact, most communication systems' customers focus most of their usage on a few other people who are friends or associates. However, they may also value the ability to communicate with merchants and people anywhere in the world in addition to their list of 'friends and family,' in which case the scope of the network effect is quite large, even universal.

Indirect network effects are almost always broad in scope, as they depend upon a complementary good, service, or asset. The value of having lots of other customers for VHS video player is that there are many video titles produced for the VHS format, which is a national or even global phenomenon. Therefore, the scope of the video player network effect is national or global, and not limited to just a handful. On the other hand, the value of purchasing a particular brand of automobile depends on how many *local* service centers can repair this auto, which depends upon the number of customers in my local area, not in the nation as a whole.

The presence of network effects would appear at first glance to be very helpful in determining market definition in antitrust analysis, and in some cases this may be true. For example, there are several providers of Internet messaging service, including Microsoft, AOL, and Yahoo! *Ceteris paribus*, interconnecting these systems would almost surely raise the value of each system to its customers, and this strongly suggests that these messaging systems are in the same market. However, network effects are usually only one piece of the market definition. For example, customers of wireless telephones benefit greatly from being able to send and receive calls to and from customers of traditional wireline telephones, suggesting strong network effects. But it is likely that wireless and wireline telephone service are distinct markets with (at present) fairly low cross-elasticities of demand[20], even though network effects between the two services are apparently strong.

[20] Only about 1–2 percent of US cellular phone customers use this phone as their only phone; most prefer to have both a wireless and a wireline telephone. However, there are apparently more customers in Europe that use wireless exclusively. It is possible that as the price of wireless is reduced, and its quality improves, it will become a true substitute for wireline telephony.

As a general rule, interconnecting competing systems with network effects adds value for customers, at least up to the limit of their scope. Network effects industries may have a 'start-up' problem; the initial networks may be so small that they are not sufficiently attractive to potential customers; so few people join, resulting in a market equilibrium with a small network. In start-up markets, competing firms may have an incentive to interconnect so that the industry as a whole is more attractive to customers. They may also have incentive to convince customers that this market is growing, and that in the future, there will be many more customers on the network. For example, if the start-up industry has several well-established firms with well-known brands that are unlikely to suddenly exit the market, customers may be more willing to sign up now in expectation of 'getting in on the ground floor' of a sure-to-be-successful network effects service.

What is a firm's incentive to interconnect? If the technology of the service in question permits converters, then interconnection is a unilateral decision for each firm: does the customer base of your competitor justify investing in a converter? In the case of PC disks mentioned above, it paid Apple to invest in a converter in the form of a DOS disk drive, as the base of DOS customers grew much larger than the Macintosh customer base. But these facts also suggest that it would not pay a PC producer of DOS machines to invest in a converter for Macintosh disks, and indeed they didn't.

However, if converters are not possible (or can be thwarted), then, interconnection can only be by consent of each firm. In telecommunication systems, physical interconnection and shared protocols are required, so that interconnection is a mutual business decision (although regulation may require such interconnection). In this case, the decision to interconnect is a crucial strategic decision for each firm, perhaps the most important decision they will make in their corporate lives[21]. It is also a decision that may determine the degree of competition in the market, and so is of great interest to regulators and antitrust authorities.

We first consider a mature industry with network effects that are broad in scope, in which new customer growth is modest or nil, with a small number of firms of roughly the same size. Is it likely that only some (or even none) of the market participants will interconnect in equilibrium? No; if two such firms interconnect, they will both offer their customers a higher value than the remaining noninterconnecting firms. These two firms will be in the enviable position of being able to charge higher prices and attracting customers from the noninterconnecting firms! In this case, an interconnection arrangement helps each firm grow and increases its

[21] Interconnection may be costly, in at least two ways: (i) it may require investment in physical assets, such as switches and routers, as well as ongoing variable costs and (ii) it may require monitoring and/or blocking of nonconforming uses. For example, if one firm produces an excessive number of abusive or obscene communications to customers of firms, with which it interconnects, then costs are imposed on both these firms and their customers, a claim made by AOL to explain its refusal to interconnect.

profitability. Noninterconnecting firms face a choice of interconnecting with the other firms, or losing their customers to the more valuable interconnected network. In this case, the only stable outcome (i.e., the market equilibrium) is for all firms to interconnect.

A variant of this case is a mature market with a small number of largish players and a large number of very small players. In this type of market, the largish players have an incentive to interconnect with each other, but not with the smaller players. They may interconnect with the smaller players but on terms much less favorable than the terms on which they interconnect with each other. It appears this is the situation in the previously referenced Internet backbone network business, in which about five firms of roughly equal size exist with a large number of much smaller players. The five firms interconnect with each other as 'peers;' since the exchange of traffic is roughly equal, they do not charge each other for carriage of Internet traffic. However, they charge 'transit' fees to smaller backbone networks (see Kende (2000) and Milgrom, Mitchell et al. (2000)).

We next consider a mature market, in which a single firm has a dominant position, such as a market share well in excess of 50 percent. It is clear that all other firms have an incentive to interconnect with the dominant firm, as its customer base adds much value to the smaller firms' networks. The dominant firm has a lesser incentive to interoperate, as adding the smaller firms' customers to its network adds much less value to the larger firm. But it also has an incentive to refuse interoperation, since it may expect that customers of the smaller firms will soon join their larger network to obtain the higher value of more network customers. This would increase the size of its network even more, making it even more attractive to the customers that chose to remain with smaller firms. It is possible that market-driven feedback loop is so compelling that the larger firm eventually drives out the smaller firm and in equilibrium monopolizes the industry, simply by refusing to interoperate. Thus, refusal to interoperate, followed by full or partial monopolization, is a possible market equilibrium in network effects markets. The short-term loss in value from refusing interconnection could be more than made up by the long-term gain in firm value from increasing market share. Refusal to interoperate can thus become a strategy for precluding competition with no offsetting benefit to customers. It is this anticompetitive outcome that is of concern to regulators and antitrust authorities.

By way of example, consider a market with one firm at 70 percent market share and two other firms at 15 percent share each, and suppose the scope of the network effect included all customers. Even if the two smaller firms interconnected, their combined share at 30 percent may not be sufficient to keep customers attracted by the value of the network effect of the much larger firm. It is true that the value of being able to reach 70 percent of the population is less than can be obtained by full interconnection with 100 percent of the population. However, if the large firm refuses to interconnect, then the appropriate comparison is between reaching 70 percent of the population vs. reaching 15 percent or even 30 percent of the population. Clearly, the larger firm creates the greater value, and as more customers switch from the smaller firms to the larger firm, the value becomes even greater. Of course, it could be that the smaller firms may have 'diehard' customers who will never desert them, even if the network effects are very large. For example, even though Microsoft Windows appears to have over 90 percent share in the OS market and the 'application barrier to entry' as found by the antitrust court is strong (indirect network effect), Macintosh computers still command the loyalty of their few remaining diehard fans. Thus, a complete 100 percent Windows monopoly is unlikely to occur.

How likely is it that the anticompetitive refusal to interconnect equilibrium will result from the interplay of market forces? Several factors determine if the market leader adopts a noninterconnection strategy:

1. The largest player must be substantially larger than its competitors, perhaps bigger than all its competitors combined (market share greater than 50 percent).
2. The network effect must be strong, so that switching to the largest provider adds substantial value for customers.
3. Customer switching costs ('stickiness') must be low, so that switching to the largest provider is not too costly for customers.
4. It must be difficult for smaller competitors to agree to interconnect amongst themselves. For example, in a market with a leader at 40 percent share and 60 small firms each with 1 percent share that don't interconnect, the leader is likely to gain customers fairly quickly. However, if the 60 firms can agree to interconnect, then their total network share is 60 percent, and the leader is likely to be the one losing share by refusing to interconnect, not the smaller firms. Even if the leader has over 50 percent market share, its competitors stand a better chance of surviving (but no guarantee), if they combine forces through interconnection.

If all four factors are in place, then the market has 'tipped,' in that the largest firm will implement a noninterconnection strategy and drive the market toward monopoly. Tipping does not imply that the market is at monopoly or near-monopoly; merely that market conditions are such that this will be the ultimate outcome.

We next consider a growing market of broad scope, in which some potential customers have joined a network but there are still many more potential customers as yet not signed up. The same factors that matter in a mature market still matter in a growing market: market share of largest firm, strength of the network effect, customer switching costs, and the ease with which smaller firms can reach a mutual interconnection agreement. In addition, several other factors come into play in growing markets:

- How costly is it to acquire customers through marketing efforts? Of particular importance is the difference in acquisition costs among firms. For example, a firm with an established brand name may find it easier to acquire customers. Also, if a firm controls a complementary good with a large market share, it may find it particularly easy to attract customers for its new service. For example, many have claimed that since Microsoft controls the Windows OS, it has an advantage in introducing new services such as its Microsoft Messenger service; others claim that since Yahoo! is the most popular Web portal (in terms of number of page views) that it has an advantage in introducing new services such as Yahoo! Messenger[22]. Firms with particularly low customer acquisition costs are sometimes able to overcome a rival's early lead in a network effects market by using its competitive advantage and 'outmarketing' its rival to surpass it in market share and thus gain the network effects advantage. In this case, it is possible that the early leader that does not possess a customer acquisition advantage may prefer to interconnect, while the 'fast follower' firm with the ability to enlist a mass of current customers of a related service into this new service may refuse to interconnect, expecting to surpass the current market leader and tip the market itself[23].
- What is the *shape* of the network effects value as a function of share of potential market? As a practical matter, there appears to be two generic types of the value function; the first is the *S-curve*, in which the value per customer is very low for network sizes up to perhaps 20–30 percent of the potential market; the value increases sharply for network sizes from 20–30 percent up to perhaps 70 percent. Adding further customers increases the value, but at a decreasing rate. Figure 1 is a graphical representation of this type of value function.

In this particular example, a noninterconnecting firm with market penetration less than 20 percent offers virtually no value to customers. If no strategy can be found to grow the network beyond 20 percent, it is likely that the industry will either die, or be limited to markets with very narrow scope. This apparently was the fate of Picturephone, a video telephony product offered by the Bell System in

[22] Reports in the business press often assess the prospects for success of software/Internet firms in new markets by reference to their current strengths or weaknesses in complementary markets. For example, Microsoft is viewed as particularly strong in that it is able to use its Windows OS platform to 'make it easy for customers to use MS Messenger.' In economic terms, these are statements about relative customer acquisition costs. When Microsoft bundled the Internet Explorer browser into Windows 98, it reduced the cost of acquiring a customer for Internet Explorer close to zero; customers could still obtain Netscape, the competing browser, free of charge, but it required a rather lengthy download, so customer acquisition costs were higher for Netscape.

[23] This is likely to be an unnecessarily risky strategy on the part of the fast follower; it may fail to surpass its rival, and thus find itself on the wrong end of a noninterconnection strategy by the leader. On the other hand, if its strength in complementary markets is strong, it may find that interconnection is the better strategy, as it can count on brand name and customer loyalty to dominate this new market, rather than relying on counting on a 'surpass and then refuse interconnection' strategy.

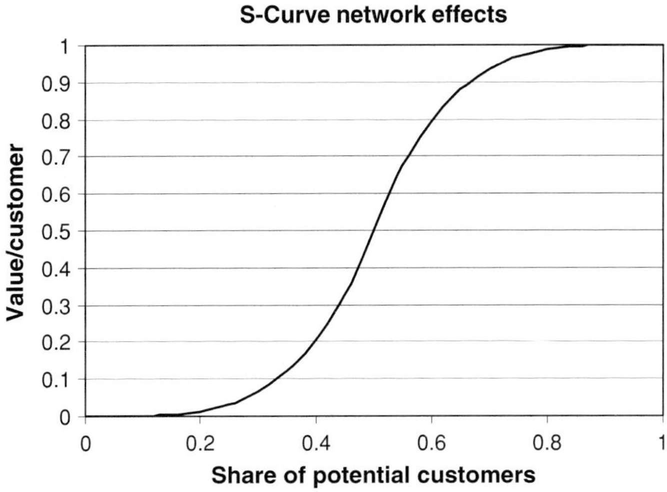

Fig. 1.

the 1970s. It never acquired a critical mass to allow it onto the steeply sloping portion of the value curve, and was withdrawn from the market (see Rohlfs (2001)). One such strategy is, of course, interconnection. If the start-up firms agree to interconnect, they could well grow the total network beyond the 20 percent, and the industry as a whole could then take off. With S-curve network effects, early interconnection agreements are likely. In fact, this occurred with Internet e-mail systems in the late 1980s.

The second generic type of value of the network effects is *concave*, in that the value as a function of share of potential market, in which the earliest growth produces the greatest value and later growth adding value but at a diminishing rate. Figure 2 is a graphical representation of a concave network effects value function.

In this case, a firm that grows rapidly gains substantial customer value through its network effects, and has potential to dominate the market at an early stage. In this example, a firm with 20 percent market penetration has already captured almost 60 percent of the network effects value. With concave network effects, refusal to interoperate at an early growth stage is likely, particularly if one firm obtains an early lead while experiencing rapid growth. This has apparently occurred with AOL's IM service and also with fax machines (see Rohlfs (2001)).

In sum, the likelihood of a noninterconnection strategy by the largest player in a growing market of broad scope depends not only on the factors identified for mature markets (market share, network effects strength, customer 'stickiness,' and ease of smaller competitors interconnecting), but also two additional factors:

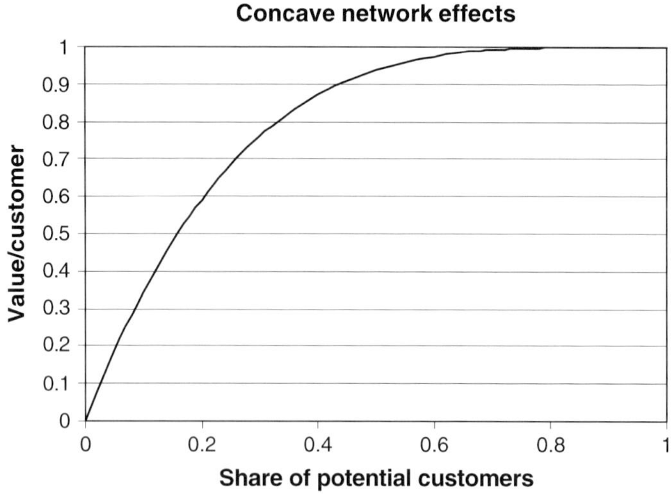

Fig. 2.

1. The relative ease of customer acquisition, taking account of competitors' positions in complementary markets; and
2. The shape of the network effects value as a function of share of potential market (S-curve or concave).

It is obvious that regulation or antitrust considerations trump strategic decisions; if regulators require interconnection (as is the case in voice telephony), then the firm has no choice (except possibly to obfuscate and sue). However, it should also be noted that the *threat* of antitrust or regulatory action may induce a dominant firm to interconnect 'voluntarily' (and thus under its own conditions), rather than face government mandates for interconnection on terms dictated by regulators.

3.2. Is tipping irrevocable? The 'serial monopoly' hypothesis

The models above suggest that a dominant firm in a market with broad network effects and favorable conditions can refuse to interconnect and thus 'tip' the market, causing customers to leave smaller suppliers and join its network until near-monopoly results. This might be taken to imply that once the dominant firm refuses to interoperate (for strategic reasons), so that the process is irrevocable and near-monopoly is a foregone conclusion.

This is not the case; even markets that have tipped to monopoly may in fact be subject to successful competitive attack by new products that are so superior that customers are willing to undertake switching costs, including network effects, to buy the superior product. Of course, this could also occur after the market has tipped and the largest provider is gaining market share. A 'killer app' changes the market dynamic and could well reverse the fortunes of the previous market leader. However, the presence of network effects almost always makes this more difficult. It is not sufficient that the new product be superior to the existing product, in which the largest provider has, or is about to have, a near-monopoly; it must be so much better as to induce customers to forego the network effect of the old product to try the new. The network effect acts as a barrier to entry; it is not insuperable, nor is the network effect irrevocable. But it requires more than simply a superior product to traverse this barrier to entry; it requires a product so superior that the added benefits of it not only overcome normal customer inertia but also the cost of foregoing network effects as well.

Several authors[24] have argued that in the new economy, this 'leapfrogging' is a natural outcome of technology-driven markets. Innovation in these markets is so rapid and revolutionary, the argument goes, that no market leader, even with strong network effects, can defend its position for long against hordes of new entrants with revolutionary products and services. The high-fixed-cost/low-variable-cost nature of the high-technology sector, coupled with strong network effects, suggests that at any point in time, monopoly may be the only sustainable market structure. Of course, this is a variant of the 'natural monopoly' argument of J. S. Mill (1848), typically used as a justification for regulated monopoly in public utilities. However, the proponents argue that the rapid pace of technical innovation in these industries will challenge and overturn the monopoly in favor of a new technology and eventually a new monopoly. The competitive struggle is *for* the market, not *in* the market. Because the monopoly is never secure in its position; it must always struggle against new entrants and new technologies by constant improvement of its product and service, and respond to constant price pressure. It does this not because of existing competitors in the market, but rather the threat of new competitors with new technologies who may enter the market. This is the 'serial monopoly' hypothesis[25]: at any time, it is likely that only one firm is in the market, but the threat that they could be overtaken at any time by any entrant disciplines their price, quality, and innovation behavior.

[24] The most recent and persuasive article in this literature is Evans and Schmalensee (2001). A popular version of the hypothesis is presented in Liebowitz and Margolis (1999), as well as earlier work of these authors.

[25] The serial monopoly hypothesis bears a strong resemblance to contestable market theory of the early 1980's. In that theory, monopoly was perfectly acceptable as long as barriers to entry were low, and new firms could enter at low cost. The monopolist would then be required to maintain low prices and high quality for fear of new entry. In the serial monopoly version of the story, high barriers to entry (in the static sense) are acceptable as long as technological innovation is fast enough so that such barriers to entry are short-lived.

This form of 'leapfrogging' should be distinguished from a 'fast follower' strategy, in which a firm carefully watches its competitors for new innovations, and follows the apparent successes to market quickly via close imitation. It may hope to learn valuable market lessons from its observation of the pioneer and possibly use complementary assets (such as another product or its marketing expertise) to overtake the pioneer. The classic case is Matsushita's introduction of a VCR less than 12 months after Sony introduced the first such product in the 1970s. Matsushita took advantage of Sony's market experience to fine-tune its fast-following product. In this case, no network effects were involved. In this strategy, the pioneer has not had time to reach a critical mass and the fast follower can overtake it without encountering strong network effects. In contrast, the leapfrogging phenomenon refers to a successful deployment of a new and superior product against an existing product, which has achieved a critical mass already (possibly a pioneer or a fast follower), and exhibits network effects. It may be argued that DVD will be the 'leapfrog' product against the VCR, perhaps overturning an existing product with strong network effects with a superior technology.

Unfortunately, there is essentially no empirical support for the hypothesis that potential entry by new, previously unknown firms either do actually occur, or its threat discipline incumbents. Evans and Schmalensee (2001); and Liebowitz and Margolis (1999) quote a number of compelling examples in the recent computer hardware and software markets. But their statements that firms dominating markets that depend upon intellectual property are 'fragile,' and subject to being overtaken on a moment's notice, has little empirical support. Of course, this is not to say their hypothesis is incorrect; it is simply unproven.

More compellingly, Evans and Schmalensee (2001) argue that traditional market definition/market share antitrust analysis is not appropriate for dynamically competitive industries characterized by high rates of innovation. They make the point that markets subject to strong economies of scale and network effects based on risky innovation investments *require* high-operating margins protected by short-term barriers to entry, or else investment would dry up. Therefore, static market power analysis based on market definition/market share is bound to find such barriers even when they are absolutely necessary to fuel investment in innovation. They argue, (correctly in our view) "Unlike price/output decisions, analysis of dynamic competition requires evidence about, among other things, the pattern of investment in developing new products (and complements thereof), the control of critical assets (particularly intellectual property and distribution channels), and the beliefs (preferably as revealed by behavior) of market participants and informed observers about the nature and pace of innovation" (p. 47).

The simplistic serial monopoly hypothesis suggests that in the 'new economy,' monopoly is the natural market structure, but technological advance ensures that all monopolies are temporary, soon to be overtaken, or at least disciplined by the Next Big Thing. In Evans and Schmalensee's (2001) more nuanced view of the serial monopoly hypothesis, regulators and antitrust authorities are not asked to have blind faith in the ability of rapidly advancing technology to solve all problems of monopoly; rather, regulators and antitrust authorities are asked to move away from blind faith in static market power/market share analysis to an approach more consonant with dynamic competition. However, developing comprehensive principles and guidelines for regulators and antitrust authorities in the area of dynamic competition is extremely difficult, and likely, beyond the current capabilities of economists. This chapter should be viewed as one small step in this much broader research program.

Inherent in the serial monopoly hypothesis is the argument that an innovator needs a period of monopoly in order to recoup its investment in innovation (including investment in failed ideas as well successful ones). Should competitors be able to immediately enter, and the market becomes fully competitive (in the static sense), then prices would drop and profits driven to zero, thus eliminating future incentives to innovate. Indeed, this is precisely the economic argument that underpins intellectual property law: a limited monopoly is granted to innovators as a reward for innovation.

Of course, this raises the question as to how best to protect both innovation and customers; should we encourage patenting such inventions as operating systems and instant messaging services, or should we stay the hand of regulators and antitrust authorities, permitting monopolies for what we hope is the short term?

There is no doubt that intellectual property protection is costly to obtain and defend, and often requires years of litigation before a truly clear title to a patent is obtained. In the fast-paced 'new economy,' one's innovation is likely obsolete long before strong patent protection can be obtained. And even once obtained, it could well be very easy to 'invent around,' so the innovator's source of rent can still be eroded by competitive forces. The usefulness of the intellectual property system varies greatly from industry to industry. For example, in pharmaceuticals, and industry known for rapid technological advance, patents are the foundation of the business models of both incumbents and start-ups, although costly to defend. In computer hardware and software as well as Internet services, practices to protect of intellectual property vary enormously, from copyrighting software to patenting business models, to the use of trade secrets, to proprietary network effects, to tying and bundling. Evans and Schmalensee (2001) argue that a dynamically competitive industry may only be sustainable via actions such as tying that appear highly anticompetitive from a static perspective.

If no intellectual property regime can protect an innovator's rent, then how should regulators and antitrust authorities treat apparent monopolization? The key to resolving this dilemma comes from intellectual property law itself, which grants an innovator a *limited* monopoly for his invention, and requires that knowledge of the technology be shared. The Department of Justice implicitly recognized this in *U.S. v. Microsoft*, objecting not to Microsoft's Windows monopoly, but only to (i) Microsoft's exclusionary contracts with OEMs; and (ii) Microsoft's attempts to *leverage* its market power from the Windows OS into new markets, such as the browser market (by tying). The government's case can be understood as attempting to limit the Windows monopoly in product space, much as intellectual property law limits innovators' monopolies in product space and in time.

The appropriate question for regulators and antitrust authorities, in addition to those posed by Evans and Schmalensee (2001), is whether or not a proposed remedy to a perceived anticompetitive practice could reasonably be expected to chill future innovation. This question has both short-run and long-run components. For example, in its review of the AOL-Time Warner merger, the FCC imposed a condition requiring AOL to interconnect its IM service with competitors prior to offering advanced services. Does this order chill future innovation? Clearly, it has two short-run effects: discouraging AOL from innovating into such advanced services, and encouraging competitors to do precisely this form of innovation, knowing that network effects will not stand in the way of their innovations. Its long-run effect is more problematic: will firms in the future be more or less likely to introduce new communications-type services knowing that the regulators, or antitrust authorities may limit the scope of the resulting monopoly by requiring interconnection? If the fairly extensive period during which AOL was able to refuse interconnection with competitors is viewed as sufficient for AOL to have realized an above-normal return on its innovation investment, then the answer to this question is that no chill on future innovation should occur. If future innovators fear that the FCC has moved too quickly, before AOL realized its innovation returns, then the answer is that the long-term effect of this order is to reduce innovation in this field.

For services with network effects, it is often difficult to assess future, or even present returns to these services, as new firms often provide the service at low or even zero price. This was the case in both *U.S. v. Microsoft*, in which Microsoft made its browser available for free to Windows customers, and in the AOL-Time Warner IM service, in which AOL made IM available free to its customers, and free to all others as an Internet download. In such cases, firms can credibly claim that since they give away the service, they earn nothing on it and so the product cannot have monopoly characteristics. This claim is disingenuous, for at least three reasons:

1. The provision of the service as a 'feature' of the firm's product, the service certainly enhances the value of the product and permits the firm to either increase its market share, raise its price, or both. For example, AOL established IM as a feature of its basic service package; the popularity of IM no doubt contributed to the rapid growth of the AOL customer base. Similarly, adding the Internet Explorer browser to the Windows OS, no doubt increased the value of the Windows OS for customers. In both cases, the firms' service introductions increased profitability, even if not via pricing the service on a stand-alone basis.

2. The critical profit impact of a network effects market occurs after the firm has rapidly built up its market. One part of the strategy to accomplish this rapid build-up is to give the product away at the early stages, thus minimizing the cost to the customer of becoming loyal to the firm. Giving away the service early, akin to introductory pricing, helps build the customer base upon which the network effects rely. Giving away the service can have a substantial effect on future profits; it is best understood as an investment in proprietary network effects, whose value is realized after the firm dominates the market.
3. In some cases, the firm gives away its service to capture a related market, from which it perceives an entry threat to its main market. In *U.S. v. Microsoft*, the government alleged that Microsoft perceived a threat to its Windows monopoly from browsers running Java programs, which could be a platform alternative to the Windows OS. In this scenario, Microsoft sought to control the browser market by giving away its Internet Explorer, because it perceived this market to be an avenue of competition to its Windows monopoly, and was anxious to shut down that avenue.

In sum, the fact that a firm gives away a service in no way removes the need for regulatory or antitrust scrutiny. Indeed, it suggests that heightened scrutiny is called for, especially in a market with very limited extant competition. It is unlikely that the firm is acting out of charity, and unless existing or likely entrants are forcing the firm to offer these services via market forces, it is likely that such actions signal anticompetitive intent, particularly in the presence of network effects[26].

3.3. The FCC's instant messaging condition in the AOL-Time Warner merger

The FCC's analysis of the merger was principally focused on AOL's IM service; in particular, the FCC was concerned that merging AOL's IM service, in which the market is (likely) tipped with Time Warner's broadband conduits to 20 percent of U.S. cable homes would enable the merged firm to leverage these assets into next generation IM services to obtain an 'instant monopoly' in these new IM services. The FCC (2001a) therefore imposed a condition that required AOL-Time Warner to interoperate with other advanced IM competitors prior to offering next generation IM services.

[26] Giving away a product or service for free can be viewed as the ultimate in predatory pricing, long a staple of traditional antitrust analysis. Evans and Schmalensee (2001) question the government's definition of predation in the Microsoft case, which relies on the increase in profits from driving one's competitors from the market. They note that in a network effects business, this is precisely the effect of vigorous (and legal) competition. Making this an antitrust violation risks placing a chill on legitimate competitive actions. It is clear that more research on predatory pricing in dynamically competitive industries is urgently required in order that regulators and antitrust authorities have the tools they need to deal with the 'new economy.'

Specifically, the FCC imposed the condition that AOL cannot offer advanced IM services until it satisfies one of the following three 'grounds for relief:'

1. AOL demonstrates that it has adopted a public standard of interoperability established by IETF, or official standard-setting body (interoperability by standard).
2. AOL enters into interoperability contracts with at least two IM providers, and stands ready to negotiate in good faith with other IM providers under the identical technical interoperability standards (interoperability by contract).
3. AOL demonstrates that it is no longer dominant in IM, either using market share data or other evidence.

Advanced IM services are defined to entail the transmission of one- or two-way streaming video over Time Warner facilities using AOL's names and presence directory, the key resource required to route IMs, and in addition, this condition sunsets in 5 years.

The condition suggests the struggle of attempting to deal with future problems in a fast-moving industry. The FCC had no wish to insist that AOL adopt an industry standard when such a standard does not exist now, and may never exist, not least as competitors may use the standard-setting process strategically. The FCC did not establish any technical standards of interoperability, preferring to leave that to independent technical standards boards, or to the market. Additionally, the FCC wished to account for changing conditions; it could well be that a competitor could come to dominate the IM market as advanced IM services became available, and AOL would not then have market power. If this could be proved, then AOL would be relieved of the interoperability requirement[27].

Perhaps most important, if there really were no next generation IM services (or, none that AOL wished to offer), then there was no interoperability requirement. The FCC gave tacit assent to the 'earned monopoly' hypothesis, that AOL was entitled to the rents from its text-based IM innovation.

In contrast to the FTC, the FCC directly addressed this as a 'new economy' merger, in which interoperation/interconnection of a network effects communications service was central. It therefore faced all the problems discussed in Evans and Schmalensee (2001), plus some even more basic evidentiary issues:

1. What evidence is required to prove the service is imbued with network effects?
2. What evidence is required to prove the market has tipped?
3. What evidence is required to prove that the firm refuses to interoperate/interconnect for strategic reasons, rather than cost or operational reasons?

[27] Of course, if AOL really were not dominant, then economic theory would predict that its optimal strategy would be to interoperate. Should they apply to the Commission for relief from interoperability on the basis that they are not dominant, the act of applying appears to refute the premise of the application: they must still be dominant if they wish to continue to refuse interoperation.

4. If the remedy is oriented toward future services, then, how can the remedy be structured so that it is effective, if the forecasted conditions obtain and has no effect if these conditions do not obtain.

4. Lessons of the AOL-Time Warner case

Without question, the two salient issues of this merger were open access (at the FTC), and advanced IM services (at the FCC). Both represent challenges to regulatory/antitrust analysis that derive from the 'new economy': should we mandate access to new economy essential facilities, or do we let the market sort it out? Should we seek to negate a network effects barrier to new entry, or recognize such barriers as a necessary part of investments incentive in the new economy?

The FTC answered the first in the affirmative; it accepted the prima facie case that more competition must be good, and adopted what appears to be an aggressively regulatory remedy. The remedy was a popular one; it was well received by consumer groups and the press. Whether customers are now better off than they otherwise would be remains unknown. The costs of this regulatory approach are not likely to be large; however, the precedent of having a monitor trustee provide ongoing supervision of an antitrust conduct remedy is somewhat disturbing. It is reminiscent of Judge Harold Greene's decade-long supervision of the AT&T breakup, essentially creating another layer of regulation on the telephone industry that quickly became tiresome to even the most ardent supporter of the Court.

The FCC's answer to the second question was in the affirmative to both parts. While there was economic analysis aplenty, the case raised troubling issues more generally for antitrust in the new economy.

4.1. Proactive vs. reactive

A merger case is necessarily forward-looking (as compared to a conduct case), but in this case, the FCC appeared to be reaching to forestall potential harms in as-yet-unimagined service markets. Such a strategy always runs the risk of hyperactive government intervening to prevent imaginary villainy; and some critics thought this was the case here. The FCC was aware of this problem and attempted to structure the remedy so that it would have no effect (and hence no cost), if the hypothesized problems never materialized. Some critics thought that this resulted in a very weak remedy with too many loopholes. This is a trade-off that regulators and antitrust authorities acting proactively must resolve: heading off future problems without causing costs if their predictions are wrong, without gutting their remedies.

The alternative is to only act on problems that are immediate and plainly evident. The FTC's remedy on open access is clearly in this vein. But often in fast-moving markets, by the time the antitrust agency recognizes a problem and gears up to solve it, the market has moved on, the remedy based on past actions is no longer appropriate, and the damage is already done. Clearly, antitrust agencies must strike this balance as well.

4.2. Serial monopoly and the new economy

Scholars working in the area of antitrust in dynamically competitive industries have argued that the competitive norm in such industries is temporary 'winner take all' innovators, succeeded in rapid order by a new 'winner take all' innovator, and so forth. The period of monopoly for each innovator is in fact a reward to such innovators, and the temporary monopoly rents are merely the quasi-rents to a social beneficial activity. Imposing 'old economy' antitrust to break that monopoly in fact will lessen innovation as it lessens the legitimate rewards to that innovation. In this view, temporary monopolies generate the quasi-rents to innovation, much as patent protection helps generate quasirents for a limited period of time. The unstated assumption of the serial monopoly theorists is that, somehow intellectual property protection is not available in these fast-moving markets. But even these scholars admit that anticompetitive behavior can still exist: for example, exclusive distribution contracts with today's monopolist can inhibit the rise of the next monopolist. The current serial monopolist, in this view, cannot use its temporary market power to forestall new innovators, and thus turn itself into a permanent monopolist.

The FCC specifically recognized that AOL had earned whatever rents it could garner from its IM innovation, and that network effects barriers to entry could well be part of this rent generation. It also recognized that a temporary monopoly should not be used to foreclose new innovators, and thus create a permanent monopoly for all future generations of this service. Its judgment was that the merger would create just such conditions and sought to forestall such actions. Its remedy was designed to ensure that the next-in-line serial monopolist would not be foreclosed by the use of merger assets. In essence, it argued that network effects combined with merger assets that dramatically improve deployment capabilities is similar in effect to exclusive distribution contracts; it makes entry by the next serial monopolist almost impossible. Whether the factual case supports the remedy can be argued; but the theory of the case seems well within the confines of antitrust in the new economy. It also illustrates the difficulty of conducting the analysis and the case with those confines.

It also illustrates the dilemma of substituting antitrust enforcement (or, non-enforcement) for intellectual property law. It may well be true that IP affords little protection to innovators in fast-moving markets, and perhaps antitrust enforcement needs to recognize the need for temporary quasirents to innovators. But the problem appears to be with the inability of IP law to afford protection in important fast-moving markets. This suggests the appropriate policy conclusion is not to alter antitrust enforcement, but rather to fix intellectual property law to provide such protection.

5. Conclusions

The AOL-Time Warner case presented a textbook case of both bottleneck access and bandwagon access, and how two different agencies reacted to each other. The FTC took an essential facilities-like approach to the cable conduit, requiring open access as a condition of the merger. The FCC took a network effects approach to AIM, requiring (future) interoperation as a condition of the merger. Both conditions can legitimately be called 'access' conditions; both problems exist simultaneously in the merger firm. Yet, the problems are very different and their solutions are very different as well.

In the technology-intensive industries of the future, it is likely that essential facilities cases will be rare; technology has a way of finding alternatives to bottlenecks. It is also likely that network effects will play a much larger role than in previous cases. Prior to the FCC's findings in AOL-Time Warner, there were few cases, in which network effects played a role (see Melamed (1999)). In order to support antitrust enforcement in network effects industries, continued research is essential.

References

Areeda, P., 1990, Essential facilities: An epithet in need of limiting principles, Antitrust Law Review, 58, 841–869.

Armstrong, M., 2002, The theory of access pricing and interconnection, in M. E. Cave, S. K. Majumdar and I. Vogelsang, eds., Handbook of Telecommunications Economics, Volume 1, 295–386.

Atkinson, J. and C. Barnekov, 2000, A Competitively Neutral Approach to Interconnection, OPP Working Paper 34, December, at http://www.fcc.gov/Bureaus/OPP/working_papers/oppwp34.pdf.

Blue Cross et al. v. Marshfield et al., 65 F.3d 1406 (7th Cir. 1995).

Brock, G., 2002, "Historical overview," in M. E. Cave, S. K. Majumdar and I. Vogelsang, eds., Handbook of Telecommunications Economics, Volume 1, 43–74.

DeGraba, P., 2000, Bill and Keep at the Central Office as the Efficient Compensation Regime, OPP Working Paper 33, December, at http://www.fcc.gov/Bureaus/OPP/working_papers/oppwp33.pdf.

Evans, D. and R. Schmalensee, 2001, Some Economic Aspects of Antitrust Analysis in Dynamically Competitive Industries, NBER Working Paper No. W8268, May, at http://www.nber.org/books/innovation2/evans5-1-01.pdf.

Faulhaber, G., 1987, Telecommunications in Turmoil, Boston: Ballinger.

Faulhaber, G., 2002, Access $\neq$ Access$_1$ + Access$_2$, The Law Review of Michigan State University - Detroit College of Law, 677–708.

Faulhaber, G., 2003, Policy-induced competition: The telecommunications experiments, Information Economics and Policy, forthcoming, at http://rider.wharton.upenn.edu/~faulhabe/Policy-Induced%20Competition.pdf.

Federal Communications Commission, 2000,00-253, Interconnection and Resale Obligations Pertaining to Commercial Mobile Radio Services CC Docket No. 94–54, at http://www.fcc.gov/Bureaus/Common_Carrier/Orders/2000/fcc00253.doc.

Federal Communications Commission, 2001a, 01-12, Memorandum Opinion and Order, CS Docket No. 00-30, at http://www.fcc.gov/Bureaus/Cable/Orders/2001/fcc01012.pdf.

Federal Communications Commission, 2001b, 01-132, Developing a Unified Intercarrier Compensation Regime CC Docket 01-92, at http://hraunfoss.fcc.gov/edocs_public/attachmatch/FCC-01-132A1.pdf.

Federal Trade Commission, 2000a, Decision and Order, Docket No. C-3989, at http://www.ftc.gov/os/2000/12/aoldando.pdf.

Federal Trade Commission, 2000b, Analysis of Proposed Consent Order to Aid Public Comment, 2; at http://www.ftc.gov/os/2000/12/aolanalysis.pdf.

Hausman, J., G. Sidak and H. Singer, 2001a, Residential demand for broadband telecommunications and consumer access to unaffiliated Internet content providers, Yale Journal on Regulation, 18, 129–173.

Hausman, J., G. Sidak and H. Singer, 2001b, Cable modems and DSL: Broadband Internet access for residential customers, AEA Papers & Proceedings, 91, 302–307.

Hovenkamp, H., 1994, Federal Antitrust Policy: The Law of Competition and Its Practice §5.6c.

Kaserman, D. and J. Mayo, 2002, Competition in the long distance market, in M. E. Cave, S. K. Majumdar and I. Vogelsang, eds., Handbook of Telecommunications Economics, Volume 1, pp. 509–561.

Katz, M. L. and C. Shapiro, 1994, Systems competition and network effects. Journal of Economic Perspectives, 8, 93–115.

Kende, M., 2000, The Digital Handshake: Connecting Internet Backbones, FCC Office of Plans and Policy Working Paper 32, at http://www.fcc.gov/Bureaus/OPP/working_papers/oppwp32.pdf.

Liebowitz, S. and S. Margolis, 1999, Winners, Losers and Microsoft, Oakland, CA: The Independent Institute.

MacAvoy, P., 1996, The Failure of Antitrust and Regulation to Establish Competition, . MIT Press.

MCI Communications Corp. v. AT&T Corp 708 F.2d 1081, 1132–1133 (7th Cir. 1983).

Melamed, A. D., 1999, Network industries and antitrust, 23 Harvard Journal of Law & Public Policy, 147.

Otter Tail Power Co. v. United States 410 U.S. 366 (1973).

Mill, J. S., 1848, Principles of Political Economy, London: John W. Parker.

Milgrom, P., B. Mitchell and P. Srinagesh, 2000, Competitive effects of Internet peering policies, in I. Vogelsang and Compaine, eds., The Internet Upheaval, MIT Press, 175–195.

Rohlfs, J., 1974, A theory of interdependent demand for a communications service, Bell Journal of Economics and Management Science, 5(1), 16–37.

Rohlfs, J., 2001, *Bandwagon Effects in High-Technology Industries*, Cambridge, MA: MIT Press.

Spulber, D. and C. Yoo, 2003, Access to networks: Economic and constitutional connections, Cornell Law Review, 88, 885–1024.

Telecommunications Act, 1996, Pub. L. 104–104, 110 Stat. 56 (1996) (codified at 47 U.S.C. § 151 et seq.)

United States v. Terminal Railroad Association 224 U.S. 383 (1912).

US v. AT&T 552 F. Supp. 131 (D.D.C., 1982).

Chapter 13

EUROPEAN AND AMERICAN APPROACHES TO ANTITRUST REMEDIES AND THE INSTITUTIONAL DESIGN OF REGULATION IN TELECOMMUNICATIONS

DAMIEN GERADIN

University of Liége and College of Europe

J. GREGORY SIDAK

Visiting Professor of Law, Georgetown University Law Center

Contents

Handbook of Telecommunications Economics, Volume 2, Edited by S. Majumdar et al.

DOI: 10.1016/S1569-4054(05)02013-0

1. Introduction

In the U.S. and the European Union, the topic of remedies in network industries cuts across antitrust law and sector-specific regulation, including telecommunications. The legal and economic understandings of a 'remedy' are not always synonymous. In both legal systems, a remedy is the corrective measure that a court or an administrative agency orders following a finding that one or several companies had either engaged in an illegal abuse of market power (monopolization in the U.S. and abuse of dominance in the EC) or are about to create market power (in the case of mergers). With the exception of merger control where remedies seek to prevent a situation from occurring, legal remedies are retrospective in their orientation. They seek to right some past wrong. They may do so through the payment of money (whether that is characterized as the payment of damages, fines, or something else). Or they may seek to do so through a mandated change in market structure ('structural' remedies), as in the case of divestiture, or in the imposition of affirmative or negative duties ('behavioral' remedies). *United States v. Microsoft Corp* (U.S. Court of Appeals for the D.C. Circuit, 2001). presented the trade-off between these various remedial alternatives (Shelanski and Sidak, 2001, p. 1) .

Industry-specific regulation is an alternative to reliance on antitrust-based remedies. Both the U.S. and the EC have heavily relied on sector-specific regulation. A leading example is the telecommunications field, in which highly detailed legislation has been adopted. This legislation is further developed through additional regulations adopted by the Federal Communications Commission (FCC) in the U.S. and the national telecommunications authorities in the EU. In addition, state public utilities commissions (PUCs) in the U.S. regulate retail rates for landline services and implement on a localized basis the general framework established by Congress and the FCC for the pricing of unbundled access to the incumbent's network. The state PUCs sometimes retard deregulation, and the EU, in contrast, has attempted to end retail regulation and totally replace it with 'bottleneck' regulation of the wholesale (input) market.

We put 'bottleneck' in quotes because of a potentially major difference between the U.S. and EU approach in this respect. The U.S. starts from the essential facilities doctrine and only differentiates via the 'necessary and impair' standard, to be discussed below. However, duplication usually voids the bottleneck property. In contrast, the EU considers markets for bottleneck services and lets NRAs analyze whether firms have significant market power ('SMP') positions in such a market. Thus, one could have duplication and still trigger regulation. The two approaches may not be so different in their outcome because the EU market definition includes items like the persistence of entry barriers and the impossibility to overcome the bottleneck problem through competition policy and because the U.S. 'necessary and impair' standard could conceivably be applicable under (imperfect) duplication.

In contrast to these legal connotations of a remedy, the economic meaning of a remedy emphasizes market failure. The market failure may result from the unchecked exercise of market power, or from the uncompensated generation of an external cost or benefit, or from an insufficiency of information with which to make efficient choices concerning consumption, production, or investment. Whereas lawyers think of a remedy as what to do after a finding of illegal conduct[1], economists think of a remedy as what to do after a finding of market failure. The two approaches overlap perfectly if legislators and courts make liability rules that are triggered only after a finding of market failure. Of course, if legislators and courts actually did so, the *Journal of Law & Economics* would be a very slim volume that would have ceased publication years ago.

The difference between the legal and economic conceptions of remedy highlights another important distinction, namely, the difference between *ex ante* and *ex post* interventions in the market. Under the *ex post* approach, a remedy is imposed if and only if an illegal conduct is first proven. And it is the government or a private plaintiff that bears the burden of proving their case. This arrangement describes the operation of monopolization law under the U.S. Sherman Act or the concept of abuse of a dominant position in the EU.

In contrast, the *ex ante* approach imposes a remedy before any specific finding of illegal conduct. The rationale for this prophylactic approach may be one or more of the following considerations:

- The probability of anticompetitive behavior in the absence of the prior restraint is high;
- The magnitude of the harm from such behavior would be great;
- The likelihood and magnitude of offsetting efficiency justifications for the behavior are low; and
- The danger of false positives is small.

Although *ex ante* remedies are generally imposed through sector-specific regulation, such remedies may also be imposed on the basis of antitrust rules. This is, for instance, the case where remedies are imposed as a condition for clearance of a

[1] We refer to the term 'illegal conduct' rather than 'liability' as it is more neutral and is more adapted to competition law discussions in the EC legal order where antitrust actions often take place in the context of administrative proceedings where the term liability is not used.

merger between telecommunications operators. In both the U.S. and the EU, antitrust enforcement authorities have used merger control procedures as a way to extract significant concessions from the merging entities. As far as institutional design is concerned, remedies in network industries can thus come from two main sources: *ex ante* or *ex post* enforcement, and/or sector-specific regulation. As one of us has written elsewhere, there are interactions between antitrust and sector-specific regulations and these have to be taken into account when addressing the issue of remedies. Moreover, as will be seen below, both antitrust and sector-specific remedies can be influenced by decisions or, more generally, analytical tools developed in other sectors (see Geradin and Kerf, 2003).

Against this background, this chapter seeks to explore the issues of remedies and institutional design of regulation in a comparative manner by reference to U.S. and EU law. Section 2 analyzes the U.S. model and Section 3 the EU model. Section 4 contains a conclusion.

2. The U.S. model

This section is divided into four subsections. Subsection 2.1 provides illustrations of *ex ante* and *ex post* remedies in telecommunications. As will be seen, the reliance on consent decrees makes the border between *ex ante* and *ex post* remedies thin as such decrees can be seen as an amalgam (hence, the concept 'hybrid' remedies) between the *ex ante* and the *ex post* approaches. Subsection 2.2 argues that, over the last two decades, antitrust law has evolved into another form of regulation as it now relies on numerous policy statements and guidelines that resemble the type of prospective rulings made by regulatory agencies. Subsection 2.3 outlines the risk that the approach followed by the FCC in its Local Competition Order and recently vindicated by the Supreme Court in the Verizon case could affect the development of antitrust-based remedies in network industries. Finally, Subsection 2.4 argues that, by attempting to impose the TELRIC model to U.S. trading partners, the U.S. Trade Representative has turned itself into a telecommunications regulator. We also argue that this process lacks legitimacy and could have an unintended 'boomerang' effect on U.S. telecommunications operators.

2.1. Ex ante, ex post, *and hybrid remedies*

The U.S. epitomizes the use of heavy-handed *ex ante* regulation, which originates from the adoption of the 1996 Telecommunications Act and a large number of implementing provisions.

For instance, this Act seeks to overcome the obstacles raised by the lack of competition in the local loop by imposing a series of obligations on telecommunications carriers involved in local exchange. Section 251 of the 1996 Act imposes a series of duties on telecommunications carriers involved in local exchange. Section 251 requires each 'telecommunications carrier' to interconnect with other carriers[2]. In addition to interconnection, all local exchange carriers (LECs) are barred from either prohibiting or imposing discriminatory conditions on the resale of telecommunications services[3]. They are also required to provide 'number portability' and 'dialing parity[4],' as well as access to their poles, conduits, and other rights of ways to competing providers of telecommunications services[5].

Additional obligations are imposed on incumbent local exchange carriers (ILECs)[6]. In addition to all of the duties listed in the preceding paragraph, these ILECs are required to provide, at just and reasonable rates, interconnection 'at any technically feasible point with the carrier's network[7].' They must also provide competitors 'unbundled' network elements upon request[8]. In addition, the Act requires ILECs to offer for resale 'at wholesale rates any telecommunications service that the carrier provides at retail to subscribers[9].' Finally, an ILEC must permit firms seeking interconnection to locate their equipment on the ILEC's premises (collocation)[10].

In addition, Sections 271 and 272 of the Telecommunications Act address the BOCs' entry into the long-distance market[11]. Section 271 allows the BOCs to provide long-distance services to their own customers provided they have concluded, with one or more competitors, interconnection agreements that satisfy

[2] Section 251 defines 'telecommunications carriers' to include incumbents and new local exchange carriers. 47 U.S.C. § 251(a)(1).

[3] 47 U.S.C. § 251(b)(1).

[4] 47 U.S.C. § 251(b)(2) and (3).

[5] 47 U.S.C. § 251(b)(4).

[6] 'Incumbent local exchange carriers' denote the local carriers in existence when the Act was adopted.

[7] 47 U.S.C. § 251(c)(2)(B). In practice, this means that incumbents must interconnect with all carriers upon request, at the locations they specify, while other carriers may interconnect with each other indirectly (i.e., by each carrier connecting to the incumbent).

[8] 47 U.S.C. § 251(c)(3). 'Unbundled access' means that the availability of access to distinct parts of the incumbent's network, at an appropriately lower cost than access to all elements of the network. Thus, a competitor can purchase only those network components and functions that it needs to offer its services.

[9] 47 U.S.C. § 251(c)(4)(a).

[10] 47 U.S.C. § 251(c)(6).

[11] Generally on Section 271, see Sloan (1998).

the requirements of Section 271(c)(2)(B), the so-called 'competition checklist[12].' The requirements contained in this list essentially relate to the interconnection obligations imposed in Section 251. The BOCs' ability to offer long-distance services is thus conditioned on meeting their interconnection obligations, thereby giving them an incentive to open their local service areas to competitors[13].

The obligations imposed by Section 251 were made even more intrusive by the implementing orders adopted by the FCC. In its first Local Competition Order, the FCC determined, which network elements ILECs had to make available to their competitors on an unbundled basis. In addition, it sets forth a methodology to be used by state utility commissions in establishing rates for interconnection between competitors, which have each established their own facilities and for the purchase of unbundled elements (when one competitor has not established full-fledged facilities). The Order concludes that this pricing methodology must be based on the incumbent's total element long-run incremental cost (TELRIC).

Both the determination of the elements that need to be offered of an unbundled basis and the TELRIC pricing methodology have been criticized by one of us in a series of papers. Hausman and Sidak (1999) argue that regulators should ask whether an ILEC could exercise market power over end users by restricting competitors' access to a requested telecommunications network element in a particular geographical market. If the market for services to end users is competitive, then the justification for mandatory unbundling could not be to increase consumer welfare. Because the Hausman–Sidak proposed approach would focus on the effectiveness of competition in the end-user services market, rather than on the ability of a particular competitor to earn profits, it would use consumer welfare, rather than competitor welfare, as its touchstone. Once regulators order a network element to be unbundled, Hausman and Sidak (1999) argue that its price should be based on the element's real option value.

[12] 47 U.S.C. § 271(c)(2)(B). Alternatively, if a BOC has not received a qualifying interconnection request within a designated period of time, the BOC can satisfy this requirement by providing a statement of generally-available terms and conditions that comply with the competition checklist and that 'has been approved or permitted to take effect by the (relevant) state commission.' *Id.*

[13] There are two additional conditions: (i) Section 271(d)(3)(C) states that the FCC may not approve a BOC's application unless it determines that 'the requested authorization is consistent with the public interest, commerce and necessity'; and (ii) even with the competitive checklist in place, Section 272 requires that a BOC creates a separate affiliate to provide long-distance services. 47 U.S.C. § 272(a)(1). This separate affiliate must operate independently from its BOC parent, keeping separate books and records and having separate offices, directors, and employees. 47 U.S.C. § 272(b)(2) and (3). In order to prevent illegitimate subsidies, all transactions between an affiliate and its BOC parent must be 'on an arm's length basis' *Id.* at § 272(b)(5).

In essence, whether the FCC admits it or not, it has interpreted the Telecommunications Act to create a competitor-welfare standard rather than a consumer-welfare standard for deciding what must be unbundled and how it must be priced (Hausman and Sidak, 1999).

Reliance on a competitor-welfare standard can also be observed with respect to the imposition of a resale obligation on an ILEC's provision of digital subscriber line (DSL) service, despite the fact that cable modem service offered by cable operators holds twice the market share as DSL (Crandall et al., 2002). In this case, *ex ante* remedial duties are imposed in the absence of any dominance on the part of the ILEC. As will be seen below, this questionable approach could not have been implemented under the new EC framework on electronic communications where dominance must be established before intrusive regulatory requirements are imposed on an incumbent.

Besides sector-specific regulation, *ex ante* remedies can also be imposed in the case of mergers, which need to be cleared by both the FCC and the DoJ. Over the last 10 years, many mergers took place in the telecommunications and cable sectors[14]. Most of these transactions were cleared provided that the parties accepted to comply with conditions, both of 'structural' and of 'behavioral' nature. For instance, in June 2000, the FCC approved the Bell Atlantic–GTE merger (FCC, 2000), subject to the obligation for the merged company to transfer substantially all of GTE's Internet business into a separate public company[15]. The merged entity had also to comply with 25 merger conditions designed to enhance local phone competition in the markets, in which Bell Atlantic or GTE is the ILEC, strengthen the merged company's incentives to enter local phone

[14] They include mergers between: (i) BOCs (e.g., the mergers between Bell Atlantic and NYNEX, SBC and Pacific Telesys, and SBC and Ameritech); (ii) BOCs and non-BOCs local exchange carriers (e.g., the mergers between SBC and SNET and Bell Atlantic and GTE); (iii) long-distance operators (e.g., the merger between WorldCom and MCI and the failed merger between WorldCom/MCI and Sprint); (iv) long-distance and/or international operators (e.g., the mergers between WorldCom and MCI and between AT&T and BT); (v) long-distance operators and cable companies (e.g., the merger between AT&T and TCI and AT&T and Media one); (vi) long-distance operators and competitive local access providers (e.g., the merger between AT&T and Teleport), (vii) wireless carriers (e.g., the merger between Airtouch and Vodafone); and (viii) a BOC and a long-distance operator (e.g., the merger between Qwest and U.S. West). For a discussion of these mergers, see Waters (1999).

[15] This condition was necessary to guarantee compliance with Section 271 which forbids a BOC company, such as Bell Atlantic, from providing long-distance voice or data services to customers in its service territory before it demonstrates that its local market is open to competition. At the time the transaction was cleared, Bell Atlantic had only received authorization to offer long-distance services in New York State.

markets outside of its territories, and promote equitable and efficient advanced services deployment[16].

Ex post remedies can be imposed on telecommunications operators by the antitrust enforcement authorities. The Antitrust Division (and to a lesser extent, the Federal Trade Commission) can sue a company or group of companies for violating the antitrust laws. If the violation of such laws is established, a deferral court can then impose remedies taking the form of fines and monetary damages, as well as behavioral and structural remedies.

The frontier between *ex ante* and *ex post* remedies is, however, thin as antitrust cases are often settled pursuant to a consent decree. In other words, issue-specific litigation leads to a negotiated, prospective regime of company-specific regulation. If a single firm is the object of the antitrust case, and if it is prominent enough in its industry (we will avoid using the loaded term 'dominant'), then the consent decree becomes the de facto asymmetric regulation of the entire industry. The most obvious example is the Modification of Final Judgment[17], by which the federal judiciary governed the telecommunications industry after the antitrust breakup of the Bell System in January 1982 until Congress enacted the Telecommunications Act in February 1996[18]. A more recent example, of course, is the *Microsoft* case, whose remedial structure following a settlement between the Antitrust Division and Microsoft remains the subject of continuing litigation in federal court.

The consent decree is an amalgam of the *ex post* and *ex ante* approaches. This characteristic explains why more than a decade ago Professor (now Circuit Judge) Michael McConnell questioned the constitutionality of consent decrees (McConnell, 1987). He regarded them as a commingling of essentially *ex post* law enforcement powers belonging to the Executive Branch and *ex ante* legislative powers belonging to Congress, which then were handed over to the Judiciary to oversee.

American telecommunications deregulation provides other current examples of the combination of *ex ante* and *ex post* regulatory models. We mentioned earlier the process under Section 271 of the Telecommunications Act, by which a Bell operating company may apply to enter the inter-LATA market. Such applications are reviewed by the FCC and the relevant state PUC, obviously under an *ex ante* approach. For these regulatory commissions, the status quo is the continuation

[16] For a summary of these conditions, see http://www.fcc.gov/ba_gte_merger/conditions.txt.

[17] Modification of Final Judgment, *reprinted* in United States *v. AT&T Co.*, 552 F. Supp. 131, 226–34 (D.D.C. 1982), *aff'd sub nom. Maryland v. United States*, 460 U.S. 1001 (1983).

[18] Pub. L. No. 104–104, 110 Stat. 56 (codified as amended in scattered sections of 15, 18, and 47 U.S.C.).

of an entry barrier. That is a kind of the prospective remedy, though an inadequate one in our opinion, because it hurts consumers in one part of the telecommunications sector in order to enhance competition in another part.

The tension between *ex post* antitrust remedies and *ex ante* telecommunications regulation also has arisen in a set of cases known as the *Goldwasser* cases (U.S. Court of Appeals for the 7th Circuit, 2000)[19], named for the first case in a series of conflicting lower court rulings. The Supreme Court has granted certiorari in one such case, *Trinko*, for the October 2003 Term[20]. The issue in these cases is whether a Sherman Act claim for monopolization is available to a competitive local exchange carrier (CLEC) that alleges that the ILEC has failed to comply with the FCC's unbundling and pricing regulations. It is a fair question to ask why it is necessary to have both *ex ante* and *ex post* remedies to address the perceived market failures that motivated passage of the Telecommunications Act of 1996.

Finally, the tension between antitrust and sector-specific regulation is also collapsing at the institutional level where, on the one hand, the Antitrust Division of the Department of Justice is taking care of issues that should normally be taken care of by a regulatory agency and, on the other, the FCC is ruling on issues that should typically be dealt with by antitrust authorities only. The first issue can be illustrated by the DoJ's involvement in the Section 271 process described above. Setting aside the wisdom or folly of the remedy, it seems odd that the Antitrust Division participates in this process. Although the Division has expertise in telecommunications, it is an enforcement body. It executes laws on an *ex post* basis—by applying an existing legal standard to a set of facts that have already occurred. The Antitrust Division is not a legislative body that exercises the power of adopting rules regulating pricing, entry, and other terms and condition of competition in network industries.

The second issue can be illustrated by the intervention of the FCC is merger proceedings (Geradin and Kerf, 2003). The Antitrust Division is the most experienced authority when it comes to analyzing whether a merger risks substantially lessening competition. Moreover, unlike the FCC, it has a horizontal view of merger control and it is important to maintain a minimum coherence across merger decisions across sectors. On the other hand, the FCC's intervention in telecommunications merger cases increases the length of the review process and

[19] For representative decisions that show the divergence of opinion on this legal question, see *Covad Communications Co. v. BellSouth Corp.*, 299 F.3d 1272 (11th Cir. 2002); *Law Offices of Curtis V. Trinko v. Bell Atl. Corp.*, 294 F.3d 307 (2d Cir. 2002), *cert. granted*, 123 S. Ct. 1480 (2003); *Cavalier Tel., LLC v. Verizon Va., Inc.*, 208 F. Supp. 2d 608 (E.D. Va. 2002).

[20] Supreme Court (2003).

involves a great deal of duplication (and thus costs). It also raises the risk of contradictory decisions. In the Bell Atlantic/NYNEX merger, for instance, the Antitrust Division did not raise any objections (U.S. Department of Justice, 1997). By contrast, the FCC only allowed the merger after imposing several conditions (Federal Communications Commission, 1997). This increases uncertainty in the market place. In our opinion, it would be preferable to concentrate merger review in the hands of the Antitrust Division, which seems best placed to handle merger review in a consistent fashion. Our position is not isolated. It is supported by at least one former FCC Commissioner and by several influential Congress members. In a series of speeches and articles in the press, former FCC Commissioner Harold Furchtgott-Roth argued that the review of mergers between telecommunications firms should be left to the Antitrust Division (Furchtgott-Roth, 1999, 2000).

Let us shift the subject slightly. So far we have discussed only private firms in network industries, but many network industries, such as postal services, still have state-owned enterprises dominating them. With public enterprises in network industries, the causes of competitive concern and the range of remedial policy instruments are different. State-owned enterprises have a greater incentive than private, profit-maximizing firms to engage in predatory pricing and anticompetitive network discrimination (Sappington and Sidak, 2000, 2003a,b). In principle, state ownership of enterprise is supposed to internalize regulatory decisions within managerial decisions. At a stylized level, the state-owned enterprise is assumed to maximize some specification of social welfare, which presumably would include consumer welfare.

With respect to state-owned enterprises, the feasible set of remedies in cases of market failure gets truncated because of at least three factors. First, the state's conflicting interest in maximizing the firm's value in anticipation of its privatization may impose practical political constraints on the intensity and invasiveness of potential remedies designed to increase competition. Second, where independent regulators do exist, as in the case of the Postal Rate Commission in the U.S., the regulator may be weak, both legally and politically, especially given the political influence of the large work force that a state-owned enterprise often employs. Third, at least in the U.S., the doctrine of sovereign immunity may bar private parties from pressing antitrust claims against the state-owned enterprise[21]. For these reasons, it is important to keep state ownership in mind when examining the

[21] The Supreme Court granted certiorari for the October 2003 Term in a case in which the Ninth Circuit had denied the U.S. Postal Service sovereign immunity, see U.S. Court of Appeals for the 9th Circuit (2003).

feasible set of remedies in network industries. We will see that the situation is different in the EU.

2.2. Antitrust as a new form of ex ante *regulation*

Let us return to the distinction between *ex ante* and *ex post* remedies in network industries. Given the choice between *ex ante* dominant firm regulation and *ex post* antitrust litigation, which approach has been more intellectually forceful in shaping what I will broadly call the 'remedial orientation of competition policy'? Twenty years ago, it was clearly the case that its embrace of economic analysis made antitrust law intellectually dominant over industry-specific regulation in the U.S. More than any of the FCC proceedings that preceded it, the antitrust case against the Bell System is considered (sometimes for the wrong reasons) the defining moment in reorienting the telecommunications industry toward deregulation. The diffusion of ideas flowed from antitrust to the regulatory agencies.

Then something happened, and the direction of policy innovation reversed. Today, American antitrust law and its notions of feasible remedies in network industries are influenced by the theories of market failure predicated on network effects (Farell and Saloner, 1985, 1986; Katz and Shapiro, 1985, 1986)[22]. Those theories were developed at Berkeley and Stanford in the 1980s. They began influencing thinking on telecommunications regulation, and by the early 1990s they dominated policy formation at both the FCC and the Antitrust Division, when the Berkeley and Stanford theorists came to Washington.

Even the practice of antitrust law evolved over that period into more of an administrative practice, characterized by numerous policy statements and guidelines issued by the Antitrust Division and Federal Trade Commission that resembled the prospective rulemakings at the FCC (Melamed, 1995). In relative terms, antitrust became less a body of actual law written by courts deciding specific cases on an incremental basis, and more a body of regulation taking the form of generalized statements of abstract principles, promulgated by a bureaucracy. As will be seen, a similar evolution can be observed in the EU. The culmination of that process was the *Microsoft* antitrust case, by which the Antitrust Division installed itself, whether it intended to or not, as overseer of a regime of dominant firm regulation of the software industry. Given the rapid technological change in software, that de facto regulation was necessarily prospective and hypothetical.

22 This phenomenon has received little attention from scholars. For two timely and thoughtful exceptions, see Geradin and Kerf (2003) and Farrell and Weiser (2003).

2.3. The shifting balance of influence between antitrust and sector-specific regulation: telecommunications law's potential to shape antitrust remedies in network industries

So it is now natural to speculate about how the FCC's crowning achievement since 1996—namely, the Supreme Court's vindication of the agency's TELRIC pricing rules in 2002 in the *Verizon* case (Supreme Court, 2002)—will influence the development of antitrust law concerning remedies in network industries. How, for example, will TELRIC pricing affect the development of antitrust law concerning Microsoft? The influence may prove to be substantial.

There is an obvious relationship between an *ex ante* regulation requiring unbundling of network elements and an *ex post* antitrust rule penalizing the failure to offer a product or functionality on an unbundled basis. The latter is the antitrust doctrine concerning tying arrangements, which was so contentious in the *Microsoft* case. When read together, *Verizon* and *Microsoft* have potentially broad implications for antitrust remedies relating to bundling and unbundling of products having substantial sunk costs and network complementarities, including intellectual property. The traditional antitrust case law on tying is not much help in the context of intellectual property and other sunk-cost investments that exhibit network effects. In this respect, such sunk-cost assets cannot really be treated the same as widgets in bundling cases. We have three observations in this regard.

First, to repeat the obvious, after the D.C. Circuit's 2001 decision in *Microsoft*, the economic subtleties of product bundling in network industries lend themselves better to analysis under the monopolization principles embodied in Section 2 of the Sherman Act[23] than to the more linguistic formulations of liability in Section 1 of the Sherman Act[24] and Section 3 of the Clayton Act[25]. Along these lines, the separate-product analysis in tying cases is less likely to be fruitful in cases involving intellectual property, such as computer software, than in cases involving widgets. The strategic motivation for bundling may have nothing to do with conventional theories of tying predicated on leveraging or price discrimination. Furthermore, the attempted preservation of a monopoly over the tying product—whether it is an operating system, a primary patent, a broadband Internet conduit, or the like—is hard to evaluate in economic terms when forced into traditional tying law.

[23] 15 U.S.C. § 2.
[24] *Id.* § 1.
[25] *Id.* § 14.

Second, although certainly critical of the *Microsoft* case, we encourage scholars, enforcement agencies, and courts to refine David Sibley's theory of 'partial substitutes,' which was essential to the government's theories of liability and remedies in that case (Sibley, 1998)[26]. Much of the government's economic theory in the 1999 trial of Microsoft focused on an elaborate version of this theory of anticompetitive tying. This economic theory was first presented in detail in Sibley's pretrial declaration on behalf of the government in May 1998[27]. There, Sibley proposed that Microsoft's actions to put in place contracting restrictions and to distribute the Internet Explorer (IE) browser tied to its Windows operating system (OS) for free were an attempt to preserve its OS monopoly[28]. To support his argument, Sibley pointed to the case of a monopoly with allegedly exclusionary practices in a complementary market that it serves, where the general conclusion has been that:

> if the price level in the complement's market is limited by competitive forces, then in the absence of efficiency justifications . . . , the monopolist's control over the bottleneck input does not give it any profit incentive to restrict or exclude a competitor's product in the complement's market . . . [because] . . . control over the bottleneck input allows the monopolist to extract value from consumers no matter whose version of the complementary good the consumer buys[29].

Applied to the Microsoft case, Sibley stated, the bottleneck input is its OS, while the complementary product is the browser[30]. He maintained that the threat to Microsoft's alleged OS monopoly arose because browsers expose their own applications programming interfaces (APIs), a condition, which enables browsers to serve as a software applications platform independent of the underlying OS; in turn, the existence of a competing platform would break down the so-called 'applications barrier to entry' in the PC OS market. A new entrant in the OS market, Sibley reasoned, 'would not have to then create an installed base of software applications complementary to its OS and comparable to Microsoft's in its size and use in order to succeed[31].' Instead, applications that were written to the browser platform would be accessible to a user employing any OS that supported that browser.

[26] See also Sidak's (2001) critique of Lawrence Lessig's application of Sibley's theory of partial substitutes.

[27] See Sibley (1998).

[28] *Id.*

[29] *Id.* at 44.

[30] *Id.*

[31] *Id.*

Sibley provided perhaps the most innovative theory of antitrust liability since the raising rivals' cost literature emerged more than a decade earlier. But the theory's eventual exposition in Franklin Fisher's testimony, and in the government's subsequent briefs, left the impression that a formal economic model has yet to be presented[32]. We do not have a formal explanation in consumer demand theory for how a complement turns into a substitute. Yet this metamorphosis is a recurring theme in the discussion of remedies in network industries. In telecommunications, for example, the leasing of selected unbundled elements at regulated prices is vigorously defended by CLECs and regulators as a complement to subsequent facilities-based investment, not a substitute for it.

Third, if we take tying law seriously in the context of network industries, we arrive at a serious pricing problem at the stage of fashioning a remedy. This pricing problem is likely to be much more challenging when the bundled products consist almost entirely of intellectual property, because of its zero marginal cost. Presumably, a prohibition against tying does not mean that a firm may not offer *A* and *B* in a bundle. Presumably, the prohibition means only that the firm must also offer *A* and *B* separately. Call *A* the tying product, which is a bottleneck of some sort. Call *B* the tied product, which is competitively supplied. How much of a discount off the bundled price must the firm therefore offer when it is compelled by antitrust law to sell *A* on an unbundled basis? When a high price is demanded for an unbundled version of *A*, does that price itself become an antitrust violation?

This question is closely related to the one that the FCC and the Supreme Court addressed in the *Verizon* case concerning pricing of unbundled network elements based on total TELRIC (see Sidak and Spulber, 1997a,b; Hausman and Sidak, 1999;). If TELRIC-based pricing is reasonable to impose on a former statutory monopolist subject to rate regulation that has not committed any antitrust violation, then it is doubtful that a court in an antitrust case would have qualms about applying TELRIC to an unregulated monopolist found to have violated Section 2 of the Sherman Act by its unlawful bundling of software. There are, of course, many alternative pricing rules that might be employed to fashion the remedy in such a tying case, but surely TELRIC rules the day and will be pursued by plaintiffs and prosecutors because it is most favorable to their cause.

How then would antitrust law implement a TELRIC approach to fashioning the unbundling remedy in a case of software integration? One approach is the

[32] See Fisher (1998, 1999, 2000). ('Microsoft's bundling of IE with the Windows software it distributes through retail channels is a similar effort to weaken Microsoft's browser competition in order to protect Microsoft's dominance in operating systems.')

top-down, avoided-cost calculation: What is the long-run average-incremental cost (LRAIC) of *B* that is avoided when *A* is unbundled? Subtract that LRAIC from the previous bundled price to determine the permissible unbundled price of *A*. But, if the telecommunications experience is any guide, the objection will be raised that the bundled price incorporates monopoly rent and inefficiency, and that these components must be subtracted also. It will also be argued that product *B* should contribute substantially to the recovery of the defendant's common costs.

The defendant in such a case will argue in rebuttal that the cost that it avoids when selling *A* without *B* bundled to it is trivial if the provision of *B* exhibits economies of scale—since, by assumption, it will still be lawful for the firm to offer a bundled version of *A* and *B*. The defendant can further be expected to argue that there may be new incremental costs of unbundling (perhaps making the net avoided cost negative), and naturally there will be a dispute over who shall pay those incremental costs of unbundling.

The other remedial approach is a bottom-up calculation of the LRAIC of product *A*, in addition to which the defendant should be allowed to recover a reasonable share of common costs, including a competitive return on capital. In principle, the top-down and bottom-up approaches should yield equivalent results. However, if they do not practice, obvious strategies will emerge between plaintiffs and defendants (over which approach is the proper test.) The experience in telecommunications is that regulators implement the two pricing calculations in ways that permit divergent results, and that there is no acknowledgment by regulators or courts of the strategic behavior that such a divergence induces. The controversy over whether ILECs have a duty to offer all network elements as a platform, priced at the sum of the TELRIC prices, would not exist if not for this methodological inconsistency tolerated by regulators (see Ingraham and Sidak, 2003).

In short, the *Verizon* case concerning TELRIC pricing will likely influence the shape of antitrust remedies in product integration cases. In the intellectual property area, we can expect to see more monopoly-preservation tying cases, relying on Sibley's theory of partial substitutes. These cases will immerse the litigants and the courts in TELRIC-like questions of the pricing of the tying product on an unbundled basis. The sunk-cost character of intellectual property will make these remedial proceedings highly contentious and highly consequential, for the desired remedy may succeed in appropriating quasi-rent rather than preventing the defendant from earning true economic rent. Nonetheless, the remedial experience in American telecommunications regulation since 1996 suggests that plaintiffs and prosecutors will prevail at the end of the day.

2.4. The U.S. trade representative as regulator

A final regulatory design takes the form of bilateral or multilateral trade agreements (Rohlfs and Sidak, 2002). On February 15, 1997, 70 countries working within the framework of the World Trade Organization (WTO) agreed on a multilateral reduction of regulatory barriers to competition in international telecommunications services (WTO, 1997)[33]. At the time, the signatory nations to the WTO agreement on telecommunications services represented markets generating 95 percent of the $600 billion in global telecommunications revenues (Andrews, 1997; Swardson and Blustein, 1997). Beginning on January 1, 1998, those nations started a phased process to open their telecommunications markets to competition. Since 1997, the U.S. government has attempted to use the WTO agreement on telecommunications services as a vehicle for 'exporting' American principles of telecommunications regulation to other nations.

In 1997 the U.S. took the position that the WTO agreement on telecommunications services requires signatory nations to follow the FCC's practices on interconnection pricing under the Telecommunications Act of 1996[34]. That effort has culminated in the initiative by the Office of the United States Trade Representative (USTR) to use the implicit threat of trade sanctions to influence Japan's domestic regulatory policy on the pricing of mandatory competitor access to the unbundled elements of the local network belonging to the operating companies of Nippon Telegraph and Telephone Corporation (NTT) (see Rohlfs and Sidak, 2002). The USTR's efforts against Japan have not been an isolated case. The USTR has sought to place detailed interconnection requirements in a bilateral treaty with Singapore, and it has initiated a WTO arbitration proceeding against Mexico over telecommunications pricing issues in what is the very first WTO case of any sort under the General Agreement on Trade in Services.

The USTR's expertise lies in negotiating trade agreements. 'The Trade Representative shall have primary responsibility . . . for developing, and for coordinating the implementation of, U.S. international trade policy' and 'shall serve as the principle [*sic*] advisor to the President on the impact of other policies of the U.S. government on international trade[35].' The USTR's expertise is not access pricing, telecommunications economics, antitrust law, or industrial organization. It appears that the USTR was, and may still be, unaware that almost continuously

[33] For an analysis of the WTO agreement on telecommunications services (see Harwood et al., 1997; Sidak, 1997; Edward and Richardson, 1999).

[34] Pub. L. No. 104-104, 110 Stat. 56. (Feb. 8, 1996).

[35] Reorganization Plan No. 3 of 1979, *reprinted* in 19 U.S.C. § 2171(b)(1) (1982).

since 1996, many American experts on telecommunications policy have doubted that American consumers have benefited from the very FCC policies that USTR would have Japan, Singapore, Mexico, and other nations emulate. Commenting on the applicability of the U.S. model of telecommunications liberalization to other nations, Robert Crandall wrote in 1997 that '[t]he most contentious single issue in implementing the 1996 Telecommunications Act in the U.S. is the measure of cost to be used in setting rates for wholesale unbundled elements (Crandall, 2000).' Not surprisingly, the FCC's policy in this area has generated continuous litigation since 1996, including two Supreme Court cases, and is too unresolved for the U.S. to force on its trading partners. Yet, despite that irresolution, interconnection pricing is today the very aspect of the Telecommunications Act of 1996 that the USTR aggressively seeks to impose on other nations in the name of enforcing the WTO agreement on telecommunications services.

It is unlikely that the USTR has the detailed knowledge, the expertise, and the proper incentives to negotiate trade agreements on interconnection pricing. The public policy issues associated with telecommunications regulation are far more complex than those associated with steel or bananas. One should question the propriety of using the USTR to influence the domestic regulatory policy of another country on a topic as complex as the efficient pricing of mandatory access to unbundled network elements. The USTR's power to formulate trade policy on this subject resides in officials who are unlikely to possess the economic expertise and resources necessary to evaluate the consumer-welfare implications of the policies that they would have Japan and other nations adopt. For these reasons, the USTR cannot credibly make the interconnection pricing policies of another nation a legitimate concern of U.S. trade policy.

Moving from process to substance, the USTR's negotiating positions implicitly espouse a competitor-welfare approach to telecommunications regulation rather than a consumer-welfare approach. It is understandable that USTR would want to promote the interests of American companies. But in this case, it is promoting the interests of a subset of American carriers while ignoring the interests of other American telecommunications carriers as well as American producers of telecommunications equipment.

No American carrier will want to invest in building a network in a less-developed country if it knows that it will immediately have to lease unbundled network elements to a competitor at a price calculated, after considerable debate, on the basis of long-run average incremental cost. The disincentive to investment will not produce any sales of telecommunications equipment by American producers. How is that outcome a good trade policy for *any* constituency in the U.S.? It certainly does not help consumers in the less-developed country.

Congress, the Administration, and the FCC should beware of the USTR boomerang. Section 252(i) of the Communications Act provides:

> 'A local exchange carrier shall make available any interconnection, service, or network element provided under an agreement approved under this section to which it is a party to any other requesting telecommunications carrier upon the same terms and conditions as those provided in the agreement[36].'

It will surely be argued, on the basis of Section 252(i), that treaty obligations that the U.S. undertakes pursuant to a bilateral agreement apply to domestic carriers as well. In other words, uncompensatory pricing policies for unbundled network elements that USTR succeeds in imposing on Singapore, for example, will become the new standard that U.S. competitive LECs seek to have imposed by domestic regulators on U.S. incumbent LECs. Suddenly, a career bureaucrat in USTR will have overridden Congress and the FCC and the federal courts. To make matters worse, judicial review of USTR actions seems difficult if not impossible under D.C. Circuit precedent (Rohlfs and Sidak, 2002).

3. The EC model

For most of the 20th century, network industries were considered to be natural monopolies, a situation, which justified the granting of exclusive rights to an operator (Geradin and Kerf, 2003). Unlike in the U.S. where such monopolies were (with the exception of postal services) run by private companies regulated by independent agencies, European States opted for a public monopoly model (Frieden, 2001).

In the early 1980s, the public monopoly model was challenged on the ground that, even if network industries contained some bottlenecks (e.g., the local loop), many products and services were potentially competitive and, thus, should be open to competition[37]. Moreover, European governments were under pressure from industry and consumer associations to liberalize these sectors, as it was felt that competition would generate better products and services at lower prices. Finally, the European Commission considered that public monopolies organized on a national basis were impeding the flow of goods and services and were incompatible with the internal market. The Commission thus engaged into a wide

[36] 47 U.S.C. § 252(i) (2000).

[37] Geradin and Kerf (2003). Communication by the Commission of June 30, 1987, Toward a Dynamic European Economy: Green Paper on the Development of the Common Market for Telecommunications Services and Equipment, COM(87) 290.

program of market opening reforms in the aviation, energy, postal services, and telecommunications sector (Blum et al., 1998). The Commission's liberalization strategy essentially relied on three pillars: (i) removal of monopoly rights through liberalization directives; (ii) adoption of 'pro-competition' regulatory frameworks through harmonization directives; and (iii) application of antitrust rules.

A major difference between the U.S. and the EU approaches to telecom regulation is the unsystematic nature of the U.S. approach compared to the systematic EU framework (Marcus, 2003). This difference is largely based on the history (path dependence) and institutions, which make it hard to draw realistic policy consequences. The EU had the luxury that it was driven by the vision of a common market for services that required liberalization and harmonization. Also, it was developed with less political—and, in particular, procedural—interference than the U.S. approach. This difference explains, for example, why the U.S. has such a hard time in developing a unified approach to all parts of the telecommunications sector, including Internet access, cable television, and the like.

The following discussion is divided into four subsections. Subsection 3.1 explores *ex ante* and *ex post* remedies in telecommunications. As in the U.S., we will see that the border between *ex ante* and *ex post* remedies is thin and that some remedies appear as an amalgam between these two kinds of remedies. Subsection 3.2 argues that, as in the U.S., EC antitrust law has taken a regulatory tone with the multiplication of notices and guidelines, which resemble the prospective rulings made by regulatory agencies. Subsection 3.3 shows that antitrust concepts and principles play a central role in the new EC framework on electronic communications. Finally, Subsection 3.4 observes that, unlike the U.S. Trade Representative, DG Trade does not behave like a telecommunications regulator. However, the enlargement process allows the EC to progressively expand the number of nations, to which its regulatory principles in the area of telecommunications and in other network industries will apply.

3.1. Ex ante, ex post, *and hybrid remedies*

As in the U.S., the EC combined the removal of monopoly rights with the adoption of pieces of legislation designed to address the 'bottlenecks' that would prevent the arrival of competition in the telecommunications sector[38].

[38] Some pieces of legislation were also aimed at addressing universal services and consumer protection issues. See, for instance, Directive 97/51 of October 6, 1997 amending Directives 90/387 and 92/44 for the purpose of adaptation at a competitive environment in telecommunications, (1997) O.J. L 295/23; Directive 98/10 of February 26, 1998 on the application of ONP telephony and on universal service for telecommunications in a competitive environment, (1998) O.J. L 101/24.

For instance, Directive 97/33 on interconnection provides that telecommunications operators, which have significant market power (hereafter, the 'SMP operators')[39] have to meet all reasonable requests for access to the network, including at access points other than the network terminations points offered to the majority of end users[40]. In addition, Regulation 2887/2000 on local loop unbundling provides that all SMP operators had to meet all reasonable requests for unbundled access to their local loops and related facilities under transparent, fair, and nondiscriminatory conditions[41]. This Regulation also states that the prices charged by these operators for unbundled access to the local loop and the related facilities should be set on the basis of cost orientation[42]. These obligations are good examples of asymmetric regulation, as they only bear on SMP operators defined in the interconnection directive as operators that have 'a share of more than 25 percent of a particular telecommunications market in the geographical area in a Member States, in which it is authorized to operate[43].' Although the SMP threshold was low, this legislation essentially aimed at the incumbents the market power, of which had to be controlled to facilitate entry in liberalized telecommunications markets.

This regulatory framework played a major role in helping to create competition in the telecommunications market, although lack of competition can still be witnessed in several market segments[44]. Unlike the U.S. telecommunications regulatory framework, the EC model proved to be quite flexible as it is based on a decentralized approach whereby the implementation of EC legislation is carried out by the national regulatory authorities (hereafter, the 'NRAs'), rather than by a European Telecommunications Agency (Geradin, 2001). In the EC, there is thus no equivalent of the FCC, which would adopt painfully detailed regulatory requirements to be implemented by telecommunications operators located in the EC. In addition, EC legislation never formally required NRAs to rely on the

[39] Under the former regulatory framework (i.e., the regulatory framework, which preceded the new regulatory framework on electronic communications), considered as having SMP all operators, which held 25 percent or more market shares in a given market. Under the new framework, SMP is assimilated to the concept of 'dominance' as understood in EC competition law.
[40] Article 4.2 of Directive 97/33 of June 30, 1997 on interconnection in telecommunications with regard to ensuring universal services and interoperability through application of the principles of Open Network Provision (ONP), (1997) O.J. L 199/32.
[41] *Id.* at Article 3.2.
[42] *Id.* at Article 7.2.
[43] *Id.* at Article 4.3.
[44] See Larouche (2000), Communication from the Commission of November 19, 2003, Ninth Implementation Report, COM(2003) 715.

LRIC model (we have seen above, that it prescribes pricing regimes be 'cost oriented'), although it has recommended that NRAs opted for such a model[45]. The flexibility of this model is to be valued, but some observers argue that the national variations it allowed could compromise the completion of an integrated telecommunications market spanning across Europe.

In the late 1990s, the Commission engaged in the process of reviewing this regulatory framework and came up with the so-called 1999 Review, a report outlining the future of EC telecommunications (now referred to as 'electronic communications') regulation[46]. This report contains broad sets of proposals to adapt the existing regulatory framework to new technological (the process of convergence) and market (growing competition) circumstances. Interestingly, the Commission appears to range on the side of the growing consensus that competition law and principles should progressively replace sector-specific regulation. The Commission observes that the aim of the Review is 'to create a regulatory regime, which can be rolled back as competition strengthens, with the ultimate objective of controlling market power through the application of Community competition law[47].' Eventually, the 1999 Review led to the adoption of the new EC regulatory framework on electronic communications. This new framework will be analyzed in Subsection 3.3.

As already illustrated in the U.S. section of this chapter, antitrust rules (hereafter referred to as 'competition rules,' as this is the terminology used in the EC) can also be applied to impose *ex ante* remedies. This is the case when the Commission is asked to clear joint ventures (hereafter, "JVs") or mergers, which raise competition law concerns. Merger control is an area where the Commission has been particularly active this last decade as it has been asked to clear a large number of transactions involving telecommunications and media operators (Garzaniti, 2003). Most of these transactions were cleared without difficulty, but some of them required more detailed attention as they raised competition law concerns. Several transactions were prohibited on the ground that they would

[45] Commission Recommendation of January 8, 1998 on interconnection in a liberalized telecommunications market (Part 1-Interconnection Pricing), (1998) O.J. 73/42, last amended by Commission Recommendation of February 22, 2002, (2002) O.J. L 58/56, at §3; Commission Recommendation of May 25, 2000 on unbundled access to the local loop: Enabling the competitive provision of a full range of electronic communications services including broadband multimedia and high speed Internet, (2000) O.J. L 156/44 at Article 1(6).

[46] European Commission, Toward a New Framework for Electronic Communications Infrastructure and Associated Services: The 1999 Review, Com (1999) 539.

[47] *Id.*

create irreparable damage to competition[48]. However, most of the problematic mergers were eventually cleared after the parties offered substantial remedies to the Commission. The measures took the form of both structural remedies that stimulate network competition (e.g., cable divestiture) and behavioral remedies that ensure competitors to the merging entities will have sufficient access to key inputs, such as content, local loop, or set top boxes (de Streel, 2003). For instance, in *Vodafone/Mannesman*, the Commission only cleared the merger after the parties submitted commitments to demerge Orange Plc and to give other mobile operators access to their inter-operator roaming tariffs and wholesale services[49].

Competition law authorities can also intervene *ex post*, which is the core business, as *ex ante* analysis is limited to the clearance of JVs and mergers. Like in other industries, the European Commission engaged many procedures on the basis of Article 82 of the EC Treaty (which prohibits abuse of dominance). For instance, it initiated a variety of proceedings, including cases of exclusionary abuses (e.g., refusal to provide access to the fixed local infrastructure to competitors)[50], as well as exploitative abuses (e.g., excessive pricing of retail services)[51]. Another illustration of the *ex post* approach can be found in the Commission's reliance on the so-called sector enquiries, a rarely used procedure pursuant, to which the Commission enquires on a whole market, rather on specific companies. The Commission launched a sector enquiry in 1999 with the purpose of successfully examining three markets: the provision and pricing of leased lines[52], mobile roaming services[53], and the provision of access to and use of the residential local loop[54].

Yet, as we have seen in the U.S. section, the border between the *ex ante* and *ex post* approaches may be quite thin. For instance, in the competition cases examined above, the Commission generally chose to transfer when possible these

[48] In total, six transactions were prohibited by the Commission: Case No. M.469 *MSG Media Service* (1994) O.J. 1994, L 364/1. Case No. M. 490 *Nordic Satellite Distribution* (1995) O.J. 1996, L 53/20. Case No. M 553 *RTL/Veronica/Endemol (I)* (1995) O.J. 1996, L 134/32. Case No. M.993 *Bertelsmann/Kirch/Premiere* (1998) O.J. 1999, L 53/1. Case No. M. 1027 *Deutsche Telekom/BetaResearch* (1998) O.J. 1999, L 53/31. Case No. M. 1741 *MCIWorldCom/Sprint*.

[49] Case No. M. 1795 (2000) *Vodafone/Mannesmann*, IP/00/373 of April 12, 2000.

[50] See, Commission Decision of May 21, 2003, *Deutsche Telekom*, O.J. 14.10.2003 L 263/9 (price squeeze between wholesale and retail local access charges); Commission Decision of July 16, 2003, *Wanadoo*, not yet published IP/03/1025 (predatory price in the retail market for broadband Internet access).

[51] IP/98/707 of July 27, 1998 (Commission launching four cases for excessive and discriminatory fixed termination charges originated from mobile and eight cases for excessive fixed retention charges).

[52] IP/99/786 of October 22, 1999; IP/00/1043 of September 22, 2000.

[53] IP/00/111 of February 4, 2000; MEMO/01/262 of July 11, 2001.

[54] IP/00/765 of July 12, 2000.

cases to the NRAs on the ground that they would be able to handle these matters by enforcing sector-specific legislation[55]. *Ex post* intervention based on competition rules has thus been prolonged by an application of *ex ante* sector-specific regulatory regimes. This suggests that *ex post* and *ex ante* regimes are seen by the Commission as alternative tools to control market power in telecommunications. However, where a matter can be properly dealt with by the NRAs, competition cases should be passed on to them.

If one goes a step further, we could even say there is a blurring of the distinction between sector-specific regulation and competition law. Clearance of JVs and mergers has often been used by the Commission to further regulatory objectives. For instance, in *Atlas*, a JV between France Telecom and Deutsche Telekom, the Commission granted a 5-year exemption that provided inter alia that France and Germany liberalize alternative infrastructures, a requirement that was not yet imposed by EC telecommunications legislation[56]. Similarly, the Commission cleared the *Telia/Telenor* merger after the Swedish and Norwegian governments committed to introduce local loop unbundling in their countries—that is, more than a year ahead of Regulation 2887/2000, which imposes local loop unbundling to the 15 Member States[57]. It is subjected to question whether the Commission should use merger control to advance its regulatory agenda[58]. In any event, these cases show the close ties between governments and business in network industries as the Member States in question agreed to anticipate EC regulatory requirements in order to help national champions to obtain clearance from the Commission.

This naturally leads us to talk about the situation of state-owned enterprises active in network industries. This issue is particularly important in the EC since European nations have historically opted for the public monopoly model in most network industries. Although privatization programs have significantly reduced public participation in sectors, such as telecommunications and air transport, state ownership remains dominant in postal services and rail transport. In the U.S. section, we argued that several factors may constrain the ability to impose remedies on state-owned companies. In the EC, the close links between postal and rail operators and their governments made it difficult for the EC to engage into market-opening reforms combined with regulatory remedies, although the most

[55] See, for instance, IP/98/763 of August 13, 1998 and IP/99/279 of April 29, 1999 (Commission launching seven cases for excessive accounting rates, but subsequently passed them to the NRAs, which imposed substantial price reductions).

[56] Case no. 35.337, *Atlas* (1996) O.J. 1996, L 239/23.

[57] Case no. M. 1439, *Telia/Telenor* (1999) O.J. 2001, L 40/1.

[58] For a discussion of this issue in the electricity sector, see Piergiovanni (2003).

recent legislation shows encouraging signs[59]. However, the European Commission initiated a large number of cases against postal, as well as to a lesser extent rail, operators to place an end on abusive practices, including cross-subsidization, discrimination, predatory pricing, excessive pricing, tying, refusal to supply, etc. The numerous proceedings against Deutsche Post illustrate the tough line taken by the Commission in the postal sector.

3.2. *Antitrust as a new form of regulation*

The prior section concluded that on several occasions, the European Commission used its powers to review JVs and mergers to achieve regulatory objectives. Looking at this from a different angle, one could also say that the Commission decisions over telecommunications JVs and mergers appear very much like catalogues of regulatory requirements listing things that the merging entities should or should not do in the future. Merger decisions thus very much look like prospective regulatory regimes.

Another illustration of this prospective approach can be found in the various guidelines and notices adopted by the Commission in the area of telecommunications. In 1991, at the outset of the liberalization process, the Commission adopted guidelines on the application of EC competition rules to the telecommunications sector[60]. In 1998, at the time the telecommunications market was completely liberalized, the Commission issued a Notice on the application of competition rules to access agreements in the telecommunications sector[61]. In 2000, at a time when the pervasive domination of the incumbent on the local loop was increasingly seen as impeding a rapid development of broadband access in Europe, the Commission issued a Communication on the application of competition rule to the unbundling of the local loop[62]. In these instances, telecommunications

[59] Directive 2002/39/EC of the European Parliament and of the Council of June 10, 2002 amending Directive 97/67/EC with regard to the further opening to competition of Community postal services, (2002) O.J. L 176/21; Directive 2001/12/EC of the European Parliament and of the Council of February 26, 2001 amending Council Directive 91/440/EEC on the development of the Community's railways, (2001) O.J. L 85/13.

[60] Commission guidelines on the application of EEC competition rules in the telecommunications sector, O.J. 6.9.1991, C 233/2.

[61] Commission Notice on the application of competition rules to access agreements in the telecommunications sector O.J. 22.8.1998, C 265/2C.

[62] Communication from the communication of April 26, 2000 on the unbundled access to the local loop, O.J. 23.9.2000 C 272/55.

operators are prospectively informed of the way the Commission intends to apply competition rules to certain practices.

One could, of course, argue that a basic distinction between guidelines and regulatory requirements is that the former do not have the binding of the latter. This distinction is more apparent than real as operators are generally well advised to follow Commission guidelines on pain of being subject to competition law investigations. Commission guidelines may thus have a greater legal significance that their 'soft law' status would tend to suggest. Let us now turn to this new regulatory framework.

3.3. The shifting balance of influence between antitrust and sector-specific regulation: competition law concepts penetrating sector-specific regulation

In the U.S. section, we saw that antitrust law was probably no longer the driving force in terms of shaping the telecommunications sector. In fact, the Government did not initiate any antitrust proceeding in this sector since the adoption of the 1996 Telecommunications Act and virtually all the proceedings initiated by competitive local exchange providers against the BOCs failed. We also observed that the approach followed by the FCC in its Local Competition Order and vindicated by the Supreme Court in 2002 in *Verizon* could influence the development of antitrust-based remedies in network industries.

The reverse situation is currently taking place in the EC. First, since the complete opening of the telecommunications sector on January 1, 1998, EC competition law has played a major role in shaping the telecommunications sector (Ungerer, no date). Moreover, competition law concepts and principles are at the core of the new regulatory framework on electronic communications adopted by the EC in 2002 and which entered into force in July 2003. As we have seen in Section A , the 1999 Review proposed a vision whereby sector-specific regulation would be progressively rolled-back, market power being exclusively controlled by competition law. This vision is to a large extent translated into the new framework, although we will see that sector-specific regulation will retain a significant importance, at least in the next few years.

The new regulatory framework is composed of five directives: one framework directive[63] and four specific directives respectively dealing with authorizations[64],

[63] Directive 2002/21 on a common regulatory framework for electronic communications networks and services, 2002 O.J. L 108/33.

[64] Directive 2002/20 on the authorization of electronic communications networks and services, 2002 O.J. L 108/21.

universal service[65], access and interconnection[66], and data protection and privacy in the telecommunications sector[67]. In order to take into account the so-called 'convergence' between the telecommunications, media, and information technology sectors, this framework not only covers telecommunications, but all forms of electronic communications and services to the exclusion of content-related aspects, which are dealt with by separate legislation.

The new framework is based on the so-called SMP regime[68]. Pursuant to this regime, regulatory obligations can be imposed on operators holding SMP in one given market after a four-step analysis, which requires coordinated efforts of the European Commission and the NRAs.

First, the Commission periodically adopts a Recommendation[69], which identifies, in accordance with the principles of competition law, the products and services markets within the electronic communications sector, the characteristics of which may be such as to justify the imposition of regulatory obligations set out in specific directives[70]. In its Recommendation, the Commission identifies three criteria that have to be taken into account for a market to be selected to be analyzed by the NRAs: (i) the presence of high and nontransitory entry barriers whether of structural, legal or regulatory nature; (ii) the presence of a market structure such that the market does not tend towards effective competition within the relevant time frame horizon; and (iii) the application of competition law alone would not adequately address the market failure(s) concerned[71].

The first criterion appears to be in full line with economic theory as we know that high barriers to entry are one of the key factors when it comes to assessing the presence of market power. As interpreted in the Recommendation, the second criterion can be used to narrow down the scope of markets falling under the first criterion. Indeed, the Recommendation explains that '[e]ven when a market is

[65] Directive 2002/22 on universal service and users' rights to electronic communications networks and services, 2002 O.J. L 108/51.

[66] Directive 2002/19 on access to, and interconnection of, electronic communications networks and associated facilities, 2002 O.J. L 108/7.

[67] Directive 2002/58 concerning the processing of personal data and the protection of privacy in the electronic communications sector, 2002 O.J. L 201/37.

[68] For a good discussion of this regime, see (de Streel, 2003).

[69] Commission Recommendation of February 11, 2003 on relevant product and service markets within the electronic communications sector susceptible to *ex ante* regulation in accordance with Directive 2002/21/EC of the European Parliament and of the Council on a common regulatory framework for electronic communications networks and services, O.J. 8.5.2003 L 114/45.

[70] Article 15(1) of the Framework Directive.

[71] §§ 9-16 of the Recommendation.

characterized by high barriers to entry, other structural factors in that market may mean that the market tends towards an effectively competitive outcome within the relevant time horizon[72].' The Recommendation then states that this may for instance be the case 'in markets with a limited, but sufficient number of undertakings having diverging cost structures and facing price-elastic demand[73].' The third criterion will be discussed below.

The combined analysis of the three criteria suggests that markets characterized by the presence of high entry barriers (first criterion), which is not compensated by a dynamic market structure (second criterion) should generally be selected unless the situation can be dealt with adequately by competition law remedies (third criterion). The prime targets for selection will thus be markets having a natural monopoly nature or oligopolistic features, especially when there is a risk of collective dominance[74].

Second, taking 'utmost' account of this Recommendation and the Commission guidelines on market analysis, the NRAs define relevant markets appropriate to national circumstances, in particular, relevant geographic markets within their territory, in accordance with the principles of competition law (Rey, 2002).

Third, the NRAs analyze relevant markets, where appropriate in cooperation with the national competition authorities, to determine whether they are competitive or not[75]. This amounts to determining whether one or several operators hold SMP in a given market. In this context, Article 14(2) of the Framework Directive provides that an undertaking is deemed to have SMP 'if either individually or jointly with others, it enjoys a position equivalent to dominance, that is to say a position of economic strength affording it the power to behave to an appreciable extent independently of competitors customers and ultimately consumers.' This provision includes word by word the standard understanding of dominance stemming from the European Court of Justice competition case-law[76]. There is some logic in using the same tools to define markets under both competition law and sector-specific regulation, although this may raise some issues that will be discussed below. In any event, this is an improvement on the prior regime, which considered as having SMP all operators that enjoyed 25 percent or more market

[72] *Id.* At §14.
[73] *Id.*
[74] Commission guidelines on market analysis and the assessment of significant market power under the Community regulatory framework for electronic communications networks and services, O.J. 11.7.2002, C 165/6. For an analysis of collective dominance telecommunications, see [o]Rey (2002).
[75] *Id.* at Article 16(1).
[76] See Case 27/76, United Brands, (1978) ECR 207.

shares in a given market. The threshold for SMP was thus particularly low (hence, running the risk of over-regulation) and set in a rather inconclusive manner as one knows that market share is only one of the factors indicating market power.

Fourth, when the NRA concludes that the market is effectively competitive, it cannot impose or maintain any specific obligations on the operators active in this market, and in cases where sector-specific obligations already exist, the NRA must withdraw them[77]. Conversely, when the NRA determines that a relevant market is not effectively competitive, it must first identify operators with SMP on that market and impose specific obligations on them from the menu of remedies proposed by the directives[78]. For instance, when an operator is found to have SMP on a wholesale market (e.g., the provision of local loop elements), the NRA must at least impose one, but can also impose several remedies among those provided in the so-called Access Directive. The remedies include obligations of transparency, nondiscrimination, accounting separation, access to network infrastructures, or price control. When an NRA imposes one or several remedies, it must ensure that they are proportionate to the policy objectives identified[79].

This short description of the SMP regime deserves some remarks.

First, this regime requires a co-operation between the Commission and the NRAs, pursuant to which both levels of power have a key role to play. The system seems, thus, more decentralized than the U.S. system (where key regulatory decisions are entrusted to the FCC and little is left to the PUCs, which is one of the factors that led to litigation against the FCC's Local Competition Order). Under the new regime, NRAs will take decisions over market definition, identification of SMP operators, market analysis, and the choice of remedies, although in taking these decisions it will have to take into account the soft law instruments adopted by the Commission.

Second, this regime relies heavily on the principles of competition law[80]. For instance, we have seen that the NRAs must define relevant markets in accordance with the principles of competition. We have also seen that the definition of SMP that is relied on in the directives is identical to the notion of dominance as defined in EC competition law. The reliance on common principles for competition law inquiries and sector-specific regulation makes sense, especially when both sets of rules are called to apply in the same markets, sometimes in a parallel manner.

[77] *Id.* at Article 16(3).
[78] *Id.* at Article 16(4).
[79] §118 of the Commission guidelines on market analysis.
[80] For a discussion of the alignment of sector-specific regulation on competition rules, see Larouche (2002).

Some authors have, however, noted that when identifying SMP for the purpose of assessing the competitiveness of a market, NRAs are working on a different set of assumptions than competition authorities acting in a competition case (Bak, 2003)[81]. Indeed, a finding of SMP on the basis of Article 14(2) of the Framework Directive will lead to the automatic imposition of one or several remedies, whereas a finding of dominance in an Article 82 case does not lead by itself to the imposition of remedies, as remedial action will only take place when an abuse has been found. The impact of a finding of SMP/dominance is thus different. In addition, an investigation of abuse of dominance is based on a set of findings reflecting the undertaking's position and behavior in a market over a given period of time in the past, while under the new regulatory framework *ex ante* obligations arise (and are withdrawn) as a result of forward-looking market analysis based on existing market conditions. Moreover, given its forward-looking nature, the analysis of dominance will often work on the basis of predictions or speculations than on existing evidence. Potential competition and supply substitutability will thus be central to such analysis, while demand substitutability may play a reduced role, although this is a key criterion for EC competition law.

Although this line of argument is interesting, it only reassesses something that we could already observe before the adoption of the new regulatory framework just by looking at the *ex ante* and *ex post* analyses carried out by competition law authorities. In fact, every time a competition law authority has to adopt *ex ante* remedies in relation to a dominant operator (e.g., in the context of a merger), the analysis of dominance will be different than from the dominance analysis is carried out in the context of an *ex post* inquiry (e.g., in the context of an Article 82 case). This seems to suggest that, when competition law authorities or NRAs are involved in the imposition of *ex ante* remedies, they should err on the side of prudence, as the analysis of dominance, on which it is based will often be speculative and lead to dramatic consequences for the operators involved whereas they have not committed any competition law infringement.

Third, under the new regime, sector-specific regulation has a subsidiary role—that is, it should be adopted only when competition law remedies would not suffice to address the identified market failure. For instance, we have seen that in its Recommendation on relevant markets, the Commission identifies among the criteria that have to be taken into account for a market to be selected to be analyzed by the NRAs, the relative efficiency of competition law remedies alone to address the market failure identified compared to the use of *ex ante* regulation.

[81] See also Commission guidelines on market analysis, §§ 24–32.

To impose regulatory requirements, the NRAs must demonstrate that regulation is better able to address the market failure in question than competition law. Thus, the SMP regime gives preference to *ex post* competition law, which is less intrusive than sector-specific regulation.

Fourth, the SMP regime provides for mechanisms to ensure that sector-specific regulatory requirements will remain only as long as they are necessary to correct a market failure. For instance, we have seen above that when an NRA discovers that a relevant market is effectively competitive it cannot impose or maintain any sector-specific obligations, but it should also withdraw all existing obligations. This creates a process whereby, as markets become competitive, regulation will progressively fade away. This system of market-by-market sunset clauses seems more effective than Section 160 of the 1996 U.S. Telecommunications Act, which operates the other way round as it enables the FCC to forbear from applying provisions of this Act if it determines that forbearance from enforcing these provisions 'will promote effective competition.' Use of Section 160 is, however, restricted with respect to some of the provisions of the Act, and it also requires that a series of strict conditions be met[82]. Moreover, this provision leaves a lot of discretion to the FCC. The EC is much more radical since, as soon as a market is effectively competitive (essentially a question of fact rather than of appreciation), the NRA has no choice but to withdraw from sector-specific regulation. Section 160 of the 1996 Act also requires the FCC to review, every 2 years, all of its rules that apply to telecommunications service providers and determine whether any are no longer necessary in the public interest[83]. Section 161 then directs the FCC to repeal or modify unnecessary rules. However, the 2-year time span between each revision is too long, and the public interest criterion is too vague. In practice, the FCC does not really attempt to review all of its regulations pursuant to Section 160 , for the task, if taken literally, would consume virtually all of the agency's resources.

Although the above-described European regime presents many attractive features, it remains to be seen how it will be applied in practice. The definition of the relevant markets, the determination of whether such markets are competitive or not, the identification of operators with SMP, and the imposition of remedies on such operators, is done by the NRAs. Although the NRAs will have to take into account the Commission guidelines on market analysis and the Commission Recommendation on relevant markets, national variations will thus occur.

[82] See 47 U.S.C. § 160 (a–d).

[83] For a discussion of § 161, see Furchtgott-Roth (1998).

As the set of *ex ante* remedies, among which the NRAs will have to make their choice to correct market failures is set in specific EC directives, it is worth having a look at such directives to see whether the obligations they contain follow a competitor-welfare or a consumer-welfare standard, the latter we believe being preferable to the former. In this respect, Article 12 of the Access Directive is somehow troublesome as it provides that an NRA could impose on SMP operators to grant access to specific facilities and/or associated services, inter alia in situations where the NRA considers that 'denial of access would *hinder* the emergence of a sustainable competitive market at the retail level, or would not be in the end user's interest[84].' While the second part of the sentence places emphasis on consumer interest, the first part seems to impose a softer (i.e., easier to meet by access seekers) test than the one imposed under competition law, which requires that access be only granted to 'essential' facilities[85]. Under sector-specific regulation, SMP operators could thus be forced to grant access to nonessential facilities on the ground that such access would be desirable to stimulate competition in retail markets. We very much hope that the NRAs will interpret the test contained in Article 12 in a manner compatible with the requirements of competition law.

3.4. DG trade as a regulator?

In the U.S. section, we argued that, by attempting to impose the TELRIC model to U.S. trading partners, the U.S. Trade Representative has turned itself into a telecommunications regulator. The situation is different in the EC as, to the best of our knowledge, the Trade Directorate of the European Commission has never used bullying tactics on the EC's trading partners to impose the EC telecommunications framework on them. The EC's more relaxed attitude can be explained by several reasons. First, the U.S. has generally been more concerned than the EC about the deficits created by the regime of international accounting rates. The U.S. suffers from huge accounting rate settlement imbalances with a number of countries[86]. The WTO case initiated by the U.S. against Mexico is thus more aimed at addressing such settlement imbalances than influencing another nation's policy. Second, some U.S. carriers have been particularly aggressive in terms of gaining market shares in key parts of the world, such as Asia. High interconnection

[84] Emphasis added.
[85] See Case C-7/97, Bronner 1998 ECR I-7791. Generally on the essential facilities doctrine, see Lang (2000).
[86] Cowhey (2004).

fees unavoidably compromise their business ambitions. The decentralized implementation approach that is followed in the EC may also be a reason explaining the different attitudes between the two trade blocks. As the European Commission does not impose any pricing mechanisms on the Member States, but only sets the principles that have to be complied with by the NRAs in their decisions (e.g., cost orientation), it would be odd for the Commission to impose LRIC or other pricing models on its trading partners.

On the other hand, the enlargement process gives the European Commission another means to influence regulatory policies in other countries. On May 1, 2004, 10 new nations will become members of the EU and, as result, will have to implement in their national law, if it is not already done, the whole body of EC legislation, including the new regulatory framework on electronic communications. Moreover, aspiring candidate nations, such as Turkey, will also have every incentive to adopt the EC's new regulatory framework, as regulatory convergence is a precondition for joining the EU. Finally, in its recent 'Wider Europe' Communication, which seeks to define the EU's line of action with a series of neighboring nations in Central and Oriental Europe and the Middle East and North African region, the Commission also expresses its desire for a greater degree of regulatory convergence between these nations and the EU. This does not mean that the Commission will seek to impose specific regulatory choices, such as a model of interconnection pricing on these nations, as uniformity over such choices is not even imposed on the NRAs of the Member States, but it will allow EC electronic communications law to influence foreign regimes in a deeper and more-lasting wsense[87].

4. Conclusions

This chapter has examined remedies and the institutional design of regulation in a comparative manner by reference to U.S. and EC law. In both systems, a variety of remedies has been imposed on 'dominant' telecommunications operators to control their market power and facilitate the arrival of competition. These remedies have often been controversial, as they were seen as too harsh by some and too lax by others. Few would dispute that these remedies have played a major role in shaping, for good or for worse, the telecommunications sector on both sides of the Atlantic.

[87] Communication of the Commission, 'Wider Europe-Neighborhood: A New Framework for Relations with our Eastern and Southern Neighbors,' COM(2003)104 final.

There are several similarities in the approaches followed by the U.S. and the EC in their remedial efforts. First, the U.S. and the EC have relied on a combination of *ex ante* and *ex post* remedies to control market power in the telecommunications. Yet, both regimes have also developed 'hybrid' remedies, which represent an amalgam between the *ex ante* and the *ex post* approaches. For instance, as illustrated by the MFJ, consent decrees have been used by the Antitrust Division as a way to regulate the telecommunications sector. In both U.S. and EC, the remedies imposed as a condition for merger clearance often take the form of long lists of behavioral requirements, which can be hardly distinguished from prospective regulatory requirements. Although we agree that antitrust rules be used to maintain a competitive market structure, we question the use of antitrust remedies to reshape the telecommunications sector or achieve specific regulatory objectives. Very often, operators are under no position to negotiate, and the clearance process turns into a game of regulatory extortion.

Second, the growing amalgam between antitrust and regulation can also be illustrated by the increasing reliance by antitrust authorities on guidelines, policy statements, notices, and other tools containing abstract statements of the way these authorities plan to address anticompetitive conduct, which may arise in the future. In this context, they very much act like a bureaucracy adopting prospective rulings than as antitrust authorities deciding cases on the basis of past events. In the future, antitrust could be much less of a litigation practice, than a regulatory compliance exercise whereby adepts go through checklists of predetermined regulatory interpretations.

There are also significant differences between the U.S. and EC regimes of remedies in telecommunications. First, regulatory remedies have generally been more intrusive in the U.S. than in the EC. The combination of the 1996 Telecommunications Act and the FCC's implementing orders has produced an extremely dense regulatory framework regulating certain categories of operators' behaviors in the most excruciating detail. Although the EC regulatory framework adopted in the EC in the 1990s was also heavy-handed, the new regulatory framework has clear deregulatory features, as it relies on a market-by-market system of sunset clauses and allows regulation only when antitrust law remedies are not sufficient to address the identified market failure(s). The new EC regulatory framework no longer imposes remedies on predetermined categories of operators, but on operators holding 'SMP,' this latter concept corresponding to the notion of 'dominance' under EC competition law. The EC system seems thus better equipped to limit the imposition of *ex ante* remedies to circumstances where market failures can be identified. No other circumstance should warrant *ex ante* regulatory intervention.

Second, in recent years, antitrust rules have played a greater role in the EC than in the U.S. in telecommunications. Although the U.S. government has not initiated any major telecommunications antitrust lawsuit since the MFJ, the European Commission has launched proceedings to address a variety of anticompetitive behaviors in telecommunications. Moreover, although it is true that U.S. antitrust increasingly takes a regulatory tone, antitrust principles have not much penetrated sector-specific regulation. On the contrary, there is a risk that regulatory models developed by the FCC (such as the controversial TELRIC pricing methodology) could influence the design of antitrust remedies in the future. By contrast, EC antitrust principles play a crucial role in the new regulatory framework on electronic communications: key regulatory decisions, such as market definition, identification of SMP operators, and the adoption of remedies must be adopted in conformity with antitrust principles. There is also a clear understanding in the EC that sector-specific regulation is to progressively give way to antitrust law.

Acknowledgments

The authors thank Alexandre de Streel for helpful comments. They also thank the financial support provided by the PAI 4/04, a program initiated by the Belgian State, Prime Minister's Office, Federal Office for Scientific, Technical, and Cultural Affairs.

References

Andrews, E. L., 1997, U.S. remains odd man in global push for phone deal, New York Times, February 14, D1.

Bak, M., 2003, European electronic communications on the road to full competition: The concept of significant market power under the new EC regulatory framework, Journal of Network Industries, 4, 293–324.

Cowhey, P., 2004, Accounting rates, cross-border services, and the next round on basic telecommunications services, in D. Geradin and D. Luff, eds., The WTO and Global Convergence in Telecommunications and Audio-Visual Services, Cambridge: Cambridge University Press, 51–82.

Crandall, R. W., 2000, Telecommunications liberalization: The U.S. model, in T. Ito and A. O. Krueger, eds., Volume 8, 415–428.

Crandall, R. W., J. G. Sidak and H. J. Singer, 2002, The empirical case against asymmetric regulation of broadband Internet access, Berkeley Technology Law Journal, 17, 953–987.

de Streel, A., 2003, The integration of competition law principles in the new European regulatory framework for electronic communications, World Competition, 26, 489–514.

Edward, M. G. and J. D. Richardson, 1999, Global Competition Policy, Washington DC: Institute for International Economics.

Federal Communications Commission, 1997, FCC approves Bell Atlantic/NYNEX merger subject to market-opening conditions, Federal Communications Commission Press Release.
Federal Communications Commission, 2000, Federal Communications Commission approves Bell Atlantic-GTE merger with conditions, http://www.fcc.gov/Bureaus/Common_Carrier/News_Releases/2000/nrcc0031.txt.
Fisher, F. M., 1998, Declaration of Franklin M. Fisher, United States v. Microsoft Corp., Civil Action No. 98–1233, (filed D.D.C. May 12, 1998).
Fisher, F. M., 1999, Direct testimony of Franklin Fisher, United States v. Microsoft Corp., (D.D.C. 1999) (No. 98–1233).
Fisher, F. M., 2000, The IBM case and Microsoft: What's the difference?, American Economic Review. Papers and Proceedings, 90, 180–183 (May).
Frieden, R. M., 2001, Managing Internet-driven Change in International Telecommunications, Artech House Publishers.
Furchtgott-Roth, H. W., 1998, Report on implementation of Section 11 by the Federal Communication Commission, http://www.fcc.gov/commissioners/furchtgott-roth/reports/sec11.hmtl.
Furchtgott-Roth, H. W., 1999, The FCC racket, Wall Street Journal, November 5.
Furchtgott-Roth, H. W., 2000, Testimony before the House Committee on Commerce, Subcommittee on Telecommunications, Trade, and Consumer, http://www.fcc.gov/Speeches/Furchtgott_Roth/2000/sphfr004.html.
Garzaniti, L., 2003, Telecommunications, Broadcasting and the Internet: EU Competition and Regulation 2nd edn., London: Sweet & Maxwell.
Geradin, D., 2001, Institutional aspects of EU regulatory reforms in the telecommunications sector: An analysis of the role of National Regulatory Authorities, Journal of Network Industries, 5, 18–20.
Geradin, D. and M. Kerf, 2003, Controlling Market Power in Telecommunications: Antitrust vs. Sector-specific Regulation, Oxford: Oxford University Press.
Hausman, J. A. and J. G. Sidak, 1999, A Consumer-welfare approach to the mandatory unbundling of telecommunications networks, Yale Law Journal, 109, 417–505.
Harwood, J. H., II, W. T. Lake and D. M. Sohn, 1997, Competition in international telecommunications services, Columbia Law Review, 874, 881–884.
Ingraham, A. T. and J. G. Sidak, 2003, Mandatory unbundling, UNE-P and the cost of equity: Does TELRIC pricing increase risk for incumbent local exchange carriers?, Yale Journal on Regulation, 20, 389–406.
Katz, M. L. and C. Shapiro, 1985, Network externalities, competition, and compatibility, American Economics Review, 75, 424–440.
Katz, M. L. and C. Shapiro, 1986, Technology adoption in the presence of network externalities, Journal of Political Economy, 94, 822–841.
Larouche, P., 2000, Competition and Regulation in European Telecommunications, Oxford: Hart Publishing.
Larouche, P., 2002, A closer look at the assumptions underlying EC regulation of electronic communications, Journal of Network Industries, 3, 129–149.
Marcus, J. S., 2003, The potential relevance to the United States of the European Union's newly adopted regulatory framework for telecommunications, in L. F. Cranor and S. S. Wildman, eds., Rethinking Rights and Regulations: Institutional Responses to New Communications Technologies, Cambridge, MA: MIT Press.
McConnell, M. W., 1987, Why hold elections? Using consent decrees to insulate policies from political change, University of Chicago Legal Forum, 295.
Melamed, D., 1995, Antitrust: The new regulation, Antitrust, 10, 13–15.

Piergiovanni, M., 2003, EC merger control regulation and the energy sector: An analysis of the European Commission's decisional practice on remedies, Journal of Network Industries, 4, 227–269.
Rohlfs, J. H. and J. G. Sidak, 2002, Exporting telecommunications regulation: The United States–Japan negotiations on interconnection pricing, Harvard International Law Journal, 43, 317–360.
Sappington, D. E. M. and J. G. Sidak, 2000, Are public enterprises the only credible predators? University of Chicago Law Review, 67, 271–292.
Sappington, D. E. M. and J. G. Sidak, 2003a, Incentives for anticompetitive behavior by public enterprises, Review of Industrial Organization, 22, 183–206.
Sappington, D. E. M. and J. G. Sidak, 2003b, Competition law for state-owned enterprises, Antitrust Law Journal, 71, 479–523.
Shelanski, H. A. and J. G. Sidak, 2001, Antitrust divestiture in network industries, University of Chicago Law Review, 68, 1–99.
Sherman Act, United States Code §§ 1 et seq.
Sidak, J. G., 2001, An antitrust rule for software integration, Yale Journal on Regulation, 18, 1–83.
Sidak, J. G., 1997, Foreign Investment in American Telecommunications, Chicago: University of Chicago, 367–394.
Sidak, J. G. and D. F. Spulber, 1997a, The tragedy of the telecommons: Government pricing of unbundled network elements under the Telecommunications Act of 1996, Columbia Law Review, 97, 1081–1161.
Sidak, J. G. and D. F. Spulber, 1997b, Deregulatory Takings and the Regulatory Contract: The Competitive Transformation of Network Industries in the United States, Cambridge, UK: Cambridge University Press.
Sloan, T., 1998, Creating better incentives through regulation: Section 271 of the Communications Act of 1934 and the promotion of local exchange competition, Federal Communications Law Journal, 50, 312–398.
Supreme Court, 2002, Verizon Communications Inc. v. FCC, 122 S. Ct. 1646.
Supreme Court, 2003, Law Offices of Curtis v. Trinko, 294 F.3d 307.
Swardson, A. and P. Blustein, 1997, Trade group reaches Phone Pact, Washington Post, February 16, A33.
U.S. Court of Appeals for the 7th Circuit, 2000, Goldwasser v. Ameritech Corp., 222 F.3d 390.
U.S. Court of Appeals for the 9th Circuit, 2003, Flamingo Indus. (USA) Ltd v. United States Postal Service, 302 F.3d 985.
U.S. Court of Appeals for the D.C. Circuit, 2001, United States v. Microsoft Corporation, 253 F.3d 34.
U.S. Department of Justice, 24 April 1997, Antitrust division statement regarding Bell Atlantic/NYNEX Merger, US Department of Justice press release.
Waters, R., 1999, No sign of let-up as wannabes follow in WorldCom's wake, Financial Times, 25.
World Trade Organization, 1997, The WTO negotiations on basic telecommunications, unofficial briefing document.

Further reading

Blum, F., A. Logue, and G. Henzold, State Monopolies under EC Law, New York: John Wiley & Sons.
de Streel, A., Remedies in the electronic communications sector, in D. Geradin, ed., Remedies in Network Industries: EC Competition Law vs. Sector-specific Regulation, Antwerp: Intersentia, 67–124.

Farrell, J. and G. Saloner, 1986, Installed base and compatibility: Innovation, product preannouncements, and predation, American Economic Review, 76, 940–955.

Farrell, J. and G. Saloner, 1985, Standardization, compatibility, and innovation, Rand Journal of Economics, 16, 70–83.

Farrell, J. and P. J. Weiser, 2003, Modularity, vertical integration, and open access policies: Towards a convergence of antitrust and regulation in the Internet age, Department of Economics, Berkeley: University of CaliforniaWorking Paper No E02–325.

Lang, J. T., 2001, The principle of essential facilities in European community competition law—The position since Bronner, Journal of Network Industries, 1, 375–405.

Rey, P., 2004, Collective dominance in the telecommunications industry, in P. Buiges and P. Rey, eds., The Economics of Antitrust and Regulation in Telecommunications, Cheltenham: Edgard Elgar.

Sibely, D. S., 1998, Declaration of David S. Sibley, United States v. Microsoft Corp., Civil Action No. 98–1233.

Ungerer, H., no date, Use of EC competition rules in the liberalization of the European Union's telecommunications sector, http://europa.eu.int/comm/ competition/speeches.

Chapter 14

TELECOMMUNICATIONS AND ECONOMIC DEVELOPMENT

BJÖRN WELLENIUS*

DAVID N. TOWNSEND†

Contents

* Björn Wellenius is an independent consultant on telecommunications policy, regulation, and economics in developing and transition economies (*wellenius@attglobal.net*).
† David Townsend is the president of David N. Townsend & Associates, a consulting firm on communications economics, policy, and regulation (*DNT@dntownsend.com*).

Handbook of Telecommunications Economics, Volume 2, Edited by S. Majumdar et al.

DOI: 10.1016/S1569-4054(05)02014-2

1. Introduction

Economic development is a concept that applies worldwide. The matters of concern, including facilitating growth, responding to changing market and technological opportunities, and reaching the least advantaged members of society, are pertinent one way or another in every country. The urgency of economic development, however, is greatest where income is lowest. In this chapter, we focus on countries having low, lower-middle, or upper-middle incomes as listed in the International Telecommunication Union's *World Telecommunication Development Report 2002* (Table 1). These countries account for 85 percent of the world's population but only 20 percent of global GDP. We refer to these countries loosely as developing countries, developing economies, or emerging markets. This group includes countries in the former Soviet Union and Eastern bloc, moving from centrally planned to market economies, which we sometimes refer to as transitional economies. We occasionally compare developing countries with countries classified as having high income, which include all those usually considered to be industrialized, postindustrial, mature, or developed.

In using these categories, readers must bear in mind the huge diversity among developing countries within a category. Middle-income countries range in size from China, with 1.3 billion people, to several Pacific islands, with fewer than 100,000. There is also much heterogeneity within countries. Modern economic sectors, including telecommunications in a growing number of countries, may perform to world standards, yet coexist with a subsistence agricultural economy. More prosperous groups that could fit anywhere in the developed world live next to large segments of the population in abject poverty. Moreover, the distinction between developing and developed countries, while convenient to focus our discussion, does not mean that these form two separate worlds. The developing and developed worlds depend on each other for flows of capital, technology, goods and services, people, and communication. Developing countries, with smaller sunk costs and fewer entrenched economic and political interests than developed countries, sometimes implement sectoral changes faster and more boldly. Yet the experience of developing countries offers some lessons that apply equally well to developed countries.

This chapter focuses on the transformation in developing countries from mostly state-owned telecommunications monopolies to increasingly competitive, private-led markets. This process began in the late 1980s and is still under way. As background, we briefly outline the importance of telecommunications for development, move quickly on to explore how policies and markets have been changing and how this has been accomplished, then establish what lessons can be learned from the experience that apply to countries still in the early stages of sector reform. As we do this, we note that service innovation has played a large role in the process of change. The advent and fast growth in mobile service not only added mobility but, more important in many developing countries, made quickly

Table 1
Countries by income level, 2001

Income category	Countries	Total population (millions)	GDP[a]		Fixed and mobile phones		Examples
			Total (US$ billions)	Per capita (US$)	Total (millions)	Per 100 people	
Low	59	2440	996	432	94	4	Ghana, India, Indonesia, Moldova, Nicaragua
Lower middle	53	2073	2470	1203	484	23	Bolivia, China, Kazakhstan, Morocco, Namibia, Sri Lanka
Upper middle	35	658	3155	4866	316	48	Chile, Malaysia, Poland, South Africa, Turkey
High	49	910	24,508	27,424	1091	120	Australia, Greece, Singapore, Slovenia, United Kingdom, United States
All	196	6080	31,130	5274	1985	33	

[a]Data are for 2000.
Source: ITU, 2002b.

available an alternative to perennially scarce fixed service, created competitive pressures on incumbents to improve, and contributed to moving the sector reform agenda forward. The Internet, another major innovation during the period we examine, provides cost-effective technological alternatives and business models for delivering traditional services, and also underpins the transition from voice communications to the use of more advanced information and communications technologies and services for development. The mobile revolution is well documented in the literature, and we give only a cursory outline of what happened. By contrast, we go in more detail into the Internet, for in most developing countries it is still in its infancy, the policy and regulatory issues and options are still evolving, and the potential benefits are linked to sectors well beyond telecommunications. Finally, we examine how the telecommunications infrastructure can be used to underpin the broad range of new activities unfolding across sectors, especially electronic forms of commerce, learning, and government, which shape a qualitative change in the ways telecommunications influences development.

Although we at least touch on the most important points within the scope outlined here, we make no claim to having given fair attention to all, nor to having produced a comprehensive checklist of issues and options. The reference list includes both printed material and Websites that were current at the time of writing.

2. Development significance of telecommunications

Telecommunications is essential for economic and social development. Research in the 1960s and 1970s showed that telecommunications services, sometimes regarded as superfluous consumer goods for the rich, were actually used mainly in connection with economic production and distribution activities, delivery of social services, and government administration. They also contribute to the quality of life and to social, political, and security objectives. Where available, telecommunications benefits a broad cross-section of the urban and rural population by income, education, and occupation (Saunders et al., 1983).

Moreover, telecommunications became essential infrastructure underpinning the growing information economy. In the 1980s, information was recognized as a fundamental factor of production, along with capital and labor. The information sector accounted for one-third to one-half of GDP and employment in high-income countries in the 1980s, and this share was expected to reach 60 percent for the European Union by 2000. Information also accounted for a substantial share of GDP in newly industrialized economies and in the modern sectors of developing countries. The increasing information intensity of economic activity, coupled with the globalization of trade, manufacturing, and capital flows, resulted in strong demand for better, more varied, and less costly information and communications services. Demand growth became intertwined with rapid changes in telecommunications technology fueled by advances in optics, software, and microelectronics. These changes greatly reduced the cost of information transmission and processing, altered the cost structures of telecommunications and other industries, made possible new ways of meeting a wider range of communications needs at lower cost, reduced user dependence on established operating entities, and led to increasing convergence among previously distinct media, computer, and telecommunications technologies and services (Saunders et al., 1994; Bond, 1997).

Demand for telecommunications continued to rise in the 1990s as mobile service expanded globally, adding mobility as well as filling gaps in fixed service, and as the advent of the World Wide Web (WWW) made the Internet accessible to the broad public. Great expectations, especially about how the Internet would transform the way we live and do business, coupled with an abundance of capital seeking attractive investment opportunities, contributed to an unprecedented surge in entrepreneurial initiatives in developed countries. They also led to rising concerns in the developing world that a new gap—the so-called digital divide—was emerging between rich and poor countries. In hindsight, the expectations were exaggerated, and many investments were overly optimistic if not plainly speculative and imprudent. Many of the new ventures floundered with the collapse of the technologies markets at the end of the decade. The underlying fundamental transformation of global communications and commerce fostered by technological and market advances, however, continues.

Today, there is a fairly broad consensus that information and communications provide key inputs for economic development, contribute to global integration while helping retain the identity of traditional societies, and enhance the effectiveness, efficiency, and transparency of the public sector (World Bank, 2002). Access to information and communications is regarded as playing a crucial role in sustainable poverty reduction, mainly by increasing the efficiency and global competitiveness of economies, enabling better delivery of health and education services, and creating new sources of income and employment for poor populations (Navas-Sabater et al., 2002). A body of experience is building up in using information and communications technologies to make a difference in the delivery of services and products in rural areas of poor countries. Applications in rural India, for example, expand private participation in development, empower people through access to information and knowledge, improve and add transparency to the delivery of government services, and help public administrators improve the planning and monitoring of development programs (Bhatnagar and Schware, 2000).

The collapse of the global technologies markets has had a substantial impact on developing countries. New business opportunities that in the 1990s might have attracted important investors and operating companies, suddenly found few or no takers. Reluctance to invest in telecommunications and in developing countries has been compounded by the fact that new offerings often are inherently less attractive than those of the past. Many of the largest and most profitable, especially those in the better-off and politically less risky countries, have already been taken. Governments expecting that almost anything placed on the market would sell and at high prices are finding, sometimes repeatedly, that those times are gone and that they need to give much more attention to crafting attractive business opportunities, containing regulatory risk, and improving the domestic investment climate. Nonetheless, in developing countries, unlike in developed markets, demand for telephones, mobile service, and Internet access has continued to grow at a fast clip. In seeking to meet this demand, developing countries are shifting away from major global operators that invest substantial capital of their own (no longer an option) to smaller firms and unorthodox consortia of operators, consulting firms, and local or regional investors that might have a better strategic fit with the new projects and better knowledge of how to manage the political risk in their territory and take advantage of and nurture these expanding markets.

3. Evolution of policies and markets

Driven by growing demand and technological innovation, a wave of structural change swept the telecommunications sector in high-income countries from the early 1980s[1]. By the early 1990s, virtually all were at some stage of major sector restructuring. Sector reforms soon extended to middle- and low-income countries, where they sometimes progressed faster and further than in mature economies. Privatization of state telecommunications enterprises, initiated in Chile (1987), Argentina (1990), Mexico (1990), and Venezuela (1991), spread throughout much of Latin America, in most cases followed much later by the opening of markets to new entrants and competition (Wellenius, 2000b). Rapid progress was also achieved from the mid-1990s in Hungary, Poland, Estonia, and some other Central and Eastern European countries (Bruce et al., 1999). Asia and sub-Saharan Africa got off to a slower start, often focusing on new entry rather than on changing the ownership of existing state enterprises (Wellenius and Stern, 1994). Reform in North Africa began in the late 1990s, notably in Morocco (1996). Even so, by 2003, about one-half of all developing countries, including most of the oil-exporting countries in the Middle East, had reached only the very early stage of telecommunications reform, and some not even that.

[1] The progressive opening in the United States of the markets for telecommunications equipment and networks, beginning in the 1960s and culminating in the breakup of the Bell System in 1982 (see Geller, 1989), was followed by privatization and the beginnings of competition in the United Kingdom (1984), Japan (1985), Australia (1991), and New Zealand (1991).

3.1. Directions of change

Telecommunications reform in the developing world, as in high-income countries, was driven primarily by global technological innovation coupled with expanding market demand. Several other factors also pushed the reform:

- Call congestion, inefficient production, overall high prices, unreliable and antiquated services, and decades-old major shortfalls in meeting demand for connections.
- The inability of state enterprises to finance the large investments required to catch up with demand and develop new and better services.
- The adoption by governments of more liberal and proprivate economic strategies, coupled with public dissatisfaction with services and willingness to change.
- Awareness of new sector models being tried out in some high-income countries.
- The aggressive search for new business opportunities abroad by telecommunications companies in mature economies facing challengers in their own markets.
- The shift in international commercial banks' exposure in highly indebted countries from nonperforming loans to new investment opportunities.

Reform in developing countries also faced additional constraints:

- Incomplete telecommunications infrastructure, which reached only the larger cities and excluded populous rural and low-income urban areas.
- A scarcity of human resources, especially experienced managers, accountants, and information systems specialists needed to run telecommunications as commercial operations.
- Limited or poor-quality enterprise information, especially on accounting, finances, physical assets, network utilization, and customer demand.
- Undeveloped local capital markets, including small or nonexistent stock markets, costly short-term debt financing, and lack of long-term debt instruments.
- Weak legal and institutional frameworks, including antiquated laws, weak judiciaries, and ineffective and sometimes corrupt public administrations.
- Limited interest from foreign investors, deterred by political and regulatory risk, discretionary taxation, exchange controls, and restrictions on capital repatriation.

Sector reforms typically have comprised changes in three main directions: opening markets to new entrants and competition, increasing private sector participation in ownership and management of the operating companies, and separating policy and regulation from operations. The paths followed in each of these directions, as well as the sequencing among them, vary widely from one country to another and have tended to change over time. Box 1 summarizes the experience of Sri Lanka, which pursued changes in all three directions.

Box 1 A decade of telecommunications reform in Sri Lanka, 1992–2002

In 10 years, Sri Lanka transformed its telecommunications sector from a stagnant government operation into a dynamic, private-led, increasingly competitive market. Between 1992 and 1996, the government issued licenses to four private cellular mobile phone companies, six facilities-based data communications companies using terrestrial and satellite technologies, and two fixed wireless telephone companies. Sri Lanka Telecom, a state-owned corporation established in 1991 to take over the operating assets of the government's Telecommunications Department, was incorporated in 1996 as a public limited liability company, Sri Lanka Telecom Limited (SLTL), and partially privatized in 1997. At the same time, the Telecommunications Regulatory Commission was created.

By 2002, about 30 new companies had entered the market and were providing Internet, paging, pay phone, leased circuits, trunk mobile radio, and other services. These structural changes brought rapid growth and improved performance in telecommunications. The SLTL expanded sixfold, from 0.13 million phone lines in 1991 to 0.77 million in 2002, while new mobile and fixed wireless operators added more than 1 million connections.

The government published a new national policy for the information and communications technology sector in 2002, aimed at making affordable and effective communications choices available to everyone. It issued new interconnection rules, overhauled spectrum management, issued a class license for international gateways when SLTL's exclusivity expired, and authorized about 30 gateway operators in the next few months. It also launched public consultations on draft legislation to establish a common regulator for telecommunications, broadcasting, and other electronic communications networks and to set up a fund to accelerate the development of communications infrastructure, including broadband.

In 1997, following international competitive bidding, the government sold 35 percent of the shares in SLTL to Nippon Telegraph & Telephone Corporation (NTT). It granted SLTL employees 3.5 percent of the company shares, of which about 0.2 percent were later sold to NTT. A shareholders agreement between NTT and the government gave NTT effective management control of SLTL by allowing it to appoint the chief executive officer, who had broad authority to run the company. A services agreement, under which NTT provided management and technical support to SLTL, expired in 2002, but NTT continued to provide the chief executive officer and chief financial officer under a secondment agreement. In 2002, the government sold another 12 percent of state-owned SLTL shares through an initial public offering on the Colombo and Luxembourg stock exchanges. The offering was oversubscribed, mainly by retail domestic investors, and another 3 percent of the SLTL shares were sold. After the offering, the government held 46.5 percent of the SLTL shares, NTT 35.2 percent, the public 15.0 percent, and employees and others 3.3 percent.

3.2. Entry and competition

Policymakers increasingly recognized that a single operator, whether state owned or private, is unable to meet equally well the large, varied, and rapidly changing demands of all types of users. Opening markets to new entrants and promoting competition thus gradually became the main thrust of sector reform. In some cases, the new operators only complement the incumbent, for example, by providing mobile, or Internet services that the established operator does not offer. Generally, however, new entry is accompanied by increasing competition between the new entrants and the incumbent and among the new entrants.

The argument for new entry and competition is that they spur growth, innovation, and productivity, resulting in more service, better service, new services, and less-expensive services. Competition, or a credible threat of competition, forces established operators to focus on customers, improve service, accelerate network expansion, reduce costs, and lower prices. As new technologies change network cost structures and reduce the minimum sustainable scale of operations, entry in all market segments becomes possible. The traditional idea of natural monopoly loses currency.

The prospect of letting in new players and allowing them to compete with the incumbent and among themselves, however, upsets a long-standing situation and can raise objections or concerns among stakeholder groups. New entry, some fear, will undermine the financial viability and even survival of the incumbent. Fiscal revenue and foreign exchange earnings from telecommunications will decline. Foreign investors will demand exclusive rights. Rebalancing the incumbent's tariffs toward cost so that it can face competition will take time and may hurt low-income users. Network rollout and universal access will suffer as operating surpluses and cross-subsidies are competed away. New entrants will limit their business to the most profitable urban and business market segments. Competition will result in wasteful duplication of networks. Small markets will be unable to sustain multiple operators. In practice (as we shall see in this chapter) as well as in principle, these concerns either have proved unwarranted or have workable solutions. Still, these arguments often figure prominently in the policy and political debate surrounding sector reform.

The strategies pursued by developing and transitional economies moving from monopolies to fully competitive markets have thus ranged widely. Early reformers were driven mainly by the need to improve the performance of existing enterprises and by fiscal considerations. Latecomers have been more interested in benefiting users and have pursued competition as an objective in itself. They have had at their disposal more technological options, especially fast-growing mobile and other wireless solutions and the Internet, as well as proven policy tools and the examples of other developing countries (Rossotto et al., 2003). Reformers typically have not opened all service and network market segments simultaneously, instead often opening new services (mobile phones) first and delaying to the last those, from which the incumbent derives most of its rents (international gateways).

Yet some countries have chosen to open markets quickly, imposing minimal restrictions on services, networks, and business models, and no obligations beyond those undertaken on commercial terms between companies and customers. Notable examples include these:

- Alone among the precursors of telecommunications reform, Chile never granted exclusive rights to any operating company. Its former state enterprises, privatized in the late 1980s, always faced the threat of competition. Arguably, this was a major factor in enabling the Chilean telecommunications sector to become, and to remain, the best-performing one in Latin America and highly regarded worldwide. Not until 1994, however, did Chile finally dismantle the remaining legal and regulatory barriers to full competition, after repeated rounds involving the sector authorities, the competition authorities, and the courts of justice.
- El Salvador opened all segments of its telecommunications market to new entry and competition in 1997, with automatic licensing of network operators on application, no limits on the number of operators, no distinctions among or restrictions on lines of business or technology, and no unprofitable service obligations.

Most countries, however, have chosen to liberalize only gradually, opening specific market segments and imposing important restrictions for fairly long periods, typically 5 years and sometimes as many as 10 years. These market restrictions often include limiting the number of operators, distinguishing among lines of business and technology, reserving public voice service for the incumbent, prohibiting the transmission of voice over public data networks and the interconnection of private and public networks, limiting new entrants' right to build their own networks, or use infrastructure other than that owned by the incumbent, and limiting foreign ownership. Competition in international service has often arrived last, slowed by the incumbent operator's interests and political power, the government's reliance on monopoly rents to finance budget deficits, and inaccurate assessment of the fiscal impact (Rossotto et al., 2004). The following are some examples of gradual liberalization strategies:

- Argentina divided the countrywide monopoly into two regional domestic companies in 1990, cutting across the major Buenos Aires market, which created comparative competition between the companies and the potential for direct competition later. It granted the two companies 7 years exclusivity with a possible extension to 10 years for domestic telephone service and, through a jointly owned subsidiary, for international service.
- Mexico opened its domestic and international voice markets in 1996 without limits on the number of licenses. But it prohibited international simple resale,

required all operators to adopt uniform international settlement rates negotiated by the dominant operator, and allocated incoming international traffic among operators in proportion to their outgoing traffic. These measures excluded competition for terminating incoming calls and limited competition for outgoing calls.

- Malaysia licensed five new operators in 1995 to compete with the incumbent in all market segments, including mobile and international. The government resisted subsequent pressures to reduce the number of players, but it has not yet fully opened the market.
- Ghana opened its mobile market around 1993 to several independent operators and in 1996 licensed a second national operator, resulting in a 5-year duopoly in fixed domestic service and all international traffic.
- The Philippines, with a nominally competitive market but continued dominance by the largest operator and sustained service shortages, licensed eight additional international gateways and mobile operators in 1995, subject to substantial obligations to build out fixed service in underserved areas.
- Morocco issued a second mobile license to an operator authorized to build its own domestic infrastructure in 1999 and an international gateway 2 years later. This was followed by several satellite (VSAT and GMPCS), and Internet service provider (ISP) licenses (though with service and network restrictions). The government established a timetable for issuing additional licenses aimed at having at least three operators competing in each major market segment by 2003.
- Honduras issued a second mobile license in 2003 and committed in a bilateral trade agreement to end by 2005, the exclusive rights it had granted its state operator by law in 1995.
- Saudi Arabia in 2002 announced plans to liberalize its telecommunications market by 2010.

3.3. Private sector participation

In the beginning, telecommunications reform in developing countries focused primarily on privatizing state enterprises. Access to capital, management, and technology to help overcome sustained shortfalls of the state enterprises, coupled with the prospect of windfall revenues from their sale and in some cases reduction of sovereign debt, became the principal drivers of early initiatives to transfer ownership and control of state enterprises to the private sector. New entry and competition were left largely to the future. Some countries joining the reform movement later, however, found that sectoral change could begin by opening the markets to new entrants to complement, rather than compete with, the incumbent. For these countries, privatizing state enterprises, often regarded as politically more difficult, could wait. A range of reform strategies thus evolved, with different methods and sequences of privatization and liberalization.

The case for placing the provision of telecommunications services squarely in the private sector stems from the constraints on performance imposed by public ownership and management. State enterprises have limited access to capital markets, technology, and management experience. Their procurement, personnel, and administrative practices, designed for government administration rather than business, handicap management. The enterprises are vulnerable to political interference and to having their coffers raided by the treasury. And they are shielded from market incentives to perform and change.

Some progress in overcoming these constraints under state ownership was achieved in several countries, especially from the mid-1960s to the end of the 1970s. Operating arms of government departments were transformed into state enterprises with somewhat greater autonomy (as in Costa Rica in 1965). Investment was accelerated with financing from bilateral and multilateral development organizations (as in India, which received seven World Bank credits and loans in 1965–1980). Technical assistance was provided by consultants financed by these sources. Human resource development was supported by international organizations, particularly by the International Telecommunication Union (ITU) with financing from the United Nations Development Programme (UNDP). Some enterprises were twinned with experienced foreign operators (as in Ethiopia, with the Swedish telecommunications administration in 1960–1980). During this period, quite a few state enterprises achieved significant improvements in internal organization and management, including strengthening technical planning, reorganizing along functional lines, improving billing and collection, and establishing or improving commercial accounting and management information systems.

By the mid-1980s, however, it had become clear that the model of state monopolies had run its course. New approaches, involving the private sector and competition, were becoming viable alternatives (Wellenius et al., 1989). Gradually it was accepted that telecommunications operating enterprises, regardless of who owns and manages them, perform best when run as profit-driven commercial businesses in a competitive environment. It also became clear that ownership does matter and that eventually, if not up front, state enterprises must be privatized to function effectively in competitive domestic and international environments. Seldom and only for limited periods had it been possible to achieve, under state ownership conditions that approximated the freedoms, incentives, and market discipline of privately owned and managed commercial enterprises.

Nonetheless, there has often been strong opposition to moving telecommunications operations from the public to the private sector. In most developing countries, these operations were (and many still are) the exclusive preserve of a single enterprise regarded as a prime asset of the people regardless of how poorly it met service needs. When considering the prospect of privatizing these state enterprises, governments face losing discretionary control over a major source of revenue, foreign exchange, employment, and political patronage. Workers fear layoffs and reduced political power of labor organizations. Policymakers worry that private companies will not respond to broad economic and social development needs of the country, and will further social inequality by concentrating on large-business users and high-income urban consumers. They also worry that private ownership alone does not provide sufficient incentives for growth and efficiency, and that privatized monopolies may be more difficult to regulate than when under state ownership. Foreign ownership and control raise additional concerns, including security and the desire to promote domestic companies[2]. Together with new entry and competition, these concerns feature prominently in the policy and political debates on reform.

The main method of privatizing state telecommunications enterprises has been the sale of a controlling interest in the enterprise to a consortium led by an experienced foreign operating company as a strategic investor. This has often been followed by an initial public offering in international markets and sometimes also in the domestic market. In addition, small numbers of shares have been sold to company employees. The transfer of a controlling interest has taken a variety of forms:

- In Chile, a majority of shares in the main local phone company were purchased in 1988 by a foreign portfolio investor, which sold to a major operator 2 years later. Shares in the main long-distance company were sold to domestic and foreign investors, one of them a major foreign operating company that eventually acquired a controlling interest. Small numbers of shares in both companies had been sold earlier to employees and the public.
- In Argentina in 1990, the assets and some liabilities of the state monopoly were transferred to the two new regional companies created by dividing the market roughly in half, and 60 percent of the shares in each of these companies were sold to different consortia led by foreign strategic investors.
- In Bolivia in 1994, a capital increase doubling the number of shares in the state-owned dominant national operator was placed with a strategic foreign investor, with the proceeds of the sale staying in the company for investment. The existing state-owned shares were used to endow a newly created national pension system.
- In Mexico in 1990, the state operating company's capital was restructured into three categories of shares, with different voting rights and ownership requirements, so as to allow wide foreign participation yet keep control in Mexican

[2] The concerns about private ownership and management mirror the main arguments used in Latin America in the 1960s to nationalize private (mainly foreign-owned) telecommunications monopolies and create new state-owned monopolies. With hindsight, it could be argued that the problem was not private ownership but lack of competition. Public utility monopolies, however, reflected established practice worldwide at the time. Getting the state into the business of providing services was in line with telecommunications practice in most countries (with the notable exception of the United States), and with the prevailing economic development doctrine promoted by influential regional agencies such as the United Nations Economic Commission for Latin America and the Caribbean.

hands. Fifty-one percent of shares with full voting rights (accounting for only 20 percent of all shares) were sold to a consortium led by a Mexican group with two foreign operating companies as minority partners. About 5 percent had been sold to employees. The rest were placed in domestic and international capital markets over several years.
- In Sri Lanka, 35 percent of shares in the state-operating company were sold to a foreign operator in 1997, and another 15 percent were placed in domestic and foreign capital markets in 2002. A shareholders agreement gave management control to the foreign operator (see Box 1).

Sale of a minority of shares in the state enterprise to private investors without management control has seldom been used. Although this method offers some potential benefits for sector development, such as subjecting the enterprise's finances to regulatory scrutiny, and may meet a demand for quality instruments to develop the local equities market, it does not fully remove the enterprise from its public sector limitations.

New entry has been the preferred option for attracting private participation in countries, where opening the market to new operators that would cooperate with and complement the incumbent, has been seen as politically more expedient than selling off the state enterprise. New entry took on increasing importance as a vehicle for private participation as the appetite in international markets for investing in telecommunications (and in developing countries) declined at the end of the 1990s. It is also the main option in countries, where privatization of the state enterprise has failed but there is investor interest in smaller opportunities through new entry. Management contracts without an ownership interest or major investment commitment have been used sparingly. The following examples illustrate the diverse methods of introducing private participation through new entry:

- Uganda licensed a private mobile operator in 1995, established an autonomous regulatory authority in 1998, and in the same year licensed a second full-service private operator. It privatized the incumbent in 2000. Uganda has become one of the fastest developing telecommunications markets in sub-Saharan Africa.
- Thailand's state-owned domestic monopoly telecommunications enterprise in 1992 granted two build-transfer-operate concessions to investment consortia, including foreign operators, to expand telephone service in Bangkok (adding 2.0 million lines), and provincial cities (1.5 million lines), more than doubling the existing numbers. The consortia paid the incumbent up to 30 percent of annual profits for these concessions.
- In Indonesia, joint operating arrangements were established in 1995 between the state telecommunications monopoly and five consortia led by major foreign operators. These give the consortia exclusive rights to manage existing fixed-telephone service in major provincial areas for 15 years while obligating them to add about 1.8 million new lines. The consortia pay the incumbent 30 percent of revenue, subject to a minimum annual amount.

- In Ecuador, following a failed attempt in 2001 to sell 35 percent of shares with management control in each of the two state-owned regional fixed-telephone operators, the government offered management contracts without major investment commitments, but had no takers. It eventually negotiated a management contract for a mobile license jointly owned by both operators.
- Honduras in 2003 announced a program for the state monopoly to award contracts to private companies that would provide, on a nonexclusive and largely unregulated basis, services for which it has exclusive rights until 2005. The private companies, which could include two mobile phone companies, several cable television operators, and data and ISPs, would market the services under their own names, develop their own networks, and be eligible to apply for concessions as stand-alone operators after 2005.

3.4. Policy and regulation

In the transition from state monopoly to private and competitive market structures, regulation is needed to contain abuse of market power, foster competition, create a favorable investment climate, and narrow development gaps (Smith and Wellenius, 1999). Under state monopoly, the incumbent operator itself was typically responsible for regulatory functions. These may have included licensing new entrants (the first mobile operator in several countries was authorized by the incumbent fixed operator), managing the radio spectrum and the numbering system, and assigning rights to exploit the country's satellite orbital positions. While operations remained fully in the public sector, the political system provided, however imperfectly, for reconciling such diverse objectives as commercial efficiency and broader national and regional development (Nulty and Schneidewinde, 1989). But as reforms sought to reduce state ownership and government control and new operators—mostly private and often in competition with the incumbent—were authorized, this arrangement became a major impediment to change.

One of the first steps in most reform processes has therefore been to distance the government's policy and regulatory responsibilities from those owning and managing the operating enterprises. Policy remained with a line ministry, and regulation was often vested in a newly created agency. The institutional solutions for sector regulation have varied among countries. Sector-specific regulatory agencies have been the norm. For example, Morocco's Agence Nationale pour la Régulation des Télécommunications (ANRT) was established by law in 1997 as a key step toward the liberalization envisaged in the law. The ANRT is a public entity with a degree of administrative and financial autonomy. The regulatory authority rests with the board of directors, consisting primarily of government officials and chaired by the prime minister. The director general, appointed by the king as head of state, runs the agency day-to-day and proposes major decisions, including annual plans for market liberalization, to the board. The success in awarding a second mobile license in 1999 was attributed largely to the existence of a credible regulatory framework and institution, the transparent tender process, and the attractive terms of the license-all closely related to ANRT's effective start-up (ITU, 2001a; Wellenius and Rossotto, 1999).

In addition, multisectoral regulatory entities are being tried as a way to use scarce human resources more efficiently, attract prime-quality top regulators, and achieve consistency in regulatory policy and practice across wider segments of the economy. The jury is still out on the relative merits of this approach, but practical experience is slowly building up (Samarajiva et al., 2002). In Latvia, the Public Utilities Commission was established in 2001 to regulate gas, posts, railways, electricity, and telecommunications. It is responsible primarily for licensing, setting tariffs, resolving disputes, controlling quality, and protecting consumer interests. The line ministry retains responsibility for formulating sector policy, drafting cabinet resolutions, specifying the general financial principles of universal service, and representing the country in international forums. A separate agency is responsible for managing the radio spectrum.

4. Results

In less than 15 years, the telecommunications scene in developing countries has changed considerably. About 40 percent of these countries are now open, or moderately open to new entry and competition (Table 2). About 50 percent have private participation in the ownership and management of former state enterprises, and all new entrants are private companies (including some owned primarily by state enterprises from other countries). About 60 percent have set up separate regulatory authorities[3]. These relatively open, private-led markets account for about 70 percent of the global population and 75 percent of the world market as measured by GDP[4].

Most markets are even more competitive than these figures suggest. Mobile phones are extensively used as substitutes for fixed phones, to the point that in many countries today there are more mobile than fixed phones in use. Low-cost alternatives to conventional international phone service, such as call-back, calling cards, and Internet telephony, even if illegal in many countries, are hard to stop and account for a growing share of traffic[5].

Not all market segments have opened at the same pace. The vast majority of developing countries now allow competition in private networks, public data services and networks, mobile phone services and networks, and Internet services. Many countries, however, continue to have monopolies in public fixed-telephone services

[3] Estimated from International Telecommunication Union data for 1999 (see ITU, 2001b).

[4] Calculated from International Telecommunication Union data for 2001 (see ITU, 2002b).

[5] Reportedly, about 20 percent of public telephone traffic from the United States to Mexico bypasses routing and settlement charges prescribed by Mexican regulations.

Table 2
Developing countries by openness to new entry and competition in telecommunications, 1999

Market status	Share of developing countries (%)	Examples
Open	17	Chile, Czech Republic, Malaysia, South Africa, Sri Lanka
Moderately open	26	Bolivia, Hungary, Morocco, Thailand, Uganda
Slightly open	34	Bangladesh, Egypt, Russian Federation, Uruguay, Vietnam, Zimbabwe
Closed	23	Algeria, Armenia, Cameroon, Haiti, Myanmar
All	100	

Source: Rossotto et al., 2003.
Note: Markets are assessed on the basis of whether there is effective competition in fixed and mobile telephone service and in leased lines, the markets are open to foreign investment, and procompetitive regulation is in place. The data are based on 151 countries in the International Telecommunication Union's regulatory survey of 1999.

and networks. Incumbents often have a stranglehold on the supply of leased lines, especially international lines. Mobile and ISPs are seldom allowed to develop their own infrastructure, or to use that of alternative providers (railways, power utilities).

The progress of competition has also been uneven among world regions, although the differences are less dramatic than one might expect given differences in income and economic development. Latin America remains more competitive than the rest, followed closely by South Asia, and more distantly by East Asia and the Pacific, the transition economies of Europe and Central Asia, and sub-Saharan Africa. The markets in the Middle East and North Africa remain, by a wide margin, the least competitive (Rossotto et al., 2003).

Sector reforms have changed the market structures in some countries substantially. In Mexico, for example, reforms between 1989 and 1996 resulted in a clear separation of policy, regulatory, and operational responsibilities; the privatization of the state operator; the licensing of four mobile operators and three facilities-based fixed service operators, including long distance and international; and a proliferation of value added, Internet, and other new public operators as well as major growth of corporate networks (Figure 1).

These structural changes have had an impressive impact on sector performance. Investment accelerated, existing services grew fast and became available on demand, technological innovation and increased labor productivity made services much more efficient, prices declined, and major new services were introduced.

4.1. Investment

Sector reforms resulted in rapid investment growth. In 1991–1998, about $488 billion were invested in telecommunications in developing countries, or about $60 billion a year—accelerating from about $30 billion a year in the early 1990s to

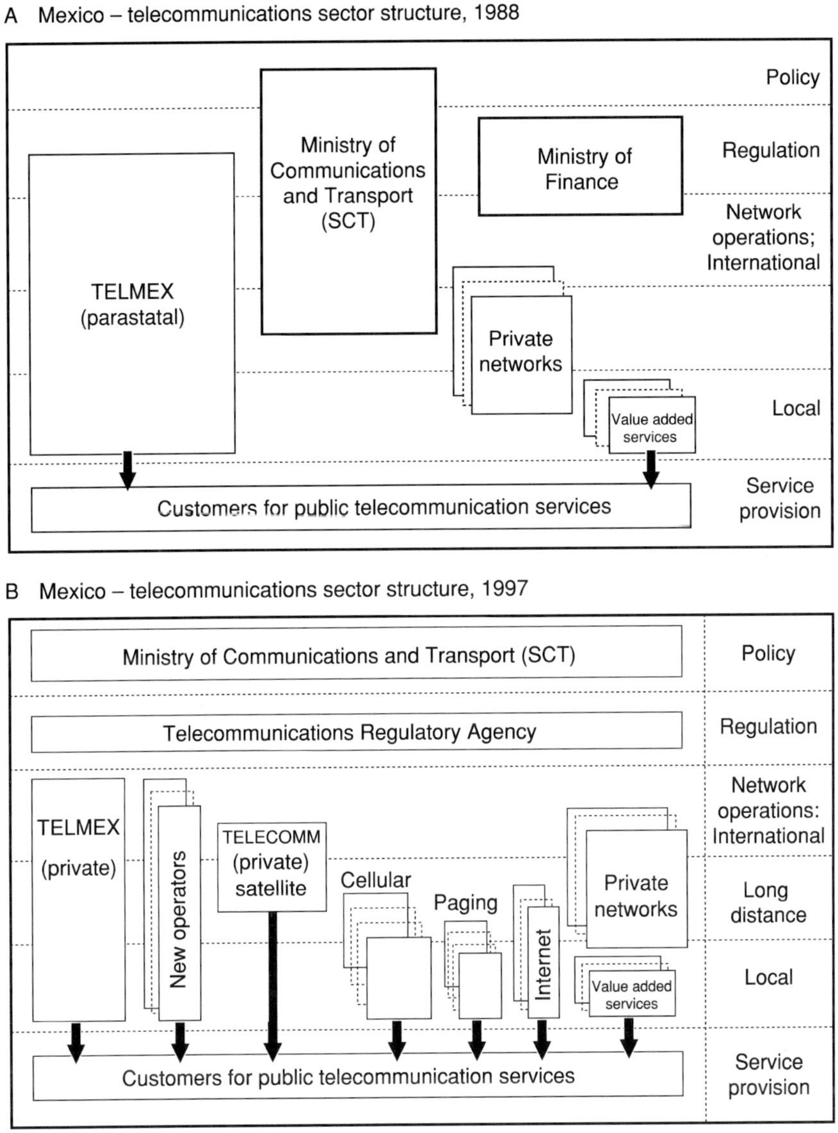

Fig. 1. Mexico – telecommunications sector structure, 1988.

about $95 billion in the late 1990s (Wellenius et al., 2000). This compares with an annual average of $17 billion in the 1980s (Pyramid Research, 1995).

About 40 percent of these telecommunications investments in the 1990s involved the private sector, increasing from about one-third in the early 1990s to about one-half in the late 1990s[6]. Investment in telecommunications projects with private participation in developing countries averaged $24 billion a year in the period 1990–1998, increasing from less than $10 billion in 1990 to more than $50 billion in 1998—for a total of $214 billion in 521 projects during this period. The bulk of these investments took place in Latin America and the Caribbean (52 percent), followed by East Asia and the Pacific (20 percent), Europe and Central Asia (15 percent), and South Asia (9 percent) (Izaguirre, 1999). After 1998, investment in telecommunications projects with private participation in developing countries declined in all regions except East and South Asia, falling by 2002 to the lowest level since 1995. The decline was due in part to the deterioration in global telecommunications markets. But it was also due to a slowdown by mobile operators after years of heavy investment in rolling out their networks and to the achievement of meeting much of the pent-up demand for fixed-line connections (Izaguirre, 2004).

The private sector invested in both mobile and fixed networks. Between 1990 and 1998, more than 300 private companies initiated mobile service on a stand-alone basis, or along with fixed service in 94 developing countries. Mobile service accounted for a large part of the investment in the years after 1998. The private sector also invested in long-distance companies in 42 countries and in local companies in 55 countries. Management contracts with major capital expenditure were used only sparingly, mainly in build-transfer-operate arrangements in Thailand and provincial revenue-sharing projects in Indonesia. Management contracts without major investment commitments were even rarer (Kiribati, Mongolia).

About one-third of the developing countries, where the private sector invested, had competitive markets (Table 3). By 1998, 28 of the 94 countries, where the private sector had invested in mobile service had between three and six competing mobile operators, and 38 had duopolies. Of the 42 countries, where private

Table 3
Developing countries with private investment in telecommunications, by market segment, 1990–1998

		Countries with private investment and competitive markets	
Market segment	All countries with private investment	Total	As percentage of all countries with private investment
Mobile[a]	94	28	30
Long distance	42	12	29
Local	55	15	27

[a]Stand-alone and together with fixed operations.
Source: Izaguirre, 1999.

[6] This is not a measure of the share of investment financed by the private sector.

investors entered the long-distance segment, 12 allowed unrestricted entry (Chile, Mexico), limited competition (Malaysia), or duopoly (Brazil, Ghana), while the rest awarded exclusive licenses to privatized incumbent operators. Of the 55 countries with private operators in fixed-local phone service, 15 allowed free entry (Chile, El Salvador, Guatemala, Mexico), limited competition (Malaysia, Sri Lanka), or transitional duopolies (Brazil, Ghana, India). The other 40 countries either awarded monopoly rights to privatized incumbents (Côte d'Ivoire, Estonia, Pakistan, Venezuela), or introduced private investment to complement the incumbent's (Bangladesh, Indonesia, Thailand) (Izaguirre, 1999). The trend of increasing competition has continued in the early 2000s.

4.2. Growth, productivity, and prices

The faster pace of investment resulted in accelerated growth in existing services, higher productivity, overall lower prices, and new services (Table 4). For developing countries as a group, the annual growth in fixed telephone main lines, which had hovered around 7–10 percent in the 1970s and 1980s, shot up to 16 percent in the second-half of the 1990s. Almost 50 million lines were added annually in 1995–2001, compared with less than 10 million annually in the 1980s. By 2001, there were half a billion phone lines in use in developing countries, almost half the world's total. Rapid expansion provided opportunities for adopting modern technologies and achieving major productivity gains. By 2001, about 92 percent of phone lines were connected to digital local switches. Labor productivity more than doubled between 1995 and 2001, from about 77 main phone lines per employee to 180.

Large reductions in domestic and international long-distance charges resulted in overall lower prices for most categories of users. The impact has been especially strong on international service prices, fueled by competitive pressures for reform of the international accounting rates system. As El Salvador prepared to open its telecommunications market to competition in 1996, the regulator reduced the maximum price for calls to the United States from more than \$1 a minute to \$0.80. Since then, the main operator's published price has come down to \$0.25, and most operators are offering even lower tariffs.

Rapid growth cleared the backlog and made it possible to meet the large new demands that appeared with increasing speed as service was seen to be available and prices declined[7]. Unmet demand was reduced from several years' growth in the 1980s to a few months' by 2000 on average. Countries that opened their markets and increased private participation fared much better than these averages.

[7] Outstanding applications for new lines underestimate demand by a wide margin, as first documented in Chile in 1968. Questions on telephone service attached to a household survey on employment in Greater Santiago showed that unmet demand was about four times the number of pending applications (see Wellenius, 1969).

Table 4
Telecommunications in developing countries, 1970–2005

Indicator	Low- and middle-income countries								High-income countries	
	1970	1975	1980	1985	1990	1995	2000[a]	2005[b]	2000[a]	2005[b]
Telephones										
Main lines										
Total (millions)[c]	22	35	55	78	124	211	502	581	543	571
Annual growth (%)	—	10	10	7	10	11	16	4	3	1
Mobile phones										
Total (millions)	0	0	0	0	0	14	393	1013	548	752
Annual growth (%)	0	0	0	0	0	—	74	27	39	8
All phone connections										
Total (millions)	22	35	55	78	124	225	895	1594	1091	1323
Annual growth (%)	—	10	10	7	10	13	26	16	13	5
Per 100 people	0.7	1.0	1.5	2.0	2.9	4.8	17.4	29.1	120	144
Outstanding applications (years)[d]	—	>10	7.0	6.5	4.4	2.0	0.3	—	0.0	—
Network digitalization (%)[e]	—	—	—	—	50	71	92	96	96	97
Productivity (lines/employee)[f]	—	—	—	—	—	77	180	220	240	280
Homes with fixed phones (%)[g]	—	—	—	—	—	14	33	—	>95	—
Internet										
Users (millions)	0	0	0	0	0	2.6	135	—	—	—
Hosts (millions)	0	0	0	0	0	0.3	6.3	—	—	—
Population (millions)	3060	3340	3640	3970	4340	4720	5150	5470	900	920

— Not available.
[a]Data are for 2001.
[b]Projected.
[c]Data for 1970–1995 are from Pyramid Research, 1995 and exclude Hong Kong, the Republic of Korea, Singapore, and Taiwan.
[d]Outstanding applications for fixed phones divided by average number of fixed main lines, plus mobile phones added in the previous 5 years. Average for high-income countries in 1996 was 2.3 days. Figures for low- and middle-income countries in 1980 and 1985 are estimates based on outstanding applications as a proportion of main lines in service in 40 countries in the period 1979–1982 (see Saunders et al., 1983, pp. 12–13).

By 2000, fixed-phone service had largely made up the past major shortfalls. Developing countries as a group realized many of the potential gains that had been blocked by outmoded sector policies and structures. Further gains are likely to result from advances in reform in countries still at an early stage rather than from additional improvements in countries that have already reformed. The aggregate growth rate is expected to decline after peaking in the late 1990s. Projections envisage another 20 million lines annually, for a total of about 580 million by 2005. Productivity may then reach the level found in developed countries today.

4.3. New services—mobile and the Internet

Competition and private participation resulted in a wide range of innovative commercial packages to meet different user needs for existing services. The greatest impact on sector development, however, came from the introduction of mobile service and, more recently, of easy access to the Internet.

Sector reforms enabled, and were propelled by, the introduction of mobile phone service from the early 1990s. Mobile networks were built, often in the regulatory vacuum of low-income countries, and quickly established themselves as an alternative to fixed service in short supply. Less important than mobility was availability on demand, even if initially at a high price.

In less than a decade, mobile service caught up with a century of fixed-telephone development. Between 1995 and 2001, 400 million mobile phones were added in developing countries. By 2001, they accounted for 44 percent of all connections (fixed and mobile), and in 24 countries, there were more mobile than fixed phones[8]. Taking fixed and mobile together, telephone growth accelerated from 7 to10 percent a year in the 1970s and 1980s to 20 percent in the 1990s. In contrast

[8] Here and elsewhere, we use phones to refer to either fixed-main lines or mobile handsets.

---

[e]Percentage of telephone main lines connected to digital local exchanges. Authors' estimates for 2000 and 2005 assume that all lines added after 1995 are digital, and that one-third of analog lines remaining at the end of the previous period are converted to digital.

[f]Number of telephone main lines divided by number of telephone company employees. Authors' estimates for 2000 and 2005 assume that the number of employees in 1996 declined annually by the average rate in 1990–1996.

[g]Authors' estimates based on data from ITU World Telecommunications Development Report, 2003 and earlier issues. Estimates assume that households with telephones have an average of 1 line in low-income countries, 1.05 lines in lower-middle-income countries, and 1.1 lines in upper-middle-income countries. Figure for high-income countries is illustrative based on International Telecommunication Union data for about 20 countries.

Source: All telecommunications data are from ITU World Telecommunications Development Report, 2003 and earlier issues, and ITU Yearbook of Statistics, 1986–1995, except where otherwise noted. Population data are taken or estimated from the World Bank Atlas, 2002.

Note: Annual growth rates are calculated for the 5-year period ending with the year shown.

with the projected leveling off of growth in fixed phones, mobile service is expected to continue to grow at a fast clip in the 2000s. By 2005, mobile phones are likely to outnumber fixed phones by almost 2 to 1. Of a total projected 1.6 billion connections in developing countries, 1 billion will be mobile. More than one-half of all mobile phones worldwide will be in developing countries.

The impact of mobile service is not just in the sheer numbers but also in the extent, to which it has quickly reached unserved population groups. The second mobile operator in Morocco reached 70 percent population coverage 1 year after launching and more than 90 percent in 2 years. In Chile, 41 percent of rural households had a mobile phone in 2002, while only 9 percent had a fixed phone (SUBTEL, 2003). Low-income customers, who are ineligible for standard telephone service because they are not creditworthy, or who are unwilling to commit to fixed-monthly payments, or need to keep close control over spending, can opt instead for prepaid mobile service. Even if they place few calls themselves, they benefit from the ability to receive calls and generate profitable new traffic from the fixed network. Prepaid customers now exceed those under monthly contract in most developing countries, and almost all new customers being added to existing networks are prepaid. In Chile, prepaid customers accounted for 75 percent of all mobile phones in 2001 and for 86 percent of all mobile customers added that year (SUBTEL, 2002). The shares are somewhat higher in lower-income countries. In Africa, 80 percent of the 34 million mobile phones are prepaid. And in South Africa, more than 90 percent of new connections are prepaid (CellularOnline, 2003). Another, unanticipated area of new service and revenue growth has been in text messaging using mobile phones. This service has become especially popular in many developing countries, particularly in Asia.

Despite the fast growth in mobile service, the bulk of telecommunications traffic and revenue continues to come from the fixed network. In Chile in 2002, mobile phones generated on average 65 minutes of traffic a month, compared with about 650 minutes for fixed phones. Although 62 percent of phones were mobile, they accounted for only 20 percent of all local traffic and 17 percent of long-distance traffic (SUBTEL, 2002).

The fixed-telephone network, through its extensive infrastructure and accelerated growth, played a fundamental part in supporting the explosion of Internet service. About 135 million people in developing countries used the Internet in 2001, 50 times as many as in 1995. There were 20 times as many hosts as in 1995. It is early still to assess the growth of the Internet in developing countries, and these figures are subject to considerable error. But there is no question that the Internet has taken off with a vigor that may be comparable to that of mobile phones a few years earlier. The Internet is making its impact on infrastructure use felt much sooner than did mobile service. In Chile by 2002, the Internet accounted for 24 percent of traffic in local networks and 47 percent in long-distance networks (SUBTEL, 2002). The growth and impact of the Internet in developing countries are discussed further in Section 6 of this chapter.

5. Lessons from reform

More than a decade's experience with telecommunications reform in developing countries suggests several conclusions that are broadly applicable to the design of reforms in countries still at an early stage of the process.

5.1. Competition as driver of change

Competition has proved to be the cornerstone of successful reform. Competitive markets grow faster. Among the Latin American precursors, countries that granted monopoly privileges of 6–10 years to the state enterprises privatized in the early 1990s, saw phone connections grow during the following 5 years at 1.5 times the annual rate achieved by state monopolies, but at only half the rate in Chile, where the government retained the right to issue competing licenses at any time[9]. An econometric analysis of 18 telecommunications privatizations in 15 countries between 1987 and 1998 found that while granting exclusivity can double a firm's sale price, it reduces growth by up to 40 percent over the exclusivity period (Wallsten, 2000). Another analysis, focusing on 86 developing countries in Africa, Asia, Latin America, and the Middle East between 1985 and 1999, showed that privatization and competition accelerated the growth in fixed-telephone service and boosted productivity, but less so when competition was introduced only after privatization (Fink et al., 2002)[10].

Lack of competition keeps prices high. In a large cross-section of developing countries in 1999, limited competition in international service resulted in call charges that were typically twice those in countries with full competition (Rossotto et al., 2004). The long exclusivity periods granted in Latin America at the time of privatization resulted in high prices for domestic and international service, affected basic telephone service, and the emergence of new services even after competition was eventually introduced. In 1995, making a phone call to the United States from Argentina (with a private monopoly), and Brazil (state monopoly) cost four to seven times as much as it did from Chile (fully open), where prices were already within the range of competitive markets worldwide. Placing a call from Argentina, Brazil, and Mexico (also with a private monopoly) to Chile cost two to three times as much as calling in the opposite direction. In Argentina in 1996,

[9] This simple comparison illustrated the importance of opening markets when privatizing state enterprises, but did not control for differences among other factors, such as income and economic growth. The findings are consistent with those of later, more formal econometric studies, however (Wellenius, 1997).

[10] While the study focused only on fixed-phone service and excluded countries that introduced competition before privatizing the state enterprise, the results nonetheless suggest that introducing competition without privatizing the state enterprise would lessen the impact of competition on the growth and productivity of service. This result probably would not apply for mobile service, where experience shows major growth following new entry regardless of what is done with the incumbent.

domestic long-distance call charges were around six times those in Chile, and 64 kbps leased circuits for Internet access providers reportedly cost more than $20,000 a month, compared with $700 in the United States for similar distances (Wellenius, 2000b).

Contrary to popular belief, investors are not inherently opposed to competition, as long as they are not burdened with regulatory uncertainty, unrealistic service obligations, and rigid tariffs and employment rules[11]. This is true even in small, low-income markets. Ghana Telecom was successfully privatized in late 1996, at the same time that a license was awarded for a second full-service national operator and three other mobile companies were already in place (Wellenius, 1997).

The licensing regime is a critical enabler of competition. Many services can be provided without license, perhaps subject only to declaration for the public record and for statistical purposes. Class licenses can be automatically granted to any applicant meeting set criteria. Individual licenses may still be necessary, however, for major operators with persistent market power (the incumbent), for any operators with special rollout or service obligations (the designated universal service provider), and for the use of large segments of scarce radio spectrum (mobile operators), which should be awarded through public competitive bidding whenever demand exceeds supply. These provisions simplify and expedite the authorization process and also reduce the opportunity for discretion, pressure, and corruption. In El Salvador, the telecommunications law of 1997 requires licenses only for using the spectrum, not for operating networks or services:

> "Operators interested in providing telephone services must request ... a license ... which shall be granted automatically ... subject only to compliance with the requirements for registration ... without any limitation of number and location, being possible that more than one license exist in the same geographical area." (SIGET, 1997)

Under this law, network operators are free to establish prices and conditions for end users as well as for one another, but must grant nondiscriminatory access to essential services[12]. Most countries, however, still have in place complex licensing arrangements that discourage, or impede new entry and fragment the market along lines of service and technologies. While this fragmentation may have been customary in the early 1990s, it is inconsistent with today's growing convergence among services, technologies, and business models.

In the early transition to competition, interconnection is the most important area of regulation, and it remains a major issue as markets mature. The basic

[11] Of course, all things being equal, investors prefer protected markets. Investment bankers usually advise governments to throw in some exclusive rights to attract investors and increase the sale price, which appeals to treasuries and also normally leads to a higher fee for the investment bankers.

[12] Defined by law, essential services are interconnection, signaling, caller identification, billing data, number portability, and directory databases.

elements of a credible interconnection regime consistent with internationally accepted principles must be in place from the outset of reform. To this end, sector legislation must establish the right and obligation to interconnect, the basic principles of interconnection (including nondiscrimination, transparency, and cost orientation), the role of the regulator in enforcing these principles, and an outline of procedures for requesting interconnection and handling disputes. Also helpful is to require the incumbent, and potentially dominant new operators, to publish reference interconnection offers (RIOs), approved by the regulator, that establish default technical and commercial terms and conditions applicable to any other operator requesting interconnection. The RIOs are often published before bids are invited for privatization or new licenses, are attached to the licenses, and thus commit both parties in advance. The RIOs are much simpler than those in mature markets and may suffice in the early years. In the absence of reliable cost accounting information, interconnection prices in a RIO can be set quickly and correctly by reference to benchmarks from other competitive markets, adjusted if necessary for major differences in some costs (capital, labor). Also needed from the start are regulatory guidelines for negotiating interconnection agreements among the parties, and a detailed process for handling disputes.

Unbundling the local loop of operators with significant market power in the fixed-public telephone network market has become mandatory in many middle-income countries, especially where the fixed network is already well developed (Chile, Mexico, the Slovak Republic). In these countries, mandatory unbundling can help develop competition in new services, especially broadband Internet, while new entrants build their own networks and alternative infrastructure. More controversial is mandatory unbundling in countries with less mature telephone networks (Albania, Ecuador), where creating incentives to roll out countrywide networks may be more urgent than sharing existing networks among competitors.

Sharing of infrastructure (poles, ducts, towers, manholes, conduits, street pedestals) and colocation (two or more operators using a common space, such as in a telephone exchange to facilitate connection of cables and transmission equipment) can also lower barriers to competitive entry. In addition, they can increase revenues for the incumbent and contain environmental damage and public inconvenience.

Domestic mobile roaming is necessary where operators have limited networks, so that customers have service countrywide. In the early stages of liberalizing mobile service, new entrants are often required to build out extensive networks of their own. This effectively puts pressure on the incumbent to expand and improve service. In later stages, however, new entrants should be allowed to supplement development of their own networks with roaming agreements with other mobile network operators using similar technologies. Whether domestic roaming should be mandatory and regulated, or left to commercial negotiation among the parties is a matter to be examined case by case.

5.2. Narrowing the development gap

Competition and private participation have effectively narrowed the market gap between what the sector actually does and what it can do, if set reasonably free. A development gap may still remain, however—a gap between what the market can do and what the government deems necessary for economic development, regional balance, social equity, cultural integration, security enhancement, or other reasons. Many rural areas and, to a lesser extent, low-income urban areas remain excluded from service. In Chile, for example, fixed-telephone lines tripled to 1.8 million in the 8 years after the state telecommunications companies were privatized in 1988. The average wait for a new connection dropped from 7 years to a few weeks, and after the last legal and regulatory barriers to entry and competition were dismantled in 1994, prices declined and remained low by international standards. Mobile phone service was taking off at a brisk pace, and the beginnings of Internet service were in place. The share of households with a telephone increased from 16 to 24 percent in 1996, and was expected to continue rising quickly[13]. Yet it was estimated that perhaps as many as half a million households would be unable to afford a telephone connection in the foreseeable future. About 15 percent of Chileans lived in localities that lacked even a public telephone. Government action was required to go beyond closing the market gap towards narrowing the remaining development gap (Figure 2).

In circumstances such as these, subsidies may be warranted to extend telecommunications services beyond the market. The economic argument for subsidy is based on the existence of consumption and production externalities, network externalities, and scale economies. Moreover, affordable access to telecommunications and other rural infrastructure services (such as water, transport, and electricity) is considered essential to enable the population to participate equitably and effectively in modern society (see Wellenius et al. (2004)). But what services

[13] According to the 2002 population census, about 54 percent of households have fixed telephones and 51 percent have mobile phones (including 41 percent of rural homes). See SUBTEL, 2003.

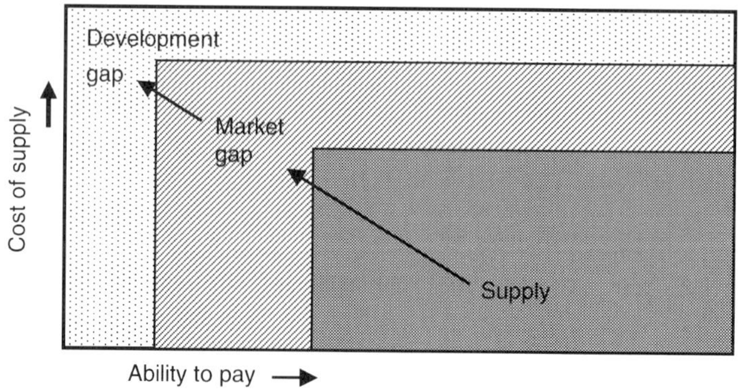

Fig. 2. Narrowing the access gap.

should become available and affordable beyond the limits of the market? Who should provide them? How much will it cost? Who should pay? These questions lie at the heart of all contemporary approaches to universal service, in developed and developing countries alike (Wellenius, 2000a).

Developing countries have sought to achieve universal access mainly to voice service, through communal facilities within reach of everyone rather than through service in all homes. In the mid-1990s, Burkina Faso, near the low end of the range of network development and with only 0.5 telephone lines per 100 inhabitants in 1994, aimed to have pay phones within 20 kilometers of most people. Chile, with about 14 phone lines per 100, sought to install pay phones in rural communities with more than 60 inhabitants (12–15 families) in the three lowest quintiles of income distribution and located at least 3 kilometers, or 30 minutes from the nearest public telephone.

More recently, an increasing number of developing countries have been adopting more ambitious targets, including public access to the Internet and broadband communications and information services. In 2000, the Philippine government announced an initiative for private companies to provide pay phone service within reach of about 30 million rural inhabitants (40 percent of the country's population) living in some 34,000 barangays (municipal subdivisions) lacking access to any telecommunications service. The initiative would also provide public access to computers, the Internet, and other information and communications technologies and services through multipurpose community telecenters in all 1525 municipalities. Extending voice or more advanced services to homes continues to be left largely to the market[14]. Improving public access for disabled users has received little attention.

A set of good practices has evolved, based primarily on experience in Latin America, for extending access to basic voice communications somewhat beyond the limits of what the market will do on its own (Box 2). Public funds are used to catalyze additional private investment as needed to reach government targets. Subsidies are limited to what is necessary to make socially desirable projects commercially viable, not to exceed initial capital costs[15]. Subsidies, typically around 1 percent of total sector revenue, come primarily from the national budget, or small mandatory contributions from the operators. Other sources of funding used or proposed, include spectrum user fees, proceeds from the sale of licenses, and international development grants and credits. Competition among operators for these subsidies, not a calculus of costs and revenues, determines how much subsidy is needed and who should receive the funds. Chile, Colombia, Guatemala, and Peru have all used this subsidy approach in rural pay phone programs

[14] A unique exception has been Colombia, where a program to provide residential telephones in low-income urban areas cross-subsidized by business and high-income households has been in place for a long time. Results have been mixed.

[15] A standard technique is to identify projects with positive social net present value (NPV)—the present value of benefits net of costs—and negative private NPV. Projects meeting these conditions can be ranked by social NPV per unit of maximum subsidy to be made available. See, for example, Wellenius (2002).

Box 2 Good practices for extending telecommunications beyond the market

Limits to the subsidy

- Service providers put their own resources at risk to build the facilities and provide the services during a given time under specified conditions.
- Government subsidies help service providers meet some investment and start-up costs. Subsidies are designed to reduce access barriers to which low-income groups are especially sensitive, such as initial connection, equipment, or installation charges.
- Customers pay for services at least as much as is needed to meet operating and maintenance costs. Consumption is subsidized only exceptionally and only up to the low level of service regarded as essential.

Steps in implementing the subsidy program

- The government defines the broad objectives, target population, and funding levels for the subsidy program. It also establishes key service conditions, such as types of services to be provided, quality standards, maximum retail prices, and the duration of service commitments.
- Prospective beneficiaries and communities largely determine service needs and choices. Economic and technical analysis is used to select and prioritize projects that are likely to be socially desirable but not commercially viable on their own, and to determine the maximum subsidy justified for each project.
- Private firms submit competitive bids for these projects. Bidders are free to develop their own business strategies, including technology choices, as long as they meet service conditions and comply with rules applying to all providers.
- Subsidies are awarded to the bidders requiring the lowest one-time subsidy. Alternatively, bids are invited for fixed subsidies, with contracts awarded on the basis of quantifiable service measures, such as the lowest price to end users or the fastest rollout of service.
- Subsidies are paid in full or in installments, linked to the completion of investments and provision of service.
- Service providers own the facilities and bear all construction and commercial risks. No additional subsidies are available downstream for the same services.

(Table 5). But Chile pioneered the approach, in a program that reduced the share of the population living in localities without even a public phone from 15 percent in 1994 to about 1 percent in 5 years (Box 3).

The principles for dealing with universal service must be outlined early in the reform process—even if implementation is to come later—as much to address concerns about serving the least favored segments of the population as to contain the risk of private companies being required to undertake unprofitable investments. For example, the incumbent's rural service obligations could be frozen at the level existing before privatization, so that investors can discount any future losses resulting from these obligations in the price bid for the company. No additional service obligations would be imposed on the incumbent, or on new entrants beyond their own business plans and what they contract with their customers.

Table 5
Results of subsidized rural pay phone programs in selected developing countries, 1994–2000

	Chile	Colombia	Dominican Rep.	Guatemala	Peru
Country characteristics					
GDP per capita, 2000 (US$)	4590	2020	2130	1680	2080
Rural population as percentage of total, 1995	15	26	—	59	28
Main line telephones per 100 people, 2001	24	17	11	7	8
Program characteristics					
Rural localities served	6059	7415	500	1253	4400
Population served					
Total (millions)	2.2	3.7	—	1.3	1.6
As percentage of rural population	>90	35	—	20	24
Subsidy					
Per locality (US$)	3600	4600	6800	4700	9600
Per inhabitant (US$)	10	9	—	5	26
Total, as percentage of sector revenue	0.3	1.4	0.8	0.3	1.0
Total investment divided by subsidy	7	—	1.7	2–4	2–3
Service obligation (years)	10	10	20	5	20
Maximum local call charge (US cents/minute)	12	11	8	4	6
Localities per project	30	—	500	—	630
Bids per project	1–5	2–7	2	2–4	—
Lowest bids as percentage of maximum subsidy offered	54	45	85	37	36
Duration of subsidy payments (years after start of service)	0	1.5	5	5	5

— Not available.
Source: Authors' compilation.
Note: The programs were initiated at various times between 1994 and 2001. Results reflect commitments for end-2002.

5.3. *Designing attractive business opportunities*

Private investors seek business opportunities that are likely to result in competitive risk-adjusted returns[16]. License design is central. Morocco's second mobile license, awarded in 1999, was particularly appealing to investors because, in addition to

[16] In particular, while investors are not opposed to competition (many, in fact, compete successfully in their home markets), they cannot be expected to also maintain large internal cross-subsidies, commit to unprofitable rollout or service obligations, and keep in place excess personnel.

Box 3 Closing the gap in access to rural communications in Chile, 1995–2000

Reforms in Chile led to rapid development of telecommunications in the 1990s, driven by the private sector. But rural areas remained largely excluded because of the high cost of providing service, low-revenue potential, and lack of strategic interest to the operating companies. To help close this gap in access, the government established the Telecommunications Development Fund in 1994. Financed by the national budget, the fund provides carefully designed subsidies to catalyze additional private investment in pay phone service in rural and urban areas with low income and low-telephone density.

The fund has been very successful. Between 1995 and 2000, it supported the provision of pay phone service to more than 6000 rural localities with about 2.2 million inhabitants, reducing the share of Chileans without access to basic voice communications from 15 percent in 1994 to 1 percent in 2002. The fund is also supporting the provision of some 25,000 individual rural telephone lines. The subsidies awarded cost the government less than 0.3 percent of total telecommunications sector revenue during the funding period, and fund administration cost about 3 percent of the money granted.

What accounts for the fund's success? The biggest factors have been reliance on market forces to determine and allocate subsidies, minimal regulatory intervention, simple and relatively expeditious processing, and effective government leadership. Competition among existing and new operators for the rural market and subsidies have kept the cost to the government far lower than in earlier public investments in similar projects. Commercial success has hinged on operators using the subsidized pay phone infrastructure to also provide individual business and residential telephone lines and, later, to offer new services (including voice mail and Internet access in some areas). Interconnection was the most important regulatory factor in commercial viability, with access charges in some cases surpassing 40 percent of rural operating revenues.

The fund's design proved to be robust, and it remains the leading example of a cost-effective solution for reducing gaps in access to basic communications in emerging economies. But questions remain about whether the services can be sustained, how to extend access to the small rural population still unserved, and whether anything needs to be done in urban areas. These questions—along with work on quality standards and monitoring and small improvements in design suggested by the fund's experience—helped guide the fund's extension into more advanced modes of communication and access to information. They can also offer guidance to other countries learning from the Chilean experience.

Source: Wellenius, 2002. Current information on the fund is available at http://www.subtel.cl.

the usual features of mobile licenses, it allowed the new operator to build a long-distance infrastructure and, after 2 years, to operate its own international gateway. This gave the new company an alternative to leasing international and domestic capacity otherwise available only from the incumbent at prices up to

10 times those in relevant competitive markets, and allowed it to directly access the lucrative international market. The license also allowed the new mobile operator to offer fixed-wireless service in rural, suburban, and industrial areas, subject to regulatory approval (Wellenius and Rossotto., 1999)[17]. By contrast, a second national operator license offered in 2002 for only fixed service and with large coverage obligations had no buyers.

Beyond making licenses attractive, enough radio spectrum must be made available to accommodate a surge in demand for wireless solutions by new entrants and incumbents. When possible, it is preferable to freely grant the use of spectrum when available, and to use market mechanisms when demand exceeds supply. Regulatory intervention should be kept to a minimum, enough to contain interference. Traditional spectrum management practices based on administrative rules should give way to modern solutions making more use of economic principles and market forces to shape the frequencies plan and assign frequencies to users (ITU, 2004). Also important in attracting investment is to ensure an adequate supply of numbering blocks and equitable allocation among operators. Misallocation and shortages of numbering blocks in Russia in 2001 forced new entrants to buy numbers from established operators at prices comparable to the monthly rental of business telephone connections.

A state enterprise can be made more attractive to private investors by undertaking internal improvements needed in any event for good management. These improvements may include updating accounts, financial statements, and external audits; preparing records of networks, and connected customers; and cleaning up stocks, legal titles (land, buildings), accounts receivable, and outstanding litigation. Hondutel, the state-owned main operator in Honduras with a monopoly in fixed domestic and international phone service, undertook most of these improvements in 2002–2003. It also negotiated a 20 percent reduction in staff and slower increases in benefits. These measures helped increase profits by 60 percent and would also aid in attracting private participation if a new attempt should be made to do so.

Rebalancing the incumbent's tariffs toward costs is also necessary to enable the company to compete effectively as well as to attract private investors. Under monopoly supply, telephone connections, monthly subscriptions, and local calls have typically been priced below costs, with the resulting deficits cross-subsidized by above-cost international and domestic long-distance call charges. As competition is brought in, the high margins of international service tend to diminish quickly, and the incumbent is increasingly left with loss-making services.

[17] The mobile company was not, however, authorized to lease its networks to other public telecommunications operators, which constrained the development of Internet service providers. Moreover, its international gateway can be used only to carry traffic from its own clients, and incoming traffic continues to be channeled by foreign operators largely through the incumbent, depriving the second mobile operator of substantial international settlements. See Wellenius and Rossotto (1999).

Table 6
Telephone tariffs before liberalization in Latvia, 1996

Tariff category	Latvia (US$)	Average of five competitive markets (US$)
Line rental, per month		
Business	7	22
Residential	1	14
Local call, per minute	0.03	0.03
Long-distance call, per minute	0.16	0.14
International call, per minute		
Near	0.36	0.38
Far	1.81	0.71

Source: Authors' compilation.

Rebalancing tariffs thus becomes urgent, not only to achieve economic efficiency but also to ensure the incumbent's financial survival and to create incentives for the new entrants to invest in a broad spectrum of networks and services, not just in the overpriced international segment. Table 6 compares the distorted telephone tariffs in Latvia in 1996, during the country's transition from a centrally planned to a market economy, with those in a basket of markets, where effective competition combined with some regulation had brought prices close to costs. This table was used to illustrate the general direction and magnitude of tariff changes needed as the company, privatized as a monopoly several years earlier, faced fierce competition from international call-back operators and prepared to liberalize all market segments toward accession to the European Union.

Leaving the responsibility for rebalancing tariffs entirely to the recently privatized companies invites difficulties. In Argentina, failure to rebalance before privatization in 1990, coupled with broad institutional weaknesses, led to more than 6 years of conflicts involving the regulatory agency, the regulated companies, the government, opposition parties, consumer associations, and judicial courts. Meanwhile, business users faced long-distance prices that were up to 50 times costlier, and international prices some four times those in neighboring countries. Price distortions encouraged the use of call-back, calling cards, and other illegal but effective forms of bypass, estimated to have siphoned off about one-fourth of international telephone revenues (Artana et al., 1998).

Outstanding labor issues in state enterprises should be resolved before inviting private participation. Most labor concerns can be accommodated as the sector's growth potential permits productivity gains with little reduction in personnel, and allows existing workers to gain from higher salaries in the private sector and new opportunities in an expanding market. The problem often is not in the number of personnel to be absorbed or reduced, but in the need to overhaul entrenched work practices and relationships with organized labor. In preparation for privatization,

Teléfonos de México (TELMEX) negotiated a major settlement with its workers union in 1989. The amended contract greatly simplified the old bargaining process by reducing the number of job categories, and gave management the flexibility to introduce new technology and allocate the labor force as required. This agreement allowed (TELMEX) to achieve significant economies by undertaking a major capital expenditure program with only modest growth in total employment (Casasús, 1994). As Ghana Telecom prepared to privatize in 1996, some 500 workers (14 percent of the combined telecommunications and postal workforce) agreed to leave with severance packages that cost the government less than 3 percent of the initial proceeds from privatization. By contrast, labor unions whose concerns—and political clout—had been ignored brought Sri Lanka's early reform program to a halt in the mid-1980s. In Colombia and Costa Rica, initial proposals to privatize the state telephone operators led to labor protests and even sabotage of the networks.

5.4. Developing regulatory capability

To give all players confidence that they have a reasonable chance to develop their business, the operations offered to private participants, as well as any remaining public sector operations, must be sufficiently anchored in laws, regulations, and main transaction documents, and supported by enough processing and enforcement capacity. This does not mean that a fully developed legal and regulatory framework needs to be in place before structural changes can be undertaken. The privatization of TELMEX in 1990 was carried out under a 1936 law on ways and means of communication (mainly transport and telegraphs). While not particularly helpful, this law was deemed to pose no obstacles to privatization. Key new provisions were established in a new concession contract for TELMEX, in the bidding documents and contract of sale, and in regulations. New legislation was later needed to prepare for opening basic services to new entrants and competition, and was enacted in 1995 (Wellenius and Staple, 1996). The Mexican Constitution was also amended at that time to allow private participation in the domestic satellite business.

Some regulatory capacity must be in place to assist and guide the early transition from monopoly to competition. A core decision-making capacity, supported by a small secretariat and outsourcing of expertise, may suffice to handle the essentials during the first 2 or 3 years. In the early stages of moving toward competition and private participation, regulators need to focus on licensing major operators, establishing and enforcing interconnection rules and procedures, and managing the spectrum and the numbering system. Other areas of competence can be gradually developed as needed (Wellenius, 1997). The skills required vary widely over time as the focus of regulatory action shifts from relationships between operators and government (licensing), to relationships among operators (interconnection) and to relationships between operators and customers (contracts, billing, quality, complaints) (Smith and Wellenius, 1999). Successive types of regulatory problems thus arise, peak, and then decline to a low simmer so that a permanent, comprehensive in-house capability may never be needed.

Countries with a sound administrative tradition, the ability to undertake durable commitments, a reasonably independent and effective judiciary, and a pool of technical and professional skills can pattern regulatory institutions along the traditional lines of ministerial regulation, or public utility commissions[18]. There seems to be no simple correlation between the structural design of regulatory entities and their performance, but the Latin American experience, longest among developing countries, suggests some lessons that may have broad validity (Box 4).

In countries with weak governance and limited administrative and professional skills, regulatory design should focus on reducing the need for action by the regulator, using resources effectively, and enhancing regulatory credibility. The usual prescriptions for institutional design may be unrealistic, or delay effective implementation well beyond the critical initial years of reform. The benefits from Ghana's privatization and initial liberalization in 1996, up to that point a leading example of reform in sub-Saharan Africa, never fully materialized because the regulatory agency failed to get off the ground. Conflicts between the two fixed operators and between the incumbent and the three mobile operators, especially on interconnection and spectrum, remained unresolved for several years. As a result of these conflicts, as well as the worse than expected performance of the economy, the fixed operators failed to fulfill the rollout and coverage plans, to which they had committed, and growth and service improvement stalled.

Reform plans often expect the regulator to do too many things too soon. A more practical approach is to reduce the need for regulatory action, especially in the early years of reform. Opening the market quickly to new entrants and competition makes the job of the regulator more manageable by providing alternative sources of information, reducing the risk of regulatory capture, and offsetting some of the dominant operator's political power. Prepackaging key regulatory rules in licenses, contracts (such as for the sale of a state enterprise), and laws limits the need for day-to-day regulatory intervention. It also adds stability to the rules of the game, reducing investors' risk of facing de facto expropriation through arbitrary changes in prices, taxes, or service obligations or the unchecked abuse of market power. The license awarded in 2000 to a second national full-service operator in Uganda included rollout obligations as

[18] The general steps in setting up effective regulation include establishing an agency with a firm foundation in law, limiting opportunity for government intervention, starting up the agency well before privatization and new entry, ensuring financial and administrative autonomy, hiring competent staff, establishing a process for appeal, giving the agency means to enforce its decisions, and setting clear boundaries and links with other authorities, including the general competition authority if one exists.

Box 4 Some lessons from early regulatory design in Latin America

- Regulation developed most effectively, where regulatory institutions were started before privatization, where they were established with a firm foundation in law rather than executive authority, and where competition arrived early.
- Regulatory institutions have tended to be closely tied to government, which has favored a good fit between regulation and policy, but made regulation vulnerable to government interference.
- When regulatory issues became political, regulators had a better chance of determining the outcomes when they reported to the highest levels of government, rather than to line ministries.
- The sharing of some regulatory functions, such as licensing, between regulator and government further tightened the excessive links between them.
- Formal separation of regulation and operation has not prevented operators from using information asymmetry, and sheer political and economic power to influence or delay regulatory decisions beyond the normal regulatory processes.
- Antitrust law and enforcement, where available, have increasingly supplemented and to some extent replaced the regulatory role in promoting and safeguarding competition.
- Participation in regional and global trade agreements and membership in regional organizations are leading to growing harmonization of regulatory principles and practice.

Source: Wellenius, 2000b

bid by the winner, price cap controls to apply for 5 years without regulatory intervention, a default interconnection agreement with the incumbent, and specific procompetitive commitments (Smith and Wellenius, 1999).

Use of external resources can augment a regulatory agency's limited skills, while also providing a basis for staff training and exposure to regulatory methods. Most regulatory functions can be contracted out, including monitoring compliance with licenses, contracts, interconnection rules, and pricing rules. Alternative mechanisms for avoiding and resolving disputes can help handle the quickly growing number of conflicts between operators and between them and the regulator, which could otherwise overwhelm the agency, and slow progress in the sector. The regulated companies can be put to work for the regulator (e.g., by giving them the responsibility for managing the numbering plan subject to public review) and the regulator's approval, or for preparing tariff proposals applying prevailing laws and regulations. Combining the regulation of telecommunications with that of other sectors, such as gas, water, or electricity, can make more effective use of scarce human resources as well as reduce the likelihood of regulation being captured by one ministry or company. Another way to make effective use of human resources is to move certain regulatory functions from governments to regional agencies. In recent years, some progress has been made in this direction by members of the Organisation of Eastern Caribbean States and the Southern African Development Community.

Trust in the regulatory function is especially critical in the early years of reform. Transparency and public consultation enhance the credibility of the regulator and the legitimacy of its decisions and make these decisions less vulnerable to change under pressure of interested parties or the government. Addressing issues important to customers, such as the terms of service contracts, billing accuracy, and service quality also helps build public confidence and support.

The stability and credibility of regulatory policies can be enhanced by support from multilateral development organizations, such as the World Bank and the African, Asian, and Inter-American Development Banks. These institutions mobilize financing and technical assistance for designing and implementing regulatory reforms. Governments receiving such assistance commit to explicit sector policies and concrete steps to establish or improve the regulatory regime. The World Bank, for example, uses a range of lending and other instruments to support governments and the private sector in telecommunications and similar sectors, as well as applications of information and communications technologies in other sectors. Creating an effective regulatory environment for private-led development is a major theme of these assistance programs (World Bank, 2003). Similarly, the ITU's Telecommunication Development Bureau has well-established programs to facilitate connectivity and access; foster policy, regulatory, and network readiness; expand human capacity through training programs; formulate financing strategies; and enable enterprises to participate effectively in the information economy. One program provides regulatory bodies with practical tools and resources for regulatory reform aimed at meeting national goals for developing information and communications technologies and increasing their accessibility and use (ITU, 2002a).

Regulatory policies can also be given added stability by locking them into regional and global trade agreements. When countries commit to specific steps and timetables for liberalization under the World Trade Organization's (WTO) General Agreement on Trade in Services (GATS), the underlying reform plans become more robust to internal political change. Countries also can commit to broad regulatory principles under the GATS, which place a floor under domestic regulatory practice. The WTO commitments are legally binding at the level of international treaty and thus take precedence over domestic laws and regulations. A WTO dispute resolution mechanism handles complaints by one country against another for alleged violation of its commitments, or general obligations under the GATS. The first formal complaint in telecommunications, and the first to deal only with services, was filed by the United States against Mexico in 2002.

Not all countries will strictly adhere to their commitments, downstream interpretation of some of these commitments may be contentious, and the dispute resolution mechanism will be triggered infrequently because of its complexity and cost to participants. But the existence of the commitments and their transparency encourage countries to stay the course and add to investor confidence. Some 80 countries (about 30 of them developing or transitional economies), together accounting for about 80 percent of the world's telecommunications traffic, have undertaken specific telecommunications commitments since 1997, and more are likely to do so in successive new rounds of WTO negotiations.

5.5. Managing the reform process

Reform is essentially a political process. Good economic and technical designs are needed to establish principles and outline solutions that are best for sector performance. But reform is largely shaped by broader political, institutional, and governance aspects. Different stakeholder groups seek different outcomes. Thus reform inevitably goes through several iterations involving cooperation, consultation, conflict, negotiation, and agreement among stakeholders in government, the private sector, and civil society. Along the way, reform designs are revisited and second-best solutions devised. What eventually emerges can differ markedly from what was initially proposed (Brinkerhoff and Crosby, 2002).

Competent management of the reform process is critical to its success. In developing countries, the planning for telecommunications reform has focused primarily on economic and technical design. The political, institutional, and governance dimensions have been brought into play indirectly through the political and business experience and skills of the people leading the process. Results have been mixed, ranging from fairly quick and successful decisions and implementation to repeated failures and total loss of credibility. Some partial lessons can be drawn from this experience.

Reform requires committed leadership at the highest level of political authority. That usually means the head of government, who then allocates responsibility for the reform to a single person with direct access to senior government officials, the freedom to cut through red tape, and the resources to assemble a small, highly qualified support team, and hire the necessary experts. That was the case in the privatization of TELMEX. The president of Mexico announced the reform in August 1989, appointed the minister of finance as chairman of the board, and gave him overall responsibility for the privatization. The finance minister handed over the chief executive's job to an experienced public administrator with a clear mandate for reform and the undivided loyalty to achieve it. The privatization was completed in December 1990. By contrast, Brazil's attempts at reform from the early 1980s failed to muster the necessary political muscle, and made no real progress until 1997.

Conflicting policy objectives are best sorted out early. The main purpose of opening markets to new entry, competition, and private participation is service—more service, better service, new services, and eventually less-expensive services. Pressures from interest groups—incumbents pushing for continued market protection, workers demanding employment guarantees, new entrants seeking special deals, treasury officials expecting to use sale revenues to reduce budget deficits, financial advisers earning success fees tied to transaction prices—can steer reform away from this purpose. Strategies that drive up the prices paid for existing companies or new licenses by creating artificial scarcities suppress growth. In India in 1996, competitive bidding for a single new fixed-service operator license offered in each region resulted in exorbitant bids that, combined with modest revenue potential, made it difficult to raise debt financing for investment. One way to increase the fiscal impact of privatization is to wait for the company to appreciate and grow under the new management. In the privatization of TELMEX, the government initially sold only 20 percent of the shares, to a strategic investor in 1990 for $1.8 billion, and then sold 31 percent more through public offerings in 1991 and 1992 for $4.5 billion—70 percent more per share (Wellenius, 1997).

Reforms should follow clearly defined processes open to participation and review by all interested parties. The public must be kept informed. The pace at which new entrants, competition, and private investors will be allowed must be known and preferably written into law or international treaty. When unique business opportunities are offered (such as buying into an existing enterprise, or purchasing a new individual license), market mechanisms, not individual negotiations, should be used to select investors and operators and to determine transaction prices. The award of licenses and contracts should strictly adhere to the evaluation criteria announced at the outset. Time is often important, but it should not be used as an excuse to cut corners, or strike deals behind closed doors. When there are no numerical limits to entry, such as in the award of general or class licenses or authorizations for the use of parts of the spectrum not in short supply, the processes should be simple, subject only to routine verification of compliance with eligibility, and allow no opportunity for discretionary intervention by the authorities. Clear rules and processes must also apply to the regulatory function. The locus and functions of the regulatory authority, and the basic procedures governing its relationships with operators, customers, and other authorities (including the general competition authority if one exists), must be defined, preferably by law (Smith and Wellenius, 1999).

6. The Internet

Not long ago many texts about the telecommunications industry virtually ignored the Internet. Although the global packet-switched network had already achieved broad familiarity and use among researchers, educational institutions, and government, it was not until the creation and wide dissemination of hypertext transfer protocol (http) and the WWW in 1994 that mainstream telecommunications industry observers, along with a rapidly growing consumer market, began to recognize the revolution under way.

6.1. Evolution of the Internet in the developing world

The story of the Internet in the developing world is only beginning to unfold. As with previous generations of telecommunications technologies, Internet service was slow to arrive in most developing countries. In 1999, only 6 percent of the world's Internet hosts were located in developing countries of Latin America, Asia-Pacific, and Africa, which contained 84 percent of the world's population and some 40 percent of its telephone lines (Männistö, 1999; ITU, 2001c). There were some exceptions. South Africa had an early start, and by 2000, had 2 million Internet users and 170,000 Internet hosts, far more than anywhere else in Africa. Several Central European countries, such as the Czech Republic and Hungary, also moved rapidly to adopt Internet technologies in the late 1990s. But even where telecommunications sector reforms had been under way for some time, and growth in basic telephone access was steadily climbing, such as in many Latin American countries, Internet service often was not widely introduced and adopted until several years after the first waves of expansion in the developed world. The share of homes with Internet access in 2000 was only 3.5 percent in Mexico, 1.7 percent in Chile, and 0.5 percent in Brazil (Caceres and Sudweeks, 2000).

The reasons for this delayed response are complex. Most fundamental are the costs in upgrading from basic voice telephone communications to Internet connectivity—costs for service providers and especially for end users. Many components of the backbone infrastructure in developing country networks require substantial enhancements to accommodate the increased data traffic of Internet transmissions, especially as more bandwidth-consuming applications are adopted. These costs often were not considered in the network expansion commitments and capital investment plans of strategic investors and others involved in privatizations of telephone operators, and doubts about the short-term return on such expenditures have made many such companies reluctant to risk already shaky profits on the unproven Internet market in developing countries. In addition, leased line charges for connection to the global Internet backbone can run to tens of thousands of dollars a month. These costs can be a major hurdle for start-up ISPs seeking to promote a fledgling Internet market and can lead to end user fees out of reach for most average consumers. Compounding the affordability problem are the much greater costs of computer hardware and software in developing countries, which can be prohibitive for consumers and many small businesses.

Perhaps in part because of such economic constraints, the international development community did not quickly embrace the potential of Internet service in its approach to telecommunications investments and policy in the way that it had with basic telephony and mobile service. In the mid- to late 1990s, most of the major donor, lending, and technical assistance institutions—such as the World Bank, ITU, and WTO—tended to consider Internet access a peripheral, luxury capability rather than a fundamental part of the strategy for developing telecommunications infrastructure. National policy bodies in developing countries echoed this perspective, and sector reform initiatives often downplayed or ignored Internet-related objectives.

This official hesitancy may also have reflected institutional inertia among national policymakers. The Internet's sudden arrival coincided with the most intensive efforts at telecommunications sector restructuring across much of the developing world (see Section 3 of this chapter). Given the effort, financing, and persuasion required to coax many governments to adopt reforms based on the standard models of the 1980s and early 1990s, it may have seemed too overwhelming a task to adjust these strategies midstream to incorporate the technological and economic implications of fostering a broad-based Internet market of uncertain future. Thus, even as those policies were helping to narrow the access gap between developed and developing countries in traditional telephone service, an even wider digital divide was rapidly opening, threatening to undermine many of the gains from market reforms.

The slow pace with which many national telephone operators introduced Internet service created an opportunity for private businesses. In many countries, some of the first and most successful ISPs were small start-up operations launched by young, computer-savvy entrepreneurs, often with experience in related ventures such as software programming, information technology consulting, or computer equipment importing and sales—and frequently with university affiliations as well. Given the absence of larger-scale competition and the general lack of experience in the market, these companies could establish themselves by offering basic low-speed dial-up Internet access to a relatively small base of business customers (Box 5). As demand for Internet service grew, demonstrating the viability of larger investments to both the business and the financing community, these early entrants could retain a leadership position in the market because of their initial customer base and name recognition along with their employees' expertise with constantly changing technology—a resource in short supply throughout the developing world.

Other Internet initiatives in developing countries have augmented the efforts of private ISPs. In response to the high costs of hardware and connectivity and the broad need for basic education and public awareness about the use and potential of Internet service, public service agencies began to launch programs to allow public access to information and communications technologies through community-based telecenters. In Peru, Red Científica, a network of independent community centers, was launched in 1994 by a nonprofit social activist organization to bring computer training, Internet access, and e-commerce services to remote rural villages. The program gained worldwide attention for its success and has been emulated in other parts of Latin America. In South Africa starting in 1997, the Department of Communications and the Universal Service Agency undertook an ambitious program to install multipurpose community telecenters in more than 500 villages. Although hindered by political and operational barriers, the telecenter program has become a model for similar projects in many other African countries. In Hungary, a national network of community 'telecottages' has been extremely successful in offering a mix of traditional and Internet-based information and

Box 5 The Africa Online story

In 1989, an ambitious Kenyan with a vision, MIT-trained Ayisi Makatiani, and two friends created a start-up venture out of his U.S. apartment, financed using credit cards. The project started out as KenyaNet, a news and communications platform for the Kenyan community in Boston. Within a year, thanks to a computer purchased with donations from appreciative mailing list users, the venture was enabling more than 2000 Kenyans around the world to use e-mail and receive news from Kenya.

Taking the name Africa Online, the venture became a full-service ISP in 1994, offering dial-up Internet access in Kenya, and eventually throughout the neighboring region. Makatiani later approached the U.S. online service company Prodigy, which bought into the company and helped lay the foundation for its rapid expansion in Africa. A series of buyouts and takeovers followed until a group of managers, including Makatiani, bought back the company in 1998 with the help of a \$2.8 million loan from African Lakes Corporation, once a British colonial trading company and now trying to establish a reputation as a technology business.

Today, Africa Online is the continent's largest ISP outside South Africa, boasting 30,000 subscribers and connecting roughly 100,000 more through E-touch, a chain of Internet cafés that people can visit to surf the Net or read their e-mail.

communications services to rural Hungarians. Begun in 1996 by some community libraries, the initiative expanded with government and donor support to establish telecottages within reach of a majority of the population by 2002.

These and many other examples form the vanguard of a global movement that has emerged in recent years to use the community telecenter as, at least an interim vehicle for bringing the Internet to rural and low-income populations in developing (and developed) countries. The key issues that this movement now confronts involve how to design economically self-sustainable and replicable telecenters, and how to tailor their services, including training and education to the needs of the populations they would serve.

The community access model has also expanded rapidly into the private sector with the recent growth in Internet cafés in the major urban areas of many developing countries. In these for-profit public sites, comparable to telecenters, consumers can buy time on computers for online access and related activities. Many Internet cafés are privately owned and operated by individuals, who may offer computer services together with (as the name implies) food and drink, as well as virtually any other retail shop or service. In other cases, chains of cafés are owned by larger companies, including ISPs and telephone operators, and share their brand name. In many large cities throughout the developing world, it is now common to find several competing Internet cafés along the same streets of downtown market centers, all of them crowded with users and waiting customers during peak business hours.

Thanks to these varied developments, Internet use has grown much more rapidly in developing countries since 2000. Precise statistics are difficult to determine in markets, where many users rely on public connections such as Internet cafés, rather than home access accounts. But estimates suggest that the number of Internet users in Africa, for example, excluding the robust South African market, more than doubled between 2000 and 2002, from 2.1 million to more than 4.8 million (ITU, 2001c). The growth in Latin America has been even more rapid (Figure 3).

The success of the varied approaches to providing Internet service in the developing world has often helped to foster a competitive consumer market for this service while reinforcing demand levels and demonstrating beyond any lingering doubts the economic value of the Internet for developing societies. The economic benefits can take many forms and are not necessarily the same as those in fully developed economies (Box 6). Yet, while Internet growth in developing countries has remained strong even in the face of the global technology sector collapse after 2000, penetration and use remain vastly lower than in the developed world. This situation suggests that the opportunities for further Internet growth remain substantial—as do the challenges in encouraging that growth and the economic benefits it would bring. Key among these challenges are fostering a competitive regime for Internet development and establishing appropriate interconnection arrangements, especially at the national level.

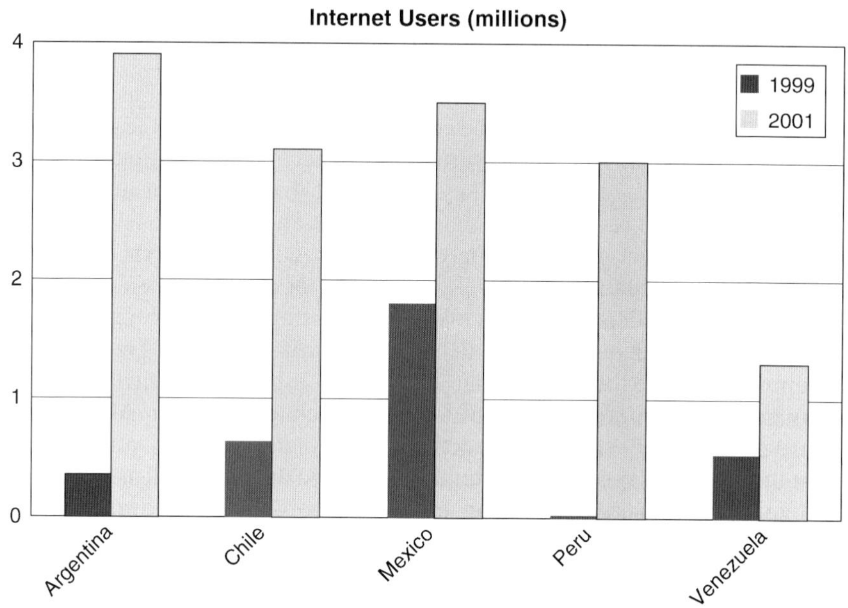

Fig. 3. Growth of the Internet in selected developing countries, 1999–2001. *Note*: Data are for December in each year specified, except Argentina (August 1999, July 2001) and Peru (April 1999, December 2001). *Source*: NUA Internet Surveys, 2004.

Box 6 From telex to fax to e-mail

For years, the standard method for transmitting long messages and documents internationally was the telex. Many international businesses and government offices routinely maintained telex access numbers even into the 1990s, although the volume of telex traffic had declined precipitously. Starting in the mid-1980s, telex began to give way to the latest technology option for transmitting messages and documents, the facsimile machine. Estimates of the growth in fax machines and traffic vary, but by the 1990s, millions of fax pages were being transmitted daily, and virtually every organization, large or small, maintained dedicated fax lines.

Fax technology offered efficiency improvements over telex, allowing users to easily transmit documents without having to type them into a telex terminal, or perform complex formatting functions. But high call charges and long holding times made international fax traffic expensive for users in developing countries.

Enter e-mail, which combines and improves on the advantages of both fax and telex in a new communications paradigm that is also vastly less expensive. It is impossible to overstate the transformative potential of e-mail for international commerce, especially for developing countries. E-mail makes it possible to transfer instantly and at virtually no cost documents that would require hours and hundreds of dollars to transmit by fax. E-mail (and such applications as chat) allows conversations between individuals or among a group, again at almost no cost. And e-mail overcomes the problem of time zone differences that inhibits international telephone communication. E-mail alone, excluding all other functions of the Internet, has brought to the developing world a vast new opportunity to participate in the global information economy.

6.2. Internet policy issues for developing countries

Successful development of the Internet as a mass means of communication and information access in developing countries will depend on a number of key factors. The most successful models of Internet growth around the world have been based on a hands-off, unregulated approach, allowing the private, competitive market, spurred largely by innovative researchers and entrepreneurs, to experiment with new ideas and applications and adopt those with popular appeal and economic viability. The developing world shouldn't necessarily emulate every step (and misstep) of countries where the Internet has become dominant, but the critical role of the market should be acknowledged in shaping a range of policy issues.

6.2.1. Competition among Internet service providers

Because private ISPs can operate entirely by utilizing the existing public telecommunications network, they properly fall into the category sometimes known as value-added networks or enhanced service providers. In many countries, this classification initially placed ISPs outside the realm of dominant incumbent telephone companies, or regulated public services, allowing the industry to flourish with minimal government intervention. As the market grew and regulatory structures became more established in the developing world, however, ISPs came under greater scrutiny and sometimes new restraints. Newly authorized regulatory agencies often instituted requirements for ISPs to obtain licenses, pay fees, and even submit to pricing and quality of service regulation. In some cases, recently privatized national telephone operators, whose foreign strategic partners had obtained blanket market exclusivity provisions in exchange for multimillion-dollar equity investments and network rollout commitments, sought to shut down private Internet operators, claiming that competition from such services violated their monopoly rights. In South Africa, where a vibrant private ISP market had been in place for several years, Telkom South Africa initiated a legal battle attempting to shut down all competing ISPs after the 1996 Telecommunications Act granted it a 5-year exclusivity in public switched telephone service.

Such attempts to restrict and regulate the Internet market risk slow the gains made by private entrepreneurs just as the industry is beginning to take off. A competitive Internet access market tends to encourage innovation and minimize prices, allowing the technology to penetrate faster and the entire information economy to grow. Regulatory intervention in this market, including licensing requirements, pricing restrictions, and favoritism toward the incumbent telephone operator, can inhibit this growth without achieving any significant benefits that a competitive market won't produce by itself. Econometric analysis of data from a survey of regulators in 60 developing countries, from the ITU, and from the World Bank, concluded that regulation is strongly correlated with lower-Internet penetration and higher Internet-access charges. After controlling for income, time trends, and the availability of telecommunications and computers, the analysis found that countries that require formal regulatory approval for ISPs to begin operations have fewer Internet users and Internet hosts than countries that do not require such approval. And countries that regulate ISP end-user prices have higher Internet-access prices than countries that do not (Wallsten, 2003).

Still, regulators clearly have an important role to play in preventing anticompetitive behavior by dominant network operators that both control Internet backbone access and compete directly with independent ISPs needing that access. The dependence of most ISPs on the incumbent's network for connectivity places them at a potentially crucial competitive disadvantage relative to the ISP owned by the telephone company if that ISP does not have to pay the same interconnection charges as its competitors. This was the case in Morocco, for example, where the more than 100 ISPs that initially registered eventually dwindled to only three. One of these, Maroc Telecom's own operation (Menara), built up a dominant position with 70 percent of client connections. The second largest, MarocConnect (a subsidiary of France's Wanadoo), complained that the prices Maroc Telecom charged for its international leased circuits were several times cheaper than those in Europe and that it used anticompetitive practices, including predatory pricing and price squeeze. The regulator found for the complainant and forced Maroc Telecom to raise the prices of its retail Internet service and lower its charges to competitors for using its local loops and transiting its network.

6.2.2. Backbone and national interconnection for Internet service providers

A more vexing policy issue involves the relationship between domestic Internet networks and services and the handful of global Internet backbone network providers (Brooks, 2000). While the long-standing arrangements for international voice telephone service at least theoretically divide the costs of country-to-country calls proportionately between carriers on each end, Internet connectivity is far more asymmetric. The ISPs in developing countries typically must connect their servers to the international backbone through one of the major access points, most often in North America, Europe, or East Asia. The cost of an international high-capacity fiber or satellite link from an ISP in, say, Africa or Latin America to such an access point is generally borne almost entirely by the connecting ISP. This arrangement is the same regardless of the direction and volume of the data traffic flowing over the connection.

Policy advocates from developing countries have complained that these arrangements amount to an unfair subsidy of Internet service in developed countries at the expense of users in developing countries. They argue that any time a U.S.-based Internet subscriber downloads information that is hosted in Africa, the U.S. ISP pays nothing toward the incremental cost of establishing the link between its customers and the African Website—while African Internet users must implicitly pay virtually all of the connection cost for any information that they access from U.S. servers. The difficulty is that while there may be some merit in suggesting more equitable cost sharing for international backbone access, there is no practical global regulatory mechanism that could impose such a system.

The Internet is fundamentally different from the international telephone system that has evolved over the past 150 years (and, before it, the telegraph system). Since there are no wholesale per-transaction or per-minute charges for traffic between network segments, but merely capacity-based connections between routers and servers along any of thousands of backbone links, there is no basis for measuring and allocating costs according to bits transmitted, or any other criterion. When a new ISP in any country 'plugs into' the vast planetary Web of circuits, switches, and servers making up the Internet, it automatically gains access to the entire public network and makes available all public data on any of its own servers to anyone, anywhere in the world. The choice to establish that new network node is entirely unilateral: existing Internet users and ISPs have no reason—or means—to underwrite the cost of extending the network to a new population of users and Websites even if they will ultimately benefit from their inclusion.

A partial solution to this dilemma that some national policymakers and Internet industry leaders in developing countries have begun to implement involves coordinating backbone access at a national or regional level. The benefits of establishing such national backbone networks and Internet exchange points are twofold: they aggregate national traffic by allowing multiple ISPs to connect over shared high-capacity links to the international backbone, saving costs and increasing efficiency; and they allow internal Internet traffic to be routed only within the country, or region rather than transiting as far as the U.S. backbone and back, as often occurs. Although coordinating private ISPs in this way is essentially an industry choice, active encouragement by national regulators can help accelerate the process of decentralizing the Internet and enhancing its cost-effectiveness in the developing world.

6.2.3. Voice over Internet protocol

For many developing country's telephone operators and regulators, another key policy challenge has been the rapidly evolving technologies and market for voice over Internet protocol (VoIP) service—that is, the use of the Internet to transmit telephone calls. The issue is not the increasing integration of Internet protocol technology into the public switched networks of major carriers as a cost-saving device, an important application of Internet technology. Instead, it is the emergence of retail VoIP services provided by third-party telephone resale businesses that offer consumers deep discounts on prevailing tariffs for international voice telephone calls.

Even where the ISP market is relatively open and competitive, this trespassing into the voice service market has led to policy and legal conflicts with licensed carriers that control international telephone service. Their claim, often supported by regulators, is that transmitting voice calls over the Internet violates exclusivity provisions of their licenses and deprives them of substantial income from both outgoing call charges and incoming international settlements. Such a dispute led to a protracted legal battle in Sri Lanka starting in 1999, as Sri Lanka Telecom claimed that illegitimate VoIP services cut deeply into its international revenues, and balked at following through on market opening and tariff rebalancing commitments as a result.

The VoIP issue illustrates some of the challenges that Internet technology introduces into the traditional dynamics of the telecommunications sector. If two users exchange messages by e-mail, or even in real-time through instant messaging, that conversation is accepted as a legitimate communication utilizing the Internet. But if those two users have the same conversation by voice using Internet-based software and transmission rather than voice circuits, this transaction is technically illicit under the regulations of many countries. In this respect, VoIP strongly resembles the call-back services that sprang up in the 1990s to exploit the rapidly growing disparity in outgoing call charges between developing countries and competitive, developed markets. The difficulty for policymakers is that these types of services are enabled by bona fide technological advances—not loopholes or theft-of-service schemes—which are inexorably driving down underlying network costs even as some administrations attempt to maintain above-cost prices. The ongoing convergence of media and markets makes it impractical and often impossible to enforce such anachronistic regulatory distinctions among service categories.

6.3. Information technology and international trade

An unavoidable consequence of the digital revolution has been the progressive merging of the telecommunications and information technology fields—into what is now commonly referred to as the information and communications technology (ICT) sector. For private businesses and governments alike, this trend has required closer integration of planning, procurement, and strategic policies relating to traditional telephone services and data processing operations. A computer is now as much a communications device as it is a data storage and analysis tool.

As a result, a wide range of economic and policy issues relating to the IT component of ICTs need to be considered in any significant review of telecommunications and development. Especially important are issues having to do with international trade in the telecommunications arena and related import, export, and transborder laws, tariffs, and treaties. Most computer hardware and software are still produced by a handful of multinational corporations based in the most advanced economies, although the actual manufacturing and programming have begun to shift to the developing world. At the same time, the ubiquitous, borderless nature of the Internet has fundamentally altered the dynamics of world trade, especially in soft products and services, and is compelling nearly every nation to reevaluate its position in the global market.

6.3.1. Personal computers in developing countries

For the Internet and any other information technology to be accessible to a broad base of consumers in developing countries, personal computers (PCs) must first be accessible and affordable. But while worldwide costs of PCs have declined precipitously in the past several years, the prices in many developing countries pose a substantial barrier for average consumers and even for small businesses. In part this simply reflects lower incomes, which can make even a few hundred dollars of hardware impossible to afford. But other factors tend to exacerbate the problem, including ill-considered high-import tariffs on IT equipment. These, combined with the inherently high costs of transport and delivery, can drive up PC prices in developing countries to as much as twice the price in developed markets.

Prospective users of basic e-mail and Web features have little need for high-end processing power and other complex features of traditional PCs, yet until recently they had few other options. New initiatives in some of the larger developing markets have begun to explore the benefits of domestic manufacture of stripped-down, lower-cost computers that would be affordable for a wide segment of the population and provide the basic functionality needed to connect to the Internet.

In India, where nearly 50 percent of the population is illiterate, a team of scientists and engineers has come up with a way for people with limited literacy to take advantage of the Internet. The team has developed a small, powerful computing device called the 'Simputer'—short for "Simple Inexpensive Multilingual People's Computer"—that offers natural user interfaces based on sight, touch, and audio. The Simputer reads out the text on Web pages in a number of India's many native languages and costs only around \$200.

In 2001, the Brazilian government announced a project intended to make stripped-down desktop computers, known as 'Popular PCs,' available for about \$300. Developers were able to save on licensing fees by using free, open-source Linux as the operating system rather than Microsoft's Windows. The government also proposed to underwrite the expansion of the Popular PC by lending low-income citizens funds to buy the machines—loans they can repay at about \$15 a month.

The approach in The Gambia is to produce low-cost computers by reconditioning old ones. When upgrading computer equipment, many companies not only discard machines and software that would be useful to others, but also pay up to \$40 per machine for disposal. A nonprofit group in The Gambia created a network that enables these computers to be donated to schools, with the schools paying the shipping and handling charges. People from government offices, health facilities, and the schools have been trained to maintain and repair the equipment.

These and other creative approaches may eventually compel the dominant international computer manufacturers to differentiate their products more according to the needs and budgets of emerging markets, especially as the spread of Internet access and other ICT services leads to a mushrooming demand for such equipment—and as the market for new PCs in the developed world becomes increasingly saturated.

6.3.2. Development of the software industry

Software development is among the fastest growing industries in the developing world, with the potential to fundamentally alter the global information technology market. The leading example is India, where the industry has achieved annual growth rates of 50 percent for several years, producing nearly \$10 billion annual exports, employing more than half a million people, and accounting for 3 percent of GDP (O'Connor, 2003). Many other countries are seeking to develop a similar industry. China, for example, has seen rapid growth in its information technology and software sector, which focuses on the potentially vast domestic market for custom software as well as export opportunities (Tschang, 2003). The global software industry appears to be following in the footsteps of manufacturing, taking advantage of lower labor costs (though much higher than those for unskilled factory workers) to shift programming and even customer service offshore. The Internet makes this transformation vastly more practical by facilitating robust collaboration and communication among geographically distant managers, programmers, marketers, and distributors and by reducing transaction costs to minimal levels.

Developing countries are also beginning to play a key role in the impending showdown between proprietary and open-source software platforms. Simply because of their limited budgets, many developing country's governments and companies are embracing free open-source software, such as the Linux operating system, to avoid paying costly license fees. They are thereby creating a base of expertise and awareness of these options that should help fuel the continued growth of the open-source movement.

6.3.3. *Intellectual property*

Much of the consumer and business trade over the Internet involve selling or licensing information, cultural products, and technology protected by intellectual property rights, yet this medium by its very nature is especially susceptible to risks of theft, or misuse of protected works. Many developing countries have been at the center of copyright violation disputes in recent years. In Bangkok, Thailand, major shopping venues sell pirated software, video games, and film and music recordings to the public openly and with impunity. Concerns about intellectual property rights have been a primary focus of international deliberations in recent years. Through the WTO, for example, countries have negotiated the Agreement on Trade-Related Aspects of Intellectual Property Rights (TRIPS), making intellectual property an integral part of the multilateral trading system. Both the TRIPS Agreement and the copyright treaties of the World Intellectual Property Organization (WIPO) seek to provide a legal framework to forestall the spread of pirating worldwide. The challenge is immense, however, as technology makes copying and distributing digital material increasingly easy, and the countries, where such practices proliferate are ill-equipped to enforce the rules.

6.3.4. *Network and data security*

Before the Internet, it was inconceivable that a single anonymous miscreant, hidden away in Brazil or the Philippines or Romania, could threaten the viability of communications networks worldwide while imposing untold millions of dollars in costs on operators and users. This scenario is now close to commonplace. The ubiquity of the Internet, and the fragility of some of its components, has created a vulnerability that can be exploited by relatively amateur programmers operating virtually anywhere. Computer viruses and worms (and hoaxes) have been proliferating across the Net for years. Many originate in developing countries, which lack the resources and laws to track down the perpetrators. And while viruses most often tend to be a malicious practical joke, other assaults on the integrity of data networks—hacking, identity theft, even cyber terrorism—carry much more ominous potential.

Arguably, however, the greatest threat to the economic viability of the Internet is an insidious offshoot of its commercial value: the unrestrained explosion of spam e-mail solicitations and related forms of exploitive electronic advertising such as spyware. Spam has been increasing exponentially for several years, despite feverish efforts by ISPs and software makers to curtail it (Gates, 2003). The global nature of the network allows such messages, like viruses, to be initiated from any location, and the antispam regulations being passed in some developed countries are unlikely to be effective in slowing the massive growth in spam. Moreover, as the Internet continues to expand in the developing world, spam traffic will probably expand along with it.

The clogging of server mailboxes and the costs of antispam software filters and other defensive measures are inhibiting the efficiencies that Internet communication otherwise accelerates. And as the industry seeks to expand further into a mobile Internet environment, the unwanted and overwhelming commercial messages threaten to place even greater roadblocks in its way. Among the international imperatives in developing a productive, responsible global electronic trade regime therefore must be coordinated legal and technological efforts to reverse this disruptive trend.

7. Economic opportunity and the future

Communications is both an economic sector in its own right, accounting for several percentage points of GDP, and an input used by most other sectors. Unlike food, housing, medical care, and human and financial resources, telecommunications and information are primarily facilitators of development and economic opportunity—means to help communities, businesses, and governments increase access to those basic goods and services. In an increasingly globalized information economy, however, the boundary between communications functions and related economic benefits—jobs, investment, exports—is blurring. Virtually every sector of a developing economy stands to benefit from the efficiency gains and broader market access that modern telecommunications brings, and these effects are far greater than those possible even a generation ago. Moreover, the ICT sector itself can magnify economic benefits for developing countries by attracting foreign investment and customers from all corners of the world. The first set of effects—efficiency and market access—offer greater potential for most countries, but the importance of the second may be growing more rapidly.

E-commerce, e-learning, and e-government—the three *e*'s—are perhaps the most common focus of short- and long-term strategies for economic opportunity linked to development of the ICT sector. These activities capitalize on the efficiency effects of information and communications technologies in the private sector, in the workforce, and in public services. All three activities are expanding rapidly throughout the developing world.

7.1. E-commerce

Electronic commerce represents a new way of carrying out traditional business activities. As long as communications networks have been available, entrepreneurs have used them to their fullest capability to create business opportunities. But the meteoric rise of the Internet and the WWW has accelerated the transformation of global commerce, allowing instantaneous, inexpensive contacts among sellers, buyers, investors, advertisers, and financiers throughout the world. The rapid integration of the Internet and other telecommunications-based functions into nearly every sphere of business is what has given rise to the recent international focus on the new world of electronic commerce.

Among the principal media contributing to global electronic commerce are traditional text, voice, and data telecommunications services; newer online services, based principally around the Internet; and a variety of subscription, or transaction-based information services and software sales. Electronic commerce can be classified into five broad categories:

- Business-to-business (B2B) (wholesale and retail);
- Business-to-consumer (B2C) (consumer retail sales);
- Business-to-government;
- Government-to-consumer;
- Consumer-to-consumer.

These activities have been concentrated in a few industry and market segments:

- Advertising and marketing services;
- Financial services and transactions;
- Tourism;
- Entertainment and information services;
- Ancillary functions contributing to business and commercial activities.

Similarly, the tangible products most likely to be traded through electronic means are concentrated in a few categories:

- Computer products (hardware and software);
- Books, music, and videos;
- Gifts and flowers;
- Apparel;
- Food and agricultural products.

Most e-commerce has been taking place in countries with advanced economies and infrastructure. For developing countries, electronic commerce presents important new opportunities to level the playing field with larger, more developed economies, because it diminishes in-place advantages of cost, communications, and information and creates huge new markets for indigenous products and services. But while many developing countries are beginning to take advantage of the potential of e-commerce, critical challenges remain to be overcome before the vision of a truly integrated and equitable world economy can be realized. Foremost among these is to develop adequate infrastructure, including technology networks, financial services, and human resources. In addition, the legal and regulatory framework often needs to be modernized to take account of the range of new issues raised by electronic transactions. And intellectual property and data security issues represent key challenges to the widespread use of e-commerce in the developing world.

Nonetheless, e-commerce has been robustly expanding around the world as technology and innovation continue to spread. The United Nations Conference on Trade and Development, in its *E-Commerce and Development Report 2002* (UNCTAD, 2002), identified several trends:

- Africa (excluding South Africa) has seen the slowest growth in online transactions, but the B2C sector is showing potential, such as in handicrafts, and B2B opportunities are beginning to appear. In Latin America, up to 70 percent of enterprises are connected to the Internet, although B2B transactions are concentrated among large transnational corporations. The Asia-Pacific region is exhibiting the strongest growth in e-commerce, particularly in manufacturing, and the Chinese sector is growing especially fast. Central and Eastern Europe are also actively expanding their e-commerce sectors, but in Central Asia and the Caucasus, e-commerce remains largely undeveloped.
- The ICT sector is creating new opportunities for women entrepreneurs in many developing countries, particularly in retail B2C commerce. It is also generating new employment opportunities for women in teleworking, data processing, and customer service.
- The explosive growth in mobile phones and personal data assistants (PDAs) is contributing to e-commerce in developing countries, both by increasing the mobility of workers and businesspeople and by allowing innovative applications of such services as text messaging.
- Information and communications technologies and e-commerce are fostering the development of new small and medium-size enterprises. A key challenge remains to develop broadly available 'e-finance' initiatives to help support these enterprises, and to simplify and expand access to credit and other financial resources.
- A range of other industries based on information and communications technologies show significant growth potential in the developing world, including insurance, electronic publishing, and export of services, including computer software and customer service.

7.2. E-learning

Like e-commerce, e-learning encompasses a broad range of activities. The best prospects combine basic and innovative technologies to enhance rather than overhaul traditional education practices. But even basic technology—such as computers, multimedia tools, and customized educational software, along with teacher training in their use—represents a major cost hurdle for most school systems. However, international donor and support programs have begun to make a dent in these infrastructure needs.

Still more challenging is the development of interactive and culturally relevant educational materials along with the means to access and share this content across far-flung locations through distance learning techniques. The SchoolNet Thailand program, sponsored by the National Electronics and Computer Technology Center, has become one of the most ambitious program in Asia, providing 5000 schools across Thailand with free (government-subsidized) Internet access, equipment, training, and content. The program includes a large, interactive digital library that allows schools and teachers to contribute curricular material. And it facilitates collaboration among schools as well as universities and public and private institutions.

The benefits of basic e-learning programs in primary and secondary education will be realized only over time, and will be difficult to measure against their investment costs. But there is ample reason to expect that the long-term payoff will be immense. These programs not only should improve the quality and accessibility of basic education; they also will increase students' exposure to information and communications technologies and thus help prepare them to succeed in an economy dependent on these technologies. Perhaps even more fundamentally, experience with advanced communications, especially for younger generations, helps fuel market demand for greater access: the customers of the booming prepaid mobile and Internet café industries are overwhelmingly young. As public education systems, even in remote rural areas, place communications technologies in the hands of youth, they will create new customers for the private ICT sector.

Beyond primary and secondary education, e-learning programs also emphasize high-tech training at the university and professional levels. Many developing countries not only face a deficit of skilled engineers and technicians in the ICT fields, but also lack adequate resources, institutions, and trainers to provide ICT training. Many of the most promising graduates and professionals in high-tech fields have left developing countries for jobs paying vastly higher salaries than are available in education, or even the private sector back home. The challenge of advanced technology training is thus to establish an institutional setting that is both attractive to trainers and affordable for trainees.

7.3. E-government

Application of advanced information and communications technologies to government functions also encompasses a variety of activities. Many of these focus on the core financial and accounting functions of government and the computerization and standardization of data records, procedures, and transactions. The goal is to save time and expense and to smooth the operation of the many disparate agencies that, together, invariably represent the largest budget and employer in any developing country. Thus planning and expenditure for e-government programs, especially in the early stages, focus on computer hardware, software, networking, and management, including the personnel and technical support to operate these systems.

Telecommunications plays a much smaller role in many of these e-government programs. But communications networks and services can enhance government performance and accessibility in at least three important areas:

- *Information services.* Much of a government's interaction with citizens involves disseminating and collecting information (and funding). In developing countries, the connections between government and the governed are often very distant and difficult to effect. When villagers need to register for programs, or obtain information about benefits, opportunities, or obligations, they may have to travel all day just to complete a brief procedure at an administrative center. Innovative ICT projects, including mobile public service centers and government information kiosks at telecenters or post offices, can vastly reduce such burdens while expanding citizens' access to and understanding of their government. For those with more regular Internet access, public agency Websites and similar online features are becoming an indispensable part of efficient modern governance.
- *Political participation.* In countries where democracy is comparatively new, encouraging and facilitating active participation in the political process is an essential goal. The ICT projects can help foster a sense of inclusion. For fledgling political parties, contact with constituents through traditional and new media can supplement grassroots organization. In South Africa, shortly after the end of apartheid, the government set up electronic voter information stations giving citizens access to video campaign statements and information about every political party's platform and candidates.
- *Decentralization and infrastructure development.* In many developing countries, there are inherent political tensions between traditional tribal and local affiliations, on the one hand, and centralized national bureaucracies and government agencies, on the other. Most government reformers acknowledge the value of local leadership and organization even while pursuing common national policy objectives. Decentralizing decision-making and policy implementation can improve the efficiency and responsiveness of many public programs, but achieving these goals still requires effective communication and coordination at the national level. Decentralization presents an especially strong opportunity to combine infrastructure development projects with government reform initiatives by meeting government telecommunications needs through private investment in public networks.

7.4. Knowledge societies and e-readiness

The economic, technological, and policy issues confronting developing countries as they seek the benefits of information and communications technologies are broad, complex, and interconnected. Efforts have been made in recent years to bring together all the issues and objectives in a comprehensive agenda that can guide analysts and policymakers in devising integrated national and global development strategies and in evaluating progress. These integrated approaches aim generally at transforming impoverished, agrarian, disconnected economies into modern 'knowledge societies' fully linked to the emerging global information economy. The strategies emphasize:

> the need to harness ICTs for development by enabling their use for empowering the poor and for scientific and technical capacity building that is consistent with development goals . . . [including] democratic decision-making, more effective governance, and lifelong learning. (Mansell and When, 1998, p. 8)

Most recently, the first phase of the World Summit on the Information Society, held in Geneva in December 2003, produced a consensus declaration of principles on the goals for global expansion of information and communications technologies. This declaration defines a shared vision:

> to build a people-centred, inclusive and development-oriented Information Society, where everyone can create, access, utilize and share information, and knowledge, enabling individuals, communities, and peoples to achieve their full potential in promoting their sustainable development and improving their quality of life. (ITU, 2003)

Consistent with these ideals, a number of countries have begun to define national ICT strategic plans and information infrastructure policies that cut across traditional sector and political boundaries, and seek to include stakeholders from government, education, private industry, and nongovernmental organizations in the process of national development. Some countries have launched formal ICT initiatives:

- Azerbaijan approved a 10-year national ICT strategy in 2003, with support from the UNDP. The project emphasizes initiatives in e-government, 'e-clerical work' (to facilitate entry of handwritten documents into computers), and promotion and use of the Azeri language in electronic media (NICTS, 2003).
- Morocco has begun implementing a program, known as 'e-Maroc', that includes establishing an intergovernmental network and promoting private sector development. Maroc Telecom, with the support of the Ministry of Industry, Commerce, and Telecommunications, has launched a program offering households a personal computer and 2 years of Internet service at a total price of less than $400. The aim is to promote use of the Internet as well as of the local network (SEPTI, 2003).
- Burkina Faso established an initial national communications development plan in 1998, under its Délégation Générale a l'Informatique, which has since been expanded and updated to cover a range of ambitious projects. These include establishing multipurpose rural telecenters, installing a government voice and data network, providing community radio facilities in rural areas, and establishing a new national technology center to coordinate training and development activities (DELGI, 2003).

Common to all these policies is a recognition that ICT development and policy cut across many sectors of government and industry, so that coordinated action is required for rapid progress. For example, education policy and technology usually fall to the education ministry, much technical worker training is the responsibility of the labor or industry ministry as well as private companies, and issues of telecommunications infrastructure and IT hardware imports are governed by a range of laws, regulations, tariffs, and business relationships. Thus achieving even a simple goal, such as increasing the training in advanced telecommunications engineering, requires shared objectives and resources among all these and other institutions. Many countries have therefore taken an integrated approach, setting up a multisectoral, or interagency committee with cabinet-level authorization to introduce collective policies and bypass many of the bureaucratic hurdles confronted by lower-level initiatives.

Another challenge in promoting national ICT development is the need to measure the progress and effectiveness of programs, both at the domestic level and through international comparisons. Toward this end, studies have proposed statistical metrics for evaluating relative conditions across countries and developed a range of methodologies for calculating a country's 'e-readiness[19].' These generally involve assessing dozens of statistical indicators along comparative scales, such as literacy levels, pricing benchmarks, and access to telephones and computers, to help identify successes as well as the areas needing most attention. One representative set of e-readiness benchmarks is that produced by the Harvard University Center for International Development, which yielded the country rankings in Table 7.

The studies tend to define the parameters of ICT development needs fairly consistently, although there have been some disputes over the country rankings they have produced. The usefulness of these studies derives not from their country scores, or rankings for a particular moment in time, however, but from what they show about how different policies and initiatives can improve a country's score or ranking over a period of years. Even in the short time that these methods have been used, the new entry in mobile markets, among other developments, has had a marked impact on the performance of several countries on these aggregate scales.

8. The path ahead

The business of telecommunications reform in developing countries is not over. About one-fourth of the developing world's telecommunications markets remain essentially closed, and about one-third have opened barely a crack. And despite the dramatic progress made in less than two decades, at the beginning of the 21st century, the majority of the world's population remains disconnected from virtually any form of modern communications.

[19] For a summary of many of these methodologies, see infoDev (2002).

Table 7
E-readiness rankings, 2001

Rank	Economy	Networked readiness index
1	United States	6.05
2	Iceland	6.03
3	Finland	5.91
4	Sweden	5.76
5	Norway	5.68
6	The Netherlands	5.68
7	Denmark	5.56
8	Singapore	5.47
9	Austria	5.32
10	United Kingdom	5.31
11	New Zealand	5.23
12	Canada	5.23
13	Hong Kong	5.23
14	Australia	5.22
15	Taiwan	5.18
16	Switzerland	5.17
17	Germany	5.11
18	Belgium	4.90
19	Ireland	4.89
20	Korea, Rep. of	4.86
21	Japan	4.86
22	Israel	4.84
23	Estonia	4.73
24	France	4.71
25	Italy	4.70
26	Spain	4.62
27	Portugal	4.57
28	Czech Republic	4.38
29	Slovenia	4.24
30	Hungary	4.14
31	Greece	4.13
32	Argentina	4.01
33	Slovak Republic	4.01
34	Chile	4.00
35	Poland	3.85
36	Malaysia	3.82
37	Uruguay	3.80
38	Brazil	3.79
39	Latvia	3.78
40	South Africa	3.71
41	Turkey	3.67
42	Lithuania	3.59
43	Thailand	3.58
44	Mexico	3.58
45	Costa Rica	3.57
46	Trinidad and Tobago	3.52
47	Dominican Republic	3.52

Table 7 (*continued*)

Rank	Economy	Networked readiness index
48	Panama	3.42
49	Jordan	3.42
50	Venezuela	3.41
51	Mauritius	3.40
52	Peru	3.38
53	Bulgaria	3.38
54	India	3.32
55	El Salvador	3.30
56	Jamaica	3.29
57	Colombia	3.29
58	Philippines	3.27
59	Indonesia	3.24
60	Egypt	3.20
61	Russian Federation	3.17
62	Sri Lanka	3.15
63	Paraguay	3.15
64	China	3.10
65	Romania	3.10
66	Ukraine	3.05
67	Bolivia	3.04
68	Guatemala	3.00
69	Nicaragua	2.83
70	Zimbabwe	2.78
71	Ecuador	2.65
72	Honduras	2.64
73	Bangladesh	2.53
74	Vietnam	2.42
75	Nigeria	2.10

Source: Harvard University Center for International Development, 2002.
Note: The Networked Readiness Index measures the potential of a country to participate in the networked world economy. The index reflects network use (defined by five variables related to quality and quantity of ICT use), and four enabling factors (network access, network policy, networked society, and networked world) each comprising a number of component indices.

8.1. The unfinished reform agenda

The countries lagging behind in telecommunications reform include most of the poorest, those least equipped to do the job but also those where the returns to reform may be the largest. Much remains to be done in promoting and supporting change in these countries, largely in line with established practice. International development organizations would be wrong to abandon efforts in these countries.

At the same time, early reformers are facing new challenges and opportunities. Although they have largely done what they set out to do, markets and technologies have in the meantime continued to develop and regulatory concerns have changed. These countries are, of course, much better equipped than at the start of the wave of reforms to deal with these matters. The wide range of players in the market ensures lively debates and a variety of viewpoints, interests, and sources of knowledge. Governments have honed their policy skills, and core regulatory competencies, in telecommunications as well as in competition, and have provided a base for moving ahead on this front as well.

Yet without a better understanding of how to manage the process of reform, taking explicitly into account political, institutional, and governance constraints, there is a risk of failure in implementing even good sectoral designs. Telecommunications reform processes have been managed largely by instinct and political savvy. In sharp contrast, work on the economic and technical design of reform in a country can draw on what is by now a well-established body of good practice, backed by a large volume of academic and professional publications specific to telecommunications. Surely it is possible to do better in managing reform as well.

Since the 1950s, economists and political scientists have worked on the question of how to implement policy reforms in general. Successive schools of thought have emerged and been applied with varying success in pursuit of economic development. Research has also looked into telecommunications. A study applying the methods of institutional economics to the analysis of telecommunications reform in six developing and developed countries yielded useful lessons on the viability of regulatory solutions given a country's institutional endowment and capacity to commit (Levy and Spiller, 1996). In addition, a toolkit for managing the process of reform was published in 2002 based on 10 years of practice in 40 developing countries, including in telecommunications (Brinkerhoff and Crosby, 2002). But this sort of knowledge and expertise has not yet been brought to bear in telecommunications reform.

8.2. Universal access

The policies and principles for market reforms in basic telecommunications apply to advanced information and communications technologies as well, but these technologies also pose additional challenges. The multipurpose community telecenter model, combined with liberalization policies, appears to be among the most promising options for bringing full-service access to information within reach of the largest number of people. The challenge is to establish commercially sustainable business models while expanding the training in e-commerce and other applications, and in communications technology in general, and increasing awareness of their potential. A growing number of examples suggest that communications services, even in rural areas, can generate a reasonable flow of income from local users. These revenues, combined with other payments—such as payments for incoming telephone traffic and government support for public service programs—offer the potential for running telecenters as profitable enterprises rather than as activities requiring permanent subsidies. As technology continues to evolve, new applications such as wireless fidelity (WiFi) may improve the cost-effectiveness of rural services even further. And as communications media converge, telecenters may be able to incorporate broadcast radio, television, and other multimedia services, expanding their market appeal and taking advantage of economies of scale (Townsend, 2002).

Still, the focus must not be on which technologies, or operational models may ultimately succeed in the marketplace, but on what economic incentives are needed to promote workable, long-term solutions. One strategy used in a growing number of countries is to establish a public funding mechanism to help provide seed capital and guidance to prospective investors in less attractive rural markets (Townsend, 2003). As discussed, experience in several Latin American countries shows that such funds, using market-based approaches, can help expand access far beyond traditional barriers.

Finally, a critically neglected need is to promote the development of culturally diverse content with information relevant for all societies. Nearly 70 percent of Websites are in English, and most software is heavily U.S.-centric in design and function. Sorely lacking are educational and cultural media emphasizing the languages, interests, and experiences of rural societies and developing cultures—from books to videos to interactive programs. As access to technology continues to spread and market mechanisms further open up the economic potential of these neglected populations, there will be an ever increasing demand to tap their knowledge, experiences, and traditions for the lasting benefit of all humanity.

References

Artana, D., F. Navajas and S. Ubizondo, 1998, Contractual adaptation in regulated utilities: A few observations from Argentina, Buenos Aires: Fundación de Investigaciones Económicas Latinoamericanas.

Bhatnagar, S. and R. Schware eds., 2000, Information and Communication Technology in Rural Development: Case Studies from India, Washington, DC: World Bank Institute.

Bond, J., 1997, The drivers of the information revolution-cost, computing power, and convergence, Public Policy for the Private Sector (World Bank, Finance, Private Sector, and Infrastructure Network, Washington, DC), September, 27–30.

Brinkerhoff, D. W. and B. L. Crosby, 2002, Managing Policy Reform, Bloomfield, CT: Kumarian Press.

Brooks, A., 2000, Internet interconnection, in ITU, Trends in telecommunication reform 2000–2001, Geneva: International Telecommunication Union.

Bruce, R., I. Kessides and L. Kneiffel, 1999, Overcoming obstacles to liberalization of the telecommunications sector in Estonia, Poland, the Czech Republic, Slovenia, and Hungary, World Bank technical Paper 440, Washington, DC: World Bank.

Caceres, M. and F. Sudweeks, 2000, The diffusion of the Internet in Chile, Perth, Australia:Murdoch University, School of Information Technology, Available: http://www.it.murdoch.edu.au/~sudweeks/papers/chile.pdf.

Casasús, C., 1994, Privatization of telecommunications: The case of Mexico, in B. Wellenius and P. A. Stern, eds., Implementing Reforms in the Telecommunications Sector: Lessons from Experience, World Bank Regional and Sectoral Study, Washington, DC: World Bank.

CellularOnline, 2003, Africa mobile statistics, Available: http://www.cellular.co.za/.

DELGI, 2003, Les TICs au Burkina, Burkina Faso, Office of the Prime Minister, Délégation Générale à l'Informatique, Available: http://www.delgi.gov.bf/.

Fink, C., A. Mattoo and R. Rathindran, 2002, Liberalizing basic telecommunications: Evidence from developing countries, World Bank, Development Research Group, Washington, DC.

Gates, W., 2003, A spam-free future, Washington Post, November 24, p. A21.

Geller, H., 1989, U.S. telecommunications policy: Increasing competition and deregulation, in B. Wellenius, P. A. Stern, T. E. Nulty and R. D. Stern, eds., Restructuring and Managing the Telecommunications Sector, A World Bank Symposium, Washington, DC: World Bank.

Harvard University Center for International Development, 2002, The global information technology report 2001–2002: Readiness for the networked world, New York: Oxford University Press, Available: http://www.cid.harvard.edu/.

infoDev, 2002, Illustrative examples of e-readiness assessments and assessment methods, Available: http://www.infodev.org/ereadiness/methodology.htm.

ITU, 2001a, Effective regulation case study, Morocco, Geneva: International Telecommunication Union.

ITU, 2001b, Trends in telecommunication reform 2000–2001, Geneva: International Telecommunication Union.

ITU, 2001c, World telecommunication indicators, Geneva: International Telecommunication Union.

ITU, 2002a, Final report WTDC–02, Istanbul, Turkey, Geneva: International Telecommunication Union.

ITU, 2002b, World telecommunication development report 2002, Geneva: International Telecommunication Union.

ITU, 2003, Declaration of principles: Building the information society—A global challenge in the new millennium, Document WSIS–03/GENEVA/DOC/4-E, Geneva: International Telecommunication Union.

ITU, 2004, Workshop on radio spectrum management for a converging world, Available: http://www.itu.int/osg/spu/ni/spectrum.

Izaguirre, A. K., 1999, Private participation in telecommunications—Recent trends, Viewpoint Note 204, World Bank, Finance, Private Sector, and Infrastructure Network, Washington, DC.

Izaguirre, A. K., 2004, Private infrastructure—Activity down by 30 percent in 2002, Viewpoint Note 267, World Bank, Private Sector Development Vice Presidency, Washington, DC.

Levy, B. and P. Spiller eds., 1996, Regulations, Institutions, and Commitment, Cambridge and New York: Cambridge University Press.

Männistö, L., 1999, Examining the opportunities and challenges for ISPs in developing countries, paper presented at ISP Forum, IIR, Amsterdam, November 30, International Telecommunication Union, Geneva.

Mansell, R. and U. When eds., 1998, Knowledge Societies: Information Technology for Sustainable Development, New York: Oxford University Press.

Navas-Sabater, J., A. Dymond and N. Juntunen, 2002, Telecommunications and information services for the poor, World Bank Discussion Paper 432, Washington, DC: World Bank.

NICTS, 2003, National Information Communication Technology Strategy Project, Azerbaijan, Available: http://www.nicts.az/english/1yearen.htm.

NUA Internet Surveys, 2004, Available: http://www.nua.ie/surveys/index.cgi.

Nulty, T. E. and E. Schneidewinde, 1989, Regulatory policy for telecommunications Restructuring and Managing the Telecommunications Sector, in B. Wellenius, P.A. Stern, T. E. Nulty and R. D. Stern, eds., A World Bank Symposium, Washington, DC: World Bank.

O' Connor, D., 2003, Of flying geeks and o-rings: Locating software and IT services in India's economic development, OECD Development Centre Technical Paper 12, Available: http://www.oecd.org/dev/technics/IndiaIT.pdf.

Pyramid Research, 1995, Analysis of telecommunications investment and financing in less developed countries, prepared for the World Bank, Telecommunications and Informatics Division, Washington, DC.

Rossotto, C., K. Sekkat and A. Varoudakis, 2003, Opening up telecommunications to competition and MENA integration in the world economy, Middle East and North Africa Working Paper 33, World Bank, Middle East and North Africa Region, Office of the Chief Economist, Washington, DC.

Rossotto, C., B. Wellenius, A. Lewin and C. G. Gomez, 2004, Competition in international voice communications, World Bank, Global ICT Department, Policy Division, Report 27671, Washington, DC.

Samarajiva, R., A. Mahan and A. Barendse, 2002, Multisector utility regulation, Technical University of Denmark, Center for Tele-Information, World Dialogue on Regulation in Network Economies, Lyngby, Available: http://www.regulateonline.org.

Saunders, R., J. Warford and B. Wellenius, 1983, Telecommunications and Economic Development, London and Baltimore: Johns Hopkins University Press.

Saunders, R., J. Warford and B. Wellenius, 1994, Telecommunications and Economic Development, 2nd edn., London and Baltimore: Johns Hopkins University Press.

SEPTI, 2003, Plan d'action e-Maroc, Morocco, Secrétariat d'Etat auprès du Premier Ministre chargé de la Poste et des Technologies des Télécommunication et de l'Information, Available: http://www.septi.gov.ma/AdminenLigne/AdminenLigne.htm.

SIGET, 1997, Ley de Telecomunicaciones, Decreto Legislativo 142, El Salvador, Superintendencia General de Electricidad y Telecomunicaciones, Available: http://www.siget.gob.sv/.

Smith, P. and B. Wellenius, 1999, Mitigating regulatory risk in telecommunications, Viewpoint Note 189, World Bank, Finance, Private Sector, and Infrastructure Network, Washington, DC.

SUBTEL, 2002, Estadísticas del sector de las telecomunicaciones en Chile, Santiago: Subsecretaría de Telecomunicaciones.

SUBTEL, 2003, Análisis de estadísticas por hogar del sector telecomunicaciones según censo de población y vivienda Chile 2002, Informe Estadístico 7, Santiago: Subsecretaría de Telecomunicaciones.

Townsend, D., 2002, The World Cup and communications development: A new vision? Paper presented at Universal Access and Rural Connectivity, Regional Workshop for Southern and Eastern Africa, Telecommunications Regulators Association of Southern Africa, International Telecommunication Union, and Commonwealth Telecommunications Organization, Dar es Salaam; and Seminar on Digital Opportunity for All: ICTs and the Fight against Poverty, Asia Pacific Telecommunity, Chiang Rai, Available: http://www.dntownsend.com/dnta/Reports/DNT%20WCup%20Com.doc.

Townsend, D., 2003, Model universal service fund: Policies and procedures, in ITU, Trends in telecommunication reform 2003, Geneva: International Telecommunication Union.

Tschang, T., 2003, China's software industry and its implications for India, OECD Development Centre Technical Paper 205, Available: http://www.oecd.org/dev/technics/ChinaIT.pdf.

UNCTAD, 2002, E-commerce and development report 2002, Geneva: United Nations Conference on Trade and Development.

Wallsten, S., 2000, Telecommunications privatization in developing countries: The real effects of exclusivity periods, Stanford, CA: Stanford University; and Washington, DC: World Bank.

Wallsten, S., 2003, Regulation and Internet use in developing countries, Policy Research Working Paper 2979, World Bank, Economic Development Vice Presidency, Washington, DC.

Wellenius, B., 1969, Hidden residential connections demand in the presence of severe supply shortages, IEEE Transactions on Communication Technology.

Wellenius, B., 1997, Telecommunications reform: How to succeed, Viewpoint Note 130, World Bank, Finance, Private Sector, and Infrastructure Network, Washington, DC.

Wellenius, B., 2000a, Extending telecommunications beyond the market: Toward universal service in competitive environments, Viewpoint Note 206, World Bank, Finance, Private Sector, and Infrastructure Network, Washington, DC.

Wellenius, B., 2000b, Regulating the telecommunications sector: The experience of Latin America, in L. Manzetti, ed., Regulatory Policy in Latin America: Postprivatization Experience, Miami: North-South Center Press.

Wellenius, B., 2002, Closing the Gap in Access to Rural Communications: Chile 1995–2002, World Bank Discussion Paper 430, Washington, DC: World Bank.

Wellenius, B. and C. Rossotto, 1999, Introducing telecommunications competition through a wireless license-lessons from Morocco, Viewpoint Note 199, World Bank, Finance, Private Sector, and Infrastructure Network, Washington, DC.

Wellenius, B. and G. Staple, 1996, Beyond Privatization: The Second Wave of Telecommunications Reforms in Mexico, World Bank Discussion Paper 341, Washington, DC: World Bank.

Wellenius, B. and P. A. Stern eds., 1994, Implementing Reforms in the Telecommunications Sector: Lessons from Experience, World Bank Regional and Sectoral Study, Washington, DC: World Bank.

Wellenius, B., C. Braga and C. Qiang, 2000, Investment and growth of the information infrastructure: Summary results of a global survey, Telecommunications Policy, 24, 639–643.

Wellenius, B., V. Foster and C. M. Calvo, 2004, Private provision of rural infrastructure services: Competing for subsidies, Policy Research Working Paper, World Bank, Economic Development Vice Presidency, Washington, DC.

Wellenius, B., P. A. Stern, T. E. Nulty and R. D. Stern eds., 1989, Restructuring and Managing the Telecommunications Sector, A World Bank Symposium, Washington, DC: World Bank.

World Bank, 2002, Information and Communication Technologies: A World Bank Group Strategy, Washington, DC: World Bank.

World Bank, 2003, ICT and MDGs—A World Bank Group Perspective, World Bank, Global ICT Department, Washington, DC.

Chapter 15

INSTITUTIONAL CHANGES IN EMERGING MARKETS: IMPLICATIONS FOR THE TELECOMMUNICATIONS SECTOR

PABLO T. SPILLER

Contents

Handbook of Telecommunications Economics, Volume 2, Edited by S. Majumdar et al.

DOI: 10.1016/S1569-4054(05)02015-4

1. Introduction

Oliver Williamson's *Markets and Hierarchies* opened a new approach to institutions. In *Markets and Hierarchies* institutions are not seen simply as organigrams or chains of command whose layers are determined by the workings of communications channels. Instead, institutions are now seen as designed to deal with a basic time inconsistency inherent to most transactions, namely opportunism coupled with the inability to write fully contingent contracts. Indeed, in explaining the differences with the prior literature, Williamson writes (1975, p. 7):

> ...(2) I expressly introduce the notion of opportunism and am interested in the ways that opportunistic behavior is influenced by economic organization. (3) I emphasize that it is not uncertainty or small numbers, individually or together that occasion market failure, but it is rather the *joining* of these factors with bounded rationality on the one hand and opportunism on the other that gives rise to exchange difficulties.

The main point of this essay is that to understand the impact of institutional change in emerging markets and its implications for telecommunications, a deeper analysis of opportunism is necessary. Paraphrasing Williamson (1975, p.7), "opportunism, in a rich variety of forms, is made to play a central role in the analysis of [institutions] herein." In this paper, the focus is mostly on the role of institutions and its implications for regulatory commitment, and in particular, for regulatory commitment in the telecommunications sector. As discussed below, the organizational problems arising from regulation (the relation between the different components of government, the regulated firms and other interested parties) are particularly amenable to analysis using a synthesis of transaction cost economics and positive political theory. I will further submit that understanding the organizational implications of opportunism is the key to understand the organization of society in its various forms, and in particular, to understand the organization of the different branches of government of which regulatory agencies are just one[1].

2. The utilities' problem

Three features characterize utilities: first, their technologies are characterized by large specific, sunk, investments[2]; second, their technologies are characterized

[1] In this sense, the earlier work of McCubbins et al. (1987, 1989) on administrative procedures in the U.S. can be seen as an attempt to show how the administrative process is designed so as to limit the potential for opportunistic behavior by the different parts of government. For further elaborations on the "process provides commitment issue, (see Macey, 1992; Shepsle, 1992)."

[2] Specific or sunk investments are those that once undertaken their value in alternative uses is substantially below their investment cost.

by important economies of scale and scope; and third, their products are massively consumed. Consider, for example, a local telephone company. Its outside plant has very little value in alternative uses (it is very expensive to bring down cables and posts, to dig out trenches, etc)[3]; network externalities and economies of density imply that it may not be economical to have multiple telephone wires deployed along the same residential street; and finally, its product is consumed by a large proportion of the city's population. Compare this situation to that of another industry characterized by large sunk investments: steel. While steel mills have very little value in alternative uses, the economies of scale and scope are trivial compared to the size of the market[4], and furthermore, while everybody indirectly consume steel products, very few individuals in society pay any attention to the price of steel. Thus, it is neither simply specific investments that characterize utilities, nor it is simply economies of scale.

Now consider newspapers. It is quite clear that there are large economies of scale and scope in the operation of city-newspapers. The increase in speed of communications and the increased use of computer design has drastically increased the extent of economies of scale in the sector. This has translated in a reduction in the number of newspapers per city while maintaining relatively constant the readership numbers. While readers are usually a relatively large portion of the population (at least of the voting population), newspapers are not utilities. The reason being, that while there may be substantial amount of sector-specific human capital (reporters' contacts with local politicians may be specific to the locality), the technology is increasingly generic. Shutting down a newspaper and moving the printing presses, their desks, computers, etc, elsewhere is becoming more and more common.

Thus, what separates the utility sector from the rest of the economy is the combination of three features: specific investments, economies of scale, and widespread domestic consumption. These features are at the core of contracting problems that have traditionally raised the need for governmental regulation of utilities[5]. In turn, they make the pricing of utilities inherently political.

The reason for the politicization of infrastructure pricing is threefold. First, the fact that a large component of infrastructure investments is sunk, implies that once the investment is undertaken the operator will be willing to continue operating as long as operating revenues exceed operating costs. Since operating costs do not include a return on sunk investments (but only on the alternative value of these assets), the operating company will be willing to operate even if

[3] Although the development of fiber optics has increased the value in alternative uses of their conducts, rights of way, and telephone poles.

[4] In some developing countries that protect the production of steel, that may not be so, as there may be just a few steel mills producing for the relatively small local market.

[5] See, among others, North (1990), Williamson (1988), Goldberg (1976), Barzel (1989), Levy and Spiller (1994) and Spiller (1993).

prices are below total average costs[6]. Second, economies of scale imply that in most utility services, there will be few suppliers in each locality. Thus, the whiff of monopoly will always surround utility operations. Finally, the fact that utility services tend to be massively consumed implies that politicians and interest groups will care about the level of utility pricing. Thus, massive consumption, economies of scale, and sunk investments provide governments (either national or local) with the opportunity to behave opportunistically vis-à-vis the investing company[7]. For example, after the investment is sunk, the government may try to restrict the operating company's pricing flexibility, including sales to competitors, may require the company to undertake special investments, purchasing or employment patterns or may try to restrict the movement of capital. All these are attempts to expropriate the company's sunk costs by administrative measures. Thus, expropriation may be indirect and undertaken by subtle means.

Expropriation of the firm's sunk assets, however, does not mean that the government takes over the operation of the company, but rather that it sets operating conditions that just compensate for the firm's operating costs and the return on its nonspecific assets. Such returns will provide sufficient *ex post* incentives for the firm to operate, but not to invest[8]. Indeed, sunk assets expropriation has been more prevalent in Latin America than direct utility takeovers or expropriation without compensation[9]. While the government may uphold and

[6] Observe that the source of financing does not change this computation. For example, if the company is completely leveraged, a price below average cost will bring the company to bankruptcy, eliminating the part of the debt associated with the sunk investments. Only the part of the debt that is associated with the value of the nonsunk investments would be able to be subsequently serviced.

[7] Observe that this incentive exists vis-à-vis public and private companies. On this issue (Spiller and Savedoff, 2000).

[8] The company will be willing to continue operating because its return from operating will exceed its return from shutting down and deploying its assets elsewhere. On the other hand, the firm will have very little incentive to invest new capital as it will not be able to obtain a return. While it is feasible to conceive loan financing for new investments, as nonrepayment would bring the company to bankruptcy, that will not however be the case. Bankruptcy does not mean that the company shuts down. Since the assets are specific, bankruptcy implies a change of ownership from stockholders to creditors. Now creditors' incentives to operate will be the same as the firm, and they would be willing to operate even if quasi-rents are expropriated. Thus, loan financing will not be feasible either.

[9] Consider, for example, the case of Montevideo's Gas Company. Throughout the 1950s and 1960s the MGC, owned and operated by a British company, was denied price increases. Eventually, during the rapid inflation of the 1960s it went bankrupt and was taken over by the government. Compare this example to the expropriation by the Perón administration of ITT's majority holdings in the Unión Telefónica del Rio de la Plata (UTRP was the main provider of telephones in the Buenos Aires region). In 1946 the Argentinean government paid US$95million for ITT's holdings, or US$677M in 2002 prices. Given UTRP's 457,800 lines, it translates at US$1372 per line in 2002 prices (deflator: capital equipment producer prices). Given that in today's prices, the marginal costs of a line in a large metropolitan city is approximately US$650, the price paid by the Perón administration does not seem unusually low. See Hill and Abdalla (1993).

protect traditionally conceived property rights, it may still attempt to expropriate through regulatory procedures.

2.1. The political profitability of expropriation

Sunk assets' expropriation may be profitable for a government if the direct costs (reputation loss vis-à-vis other utilities and lack of future investments by utilities) are small compared to the (short term) benefits of such action (achieving reelection by reducing utilities' prices, by attacking the monopoly, etc), and if the indirect institutional costs (e.g., disregarding the judiciary, not following the proper, or traditional, administrative procedures, etc) are not too large.

Thus, incentives for expropriation of sunk assets should be expected to be largest in countries where indirect institutional costs are low (e.g., there are no formal or informal governmental procedures—checks and balances—required for regulatory decision making; regulatory policy is centralized in the administration; the judiciary does not have a tradition of, or the power to, reviewing administrative decisions, etc), direct costs are also small (e.g., neither the utilities in general require massive investment programs, nor technological change is an important factor in the sector), and, perhaps, more importantly, the government's horizon is relatively short (i.e., highly contested elections, need to satisfy key constituencies, etc). Forecasting such expropriation, private utilities will not undertake investments in the first place. Thus, government direct intervention may become the default mode of operation.

Emerging markets provide counterbalancing factors in telecommunications. On the one hand, many emerging markets do not have well functioning telecommunications sectors, so the need to invest is high. In these environments opportunistic behavior should be expected late in the investment program. On the other hand, many emerging markets are characterized by substantial administrative discretion, weak judiciary systems and unstable political environments. In those environments, opportunistic behavior may be expected earlier on. Institutional changes, both in terms of regulatory structures and regulatory regimes, then, may be required to promote private investment in the sector.

2.2. The implications of government opportunism

Thus, in the presence of an important threat of opportunistic behavior, governments that want to motivate private investment will have to design institutional arrangements that will limit its own ability to behave opportunistically once the private utility undertook its investment program. Such institutional arrangements are nothing but the design of a regulatory framework. Such institutional arrangement will have to stipulate price setting procedures, conflict resolution procedures (arbitration or judicial) between the parties, investment policies, etc. In other words, regulation, if credible, solves a key contracting problem between the government and the utilities by restraining the government from opportunistically expropriating the utilities' sunk investments[10]. This, however, does not mean that the utility has to receive assurances of a rate of return nature, or that it has to receive exclusive licenses[11]. In some countries, however, such assurances may be the only way to limit the government's discretionary powers.

Unless such regulatory framework is credible though, investments will not be undertaken, or if undertaken will not be efficient. Investment inefficiencies may arise in several fronts. A first order effect is underinvestment. Although the utility may invest, it will do so exclusively in areas whose market return is very high and where the payback period is relatively short[12]. Second, maintenance expenditures may be kept to the minimum, thus degrading quality. Third, investment may be undertaken with technologies that have a lower degree of specificity, even at the cost of, again, degrading quality[13]. Fourth, up-front rents may be achieved by very high prices, which although may provide incentives for some investment, may be politically unsustainable[14]. The costs then of noncredible environments are high. Even in environments where the lack of credibility is not as high as to prevent private participation, Guasch and Spiller (1999) find that the lack of credibility may increase utilities cost of capital up to 6 percentage basis points. This is a very high cost. For example, a 5% increase in cost of capital would lead to a reduction of 30–35% in the sale price of a 25-year concession.

A noncredible regulatory framework, then by creating strong inefficiencies and poor performance will eventually create the conditions for a direct government take-over. Thus, government ownership may become the default mode of operations. Government ownership, then, reflects the inability of the polity to develop

[10] See, Goldberg (1976) for one of the first treatments of this problem (Williamson, 1976).

[11] Indeed, the Colombian regulation of value added networks specifically stipulated that the government cannot set their prices, nor that there is any exclusivity provision. Thus, regulation here meant total lack of governmental discretion.

[12] An alternative way of reducing the specificity of the investment is by customers undertaking the financing of the sunk assets to the customers. Thus, for example, in several Latin American countries—and also in Japan—the installation charge for a telephone line used to range in the initial deregulatory periods from $400 to $600. Given that the investment cost of a new line ranges from $600 to $800 in less developed countries, that initial charge essentially protects the utility from the expropriation of its sunk investments.

[13] In this sense it is not surprising that private telecommunications operators have rushed to develop wireless rather than fixed link networks in Eastern European countries. While wireless has a higher long run cost than fixed link, and on some quality dimensions is also an inferior product, the magnitude of investment in specific assets is much smaller than in fixed link networks. Furthermore, a large portion of the specific investments in cellular telephony are undertaken by the customers themselves (who purchase the handsets).

[14] The privatization of Argentina's telecommunications companies is particularly illuminating. Prior to the privatization, telephone prices were raised well beyond international levels. It is not surprising, that following the privatization the government reneged on aspects of the license, like indexation. The initial high prices, though, allowed the companies to remain profitable, even following government's deviation from the license provisions. See Spiller (1993).

regulatory institutions that limit the potential for government opportunistic behavior.

3. Sources of regulatory commitment

In Levy and Spiller (1994) we argued that the credibility and effectiveness of a regulatory framework—and hence its ability to facilitate private investment—varies with a country's political and social institutions.

Political and social institutions not only affect the ability to restrain administrative action, but also have an independent impact on the type of regulation that can be implemented, and hence on the appropriate balance between commitment and flexibility. For example, relatively efficient regulatory rules (e.g., price caps, incentive schemes, and use of competition) usually require granting substantial discretion to the regulators. Thus, unless the country's institutions allow for the separation of arbitrariness from useful regulatory discretion, systems that grant too much administrative discretion may not generate the high levels of investment and welfare expected from private sector participation. Conversely, some countries might have regulatory regimes that drastically limit the scope of regulatory flexibility. Although such regulatory regimes may look inefficient, they may in fact fit the institutional endowments of the countries in question, and may provide substantial incentives for investment.

I like to look at regulation as a 'design' problem[15]. Regulatory design has two components: regulatory governance and regulatory incentives. The governance structure of a regulatory system comprises the mechanisms that societies use to constrain regulatory discretion, and to resolve conflicts that arise in relation to these constraints. On the other hand, the regulatory incentive structure comprises the rules governing utility pricing, cross-subsidies or direct-subsidies, entry, interconnection, etc. While regulatory incentives may affect performance, a main insight from Levy and Spiller (1994) is that the impact of regulatory incentives (whether positive or negative) comes to the forefront only if regulatory governance has successfully been put into place. Now, regulatory governance is a choice, although a constrained one. The institutional endowment of the country limits the menu of regulatory governance available. Thus, regulatory commitment has two sources: the institutional endowment and regulatory governance. Both will be discussed next.

3.1. Institutional endowment

Levy and Spiller (1994) define the institutional endowment of a nation as comprising five elements: first, a country's legislative and executive institutions. These

[15] The concept of regulation as a design problem was first introduced in Levy and Spiller (1993). I use here the terminology subsequently developed in Levy and Spiller (1994).

are the formal mechanisms for appointment of legislators and decision makers, for making laws and regulations, apart from judicial decision making; for implementing these laws, and that determine the relations between the legislature and the executive. Second, the country's judicial institutions. These comprise its formal mechanisms for appointing judges and determining the internal structure of the judiciary; and for resolving disputes among private parties, or between private parties and the state. Third, custom and other informal but broadly accepted norms that are generally understood to constrain the action of individuals or institutions. Fourth, the character of the contending social interests within a society, and the balance between them, including the role of ideology. Finally, the administrative capabilities of the nation. Each of these elements has implications for regulatory commitment. The focus will be on the first two.

The form of a country's legislative and executive institutions influences the nature of its regulatory problems. The crucial issue is to what extent the structure and organization of these institutions impose constraints upon governmental action. The range of formal institutional mechanisms for restraining governmental authority includes: the explicit separation of powers between legislative, executive and judicial organs of government[16]; a written constitution limiting the legislative power of the executive, and enforced by the courts; two legislative houses elected under different voting rules[17]; an electoral system calibrated to produce either a proliferation of minority parties or a set of parties whose ability to impose discipline on their legislators is weak[18]; and a federal structure of power, with strong decentralization even to the local level[19]. Utility regulation is likely to be far more credible—and the regulatory problem less severe—in countries with political systems that constrain executive discretion. Note, however, that credibility is often achieved at the expense of flexibility. The same mechanisms that make it difficult to impose arbitrary changes in the rules may also make it difficult to enact sensible rules in the first place, or to efficiently adapt the rules in the face of changing circumstances. Thus, in countries with these types of political institutions, the

[16] For analysis of the role of separation of powers in diminishing the discretion of the executive (Gely and Spiller, 1990; McCubbins et al. (1987, 1989, and references therein).

[17] Nonsimultaneous elections for the different branches of government tend to create natural political divisions and thus electoral checks and balances (Jacobson, 1990). For an in-depth analysis of the determinants of the relative powers of the executive (Shugart and Carey, 1992).

[18] Electoral rules have also important effects on the 'effective number of parties' that will tend to result from elections, and thus, the extent of governmental control over the legislative process. For example, it is widely perceived that proportional representation tends to generate a large number of parties, while first-past-the-post with relatively small district elections tends to create bipolar party configurations. This result has been coined Duverger's Law in political science (Duverger, 1954). More generally, see Taagepera and Shugart (1993). For analyses of how the structure of political parties depends on the nature of electoral rules (with applications to the U.K.) (Cain et al., 1987, Cox, 1987).

[19] On the role of Federalism in reducing the potential for administrative discretion, see Weingast (1995) and references therein.

introduction of reforms may have to await the occurrence of a drastic shock to the political system.

Legislative and executive institutions may also limit a country's regulatory governance options. In some parliamentary systems, for example, the executive has substantial control over both the legislative agenda and the legislative outcomes[20]. In such countries, if legislative and executive powers alternate between political parties with substantially different interests, specific legislation need not constitute a viable safeguard against administrative discretion, as changes in the law could follow directly from a change in government[21]. Similarly, if the executive has strong legislative powers, administrative procedures and administrative law by themselves will not be able to constrain the executive, who will tend to predominate over the judiciary in the interpretation of laws. In this case, administrative procedures require some base other than administrative law.

A strong and independent judiciary could serve as the basis for limiting administrative discretion in several ways. For example, the prior development of a body of administrative law opens the governance option of constraining discretion through administrative procedures[22]. Also, a tradition of efficiently upholding contracts and property rights opens the governance option of constraining discretion through the use of formal regulatory contracts (licenses). This option is particularly valuable for countries where the executive has a strong hold over the legislative process. Further, a tradition of judicial independence and efficiency opens the governance option of using administrative tribunals to resolve conflicts between the government and the utility within the contours of the existing

[20] While parliamentary systems grant such powers in principle, whether they do so in practice depends upon the nature of electoral rules and the political party system. Parliamentary systems whose electoral rules bring about fragmented legislatures would not provide the executive—usually headed by a minority party with a coalition built on a very narrow set of specific common interests—with much scope for legislative initiative. By contrast, electoral rules that create strong two-party parliamentary systems—as well as some other kinds of nonparliamentary political institutions—would grant the executive large legislative powers. For an in depth discussion of the difference between parliamentary and presidential systems, and the role of electoral rules in determining the relative power of the executive (Shugart and Carey, 1992).

[21] In the U.K., regulatory frameworks have traditionally evolved through a series of acts of Parliament. For example, major gas regulation legislation was passed in 1847, 1859, 1870, 1871, 1873, and 1875. Similarly, water regulation legislation was passed in 1847, 1863, 1870, 1873, 1875, and 1887. Systematic regulation of electricity companies started in 1882, only four years after the inauguration of the first public demonstration of lighting by a public authority. The 1882 Act was followed by major legislation in 1888, 1899, 1919, and 1922, and culminating with the Electricity (Supply) Act of 1926 creating the Central Electricity Board. See Spiller and Vogelsang (1993b), for discussions of the evolution of utility regulation in the U.K., and references therein.

[22] This has traditionally been the way administrative discretion is restrained in the U.S., as regulatory statutes have tended to be quite vague. For an analysis of the choice of specificity of statutes, see Schwartz et al. (1993). Observe, however, that administrative law may not develop in a system where the executive has strong control over the legislative process. This issue is discussed in detail below.

regulatory system. Finally, it provides assurances against governmental deviation from specific legislative or constitutional commitments that underpin the regulatory system.

3.2. A tale of two countries

In the remaining part of the paper, a theory will be developed that relates the design of regulatory governance to the nature of the country's institutional environment. Essentially, two types of institutional environments will be compared: one in which governmental decisions are taken in a decentralized fashion and the other where governmental decision making is heavily centralized. The U.S. will be used as an example of the former, and the U.K. as an example of the latter category. The widely variant evolution of regulation in these two countries provides a useful contrast for illustrating and exploring the relation between institutions and commitment. Consider the case of telecommunications.

The main body of telecommunications legislation in the United States is now the Communications Act (FCA) of 1934[23], as amended by the Telecommunications Act of 1996[24]. The 1934 Act specifically directed the newly created Federal Communications Commission to regulate interstate communications so as to provide telecommunications services at 'just, fair and reasonable prices.' Nowhere in the Act were there any specific instructions about how to obtain that general goal. Furthermore, the Act presumed the existence of a monopoly supplier of long distance services. The fostering of competition was not one of the stated goals of the Act.

Even though the FCA was silent about competition, in the late 1950s the FCC started a process of partially deregulating the long distance and the customer provided equipment segments of the industry. This was a measured process. Deregulation (and entry) was allowed selectively. In long distance telecommunications, deregulation started essentially with the *Above 890* decision of 1959. In this decision, the FCC determined that any private microwave system which met proper technical criteria would be authorized to operate, even though common carrier facilities which could provide service to the private system applicant already existed[25]. This decision was followed a decade latter by allowing a small firm, MCI, to construct microwave facilities between Chicago and St. Louis. In contrast to AT&T, MCI did not propose at that time to offer end-to-end exchange service. Instead, customers had to provide their own loop service between MCI's microsites and their own facilities. Ordinarily, this would be done through lines leased from local phone companies[26]. A year later, in the *Specialized Common*

[23] 47 U.S.C. 151 (1934).

[24] The official citation for the new Act is: Telecommunications Act of 1996, Pub. LA. No. 104–104, 110 Stat. 56 (1996).

[25] 27 F.C.C. 359 (1959).

[26] 18 F.C.C. 2d 953 (1970).

Carriers Decision[27], the FCC granted other companies the right to provide similar specialized private line communications. In other decisions, the FCC allowed entry of some value-added carriers, providing more sophisticated services than those available[28].

This deregulatory process has been called by industry observers as the 'slippery slope[29],' as it eventually led to the break up of AT&T. While the FCC was a major force behind these deregulatory moves, the Courts were also active participants in deregulation. The first major judicial interventions concerning long distance services were the so called *Execunet* decisions[30]. In its 1976 *Execunet* decision, the FCC opened the private lines market for competition, while at the same restricted it entry to the standard interstate phone service[31], thus prohibiting MCI from continuing to offer its *Execunet* service[32]. MCI appealed to the U.S. Court of Appeals of the D.C. Circuit, that a year later issued its landmark *Execunet* decision, which reversed the agency. This decision was followed by two decisions (*Execunet II* and *Execunet III*)[33], which required AT&T to provide whatever interconnection was needed, and imposed similar requirements to independent local telephone companies.

The *Execunet* decisions fully opened the long distance services market for competition[34]. The Courts, however, rather than the agency or Congress, made public policy. The FCC was opposed to the Court's decision, in particular as it came just 1 year after a major FCC investigation. Finally, the main regulatory change of the 1980s, the break up of AT&T, was accomplished as a result of a settlement of the Justice Debarment's antitrust suit against AT&T, rather than by legislation or standard regulatory procedures. Furthermore, it was the Courts that managed the break-up process rather than the FCC.

Only when the settlement of the AT&T litigation (the modified final judgment) was in danger of being reversed by the Supreme Court did Congress acted and enshrined many of the restrictions of the settlement into legislation, while at the

[27] 29 F.C.C. 2d 871 (1971).

[28] See Knieps and Spiller (1983) for a detailed discussion of this process, and of the incentives faced by the FCC to engage in the partial deregulation process.

[29] See Henck and Strassburg (1988).

[30] A major previous deregulatory court decision took place in 1956 when the Court reversed an FCC decision prohibiting the use of a customer provided phone attachment. *Hush-a-Phone vs. United States*, 238 F.2d 266 (D.C. Cir. 1956).

[31] What is called in industry jargon, the Message Toll Service (MTS).

[32] 60 F.C.C 2d 25 (1976).

[33] *MCI Telecommunications Corp. vs. FCC* (*Execunet I*), 561 F.2d 365 (D.C. Cir. 1977), *MCI Telecommunications Corp. vs. FCC (Execunet II)*, 580 F.2d 590 (D.C. Cir. 1978), *Lincoln Tel. and Tel. Co. vs. FCC, (Execunet III)*, 659 F.2d 1092 (D.C. Cir. 1981).

[34] The major remaining problem was what price should the competitors pay for access to the network. This conflict was resolved through negotiations between AT&T and the specialized carriers (Temin, 1987) pp. 137–142.

same time opening local exchange markets for competition and providing the RBOCs the ability to enter into long distance markets, although under tight conditions, to be administered by the FCC[35].

Compare this process to the evolution of regulation in the U.K[36]. Untill quite recently the British telecommunications services sector could be safely identified with British Telecom (BT) and its predecessor, the telecommunications division of the British Post Office. The Post Office controlled, maintained and developed both the telephone network and controlled the supply and maintenance of terminal equipment. The Post Office operated as a Department of State under the direct control of a Minister of the Crown until 1969, at which time it was converted into a public corporation. The Telecommunications Division was separated from the Post Office in 1981 when British Telecom became a public corporation[37]. In 1984, before its privatization, BT was converted into a public limited company (BT plc)[38].

Prior to privatization, telecommunications policy, as relates to prices, investments, and technology adoption, was the result of the interaction among the Post Office (later on BT), the Secretary of State, Parliament's Select Committee on Nationalized Industries, and user groups as represented (e.g., in the Post Office Users' National and Regional Councils). During the postwar period there were substantial changes in telecommunications policy and in the controls over and the organization of the Post Office and its telecommunications division (see Boxes 1 and 2.) Although some regulatory changes involved specific legislative acts, most did not and were undertaken by cabinet, promoted through the commissioning of White Papers and implemented via executive orders.

The fluidity of governmental policy suggests that postwar governments had substantial discretion over regulatory policy and prices, although not necessarily

[35] See Spulber, this volume, for a discussion of the importance of the MFJ and the 1996 Act.

[36] For a detailed analysis of the evolution of telecommunications in the U.K. (Spiller and Vogelsang, 1993a).

[37] Prior to the separation of BT from the Post Office, the Post Office had three business centers, Posts, Giro, and Telecommunications and Data Processing, each of which had its own profit and loss accounts and balance sheet.

[38] U.K. public enterprises have usually been organized in three ways: First, as a department of a particular ministry; second, as a public corporation; and finally, as a private company organized according to the Companies' Act, only that its majority shareholder is the government. The level of involvement of the Minister, and of Parliamentary committees (in particular the Select Committee on Nationalized Industries), falls as we move from a Department of State to a public corporation, to a private company where the government is the majority shareholder. The main difference between the first and second type of organization is that in the latter the company may borrow and can maintain its own reserves. Its board, rather than being headed by a Minister, as in a Department of State, is appointed by the Minister, who is also directly involved in the company's long term strategic planning, while leaving to the Board the day to day management. In either case, though, the government has substantial discretion on the management of the enterprise. The private company form of organization, though, limits slightly more government discretion.

Box 1 Specific Telecommunications Policies from 1960 to 1980

The 1969 Post Office Act
- The Minister to make appointments to the Board
- The Minister to control investment program and borrowing
- The Post Office to provide universal service
- A Post Office Users' National Council (POUNC) and three county councils for Scotland, Wales and Northern Island to monitor the Post Office from the consumers' perspectives
- The Post Office had to consult with the POUNC before implementing any major initiative. The POUNC to make recommendations to the Minister
- The Post Office to become a Public Corporation.

The Carter Report of July 1977
- Indicated the exerciser of market power by the Post Office
- Indicated that the 1969 Act provided few incentives to lower costs
- Criticized the accounting system and data availability
- Highlighted conflict between managers and the government
- Highlighted lack of accountability and a proper framework for decision making and project evaluation
- Post Office management was too rigid and managerial salaries were too low
- Recommending separating Posts and Telecommunications into two corporations

The 1978 White Paper on the State of the Post Office
- Did not separate telecommunications from post
- Encouraged greater decentralization of decision making
- Instituted target cost reductions of 5% per annum over 5 years

Source: Spiller and Vogelsang (1993a)

over the management of the telecommunications operator. As either public corporation or a department of state, the telecommunications division of the Post Office, and later on BT under public ownership, had serious management problems, arising, to a large extent from political interference and the lack of managerial incentives (Moore, 1986). Until BT's privatization, most of the regulatory changes were attempts by the incumbent government to reduce (or formalize) either the minister's or management's discretion. The fact that several White Papers, as well as the 1969 Telecommunications Act, tried to limit the scope of ministerial interference, suggests the extent of government inherent inability to commit not to interfere with the management of the public corporations.

Even though the Tory party did not always have in its platform the privatization of state-owned enterprises, during the 1970s a movement inside the party emerged to undo the nationalizations undertaken by previous Labor Governments. While some of the supporters of privatization in the Conservative party

Box 2 Main regulatory changes and proposals prior to BT's privatization

The 1981 Beesley Report
- Recommended unrestricted resale of leased lines (private circuits)
- Recommended allowing BT to freely set prices for private circuits
- Recommended to promote network entry

The British Telecommunications Act of 1981
- Allowed some entry into VANs (to be further relaxed by 1987)
- Terminated BT's monopoly over customer premises equipment (with the exception of the first phone, to be terminated in 1985)
- Allowed licensed network entry
- Separated BT from the Post Office

The 1982 White Paper
- Proposed to sell 51% of BT
- Proposed to create the Office of Telecommunications (OFTEL) with a Director General

The Duopoly Policy of November 1983
- Government announced that, for 7 years, no nationwide competitor would be allowed besides BT and Mercury to supply fixed-link voice telephony

The 1984 Telecommunications Act
- Creates OFTEL and the position of Director General of Telecommunications
- Requires the issuing of licenses for all PTO
- Stipulates process of license amendments
- Brings the MMC into the regulation of licensees
- Silent about price setting, except that is to be determined in the license

Source: Spiller and Vogelsang (1993a).

saw privatization as a way to improve the fiscal situation, others supported privatization from a clear political perspective. For example, John Moore, the then Financial Secretary to the Treasury saw 'people's capitalism' as a way to change the political composition of the nation[39]. The privatization of BT seems to have been affected by this view, as it was done in a way that assured widespread ownership of BT's shares across the population. Whatever the reason for its privatization, though, the logic behind the process that started with the 1981 British Telecommunications Act called for competition and private ownership. Box 2 describes the process that lead to BT's privatization.

[39] Speaking at the Wider Share Ownership Council Forum, Mr. Moore said: "Our aim is to establish a people's capital market, to bring capitalism to the place of work, to the High Street and even the home." Quoted in Newman (1986, p. 41). Mr. Moore was also a strong supporter of employee share ownership schemes as they were applied to BT and other companies. See Newman (1986, p.150) and Moore (1986).

British Telecom's privatization, however, was not undertaken without the design of a particular regulatory governance structure. As Box 2 shows, an independent regulatory office (OFTEL) was created, companies were granted licenses, a license amendment process was developed, the Monopolies and Mergers Commission (MMC) was brought into the regulatory process, and price setting was instituted in the license itself rather than in the legislation. It will be discussed below that given the nature of the U.K.'s institutional environment, these institutional innovations provide private investors with some amount of protection against the threat of future policy changes, and thus provided the necessary commitment for private investment to take place.

There are four major differences in the evolution of the deregulatory processes in the U.S. and the U.K. First, is the role of judicial review of regulatory agencies. In the U.K, the courts played no role through judicial review of regulatory decisions, while they were fundamental in the U.S. Second, is the role of the legislature. While telecommunications were an important issue in the U.S Congress during the late 1970s and early 1980s (Spiller 1990), Congress did not legislate the major regulatory changes in telecommunications. In the U.K, however, Parliament passed several bills that were fundamental for BT's privatization (see Box 2). Third, is the role of regulatory agencies. Although in the U.K. the cabinet undertook major regulatory decisions (e.g., the introduction of the duopoly policy and its eventual reversal), in the U.S. the FCC, an independent regulatory agency subject to judicial review, was in charge of implementing the vague mandates of the Federal Communications Act. Finally, is the role of licenses. Although in the U.S. licenses play no major role in the regulatory process, they are the key feature of regulatory governance in the U.K.

The differences in the evolution of regulation in the U.S. and the U.K will be explained by developing a theory of the relation between the institutional endowment and the design of regulatory governance.

4. Regulatory governance: administrative process with judicial review

Regulatory governance may take very different forms. In the United States, regulatory governance consists on a complex set of administrative procedures. A major implication of this regulatory choice is that the judiciary plays a key role in the regulatory process. In the United States, administrative law stipulates that all decisions of administrative agencies can be reviewed by Federal Courts[40]. The

[40] Except for the Social Security Administration disability decisions which are heard at the Federal District Courts, most decisions of administrative agencies are reviewed directly by Federal Courts of Appeals.

U.S. Congress has enacted several pieces of legislation that set the standards for judicial review, the most far reaching of which is the Administrative Procedures Act[41]. These standards, however, are quite vague. For example, Section 706 of the Administrative Procedures Act provides that:

> the reviewing court shall: (1) compel agency action unlawfully withheld or unreasonably delayed; and (2) hold unlawful and set aside agency action, findings, and conclusions found to be: (a) arbitrary, capricious, and abuse of discretion, or otherwise not in accordance with law; (b) contrary to constitutional right, power, privilege, or immunity; (c) in excess of statutory jurisdiction, authority or limitations, or short of statutory right; (d) without observance of procedure required by law; (e) unsupported by substantial evidence in a case subject to Sections 556 and 557 of this title or otherwise reviewed on the record of an agency hearing provided by statute; or (f) unwarranted by the facts to the extent that the facts are subject to trial *de novo* by the reviewing court.

That is, there are two issues in the judicial review, substance and procedure. Both are stipulated by the APA. The courts, however, are not restricted to interpret the APA, as they can also use the due process clauses of the Constitution. The criteria for judicial review, then, are vague. While in principle courts decide questions of law and not of facts, providing deference to the agencies on the latter, the difference between a question of law and a question of fact is also vague[42].

The vagueness of the criteria under which Courts can review administrative decisions is not a legislative mistake. The APA was passed precisely with the dual intention of providing the Courts with the ability to monitor the agencies, and interest groups with the ability to participate in the regulatory process. Both aspects of the APA foster participation by interested parties, and provide members of Congress with the ability to preempt the enacting of administrative decisions they may dislike[48].

In a system, like the United States, characterized by division of powers, political control of judicial decisions, however, is limited. In a series of papers Gely and Spiller develop a basic framework to understand the role of the judiciary in what is now called 'the division of powers game[43].' The main thrust of their papers is that Courts are ideologically motivated bodies with well defined political preferences,

[41] The legislation enacting an agency will usually include specific administrative procedures the agency must follow. These may simply refer to the standards introduced by the APA, or may specify alternative ones.

[42] See, e.g., *NLRB vs. Hearst Publications*, 322 U.S. 111, where the Supreme Court held that newsboys are employees under the NLRA, reversing a previous Court of Appeals decision, thus making a statement of fact, based, however, on the interpretation of the statute.

[43] See Gely and Spiller (1990, 1992), and Spiller and Gely (1992). For other 'division of power' models see, among others, Eskridge and Ferejohn (1992a,b), Ferejohn and Weingast (1992) and Ferejohn and Shipan (1990). For a critical review of this literature see, Segal (1995). For a rebuttal, see Bergara et al. (2002).

making decisions based not on the traditional legal rules of precedent, but on the constraints imposed by the other political institutions (i.e., Congress and the Presidency)[44]. In such a framework the main constraint on the Courts' power is the potential for legislative reversal[45].

The frictionless world of Gely and Spiller has derived some useful empirical hypotheses, some of which have been already been subject to empirical tests[46]. The real world, however, is more complex. Courts, agencies and legislatures have decision costs[47], and have to make their decisions subject to binding budget constraints and substantial informational asymmetries, not the least related to the preferences of the other players[48]. In such environments, Congress can influence the Court not only through reversals, but also through impacting on their costs of making decisions. Tiller and Spiller (1999), in particular, show that process requirements as well as agency and judicial budgets, appointments and judicial expansions[49], by differentially impacting on agency and courts costs, have direct influences on regulatory outcomes, and in particular on the power of the status quo.

Judicial review in a division of powers system, however, is a double-edged sword. On the one hand judicial review may limit deviations from the status quo, but often it may be an instrument in motivating change. Gely and Spiller (1990) show how judicial review blocked the Reagan administration's attempts to deregulate through administrative fiat. Spiller (1990) shows, though, how the judiciary was the key player in the deregulation of long distance telecommunications, with Congress being active but unable to legislate a deviation from the status quo. The difference between the two scenarios is based on the relative composition of Congress vis-à-vis the Courts and the administration. The wider the disagreements among members of congress on what the policy should be, the higher the probability that the Courts will make a drastic shift in the status quo. That was not only the case in Telecommunications, but also in environmental protection (McCubbins et al., 1989) among other areas.

[44] Spiller and Spitzer (1995) analyze in detail the theoretical implications of assuming that courts are not strategic players. They show that such assumption has empirical implications not consistent with current evidence.

[45] If the decision touches on a constitutional issue, reversal has to be undertaken by a constitutional amendment. See Gely and Spiller (1992) and Spiller and Spitzer (1992). Iaryczower, Spiller and Tommasi (2002) explore political restrains on the Courts by other means than constitutional amendments.

[46] See Spiller and Gely (1992) and Bergara et al. (2002).

[47] See Spiller (1992) for the first treatment of agency discretion in a judicial hierarchy with decision costs.

[48] See, in particular, Schwartz, Spiller and Urbiztondo (1994) for an analysis of Congressional legislative action in an environment in which congressional preferences are not perfectly known by the courts, and in which the enacting congress may not be the same as the reversing congress.

[49] For a political model of strategic judicial expansions, see DeFiguereido and Tiller (1995).

4.1. Why judicial review?

A major question is, though, why is it that the United States has such a strong tradition of judicial review of administrative agencies, while most other countries do not? The answer to this question will help our understanding of whether the U.S. regulatory governance framework is applicable elsewhere in the world.

There are two conditions for a system of administrative procedures subject to judicial review to provide regulatory commitment. First, division of powers has to be real, that is, that neither the executive nor the legislative can dominate the political process. Second, that the judiciary is not easily manipulable by the political parties[50]. Under these two conditions, the Courts will naturally develop a tradition of contesting agency decisions, a legal tradition that is a basic ingredient for judicial review to provide any credibility at all. To understand why these two conditions are necessary ingredients, consider a situation where although there is a formal separation of powers, electoral rules are such that the party that wins the presidency also wins both houses of the legislature[51]. Assume, furthermore, that electoral rules are such that the regional compositions of the two houses are not too distinct[52].

Figure 1 provides such a description. Figure 1 shows three players, the House (H), the Senate (S), and the President (P), each represented by their ideal points in a generic two dimensional policy space[53]. The triangle connecting the three ideal points consists of the bargaining set among the three players. Any policy in that set is a legislative equilibrium. The Figure also represents the initial policy decision, call it x_0. x_0 is a legislative equilibrium as a change away from x_0 will make one of the players worse off and will consequently be blocked. Now, x_0, however, may not be a stable outcome. In particular, Figure 1 shows that a slight political change may render x_0 outside the set of legislative equilibria. If x_0 was implemented through an administrative process, then Congress may legislative procedural changes so that x_0 will change as well.

Figure 2 shows the implications for regulatory commitment of a system of division of powers that generates a large legislative bargaining set. The Figure depicts two legislative bargaining sets, one linking ideal points H, S and P, and another, a wider set, linking ideal points H_w, S_w, and P_w. The initial status quo, x_0, is

[50] Iaryczower, Spiller and Tommasi (2002) show that the Argentine Supreme Court was influenced by the double effect of high rotation in their membership, so that very few presidents faced Courts nominated by opposed political parties, and by the high level of concentration of power that the governing party had in Congress.

[51] Mexico, until the Fox Administration may have been such a system.

[52] This is an example of false bicameralism. For a public choice perspective on bicameralism, see Levmore (1990).

[53] The use of ideal points for multimember bodies in multidimensional policy spaces is an oversimplification, as in general, in those conditions multimember bodies may not be represented as having stable preferences.

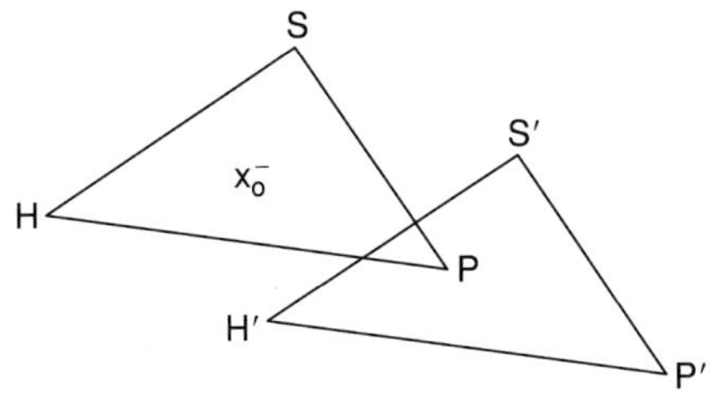

Fig. 1. Policy implications of a political shift with a small legislative bargaining set.

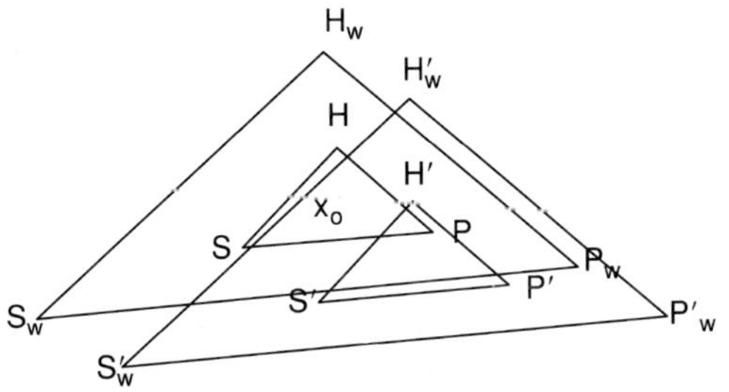

Fig. 2. Policy implications of a political shift with a large legislative bargaining set.

in both sets. Assume, now that there is a given change in public preferences that imply a shift of a given magnitude in all three main veto players. Now the relevant legislative bargaining set shifts to connect the new ideal points marked with an apostrophe[54]. With a larger bargaining set, a given preference change has a smaller probability that the initial status quo would be left outside the new legislative bargaining set, thus providing further commitment to the initial policy choice.

We have seen then that unless separation of powers generates a large legislative bargaining set, administrative procedures will not provide a credible regulatory framework as political changes will drive changes in the underlying procedures so as to modify the final regulatory outcomes. We still need to show that such type of division of powers will naturally generate a judicial tradition of challenging administrative acts.

Spiller (1996a) calls the process that generates such judicial tradition 'Pavlovian evolution.' The thrust of this argument is that judicial norms will develop differently under different legislative and executive organizations. In particular, in a true division of powers system, Courts can more easily challenge administrative

[54] The change in political preferences was assumed to move all politicians in the same general direction so that the change did not affect the size of each of the legislative bargaining sets.

and regulatory decisions, both on procedures and on substance. Such interventions will, in most cases, not trigger a legislative response, leaving the Court's decision standing[55]. On the other hand, in very centralized systems, attempts by the Courts to reverse an administrative decision may trigger a legislative or executive response overriding the Court.

Consider, for example, Figure 3. There the Court faces the three-veto structure characteristic of a bicameral presidential system with a fragmented legislature, what I call a true division of powers system. Assume that the legislation is vague and as a consequence the agency attempts to implement a particular point in the legislative set, x_A, which differs from the status quo x_0. The parties may challenge the agency decision on procedural grounds. Assume, now, that the Court has preferences over the policy outcomes. Say that there are two types of Courts, SC_H and SC_L. In the figure SC_H prefers the original status quo, while SC_L prefers the agency's choice. The Court, then, may or may not reverse the agency decision. In any case, though, the court will not be reversed as either decision is in the legislative set, and thus the court's decision is a legislative equilibrium[56].

In Presidential systems that are either false-bicameral or in two-party parliamentary systems, though, the number of veto points is substantially reduced. In these cases the Court may not easily reverse administrative decisions without facing a challenge. Figure 4 represents the legislative set for a two-party parliamentary system. There H and S represent the two branches of the legislature and

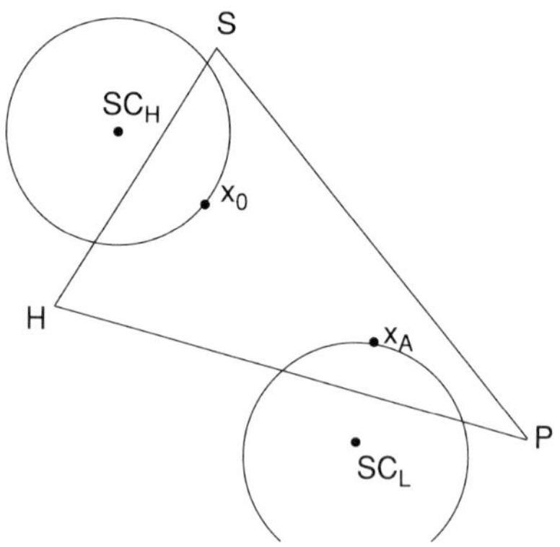

Fig. 3. Judicial choice in standard Presidential systems.

[55] That Congress pays attention to judicial decision making and often reverses the courts has been documented in Eskridge (1991). See also Spiller and Tiller (1996) for a model of Congressional reversals of judicial decisions arising from Courts having preferences not only over the policy space but also over judicial rules or other issues over which Congress cannot overrule the Courts.
[56] See Gely and Spiller (1990).

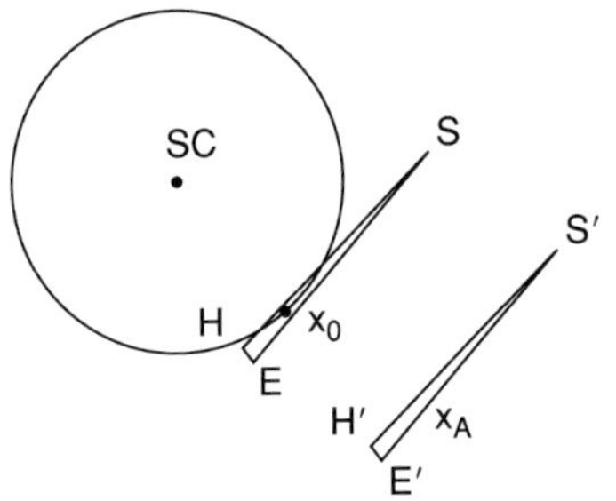

Fig. 4. Judicial choice in two party parlimentary systems of false bicameral presidential systems.

E represents the fact that the executive is drawn from the lower House[57]. In Figure 4, for example, a political move made the initial status quo untenable. An agency decision moved the policy to x_A. Attempts by the Court to reverse the agency decision (SC, in Figure 4 prefers x_0 to x_A) will directly trigger a legislative response, reversing the court and bringing the policy back to x_A[58]. This, however, may not be the case in multiparty parliamentary systems, whereby governments are formed from multiparty coalitions, and where the potential for coalition break-up is substantially bigger[59].

Here is where Pavlovian evolution may take place. A judiciary operating in a system where decision making is substantially decentralized among the different political institutions with numerous veto points will find it possible to check the executive on its administrative decisions. Such judicial actions will usually count with some legislative support and hence would be invulnerable to a political reversal. On the other hand, a judiciary operating in a political system where the legislative and the executive process are controlled by the same political group will find that such acts of judicial independence will be punished with legislative reversals, budget cuts, lack of promotion, political demotion or even nastier political tricks[60]. In such a political environment, a judicial tradition of reviewing administrative decisions will not develop. In those systems, judicial activism will only trigger a legislative reversal and political recrimination. Thus, it is not

[57] Most parliamentary system have a bicameral legislature, but in most legislative powers are rested in the lower house, who also elects the government.

[58] See Ramseyer (1994) for a fascinating discussion of cabinet control of the judiciary in Japan, and Salzberg (1991) for a similar analysis in the U.K.

[59] See, for example, Steunenberg (1995) for an analysis of judicial intervention in the euthanasia debate arising from the fragmentation of the government coalition.

[60] For example, in 1990, following clashes with the Court, the new government of President Menem, with the agreement of Congress, increased the number of Supreme Court justices from 5 to 9 and named all new justices. See Levy and Spiller (1994). See also, Iaryczower, Spiller and Tommasi (2002).

surprising that neither in the U.K. nor in Japan, Courts have developed a strong tradition of judicial review of administrative actions, developing a rudimentary administrative law[61].

To summarize, the U.S. system of judicial review of administrative agencies' decisions is based on the nature of U.S. legislative institutions, including electoral rules that have created a system with large number of veto points, and where there is substantial decentralization of decision making. In such circumstances, independent judicial review can thrive. A question that needs to be asked, though, is why is it that in such an environment Congress has chosen to delegate to agencies and Courts the interpretation of their statues, and what are the implications for regulatory commitment.

4.2. *Why delegate to independent agencies*

This question needs to be broken in two parts: first, is there such a thing as independent agencies, and second, why will Legislatures delegate to independent agencies. The first question was answered in the following way: the probability of observing independent agencies is higher in systems characterized by divided government (Spiller and Urbiztondo, 1994). We show that in systems characterized by unified government (where the preferences of the legislature and the executive are systematically aligned, as in a two party parliamentary system) control over the bureaucracy will be stronger, with a much smaller proportion of political appointees than in systems characterized by divided government (what I call here true division of power systems). The use of political appointees (including independent agencies), we claim, arises from the fact that in systems characterized by divided government the executive has a much smaller control over the professional bureaucracy, as the latter will naturally tend to be aligned with the legislature, a political institution that tends to be longer lasting than the executive. Spiller and Urbiztondo (1994) find that such characterization of divided and unified governments hold both across countries and across cities in the United States. Bambaci et al. (2002) find that such characterization fits Argentina well[62].

In an important article, Weingast and Moran (1983) raised the Congressional Dominance Hypothesis. This hypothesis, presented in a different form by McCubbins and Schwartz' (1984) 'fire alarm' framework, suggests that independent agencies are not truly independent, as they are subject to continuous congressional oversight. Actual congressional reversal is not necessary, all what is needed is the threat of such reversal. Spiller (1992) shows that independent administrative agencies' discretion in a system of division of powers depend,

[61] On U.K.'s administrative law, see Baldwin and McCrudden (1987).

[62] Bambaci et al. (2002), show that given the extreme degree of turnover in the legislative and executive, the civil service bureaucracy is essentially uncontrollable, generating the need for the development of an unusually deep "parallel bureaucracy."

among other things, on the composition of the legislature and the executive (determining the threat of congressional reversal), and also on the organization and budget of the judiciary (determining the threat of judicial reversal). In a system of division of powers, however, Congressional Dominance is a corner solution. Spiller (1990a) shows that Congressional budgetary decisions of agencies reflect an internal rather than a corner solution. Thus, agencies do not fully respond to Congressional desires. If this is the case, then, a basic question is why does Congress delegate to agencies who are not fully aligned with it, and what are the implications of such delegation for regulatory commitment?

Schwartz et al. (1993) developed a framework to understand the congressional choice between specific and vague legislation. Specific legislation is one where substantive administrative actions are specified in great detail (e.g., FCC to hold auctions for the allocation of the spectrum, EPA to organize markets for pollution rights granting utilities specific allowances), whereas vague legislation is one where congress just specifies the general principles to be interpreted by the agency and the courts (e.g., the FCC is to implement policies in telecommunications promoting the welfare of the U.S. citizens). They show that the choice of legislative specificity is a strategic one in a system characterized by division of powers and informational asymmetries. The rationale is that in such system congress can trust neither the agencies nor the courts in implementing what the current congress actually wants. Informational issues are important. In the absence of informational asymmetries, we have shown elsewhere that legislative intent plays no role[63]. There are several informational problems that may have implications for the choice. First, agencies and courts may not fully know Congress' preferences on particular issues[64]. The higher the degree of uncertainty on preferences the higher the chances that agencies and courts will err in their actions and that congress will then have to spend resources in reversing agencies and courts. In a separating equilibrium, then, specific legislation provides a signal about congressional preferences, in particular, that congress cares much about this particular issue. Second, agencies and courts may be uncertain about congress' costs of reversing their decisions[65]. In a separating equilibrium, then, specific legislation provides a signal that the enabling congress believes it will have low reversal costs in the future. If the enabling congress believed that its future reversal costs would be high, then there will be no point in drafting specific legislation as courts and agencies could then deviate and the future congress will not be able to reverse it.

[63] See Gely and Spiller (1990) showing that under full information the initial legislative act has no consequence on the final equilibrium.

[64] We differentiate between uncertainty on legislators' most desired policies and on their intensity over policies. Most desired policies relate to an individual's 'ideal point.' Intensity is related to how such individual's utility falls as the policy moves away from his or hers ideal point.

[65] See Spiller (1992) where congressional reversals arise from lack of information on congressional reversal costs.

A major result from this analysis is that legislative vagueness and agency and judicial activism are, in equilibrium, correlated. But the correlation does not arise because of courts or agencies wanting to follow the intent of congress, but rather because legislative vagueness is correlated with high reversal costs and low saliency issues, environments in which agencies, and courts, have more discretion.

Thus, delegation to independent agencies requires a system of division of powers. In this environment, legislative specificity will most probably not be the norm, as legislative costs will be high and preference homogeneity among the members of congress will most probably be low, increasing the costs of reversing agencies and courts. Under those circumstances, it is where we can expect agency independence. But, it is also here where we should expect judicial independence that, to some extent, counterbalances and limits the independence of agencies.

5. Commitment in unified government systems

The previous discussion suggests that purely administrative procedures with judicial review may not provide substantial regulatory commitment in systems characterized by unified government. The main deterrent to commitment is that government controls both the administrative and the legislative processes. Thus, political changes that bring about a change in government can also bring about legislative changes. By having few institutional checks and balances, such systems have an inherent instability that raises questions about their ability to provide regulatory commitment. Nevertheless several countries that can be characterized as having a unified form of government have developed private ownership of utilities (e.g., Japan, the U.K., Jamaica, and Mexico). Such countries have developed alternative institutional ways to provide regulatory commitment. Some, like the U.K., Jamaica, and other Caribbean countries, have based their regulatory governance structures on contract law[66]. Japan, prior to the collapse of the LDP, developed internal party structures that provided for substantial regulatory commitment[67]. Other countries (e.g., Mexico), developed private ownership of utilities by providing for substantial up-front rents[68].

[66] See Spiller and Sampson (1995) and Spiller and Vogelsang (1995) for analyses of the regulatory structures in Jamaica and the U.K., respectively.

[67] See, in particular, the volume by Cowhey and McCubbins (1995) for an analysis of the organization of the LDP and its implications for policy determination.

[68] Consider, for example, the privatization of the telecommunications company, Telmex. Prior to the devaluation of 1994, Telmex residential rates were as follows: $488 for connection, $9 per month which included 100 calls, $0.15 per a 3 minutes extra of local call, $1 for a 3 minutes national long distance call, $2.4 for a 3 minutes call to the United States, and $6.25 for a 3 minutes international call elsewhere. Obviously, these prices are orders of magnitude of comparable U.S. rates. This, in turn, provided Telmex with a very high rate of return during its initial few years. For example, in 1993 it had $7Billion in revenues and 43% profit margin. (Crespo, 1995).

A main purpose of the regulatory and institutional schemes in countries like Japan, Jamaica, and the U.K. is to provide the regulated companies with some amount of veto power over regulatory decisions[69]. Consider the case of British utilities. British utilities are regulated by different price caps methods. The distinguishing feature of these price cap methods, however, is that they are embedded in the companies' license rather than in an agency decision or piece of legislation[70]. The advantage of regulatory frameworks instituted through licenses is that since the latter usually have the power of contracts between governments and the firms, any amendment to the license will usually require the agreement of the company[71]. This feature, however, provides credibility at the cost of inflexibility. For example, if a technological breakthrough eliminates any economies of scale in a segment of the market, the regulator may have to 'bribe' the company into accepting to relinquish its legal exclusive rights over that segment. In a more flexible basic engineering choice, such decision could be taken administratively[72]. Thus, while contracts may be useful in providing assurances to the companies, they do so only by introducing rigidities in the regulatory system.

5.1. Contract-based regulation

For contracts as a governance structure to provide regulatory stability, then, they must be very specific and clearly limit what the regulator can do. A license that does not specify the regulatory mechanism in any detail, but leaves the administration free to make all regulatory decisions will fail the first criteria for regulatory stability[73]. Operating licenses in the U.S., for example, do not serve as a governance structure, as they deal mostly with eminent domain and franchise area issues. Whether specific licenses will provide regulatory credibility depends on whether the courts will see licenses as binding contracts. In particular, it must be the case that courts will be willing to uphold contracts against the wish of the administration. If courts do not treat licenses as contracts, or grant the administration substantial leeway in interpreting those contracts, then, license-based

[69] Mexico did not provide formal veto power to the company, although the fact that the company represents more than 30% of the market capitalization of the Mexican stock exchange implies that regulatory choices that will impact negatively on Telmex's profitability would not go well with the investment community nor with the government itself.

[70] Indeed, the enabling laws in the U.K. are silent about pricing schemes.

[71] In the British case, the law stipulates that in the case that the company does not agree to a license amendment proposed by the regulator, there is a process involving the Monopolies and Mergers Commission that the regulator may use to amend the license against the will of the company (Spiller and Vogelsang, 1993a).

[72] This, indeed, had been the case in Jamaica concerning the introduction of cellular telephony (Spiller, 1995).

[73] Comparing the licenses issued in Jamaica under the Jamaican Public Utilities Act of 1966 to those issued prior to 1966 or after the privatization of 1988 shows the total failure of licenses to restrain the regulators based on the 1966 Act (Spiller and Sampson, 1993).

regulatory contracts will fail as a source of regulatory stability. Observe, then, that contracts can be implemented in nations with very strong or very weak executives, with parliamentary or Presidential systems. Indeed, a basic requirement is the independence of the judiciary *and* that the judiciary sees licenses as contracts.

Contract-based regulation, however, is particularly appealing to polities whose legislative bargaining set is very narrow or small. In such cases, as in Figure 4, changes in political preferences would either bring about a new piece of legislation if the current regulatory regime is based on specific legislation or a modification of the agency's interpretation of the statute if the current regulatory regime is based on general administrative procedures. On the other hand, if x_0 was initially hard-wired through a license, then the fact that x_0 does not belong anymore to the legislative bargaining set is irrelevant. Changes in x_0 require the acquiescence of the company. Thus, by introducing the regulatory process in the license, an additional veto point is introduced, namely the company itself. Thus, the relevant set of parties required to change the status quo x_0 now includes also the company. Thus, in Figure 5, the relevant bargaining set is now the set given by the ideal points (H, S, E, and C), where C represents the company's ideal point. The Figure shows how in such circumstance regulatory credibility is enhanced even in a situation where electoral changes move against the company[74]. Thus, it is not surprising that countries like Jamaica, the U.K. and many of the other Caribbean countries have adopted license-based regulatory systems.

Informal norms may also provide some regulatory stability even if licenses can be unilaterally amended by the government. For example, the U.K. has been granting operating licenses since at least 1609 (Spiller and Vogelsang, 1993b). While initially licenses were granted by the King, eventually Parliament took over as the license granting institution[75]. Since licenses were granted by Acts of Parliament, Parliament could, in principle, change the conditions under which future licenses were to be granted. Parliament, could, furthermore, go a step

[74] For a model of regulatory decision making in the U.K., see Spiller and Vogelsang (1994).

[75] Licenses were granted in many different forms. Until 1919, the most common way was for a prospective utility to apply to Parliament for a particular license. A Select Committee would hear the case and either recommend it or not. The Select committee's recommendation would then be presented to Parliament for approval. This latter step was supposed to be a simple formality. These are called Private Bills. A license could also be granted by an Order in Council, whereby the Privy Council upon recommendation of the relevant Minister and the Cabinet will grant the license. Licenses were granted also by General Acts of Parliament. Such an example is the Public Health Act of 1875 which granted the municipalities the right to establish water and gas undertakings in their districts if no one else was empowered and willing to perform such undertaking. Provisional orders were substitutes for private bills. A Provisional Order would be granted by a Minister following a request for a license. The Minister's order would then be formalized by an Act of Parliament. Finally, following the 1919 Electricity (Supply) Act, the Electricity Commission, created by the Act, started issuing licenses through Special Orders, who had to be approved by the Minister of Transport. Special Orders did not have to obtain the formal approval of Parliament. Parliament, however, retained the right to reconsider those licenses (Spiller and Vogelsang, 1993b).

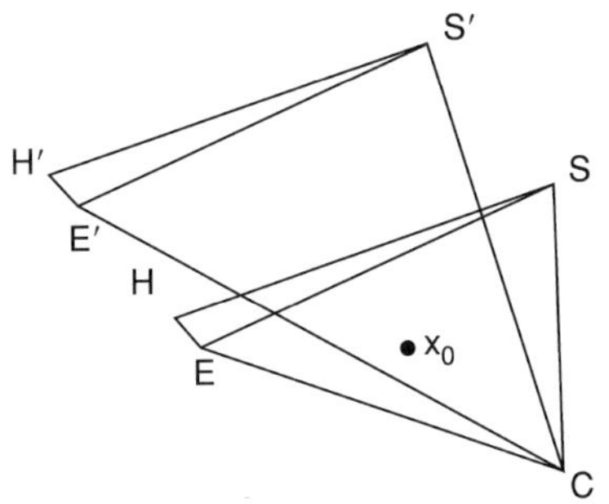

Fig. 5. Policy stability under contract based regulation.

further and unilaterally change the operating rights of current license holders. Such parliamentary power, however, made the licenses fragile instruments. Over the years, however, Parliament has developed an informal (constitutional) norm of not revoking licenses without compensation. Indeed the only case I know of when licenses were revoked was during the massive restructuring of the electricity sector in 1926. In that case, however, the license holders that were to be shut down were given very favorable transitory rights that compensated them for whatever losses could arise from the compulsory shut down[76].

These informal norms still apply, to a large extent, in the U.K., as the regulator is supposedly independent of the Secretary of State for Trade and Industry, although the latter appoints the regulator and the regulatory agency's budget is included in the Department of Trade and Industry's budget. Indeed, licenses can be amended unilaterally by the government, only following a particular procedure. Spiller and Vogelsang (1997) show that this procedure provides credibility as long as the regulator, the Monopolies and Merger Commission and the Secretary of State can be thought of as politically independent of each other[77]. They show that if the three entities are politically the same, then they can manipulate the

[76] See Spiller and Vogelsang (1993b).

[77] Indeed, on March 7th 1995 the Director General of Electricity Supply, Prof. Stephen Littlechild, considered by most the architect of the U.K. regulatory system, reversed a price-cap decision taken in August 1994 and suggested that from April 1996 distributing companies may be subject to a tighter price cap. This reversal was seen by many as showing the lack of commitment inherent in the U.K. For example, the title of The Economist article dealing with this event is "Incredible" (The Economist, March 11, 1995, p.74). This, however, fits the Spiller and Vogelsang (1995) model in that the regulator may make proposals at any time, but for those proposals to become policy they have to follow a particular procedure. In the case at hand, if the distributing companies would not agree to the regulator's proposal, then the regulator has the option of making a reference to the Competition Commission (then the Monopolies and Mergers Commission). Only if a modification of the license is undertaken against the will of the companies and without a reference to the Competition Commission can it be said that a breach of the process took place.

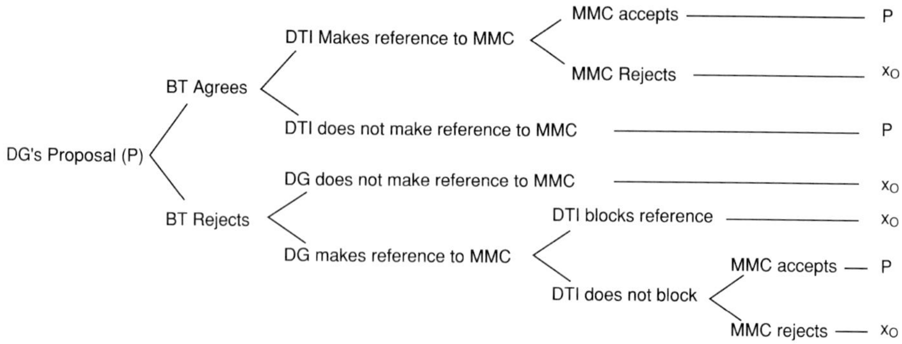

Fig. 6. BT's license amendment process.

license in almost every way, eliminating any regulatory commitment that licensing as a governance structure would provide.

To see this point, consider Figure 6. The regulatory process of telecommunications in the U.K. can be conceived as the regulator (the Director General of Telecommunications, DG) making a proposal (P) to BT to amend some of its license provisions (e.g., the price cap system, the interconnection rules, etc, with the status quo represented by x_0). If BT accepts, the secretary of trade and industry (DTI) may object and make a reference to the Competition Commission (the former MMC). If the Competition Commission (MMC) accepts the license amendment, then the regulator can sign the new license over the objections of the DTI. If, however, the MMC rejects the agreement between the DG and BT, then the status quo remains. If BT rejects the DG's proposal, then the DG may make a reference to the MMC. Such reference, though, can be blocked by the DTI, but only if the DTI claims that the DG's proposal breaches national security, a conceivable high standard. If the DTI does not block the reference, and if the MMC agrees with the regulator, then the DG may unilaterally amend BT's license. If, however, the MMC rejects the DG's proposal, then the status quo remains. Finally, if the DTI blocks the reference to the MMC, then again, the status quo remains. This discussion, then, shows that the license amendment procedures provide BT with some assurances against regulatory discretion insofar as the DG, the DTI and the CC do not have similar objectives.

Informal norms come in several ways to shore up the fragility of U.K. institutional arrangements: first, although feasible, it would be inappropriate for the DTI to attempt to influence directly the DG[78]. Second, although the government can

[78] Indirect influence, though, takes place through the media and through the lower levels of the bureaucracy.

appoint members of the Competition Commission at will, the CC is the longest lasting regulatory agency with substantial public credibility, and it would not be appropriate for the DTI to replace all its members at once. Third, it would also be unheard of for the secretary of state not to sign a license amendment that was agreed upon by BT, the regulator and the CC. Thus, although most probably BT would not be able to block any of those actions through the courts, the U.K. has developed a series of implicit understandings on how governments operate that would inflict substantial costs to a politician that drastically deviates from such norms. The emphasis on implicit norms may seem strange to those raised in the American way of government, where what is not directly illegal is proper. Issues like 'losing face' are much less important in a formalistic society as the United States. In a society like Japan or the U.K., where decision making is not public, implicit norms of conduct can be understood as an equilibrium that provides the different players in both the party in power and the private sector with incentives to invest, both in the physical sense (the private sector) as in the political sense (the politicians)[79].

Norms, however, cannot be used as analytical tools lightly. But some empirical implications can be derived from the fact that for facilitating norms to develop, political stability is a necessity. Thus, for example, countries with high political instability will not develop such supporting norms, and hence will not be able to develop regulatory structures that provide for substantial regulatory flexibility. This, indeed, is the case of Jamaica or Bolivia. These two countries have used contracts as ways to provide regulatory commitment, and have done so effectively and for long periods of time. Indeed, Jamaica regulated utilities that way from the beginning of this century until 1966 when the Jamaica Public Utilities Commission was created. Nationalization of all utilities followed in the early 1970s. In the late 1980s the telecommunications company was privatized, using again a very specific license as the basic regulatory governance instrument. Performance since privatization has been quite successful[80]. Bolivia has the longest lasting private electricity company in South America. COBEE, founded in the late 1910s, has provided electricity to La Paz and its environs. Its regulatory structure has also been based on a very specific license. Neither country's licenses, though, provide for much flexibility[81].

The U.K. case, however, differs from the others, in that the U.K. introduced a competitive environment, while Jamaica and the other Caribbean countries use of license-based regulation was in the framework of the granting of monopolies. Further use of license-based regulation in weak environments would

[79] For a similar analysis of the norms organizing the U.S. House of Representatives (Weingast and Marshall, 1988).

[80] See Spiller and Sampson (1995).

[81] See Spiller (1995).

provide further impetus to growth and investment by limiting the potential for opportunistic behavior by governments.

5.2. *Adapting contract-based regulation to unexpected shocks: renegotiation*

Because contract-based regulation introduces substantial rigidities to the regulatory process, it needs to be complemented by a well specified contract renegotiation or contract amendment processes. The U.K. example is one such process. The U.K. framework, however, may not be adaptable to other emerging markets where the types of informal bureaucratic norms described above are not present. In the absence of some types of modification process, economic or political shocks will generate too many conflicts between the regulator and the investor, eroding confidence. The regulatory process in the U.K. and in the U.S. can be thought of as a 'relational contract' process[82]. Relational contracts are informal agreements sustained by the value of future relationships (Baker et al., 2002). Relational contracts provide flexibility and adaptation to formal contracts. In the U.S., the APA provides that flexibility, while in the U.K. the license modification process does the same. Formal contracts between governments and firms, however, have an inherent flexibility, which limits the ability to adapt in an informal manner. The fact that one partner is a government entity limits adaptation[83]. Consider a cost shock that is beyond the range of expected shocks. In the U.S. or in the U.K., such cost shocks would trigger a regulatory adaptation process. In the U.S. there would be calls for rate reviews, even if the regulated firm is subject to a 'strict price cap.' In the U.K. such shock may trigger an anticipated price review. Indeed, in the U.K. price reviews are often undertaken outside the 5-year clock[84]. With contract-based regulation, such adaptation may not work. Would the regulator exempt the regulated firm from its pricing arrangement, the whiff of corruption would naturally follow. In some environments, citizens could file a suit against the regulator for not following with the letter of the law[85]. Thus, regulators are bound, both to avoid perception of corruption or influence, and to avoid being taken to the courts, to follow the letter of the contract. Thus, contract-based regulation must be designed so that they are continuously renegotiated or modified by

[82] See McNeil (1978).

[83] Normally licenses or franchise contracts are granted not by the regulatory agency but by a cabinet level entity.

[84] See Spiller and Vogelsang (1997).

[85] A recent example has been the suit filed by citizen organizations against the Argentine Government for its attempt to implement public hearings over utility price increases which did not follow the terms of the Economic Emergency Law of 2002. See, for example, La Nación, La Justicia suspendió los aumentos en servicios públicos, 25 September 2002.

common will. Without those procedures, contract-based regulation would be unworkable.

The evidence on contract-based regulation squares with this implication of transaction cost analysis. Guasch (2001) shows that, outside telecommunications sector, 60–70% of all concession contracts in Latin America were renegotiated in the first 3 years of the concession[86]. Abdala and Spiller (2000) provide evidence that concession contract renegotiation is a common feature of contract-based regulation in Argentina. They also show that contract renegotiation will be more common the higher the incidence of uneconomic investment requirements and the higher the annual concession payments to the government. The rationale is straightforward. The higher the cash flow commitments to either uneconomic investments or to the government, the higher the possibility that a shock will bring the company closer to its bankruptcy constraint. Since governments are not interested in triggering service dislocation, governments facing important economic shocks will tend to favor renegotiation over canceling the concession and relicensing. The Williamsonian fundamental transformation (from one buyer with many suppliers to one buyer to one supplier) is also present in utilities (Williamson, 1976). Thus, renegotiation is going to be a fact of life in contract-based regulation.

The fact that renegotiation is the norm more than the exception in contract-based regulation should be taken into account when granting the original contract. Concession contracts are normally granted via some type of bidding process. The bidding process may be for the lowest average price, for the highest annual—or upfront—payment, for the highest investment level or faster deployment, or for any combination (Abdala and Spiller, 2000). Not all concession contracts will be renegotiated rapidly, though. The better designed the regulatory framework (i.e., the lower the uneconomic investments, the stronger the collecting rights of the utility[87], or annual payments to the government) the lower the potential for renegotiation. If, however, the regulatory framework is designed with social policy in mind, then the bidding procedures must be adjusted[88]. Bidding for lowest price or highest annual payment may not be optimal if the constraints on the company are such that contracts will have to be renegotiated readily after the concession

[86] For a view of contract renegotiation as reflecting contracting failures and opportunistic behavior by the firm, see Guasch and Laffont (2002).

[87] In some jurisdiction the courts grant citizens a constitutional right to public utility services. In those cases, collection may suffer. Argentina is one such example. See Artana, Navajas and Urbiztondo (1999) for collection problems in water utilities in Argentina.

[88] Estache and Quesada (2001) provide a model of concession contract renegotiation in which they suggest that to avoid renegotiations, which drive prices higher, governments may consider granting the concession to a relatively inefficient firm.

was granted. In those environments, firms will strategically adjust their bids to account for the possibility of renegotiation. Indeed, companies may underbid so much that the possibility of facing negative profits may be very high right from the start. In those environments, auction procedures should be modified to reduce the need for renegotiation. Indeed, an auction method that awards the concession to the company whose present value of revenues, accounting for renegotiation, binding limit liability constraints, and renegotiation costs, is the lowest seems to substantially reduce incentives to underbid and thus would limit, but not eliminate, the need for renegotiation.

6. Final comments

Political opportunism has important organizational implications. Restraining governmental opportunism, however, is not an easy task. Countries that have succeeded in developing a healthy private sector are those that in one way or another have developed institutions that restrain governmental decision making. But such restraining is itself a political choice. Although the conditions for such choice to arise are yet to be developed, one is explored here: the nature of legislative and judicial institutions. Countries with electoral and legislative systems that bring about decentralized government have stronger chances of developing equilibria where government discretion is restrained. In the shortrun, however, countries do not have the ability to undertake drastic changes in electoral and legislative systems. Regulatory reform, in that sense, must be adapted to the country's institutional environment. This chapter highlights contract-based regulation as a way to provide some degree of credibility in environments in which administrative and judicial review are not common or effective. Contract-based regulation, however, requires a substantial rethinking of the regulatory process so as to make it adaptable to the environment. This rethinking should be at the forefront of institutional analysis.

References

Abdala, M. and P. T. Spiller, 2000, Institutions, Contracts and Regulation in Argentina (in Spanish), Buenos Aires: Temas.

Artana, D., F. Navajas and S. Urbiztondo, 1999, Governance and Regulation in Argentina, in P. T. Spiller and W. Savedoff, eds., Spilled Water: Institutional Commitment in the Provision of Water Services in Latin America, Interamerican Development Bank, Washington DC.

Baker, G., R. Gibbons and K. Murphy, 2002, Relational Contracts and the Theory of the Firm, Quarterly Journal of Economics, 117(1), 39–83.

Baldwin, R. and C. McCrudden, 1987, Regulation and Public Law, London: Weidenfeld & Nicholson.

Bambaci, J., P. T. Spiller and M. Tommasi, 2002, Bureaucracy in Weak Presidential Systems: Argentina, Mimeo, Berkeley: University of California.
Barzel, Y., 1989, Economic Analysis of Property Rights, Cambridge, MA: Cambridge University Press.
Bergara, M., B. Richman and P. T. Spiller, 2002, Modeling supreme court preferences in a strategic context, with Mario Bergara and, forthcoming, Legislative Studies Quarterly, 28, 247–280.
Cain, B., J. Ferejohn and M. Fiorina, 1987, The Personal Vote, Cambridge: Harvard University Press.
Cowhey, P. and M. McCubbins, eds., 1995, Structure and Policy in Japan and the United States, New York: Cambridge University Press.
Cox, G., 1987, The Efficient Secret, New York, New York: Cambridge University Press.
Crespo, M., 1995, Telmex: Slim's pickings, Financial World, 164(1), 20.
de Figueiredo, J. and E. Tiller, 1995, Congressional Control of the Courts: A Theoretical and Empirical Analysis of the Expansion of the Judiciary, Berkeley: University of California.
Duverger, M., 1954, Political Parties: Their Organization and Activity in the Modern State, New York: Wiley.
Estache, A. and L. Quesada, 2001, Concession contract renegotiation: Some efficiency vs. equity dilemmas, Mimeo: World Bank.
Gely, R. and P. T. Spiller, 1990, A rational choice theory of supreme court statutory decisions, with applications to the *State Farm* and *Grove City* Cases, Journal of Law, Economics and Organization, 6, 263–301.
Gely, R. and P. T. Spiller, 1992, The political economy of supreme court constitutional decisions: The case of Roosevelt's court packing plan, International Review of Law and Economics, 12, 45–67.
Goldberg, V., 1976, Regulation and administered contracts, Bell Journal of Economics, 7(1), 426–448.
Guasch, J. L. and P. T. Spiller, 1999, Managing the regulatory process: Design, concepts, issues, and the Latin America and Caribbean story, Washington DC: World Bank.
Guasch, J. L., 2001, Concessions and regulatory design: Determinants of performance—fifteen years of evidence, San Diego: Mimeo, World Bank and University of California.
Henck Fred, W. and B. Strassburg, 1988, A Slippery Slope: The Long Road to the Breakup of AT&T, Greenwood Press.
Hill, A. and M. Abdala, 1993, Regulation, institutions and commitment: Privatization and regulation in Argentine telecommunications sector, Washington, DC: The World Bank.
Iaryczower, M., P. T. Spiller and M. Tommasi, 2002, Judicial decision making in unstable environments: Argentina 1938–1998, with American Journal of Political Sciences, 46(4), 699–716.
Jacobson, G. C., 1990, The Electoral Origins of Divided Government: Competition in the U.S. House Elections, 1946–1988, Boulder, CO: Westview Press.
Knieps, G. and P. T. Spiller, 1983, Regulating by partial deregulation: The case of telecommunications, Administrative Law Review.
Levmore, S., 1990, Bicameralism: When are two decisions better than one? International Review of Law and Economics, 12(2), 145–168.
Levy, B. and P. T. Spiller, 1993, Regulations, institutions and commitment in telecommunications: A comparative analysis of five countries, Proceedings of The World Bank Annual Conference on Development Economics, 215–252.
Levy, B. and P. T. Spiller, 1994, The institutional foundations of regulatory commitment: A comparative analysis of telecommunications regulation, Journal of Law, Economics, and Organization, 10(1), 201–246.
Macey, J., 1992, A organizational design and political control of administrative agencies, Journal of Law, Economics, and Organization, 8(1), 93–110.
McCubbins, M. D., R. G. Noll and B. R. Weingast, 1987, Administrative procedures as instruments of political control, Journal of Law, Economics, and Organization, 3, 243–277.

McCubbins, M. D., R. G. Noll and B. R. Weingast, 1989, Structure and process, politics and policy: Administrative arrangements and the political control of agencies, Virginia Law Review, 75, 431–482.

McCubbins, M. and T. Shwartz, 1984, Congressional oversight overlooked: Police patrol vs. fire alarms, American Journal of Political Sciences, 28(2), 165–179.

Macneil, I., 1978, Contracts: Adjustments of long-term economic relations under classical, neoclassical, and relational contract law, Northwestern University Law Review, LCCII, 854–906.

Moore, J., 1986, The success of privatization, in Kay, J., C. Mayer and D. Thompson, eds., Privatisation and Regulation: The UK Experience, Oxford: Clarendon Press.

Newman, K., 1986, The Selling of British Telecom, London: Holt, Rinehart and Winston.

North, D. C., 1990, Institutions, Institutional Change, and Economic Performance, Cambridge, MA: Cambridge University Press.

Ramseyer, M., 1994, The puzzling (in)dependence of courts: A comparative approach, Journal of Legal Studies, 72, 1–748.

Salzberg, E. M., 1991, The delegation of legislative powers to the courts and the independence of the judiciary, University of California, Berkeley: Mimeo.

Schwartz, E., P. T. Spiller and S. Urbiztondo, 1993, A positive theory of legislative intent, Law and Contemporary Problems, 57(1–2), 51–74.

Shepsle, K. A., 1992, Bureaucratic drift, coalitional drift, and time inconsistency—A comment, Journal of Law, Economics and Organization, 8(1), 111–118.

Shugart, M. S. and J. M. Carey, 1992, Presidents and Assemblies, New York, NY: Cambridge University Press.

Spiller, P. T., 1990, Governmental institutions and regulatory policy: A rational choice analysis of telecommunications deregulation, Mimeo: University of Illinois.

Spiller, P. T., 1990, Politicians, interest groups and regulators: A multiple principals agency theory of regulation (or let them be bribed), XXXIII Journal of Law and Economics, 65–101.

Spiller, P. T., 1992, Agency discretion under judicial review, in Formal Theory of Politics II: Mathematical Modelling in Political Science, Mathematical and Computer Modelling, 16(8–9), 185–200.

Spiller, P. T., 1993, Institutions and regulatory commitment in utilities' privatization, Industrial and Corporate Change, 2, 387–450.

Spiller, P. T. and R. Gely, 1992, Congressional control or judicial independence: The determinants of U.S. supreme court labor decisions: 1949/1987, Rand Journal of Economics, 23(4), 463–492.

Spiller, P. T. and C. I. Sampson, 1993, Regulation, institutions and commitment: The Jamaican telecommunications sector, The World Bank.

Spiller, P. T. and M. Spitzer, 1992, Judicial choice of legal doctrines, Journal of Law, Economics and Organization, 8(1), 8–46.

Spiller, P. T. and M. Spitzer, 1995, Where is the sin in sincere: Sophisticated voting in the courts, Journal of Law Economics and Organization, 11(1), 32–63.

Spiller, P. T. and E. Tiller, 1996, Invitations to override: Congressional reversals of supreme court decisions, International Review of Law & Economics, 16(4), 503–521.

Spiller, P. T., 1996a, A positive political theory of regulatory instruments: Contracts, administrative law or regulatory specificity, USC Law Review, 69(2), 477–515.

Spiller, P. T. and S. Urbiztondo, 1994, Political appointees vs. career civil servants: A multiple-principals theory of political institutions, European Journal of Political Economy, 10(3), 465–497.

Spiller, P. T. and I. Vogelsang, 1993a, Regulation, institutions and commitment: The British telecommunications sector, The World Bank.

Spiller, P. T. and I. Vogelsang, 1993b, Notes on public utility regulation in the UK: 1850–1950, Mimeo: University of Illinois.

Spiller, P. T. and I. Vogelsang, 1997, The institutional foundations of regulatory commitment in the UK, (with special emphasis on telecommunications), Journal of Institutional and Theoretical Economics, 153(4), 607–629.

Steunenberg, B., 1995, Courts, cabinet and coalition parties: The politics of euthanasia in a parliamentary setting, The Netherlands: University of Twente.

Taagepera, R. and M. Shugart, 1993, Predicting the number of parties: A quantitative model of duverger mechanical effect, American Political Science Review, 87, 455–464.

Temin, P. and L. Galambos, 1987, The Fall of the Bell System: A Study in Prices and Politics, Cambridge: Cambridge University Press.

Tiller, E. and P. T. Spiller, 1999, Strategic instruments: Legal structure and political games in administrative law, Journal of Law Economics and Organization, 15(2), 349–377.

Weingast, B. R., 1995, The Economic role of institutions: Market preserving federalism and economic development, Journal of Law, Economics and Organization, 11(1), 1–31.

Weingast, B. R. and M. Moran, 1983, Bureaucratic discretion or congressional control: Regulatory policy making by the federal trade commission, Journal of Political Economy, 91, 765–800.

Williamson, O. E., 1975, Markets and Hierarchies: Analysis and Antitrust Implications, New York: Free Press.

Williamson, O. E., 1976, Franchise bidding for natural monopolies: In general and with respect to CATV, Bell Journal of Economics, 73–104.

Williamson, O. E., 1988, The logic of economic organization, Journal of Law, Economics and Organization, 4, 65–93.

SUBJECT INDEX